McGraw Hill's

NATIONAL ELECTRICAL SAFETY CODE® (NESC®) 2023 HANDBOOK

About the Authors

David J. Marne, P.E., B.S.E.E., is a registered professional electrical engineer. Mr. Marne is a nationally recognized speaker on the National Electrical Safety Code (**NESC**), and he serves on various NESC subcommittees. He is company president and senior electrical engineer for Marne and Associates, Inc. in Bozeman, Montana, where he specializes in **NESC** training and expert witness services. He has over 39 years of experience in the utility industry and is a senior member of the IEEE.

On a personal note, Dave was born and raised in Mt. Pleasant, Pennsylvania, and he is of Italian descent. His father's surname was Marmo, but was "Americanized" and changed to Marne. Dave and his wife Patty met at Montana State University in Bozeman, MT, and have been happily married for 39 years. After living in Napa, CA, and Missoula, MT, they returned to where it all began and are currently enjoying a busy life in Bozeman, MT. Outside of work, Dave enjoys spending time with his family, traveling, and spending time in the beautiful Montana outdoors.

John A. Palmer, Ph.D., P.E., C.F.E.I., has been President of Palmer Engineering and Forensics, a multidisciplinary forensic and consulting engineering firm based in North Salt Lake, Utah, since 2009 and has been performing work in the field of forensic engineering since 2001. He has worked on a wide variety of investigations for insurance claims and legal matters, including offering expert testimony, over a broad array of cases including electrical equipment failures, electrical injuries, electrocutions, fires, explosions, and intellectual property. Dr. Palmer is also an experienced teacher in the area of electric power engineering, having taught topics such as power systems analysis, power system protection, power electronics, electric machinery and electrical failure analysis over a teaching career spanning multiple decades and multiple universities. He currently teaches as an associate professor in the Electrical and Computer Engineering Department at the University of Utah. He obtained his B.S. in electrical engineering at Brigham Young University and his master's and Ph.D. degrees in electric power engineering from Rensselaer Polytechnic Institute. He is a registered Professional Engineer in multiple states and a Certified Fire and Explosion Investigator.

Looking at John from a personal perspective, you will find a guy who was born and raised in Phoenix, Arizona, and met and married the love of his life in Utah during his undergraduate studies at Brigham Young University. They are the proud parents of five children, who are now grown and raising children of their own. John enjoys cycling, hiking, and spending time with his family.

McGraw Hill's

NATIONAL ELECTRICAL SAFETY CODE® (NESC®) 2023 HANDBOOK

Based on the Current 2023
National Electrical Safety Code® (NESC®)

David J. Marne

John A. Palmer

New York Chicago San Francisco Athens London
Madrid Mexico City Milan New Delhi
Singapore Sydney Toronto

McGraw Hill's National Electrical Safety Code® (NESC®) 2023 Handbook

2 3 4 5 6 7 8 9   LCR   26 25 24 23

Library of Congress Control Number: 2022947973

ISBN 978-1-264-25718-8
MHID 1-264-25718-X

Sponsoring Editor Lara Zoble	**Indexer** Michael Ferreira
Editorial Supervisor Patty Mon	**Production Supervisor** Lynn M. Messina
Project Manager Alok Singh, MPS Limited	**Composition** MPS Limited
Acquisitions Coordinator Elizabeth M. Houde	**Illustration** MPS Limited
Copy Editor Vandana Kumari	**Art Director, Cover** Jeff Weeks
Proofreader Vandana Gupta	

Contents

Part 4.   Work Rules for the Operation of Electric Supply and Communications Lines and Equipment

Preface

McGraw Hill's National Electrical Safety Code® (NESC®) 2023 Handbook is written for the engineer, staking technician, power lineworker, communications lineworker, inspector, and safety administrator of an electric power or communication utility company, contracting company, or consulting firm. Employees involved with power substation, power transmission, power distribution, and communication network, design, construction, operation, and maintenance should have an understanding of the **NESC** to keep the public, utility workers, and themselves safe.

McGraw Hill's NESC 2023 Handbook is designed to be used in conjunction with the National Electrical Safety Code (**NESC**) published by the Institute of Electrical and Electronic Engineers (IEEE). This *Handbook* presents hundreds of figures and photos, plus examples and discussions to explain and clarify the **NESC** rules. The straightforward and practical information in this *Handbook* is intended to aid the understanding of the sometimes confusing and complicated text in the **NESC**. This *Handbook* follows the order of parts, sections, and rules as presented in the **NESC** including the general rules, electric supply station rules (Part 1), overhead line rules (Part 2), underground line rules (Part 3), and work rules (Part 4). This format ensures a quick and easy correlation between the **NESC** rules and the discussions and explanations in this *Handbook*. This *Handbook* is most effectively used by referring to the **NESC** for the precise wording of a rule and then referring to the corresponding rule number in this *Handbook* for a practical understanding of the rule. The Appendix of this *Handbook* includes numerous photos of **NESC** applications and violations.

The **National Electrical Safety Code (NESC)** is the "bible" for developing power and communication utility standards. All utility construction has some relevance to the **NESC** including items such as line design, substation design, standard drawings, material purchases, pole placement, work practices, signage, and others. The **NESC** should not be confused with the National Electrical Code (NEC) published by the National Fire Protection Agency (NFPA). The NEC is used for residential, commercial, and industrial building wiring. It does not apply to utility systems although there is some overlap of the **NESC** and the NEC at a building service. The **NESC** is written as a voluntary standard. It can be, and indeed has been, adopted as law by numerous individual states or other governmental authorities. To determine the legal status of the **NESC**, the state public service commission or public utility commission or other governmental authority should be contacted. The **Code** is written by various **NESC** committees. The organizations represented, subcommittees, and committee members are listed in the front of the **Code** book. The procedures and time schedules for revising the **NESC** are described in the **Code** book. The **NESC** has an interpretation committee that issues formal interpretations. The process for obtaining a formal interpretation is outlined in the **Code** book. The **NESC** is currently published on a 5-year cycle, although the 2023 edition was issued after 6 years due to logistical issues associated with the COVID-19 pandemic. Urgent safety matters that require a change in between **Code** editions are handled through Tentative Interim Amendments and the **NESC** Fast Track process. Original work on the **NESC** began in 1913. This *Handbook* is based on the 2023 edition of the **NESC**.

McGraw Hill's NESC Handbook is the only **NESC** *Handbook* that focuses on the practical application of the current edition of the **NESC**. The numerous figures, photos, and examples make *McGraw Hill's NESC 2023 Handbook* the most complete and useful handbook available.

Additional reference material for the interested reader includes, *The Lineman's and Cableman's Handbook* by Thomas M. Shoemaker and James E. Mack, and the *Standard Handbook for Electrical Engineers*, by Surya Santoso and H. Wayne Beaty, both published by McGraw Hill. The transmission line, distribution line, and substation design and construction standards published by the Rural Utility Service, formerly known as the Rural Electric Administration, are also invaluable references. Finally, the book titled *Overhead Distribution Lines Design and Application* by Lawrence M. Slavin contains practical engineering and design information for pole lines.

This *Handbook* should be used as a reference for understanding and applying the rules in the **NESC**. This *Handbook* is not an official **Code** document and does not contain official **NESC** committee interpretations. The information, figures, and photos in this *Handbook* are intended to be used as aids to the reader of the **Code** and are not intended to be a replacement for the comprehensive nature of the **Code** as it is written.

David J. Marne
John A. Palmer

Acknowledgments

We wish to express our sincere appreciation to many individuals who helped us bring the 2023 edition of this *Handbook* to print.

The following peer reviewers provided a technical review of our manuscript: Grant Glaus, Mickey Gunter, Keith Reese, and Lawrence Slavin. The following people helped bring the text, figures, and photographs to life: Maria Melvin and Larry Coles. The following people at McGraw Hill and their affiliates supported and guided our efforts: Lara Zoble and her hardworking associates.

Finally, we wish to thank our families for their love and support.

David J. Marne
John A. Palmer

General Sections

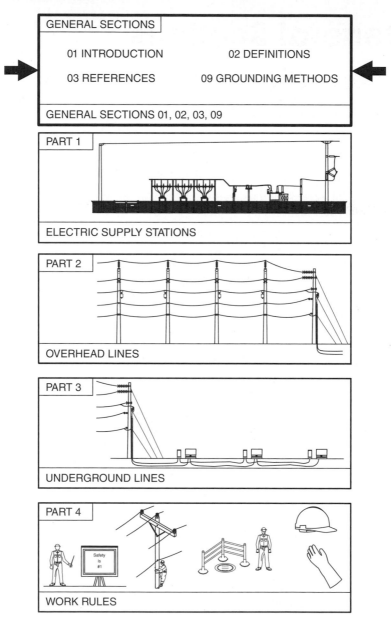

GENERAL SECTIONS

01 INTRODUCTION   02 DEFINITIONS

03 REFERENCES   09 GROUNDING METHODS

GENERAL SECTIONS 01, 02, 03, 09

PART 1

ELECTRIC SUPPLY STATIONS

PART 2

OVERHEAD LINES

PART 3

UNDERGROUND LINES

PART 4

WORK RULES

Section 01

Introduction to the National Electrical Safety Code

010. PURPOSE

The very first rule in the **NESC** outlines the purpose of the entire book. The rules contained in the **NESC** are provided for the "practical safeguarding of persons during the installation, operation, and maintenance of electric supply and communication facilities, under specified conditions." The persons the **Code** is referring to are both the public and utility workers (employees and contractors).

The **Code** uses the term electric supply for electric power. The **Code** also covers communications lines and equipment. Communications utilities include, but are not limited to, telephone, cable TV, and fiber utilities.

The **NESC** is a standard published by the Institute of Electrical and Electronic Engineers (IEEE). It is a recognized standard for utility company safety. It is adopted by utilities or by an authority having jurisdiction over utilities (e.g., a state public service commission or public utility commission) or by some other authority. To determine the specific legal status of the **NESC**, the authority having jurisdiction must be contacted. In general, utilities in 49 of the 50 United States use the **NESC** as a standard of practice or as a legal statute. Some states automatically adopt the current edition and some states have adopted specific editions. In addition, some states write additional rules to supplement or expand on the **NESC** Rules. Examples are states that require detailed inspection requirements and states that have adopted "storm hardening" rules. One state that intentionally does not adopt the **NESC** is the State of California which writes its own codes titled, General Order 95 (GO95) for overhead lines, General

Order 128 (GO128) for underground lines, and General Order 165 (GO165) for inspections.

The **NESC** contains the basic provisions necessary for safety under specified conditions. The **Code** is not intended to be a design specification or instruction manual. The **Code** specifies what needs to be accomplished for safety, not how to accomplish it. While the **NESC** is not intended as a design specification or instruction manual, many of the requirements in the **Code** impact the design, construction, operation, and maintenance of power and communications lines and equipment. The ultimate responsibility for the design, construction, maintenance, and operation of the utility systems and the safety of those systems, are the utilities that own and operate them, and they must do so in compliance with the requirements of the **Code**. Values for clearance and structure strength must be not less than the values indicated in the **Code** for safety purposes. Clearance and strength values can certainly be greater than the **Code** specifies, but greater values are usually not required for safety unless so determined by competent engineering analysis.

For example, if the **Code**-required clearance of a 12.47/7.2-kV phase conductor is 18.5 ft over the ground, the **Code** does not specify how high the poles have to be or where the wires are attached on the pole to obtain the 18.5-ft clearance. Selecting the proper pole heights and attachment heights is a design function which should take into account the type of conductor, sag, tension, operating conditions, terrain, and various other parameters. The **Code** does require that the phase wires be not less than 18.5 ft high (they can certainly be higher). Some utilities establish clearance values by using the **Code** clearance plus an adder. The adder can be thought of as a design or construction tolerance adder to maintain the required **Code** clearance over time, but not an adder for added safety. A line designed for exactly 18.5 ft of clearance over the ground could end up having problems meeting **Code**. There are several factors that could jeopardize the 18.5-ft clearance. One factor could be that the pole hole was dug 6 in deeper than it should have been dug. Another factor might be a slight rise in elevation at midspan that was not detected due to the fact that the ground line was not profiled or surveyed. Other factors that can jeopardize clearance include wire stringing issues, leaning poles, leaning crossarms, etc. Using a clearance adder (e.g., 3 ft) would require designing the line to 21.5 ft instead of 18.5 ft. The clearance adder can also help meet and maintain the required **Code** clearance over time. See Rule 230I for an additional discussion of maintaining clearances over the life of an installation.

There are many design standards available for utility companies to reference. The Rural Utilities Service (RUS) publishes transmission, distribution, and substation design manuals. Many larger utilities develop their own design manuals. Each manual written must be in compliance with the **Code**, as the **Code** is the "bible" for all utility work.

The **Code** states that rules apply to specified conditions. One example of the specified conditions is overhead line clearance for open supply conductors, 750 V–22 kV, over a road subject to truck traffic (**NESC** Table 232-1, Row 2). In **NESC** Appendix A (**NESC** Fig. A-1), the **Code** shows the reference component (of a truck) to be 14 ft. The mechanical and electrical clearance component for a conductor shown in this figure is 4.5 ft, for a total of 18.5 ft of clearance when the conductor is at the maximum sag condition. The specified conditions for the maximum sag of the conductor can be found in **NESC** Rules 230B and 232A. These are the specified conditions associated with the 18.5 ft of clearance. If an overhead line is being designed for a mining installation and the conditions are that 22-ft-high trucks are used, then this condition requires that the same line be designed with a 26.5-ft clearance ($22' + 4.5' = 26.5'$) at the maximum sag of the conductor. This concept is reflected in **NESC** Table 232-1, Footnote 26.

The **NESC** is not intended to provide design criteria for abnormal events due to the actions of others (e.g., the force of a car hitting a pole) or weather events in excess of those specified in **NESC** Sec. 25 (e.g., a tornado). See Rule 230I for additional information.

The first rule in the **Code** is a good place to discuss the format and numbering system used in the **Code**. The following are the main parts of the **Code**:

- General Sections 01, 02, 03, and 09 (Introduction, Definitions, References, and Grounding Methods)
- Part 1–Electric Supply Stations (Commonly referred to as substations)
- Part 2–Overhead Lines (Power and communications)
- Part 3–Underground Lines (Power and communications)
- Part 4–Work Rules (Power and communications, similar OSHA Standards apply)

An example of rule numbering is shown in Fig. 010-1.

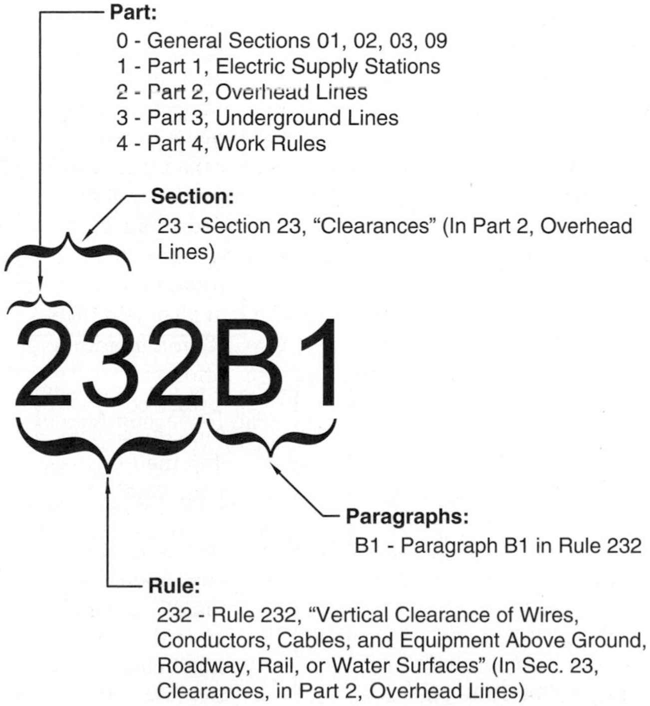

Part:
0 - General Sections 01, 02, 03, 09
1 - Part 1, Electric Supply Stations
2 - Part 2, Overhead Lines
3 - Part 3, Underground Lines
4 - Part 4, Work Rules

Section:
23 - Section 23, "Clearances" (In Part 2, Overhead Lines)

232B1

Paragraphs:
B1 - Paragraph B1 in Rule 232

Rule:
232 - Rule 232, "Vertical Clearance of Wires, Conductors, Cables, and Equipment Above Ground, Roadway, Rail, or Water Surfaces" (In Sec. 23, Clearances, in Part 2, Overhead Lines)

Fig. 010-1. Example of rule numbering (Rule N/A).

011. SCOPE

This rule defines what is covered in the **National Electrical Safety Code** and what is not covered. The **NESC** covers supply and communication facilities and associated work practices carried out by a supply or communications utility company or an entity functioning as a utility. In general, the **NESC** applies to electric supply (power) and communications utilities. Communications utilities include, but are not limited to, telephone, cable TV, and fiber utilities. An example of an organization not normally thought of as a utility but functioning as a utility could be a university campus system or a large industrial complex that owns a utility-voltage distribution system. The **NESC** covers functions of utilities including generation of energy or communication signals and transmission and distribution to the service point.

The **NESC** does not cover utilization wiring in buildings. The standard that does cover building wiring is the National Electrical Code (NEC). The NEC is the "bible" for the electrical building industry and is used primarily by engineers and electricians. The **NESC** is the "bible" used primarily by utility engineers and utility lineworkers. See Fig. 011-1.

In a few places the **NESC** and the NEC overlap or come close to overlapping. One location is at the service to a building. The **NESC** contains **NESC** Fig. 011-1 which is a one-line diagram showing the dividing line between an electric supply utility and a premises (i.e., building) wiring system. The one-line diagram does not show a meter which would normally exist at or near the service point. For an

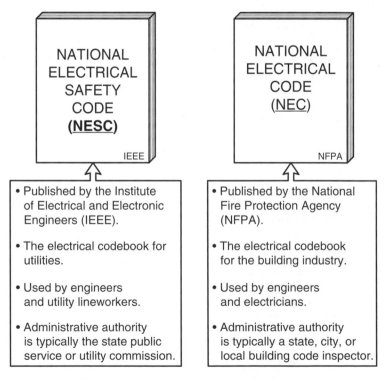

Fig. 011-1. Differences between the NESC and the NEC (Rule 011).

overhead electric service, the typical dividing point between the **NESC** and the NEC is the conductor splice at the weatherhead. This is the dividing line between the electric utility function and the electric utilization function. For underground electric services the typical dividing point can vary. Some utilities provide service to the terminals on the meter base. Others provide service to the property line and the customer provides wiring after that point. Sometimes a utility will have the customer install secondary wiring from the pad mount transformer to the meter, but then the utility will take ownership of this wiring. To determine whether the circuit is covered under the **NESC** or the NEC, the ownership of the circuit and who maintains and/or installs and controls the circuit are important factors to consider. Some utilities use a direct buried splice or a junction box under the meter base to provide a clear transition point between the two codes. The tariff that the utility has on file with the State Public Service Commission or Public Utility Commission may also help define the service point of the utility. See Fig. 011-2.

The dividing line between the **NESC** and the NEC for an overhead or underground communications service (e.g., a phone, cable TV, or fiber utility service) is typically the network interface device or demarcation box installed on the building being served. The demarcation box typically has a compartment for the communications company service wires and a separate compartment for the building communications cabling. This separates the utility communications function from the building communications function.

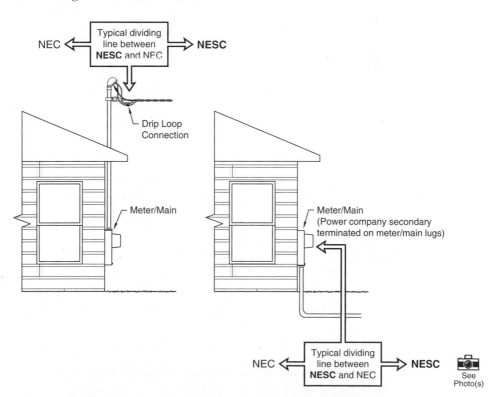

Fig. 011-2. Typical dividing lines between the **NESC** and the NEC (Rule 011).

Another item that is covered by both the **NESC** and NEC is street and area lighting. Control of the street and area lighting is the important factor to consider when determining whether street and area lighting is covered under the **NESC** or the NEC. Street and area lighting that is metered usually falls under NEC requirements. Street and area lighting that is not metered and is owned, operated, and maintained by the utility is typically covered under the **NESC**. The main differences between street and area lighting covered under the **NESC** and street and area lighting covered under the NEC are grounding methods and overcurrent protection requirements. **NESC** street and area lights are typically operated and maintained by power lineworkers, and NEC street and area lights are typically operated and maintained by electricians. Examples of street lighting systems covered by the **NESC** and NEC are shown in Fig. 011-3.

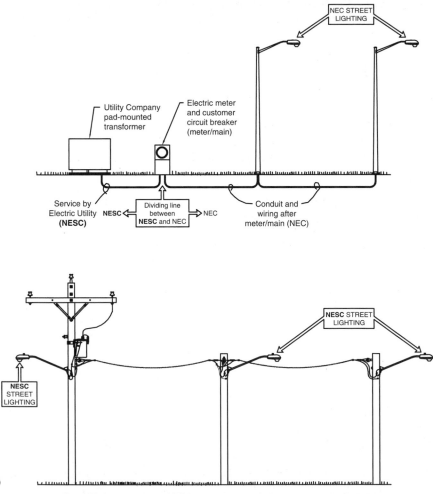

Fig. 011-3. Examples of **NESC** and NEC street lighting systems (Rule 011).

The **NESC** does not cover installations in mines (underground), ships, railway rolling equipment, aircraft, or automotive equipment. These industries have their own standards. See Rules 101 and 301 for a discussion of utilization wiring in Parts 1 and 3.

Section 02, Definitions of Special Terms, provides definitions of the terms "delivery point" for one utility delivering energy or signals to another utility and "service point" for determining the dividing line between the serving utility (supply or communication) and the premises wiring. A clear separation between the **NESC** and the NEC can be determined by applying **NESC** Rule 011 and using the definitions in **NESC** Sec. 02 and by referencing the corresponding scope rule in the front of the NEC.

Although not specifically addressed in Rule 011, the **NESC** does not cover easement conditions, environmental protection, raptor (bird) protection, FCC regulations, FAA regulations, NERC standards, electromagnetic fields (emf), settings for protective device coordination, or construction, operation, and maintenance cost issues. These items are either covered by other standards or they do not present a safety concern or they are simply not addressed in the **NESC**.

012. GENERAL RULES

This rule provides three general rules for applying the **Code**. Rule 012A requires that the design, construction, operation, and maintenance of electric supply and communication lines equipment must be in accordance with the **NESC**. Rule 012B requires that utilities or other organizations performing work for the utility, such as a contractor, are the responsible parties for meeting **NESC** requirements for design, construction, operation, and maintenance. Rule 012C acknowledges that the **Code** cannot cover every conceivable situation. Where the scope of the **NESC** applies and the **Code** does not specify a rule to cover a particular installation, design, construction, operation, and maintenance should be done in accordance with "accepted good practice" for the local conditions known at the time. This does not allow the **Code** to be ignored for a condition that is covered in the **Code**. If "accepted good practice" is needed because a specific **Code** rule does not exist, a comparable **NESC** rule or the National Electrical Code (NEC) can be referenced to find an "accepted good practice." Other standards can also be referenced, or "accepted good practice" may be determined by reviewing utility operating records to find out what safe practices are practical for the local conditions. Some rules in the **NESC** have wording such as "as experience has shown to be necessary" (Rule 214A2), or "located a sufficient distance" (Rule 231B1) or "supplemental mechanical protection" (Rule 352D2, Exception). This type of wording also necessitates the use of **NESC** Rule 012C. Rule 012 applies to both supply and communication utilities. The **NESC** applies during the initial design and construction and during the life of the installation (i.e., during operation and maintenance).

013. APPLICATION

New installations and extensions are covered under Rule 013A. This rule clearly states that all new installations and extensions must adhere to the provisions of the **NESC**. Rule 013A1 does allow the administrative authority to waive or modify the

NESC rules if safety is provided in other ways. The administrative authority refers to the public service commission or public utility commission for the applicable jurisdiction, not any employee or manager within the utility itself. An example of a situation in which such a waiver or modification may apply is listed in the **Code** to clarify this statement.

The **Code** is intended to provide guidance to utilities for the safety of their systems, but not to prevent progress and development of new technologies and methods. Thus to make it clear that it is permissible to explore new approaches, Rule 013A2 recognizes that new types of construction and methods may be used experimentally to obtain information even if they are not covered in the **Code**. This is only permitted if three conditions are met. First, qualified supervision must be provided. One example of qualified supervision is supervision under the direction of a registered professional engineer who is using engineering judgment and collecting data on the new construction. Second, equivalent safety must be provided. On joint-use facilities (e.g., poles with both power and communication lines—see definition in Section 2), all affected users must be notified in a timely manner. For example, if a new overhead power distribution line configuration is being tried on a pole that is shared by a communication company, the power utility must inform the communication utility to avoid confusion and mishaps among the communication utility personnel. Furthermore, since the **Code** is on a 5-year revision cycle, many times new construction methods arise during the 5-year period and then are included in a future edition of the **Code**.

Rule 013B discusses how the **Code** is applied to existing installations. Many people use the term "grandfather clause" when discussing this rule. The "grandfather clause" concept appears simple; however, caution must be used when applying this rule. Rule 013B1 states that if an existing installation meets or is altered to meet the current **NESC**, the installation is not required to comply with any previous edition. As an example, if an older line is being maintained or modified, no effort need be expended to ensure compliance with the **Code** that was applicable at the time of construction as long as it complies with the current edition.

Rule 013B2 states that existing installations, including maintenance replacements, that comply with a prior edition of the **Code**, need not be modified to comply with the current edition. The clause "that comply with a prior edition of the **Code**" indicates that if the current installation does not comply with a prior edition of the **Code**, it is not sufficient to bring it back to compliance with the **Code** that was in effect at the time of construction, it must be brought into compliance with the current edition of the **Code**. Additionally, Rule 013B2 states that once an existing installation is brought into compliance with a subsequent edition, the earlier editions no longer apply.

As an example, consider a pole originally set in 1970 under the 1961 **Code**. If that pole were replaced in 1999, bringing the line into compliance with the 1997 edition of the **Code**, the 1961 edition would no longer apply. After that point, it would no longer be appropriate for subsequent repairs or maintenance to later look back to clearance requirements from the 1961 edition. Rule 013B2 can be applied to changes in the **Code**. Changes from the previous edition are indicated by a vertical black bar in the left margin of the **NESC** Codebook. Applying Rule 013B2 eliminates the need to alter existing installations that do not comply with a new rule in the current edition (assuming the installation complies with the **Code** in effect at the time the line was built or a subsequent edition of the **Code** to which it was made to comply). Additionally, Rule 013B2 explains, with regard to maintenance replacements, that the new supporting structure must meet or exceed both the required strength and the appropriate height of the structure being replaced. There is one exception to the maintenance replacement part of the rule, which is that when a structure (e.g., a pole) is replaced, the current version of Rule 238C (which refers to clearances for span wires or brackets) must be met, where applicable. See Rules 202 and 238C for additional information.

Rule 013B3 specifically discusses conductors and equipment that are added, altered, or replaced on an existing structure. Rule 013B3 states that the structure or the facilities on the structure need not be modified or replaced if the resulting installation will be in compliance with either:

- The **Code** rules that were in effect at the time of the original installation, or
- The **Code** rules in effect in a subsequent edition to which the installation has been previously brought into compliance.

If either of these options is chosen, caution needs to be taken to determine if the existing installation complies with a prior **Code**. If the second bullet point above applies (bringing an installation into compliance with a subsequent edition), then the first bullet point (the rules in effect at the time of the original installation) no longer applies. If neither bullet point applies, that is, if the installation will not comply with either Rule 013B3(a) or 013B3(b), then the installation must comply with the rules in this edition.

Rule 013B4 addresses how to work on a structure (e.g., pole) that has a violation. This paragraph points to the inspection rules in Part 2 (Overhead Lines), not Part 1 (Electric Supply Stations) or Part 3 (Underground Lines). Rule 214 requires inspections. When a condition or defect is found that affects compliance with the **Code** (i.e., a violation), Rules 214A4 and 214A5 address the timeline to correct the violation. The timeline is not addressed with hard numbers like a day, a week, a month, a year, etc., the rules only provide general wording. If a violation is not corrected, it must be recorded until the correction is done. If the work to be done on the structure in itself does not create a structural, clearance, or grounding violation or worsen an existing violation, then the work on the structure can be done without first fixing the existing violation, noting that according to Rule 214A4 the violation must be recorded, even if discovered while other work is being performed. In other words, the existing recorded violation can be done subsequently in compliance with Rule 214A5. If these conditions are not met, the existing recorded violation must be corrected at the same time the other work is done or the other work must be postponed until the recorded violation is corrected. One example is adding a new service drop (power or communications) to a pole that has some other recorded violation. If adding the service drop meets the requirements in Rule 013B4 (in itself, adding the service drop does not create a structural, clearance, or grounding violation or worsen an existing violation), then the service drop can be added without fixing the other recorded violation at the same time, understanding that it will be corrected consistent with Rule 214A5.

Normally electric supply and communication engineers have the most current codebook sitting upon their desks and field workers do not have prior codebooks lying around in their trucks. Even if something as simple as adding a transformer to an existing installation or replacing an existing pole is being done, the utility performing the work must be careful not to blindly assume that the existing situation complies with a prior **Code**. Utilities may want to consult the advice of a legal professional when applying Rule 013B as interpreting this rule from a legal viewpoint can be as critical as interpreting it from a technical viewpoint. Although not a **Code** requirement, a simple solution to eliminate the complexities of applying the "grandfather clause" is to use the current edition of the **NESC** for existing installations that require maintenance replacements or additions, alterations, or replacements.

The application of a "grandfather clause" should not be confused with maintaining clearances or a change in land use under a supply or communications line. See Rule 230I for a discussion.

Rule 013C requires that inspections of new and existing facilities meet the inspection rules in the current edition of the **NESC**, not prior editions. In other words, there is no "grandfather clause" covering inspections or work practices. Inspections of electric supply stations are covered in Rule 121, inspections of overhead lines are covered in Rule 214, and inspections of underground lines are covered in Rule 313. Similarly, Rule 013C also requires that work performed on new and existing facilities meet the work rules in Part 4 of the current edition of the **NESC**, not prior editions.

014. WAIVER FOR EMERGENCY AND TEMPORARY INSTALLATIONS

In this rule the **Code** recognizes the need to waive or modify rules in cases of emergency or temporary installations. Rule 014A grants the person responsible for the installation the ability to modify or waive rules with specific requirements. Rule 014B for emergency installations applies to both overhead and underground lines. Rule 014C for temporary installations also applies to overhead and underground lines. The **Code** does not specify a time length for temporary installations or define an emergency installation. The **Code** makes it clear that temporary installations must meet the requirements for nontemporary installations. Rule 014C contains an additional requirement that temporary overhead lines be built to not less than Grade N construction. Strength reductions for temporary service are also addressed in **NESC** Table 261-1, Footnote 4. Rule 014C for temporary construction contains an exception for underground burial depths similar to the burial depth waiver in Rule 014B for emergency installations. See Fig. 014-1 for a summary of the requirements of Rule 014.

015. INTENT

This rule defines three key **NESC** words: "shall," "should," and "RECOMMENDATION." **Code** rules containing the word "shall" indicate that the rule is mandatory and it must be met. Not complying with a rule that contains the word "shall" is a direct **Code** violation. **Code** rules containing the word "should" indicate requirements that are normally and generally practical for the specified conditions.

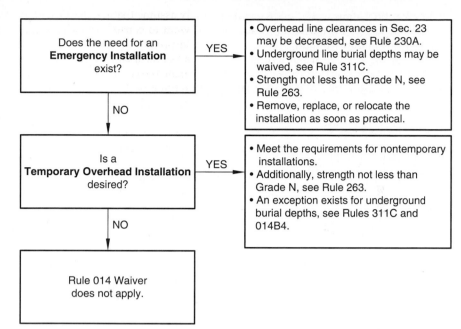

Fig. 014-1. Summary of the requirements for emergency and temporary installations (Rule 014).

The **Code** does recognize that under certain circumstances some rules may not be practical. Where this is the case, the word "should" is used. Under the definition of "should," the **Code** references Rule 012 that requires "accepted good practice" to be used. If at all possible the words "should" and "shall" should be considered the same. "Should" does not give the utility allowance to deviate from the rules simply as a matter of convenience, rather, deviations under a "should" provision are allowed only when specific local conditions make the literal implementation of the rules impractical. If it is not possible to treat "should" the same as "shall," it is prudent to carefully review the specific conditions that prohibit a utility from applying the **Code** rule that contains the word "should" and it is prudent to apply an "accepted good practice" that provides an equivalent degree of safety for the specific conditions. The **Code** uses the word "RECOMMENDATION" for provisions that are considered desirable but not intended to be mandatory. "RECOMMENDATION" is the least stringent of the three terms "shall," "should," and "RECOMMENDATION." Since the **Code** does consider a "RECOMMENDATION" desirable, it seems prudent that the utility make some effort to comply with a "RECOMMENDATION" even though it is not mandatory.

Rule 015 also contains clarifications of the words "NOTE," "EXAMPLE," and "footnote." "NOTES" and "EXAMPLES" are not mandatory, and they are provided for information and illustrative purposes but are not considered part of the **Code** requirements. "Footnotes," however, are used for tables throughout the **Code** and they carry the full force and effect of the table or rule with which they are associated.

An "EXCEPTION" to a rule has the same force as the rule itself. Exceptions are not reduced safety measures. For example, if a clearance value is reduced by an exception, some condition associated with the exception may be provided to maintain safety. Typically, the **Code** provides a larger value in the rule and smaller value in the exception.

The physical location of the words "RECOMMENDATION," "EXCEPTION," and "NOTE," and how these words are indented with respect to other text, signifies to what rule the "RECOMMENDATION," "EXCEPTION," or "NOTE" applies.

016. EFFECTIVE DATE

This rule states that the 2023 edition of the **NESC** may be used at any time on or after publication date. In addition, this edition shall become effective no later than the first day of the month after 180 days following the publication date. The effective date applies to the design and approval process, not just the construction date. If the design or approval for a new installation or extension was started before the effective date, the prior **Code** may be used. The example in this rule indicates the 2023 **NESC** publication date of August 1, 2022 and establishes February 1, 2023 as the effective date. The note in this rule explains that the 180-day (6-month) grace period allows utilities and other agencies to acquire copies of the **Code** and revise regulations, standards, and procedures. The note also clarifies that this edition is not required to be used before the 180-day period; however, it is not prohibited to use it during this period.

The **NESC** is a standard published by the Institute of Electrical and Electronic Engineers (IEEE). It is a recognized standard for utility company safety. It is adopted by utilities or by an authority having jurisdiction over utilities (e.g., a state public service commission or public utility commission) or by some other authority. To determine the specific legal status of the **NESC**, the authority having jurisdiction must be contacted. In general, utilities in 49 of the 50 United States use the **NESC** as a standard of practice or as a legal statute or regulation. Some states automatically adopt the current edition and some sates have adopted specific editions. In addition, some states write additional rules to supplement or expand on the **NESC** Rules. Examples are states that require detailed inspection requirements and states that have adopted "storm hardening" rules. One state that intentionally does not adopt the **NESC** is the State of California which writes its own codes titled, General Order 95 (GO95) for overhead lines, General Order 128 (GO128) for underground lines, and General Order 165 (GO165) for inspections.

017. UNITS OF MEASURE

The 2023 **NESC** uses the English (customary inch-foot-pound) system as the primary unit for numerical values. The metric (International System of Units or SI) system is the secondary system. In the text of the **Code**, English values are shown first with the metric system shown second and inside parentheses. Some tables in the **Code** have the metric and English system in the same table. Other more complex tables have separate tables. When separate tables are used, the table in the **Code** will be the English table. The corresponding metric table can be found in Annex 1 of the **NESC**. The values in each system are rounded to convenient numbers. An exact conversion is not used so that the values appear functional for safety purposes (e.g., rounded to the nearest half foot or tenth of a meter). Units of measure discussed in this Handbook are based on the customary (English or inch-foot-pound) system.

Rule 017B states that physical items referenced in the **Code** are in "nominal values" unless specific dimensions are provided. Other standards may set tolerances for manufacturing. An example of nominal values is shown in Fig. 017-1.

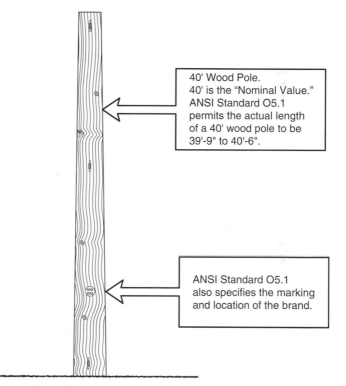

40' Wood Pole.
40' is the "Nominal Value."
ANSI Standard O5.1
permits the actual length
of a 40' wood pole to be
39'-9" to 40'-6".

ANSI Standard O5.1
also specifies the marking
and location of the brand.

See
Photo(s)

Fig. 017-1. Example of nominal values (Rule 017B).

018. METHOD OF CALCULATION

Rule 018 provides rounding requirements. In general, rounding "off" to the nearest significant digit is required unless otherwise specified in applicable rules. One rule that requires a different rounding method is Rule 230A4, which requires rounding "up" for clearance calculations as Sec. 23 deals with various overhead line clearances which are typically specified as "not less than" clearances. Rounding "off" follows the rules of traditional rounding learned in math class. An example of rounding "off" is rounding 20.02 down to 20.0 or rounding 20.66 up to 20.7. An example of rounding "up" for "not less than" clearance is rounding 20.02 to 20.1 because rounding "off" to 20.0 would not meet a clearance required to be "not less than" 20.02. See Rule 230A4 for additional information.

Section 02

Definitions of Special Terms

The **NESC** provides several terms and their definitions for use in the codebook. Section 02 is the first place to look for definitions of special terms. The **Code** text under the title of Sec. 02 references the *IEEE Standards Dictionary Online* available at https://dictionary.ieee.org for definitions not contained in this section. The *IEEE Standards Dictionary Online* should be used as a second step if the definition is not provided in Sec. 02 of the **NESC**. The third and final step for definitions not provided in Sec. 02 of the **NESC** or in the *IEEE Standards Dictionary Online* is to look the word up in a standard dictionary. These three steps are outlined in Fig. 02-1.

Occasionally, terms are defined in individual rules instead of Sec. 02. One example of this is the words "shall," "should," and "RECOMMENDATION" that are defined in Rule 015. Another example is how the word "equipment" is defined relative to a specific application. Rule 238A defines equipment relative to clearance between communication and supply facilities located on the same overhead line structure. Rule 380A provides examples of equipment relative to underground construction.

Clarifications and drawings of key terms are provided in this handbook. They are provided in the individual rules in which the terms apply instead of this section with the exception of two terms, voltage and effectively grounded, which are discussed below.

The term voltage has five definitions in Sec. 02 of the **NESC**. In some code rules and tables, voltage is specifically stated as phase to phase or phase to ground. In some locations, a voltage is stated without a phase-to-phase or phase-to-ground reference. If a voltage is stated without a phase-to-phase or phase-to-ground reference, the voltage is dependent on the type of grounding system. For example, if the code states that a clearance adder is required for lines over 50 kV (without a phase-to-phase or phase-to-ground reference) and the line is fed from a

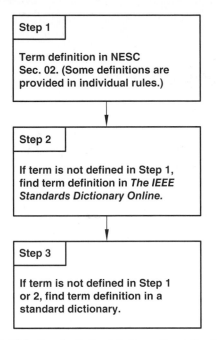

Fig. 02-1. Steps for finding definitions of terms (Sec. 02).

wye-connected single grounded system/unigrounded system with ground fault relaying that is connected to an effective ground, the clearance adder may be applied to lines over 50 kV to ground (or 86.6 kV phase to phase). If the line is ungrounded, the clearance adder must be applied to lines over 50 kV phase to phase. Voltage as defined in Sec. 02 is the effective rms voltage. See Rule 230G for a discussion and figures related to AC rms voltage, DC voltage, phase-to-phase voltage, and phase-to-ground voltage.

The terms effective ground/effectively grounded, effectively grounded neutral conductor, multi-grounded system, and single-grounded system/unigrounded system are all defined in Sec. 02 of the **NESC**. The important part of the definition of effective ground/effectively grounded is what the definition does not say. The definition of effective ground/effectively grounded does not provide a value in ohms (e.g., the **Code** does not say that an effective ground/effectively grounded system is 5 Ω or less). Multigrounded systems discussed in Rule 096C typically are considered effectively grounded, but in special cases (e.g., a very rocky 1-mile stretch of line), more than four grounds in each mile may be needed to make a multigrounded system an effectively grounded system. Rule 096D does specify a 25-Ω limit for single-grounded systems.

Section 03

References

Section 03 provides a list of standards that are referenced in the **Code**. The standards listed form a part of the **Code** to the extent that they are referenced in the **Code** rules. If a standard is cited for information purposes only, it appears in the Bibliography in Appendix E of the **NESC**. For example, ANSI O5.1, *American National Standard Specifications and Dimensions for Wood Poles,* is listed in Sec. 03 as it is referenced in **Code** Rule 261A2b(1). However, IEEE Standard 80, *IEEE Guide for Safety in AC Substation Grounding,* is listed in Appendix E of the **NESC** (Bibliography) as it is cited for informational purposes in a "NOTE" in Rule 092E. Rule 015 states that a "NOTE" is provided for information purposes only. Some standards are referenced in a **Code** rule in one location and cited in a "NOTE" in other locations. These standards with a dual reference are listed in Sec. 03. The **Code** recognizes that the standards listed in Sec. 03 provide information that does not need to be repeated in the **Code**. This helps keep the codebook from getting too wordy and utilizes standards that another agency has documented. The relationship between Sec. 03—References and Appendix E—Bibliography is shown in Fig. 03-1.

The **Code** acknowledges, with a note in Sec. 03, that current standards may be newer than the ones listed, as the **Code** is updated on a 5-year revision cycle and some standards may be updated during the middle of this process. The standards listed in Sec. 03 are an important part of a technical library of reference material. Additional notes in Sec. 03 of the **NESC** provide names and website addresses of the various standards making bodies from which the standards may be obtained.

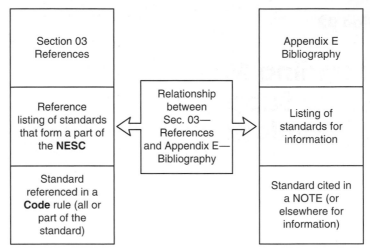

Fig. 03-1. Relationship between Sec. 03—References and Appendix E—Bibliography (Sec. 03).

Section 09

Grounding Methods for Electric Supply and Communications Facilities

090. PURPOSE

The purpose of Sec. 09 is to provide practical methods of grounding. Grounding is one of the ways to protect people from hazardous voltages. It also allows protective devices to operate during a fault condition. The basic theory behind grounding is to keep the voltage of a grounded part (e.g., equipment case, neutral conductor, and communications messenger) as close as possible to the potential of the earth so that a voltage difference does not exist between a person and a grounded metal object. The **Code** states in this rule that grounding is used as one of the means of safeguarding employees and the public from injury. Other means include, but are not limited to, guarding, adequate clearance above ground, proper burial depth, etc.

091. SCOPE

The scope of Sec. 09 is to provide the *methods* of protective grounding for supply and communication conductors and equipment. The *requirements* for grounding are listed in the other parts of the **Code**. The requirements for grounding electric supply stations are predominantly in Rule 123. The requirements for grounding overhead lines are predominantly in Rule 215. The requirements for grounding underground lines are predominantly in Rule 314. Rules 123, 215, and 314 all use the term effectively grounded. Rule 091 provides methods for effective grounding and points to the definition of effectively grounded in Sec. 02, "Definitions of special terms." The scope of Sec. 09 does not include the grounded return of electric railways or lightning protection not associated with supply and communication wires, for example, lightning protection wires connected to a lightning rod on top of a barn.

092. POINT OF CONNECTION OF GROUNDING CONDUCTOR

092A. Direct Current Systems That Are Required to Be Grounded. This rule has basic connection requirements for direct-current (DC) systems. For 750 V and less, the grounding conductor connection must be made only at the supply station. For three-wire DC circuits, the connection must be made to the neutral. For DC systems over 750 V, the grounding conductor connection must be made at both the supply and load points. The connection must be made to the neutral of the system. The ground or grounding electrode can be external or remote from each of the stations. This permits separating the electrode from areas with ground currents that can cause electrolytic damage. The **Code** permits one of the two stations to have its grounding connection made through a surge arrester as long as the other station has the neutral effectively grounded. An exception is provided for the 750-V and greater category for back-to-back DC converter stations that are adjacent to each other. For this condition the neutral of the system should be connected to ground at one point only.

**092B. Alternating-Current Systems
That Are Required to Be Grounded**
092B1. 750 V and Below. The point of the grounding connection on alternating-current (AC), wye-connected, three-phase, four-wire and single-phase, three-wire systems operated at 750 V and below is shown in Fig. 092-1.

On other one-, two-, or three-phase systems feeding lighting circuits, a grounding connection must be made to a common circuit conductor. Common examples include a 120/240-V, three-phase, four-wire center tap delta service; a 120/208-V, single-phase, three-wire service fed from a 120/208-V, three-phase, four-wire service; or a 120-V, single-phase, two-wire service fed from a 120/240-V, single-phase, three-wire service.

Wye and delta circuits that are not grounded or do not use a common (neutral) conductor for grounding cannot be used to serve lighting loads. See Fig. 092-2.

Grounding connections must be made at the source and line side of a service as shown in Fig. 092-3.

092B2. Over 750 V. Nonshielded conductors (e.g., bare neutral conductors) must be grounded as shown in Fig. 092-4.

The wording in Rule 092B2a requires unigrounding at the source (e.g., a substation transformer) and permits, but does not require, multigrounding along the line. However, various rules in Part 2, Overhead Lines, and Part 3, Underground

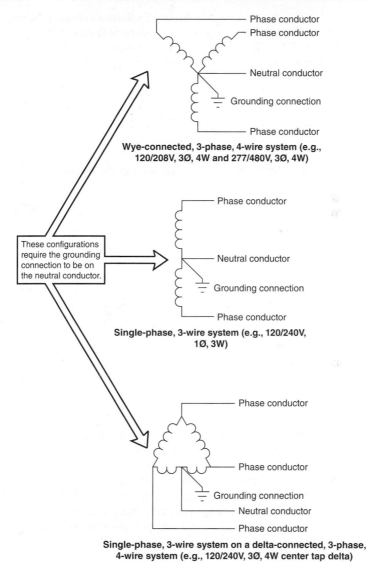

Wye-connected, 3-phase, 4-wire system (e.g., 120/208V, 3Ø, 4W and 277/480V, 3Ø, 4W)

These configurations require the grounding connection to be on the neutral conductor.

Single-phase, 3-wire system (e.g., 120/240V, 1Ø, 3W)

Single-phase, 3-wire system on a delta-connected, 3-phase, 4-wire system (e.g., 120/240V, 3Ø, 4W center tap delta)

Fig. 092-1. Grounding connection on wye-connected, three-phase, four-wire and single-phase, three-wire systems (Rule 092B1).

Lines, will require systems to be effectively grounded. Effectively grounded systems typically need to be multigrounded to provide sufficiently low ground impedance. Multigrounded systems are discussed in Rule 096C.

Shielded conductors on riser poles must be grounded as shown in Fig. 092-5.

Shielded cables without an insulating jacket must be grounded as shown in Fig. 092-6.

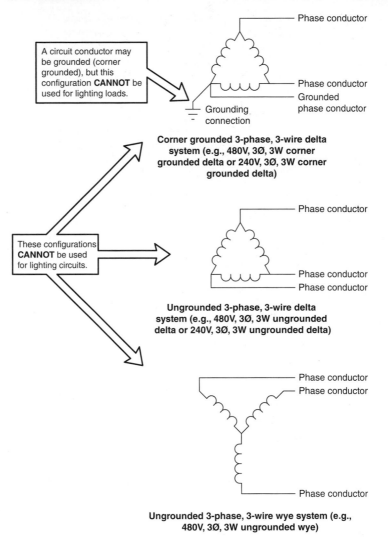

A circuit conductor may be grounded (corner grounded), but this configuration **CANNOT** be used for lighting loads.

Phase conductor

Phase conductor
Grounded phase conductor

Grounding connection

Corner grounded 3-phase, 3-wire delta system (e.g., 480V, 3Ø, 3W corner grounded delta or 240V, 3Ø, 3W corner grounded delta)

These configurations **CANNOT** be used for lighting circuits.

Phase conductor

Phase conductor
Phase conductor

Ungrounded 3-phase, 3-wire delta system (e.g., 480V, 3Ø, 3W ungrounded delta or 240V, 3Ø, 3W ungrounded delta)

Phase conductor
Phase conductor

Phase conductor

Ungrounded 3-phase, 3-wire wye system (e.g., 480V, 3Ø, 3W ungrounded wye)

Fig. 092-2. Wye and delta systems not to be used for lighting loads (Rule 092B1).

Shielded cables with an insulating jacket must be grounded as shown in Fig. 092-7.

Shielded cable without an insulating jacket that is buried in direct contact with the earth has an advantage of being grounded all along its length. However, direct-buried shielded cable without an insulating jacket is susceptible to corrosion. The insulating jacket can prevent corrosion of the shield or concentric neutral. See Rule 096C for multigrounding requirements and special exceptions.

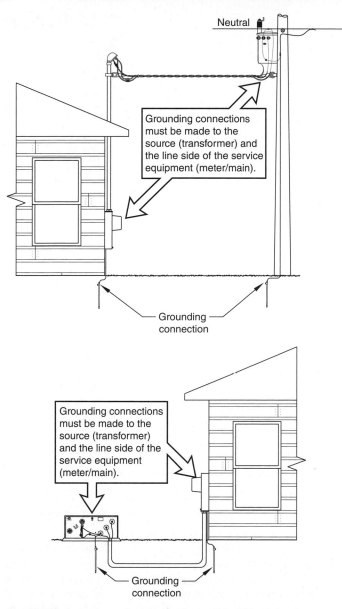

Neutral

Grounding connections must be made to the source (transformer) and the line side of the service equipment (meter/main).

Grounding connection

Grounding connections must be made to the source (transformer) and the line side of the service equipment (meter/main).

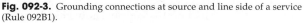

Grounding connection

Fig. 092-3. Grounding connections at source and line side of a service (Rule 092B1).

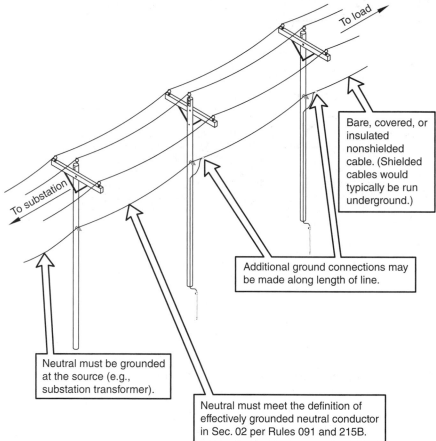

To load

To substation

Bare, covered, or insulated nonshielded cable. (Shielded cables would typically be run underground.)

Additional ground connections may be made along length of line.

Neutral must be grounded at the source (e.g., substation transformer).

Neutral must meet the definition of effectively grounded neutral conductor in Sec. 02 per Rules 091 and 215B.

See Photo(s)

Fig. 092-4. Grounding connections for nonshielded cables over 750 V (Rule 092B2a).

092B3. Separate Grounding Conductor.   If a separate grounding conductor is used on an AC system to be grounded as an adjunct (joined addition) to a cable run underground, there are several conditions that apply. The separate grounding conductor must be connected directly or through the neutral to items that must be grounded. The conductor must be located as shown in Fig. 092-8.

Adjunct (joined addition) grounding conductors are typically used with shielded supply cables. If the shield on the supply cable is not a sufficient size to carry neutral current or fault current, an adjunct grounding cable can be used. An adjunct grounding conductor should not be used to replace a corroded concentric neutral conductor in a direct-buried cable. Rule 350B requires that a direct-buried cable operating above 600 V have a continuous metallic shield,

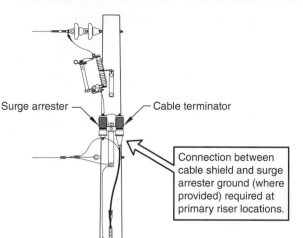

Surge arrester — — Cable terminator

Connection between cable shield and surge arrester ground (where provided) required at primary riser locations.

See Photo(s)

Fig. 092-5. Surge arrester cable—shielding interconnection (Rule 092B2b(1)).

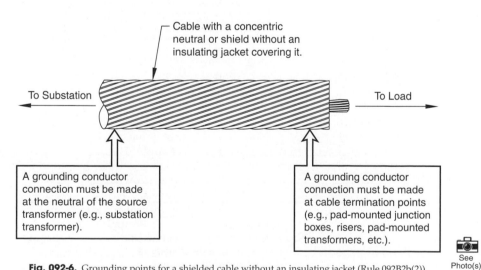

— Cable with a concentric neutral or shield without an insulating jacket covering it.

To Substation ◄

To Load ►

A grounding conductor connection must be made at the neutral of the source transformer (e.g., substation transformer).

A grounding conductor connection must be made at cable termination points (e.g., pad-mounted junction boxes, risers, pad-mounted transformers, etc.).

See Photo(s)

Fig. 092-6. Grounding points for a shielded cable without an insulating jacket (Rule 092B2b(2)).

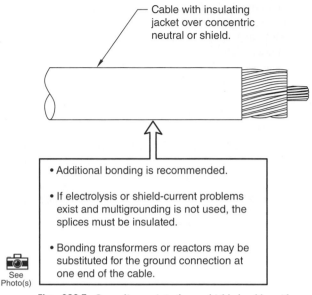

Cable with insulating
jacket over concentric
neutral or shield.

• Additional bonding is recommended.

• If electrolysis or shield-current problems
exist and multigrounding is not used, the
splices must be insulated.

• Bonding transformers or reactors may be
substituted for the ground connection at
one end of the cable.

See
Photo(s)

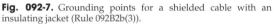

Fig. 092-7. Grounding points for a shielded cable with an
insulating jacket (Rule 092B2b(3)).

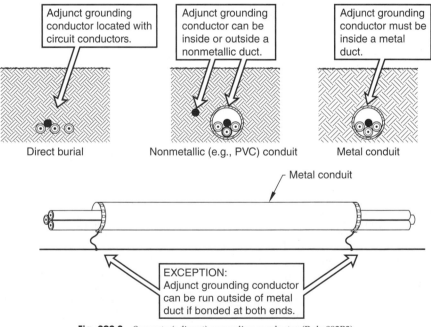

Adjunct grounding
conductor located with
circuit conductors.

Adjunct grounding
conductor can be
inside or outside a
nonmetallic duct.

Adjunct grounding
conductor must be
inside a metal
duct.

Direct burial   Nonmetallic (e.g., PVC) conduit   Metal conduit

Metal conduit

EXCEPTION:
Adjunct grounding conductor
can be run outside of metal
duct if bonded at both ends.

Fig. 092-8. Separate (adjunct) grounding conductor (Rule 092B3).

sheath, or concentric neutral. The adjunct grounding conductor can be used to supplement the concentric neutral but not replace it if it has corroded away.

092C. Messenger Wires and Guys

092C1. Messenger Wires.   The point of connection of the grounding conductor to messenger wires that are required to be effectively grounded by other parts of the **Code** is shown in Fig. 092-9.

Communications messenger wires are required to be effectively grounded in Part 2, Overhead Lines, Rule 215C. Communications messenger wires on joint-use (power and communication) poles are required to be grounded in Secs. 23 and 24 to meet certain clearance and grade of construction requirements. The messenger must meet certain ampacity and strength criteria defined in Rules 093C1, 093C2, and 093C5. For the messenger (on a joint-use

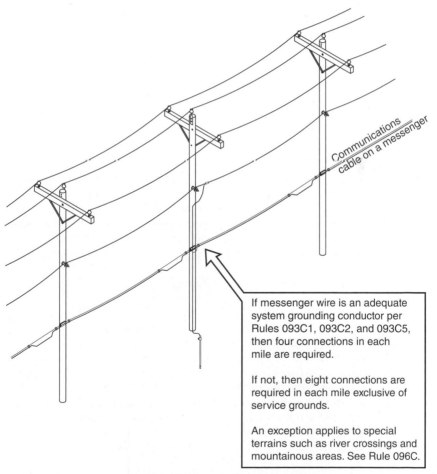

Communications cable on a messenger

If messenger wire is an adequate system grounding conductor per Rules 093C1, 093C2, and 093C5, then four connections in each mile are required.

If not, then eight connections are required in each mile exclusive of service grounds.

An exception applies to special terrains such as river crossings and mountainous areas. See Rule 096C.

See Photo(s)

Fig. 092-9. Grounding of messenger wires (Rule 092C1).

power and communication structure) to meet the ampacity requirement in Rule 093C2, the messenger wire ampacity must be rated not less than one-fifth of the neutral wire ampacity. It is sometimes difficult to find the ampacity rating of messenger wires as many communications messenger wires are actually guy wires. Manufacturers of guy wires typically provide mechanical strength ratings, not electrical ampacity ratings. The four grounds in each mile rule appears here for the first time in the Code. It is discussed in detail in Rule 096C.

092C2. Guys. The point of connection of the grounding conductor to guys that are required to be effectively grounded by other parts of the Code is shown in Fig. 092-10.

Guys (both supply and communication) must be either grounded (per Rule 215C2) or insulated (per the exception to Rule 215C2). If guys are grounded, they must be grounded using the methods in this rule.

092C3. Common Grounding of Messengers and Guys on the Same Supporting Structure. When messengers and guys are on the same supporting structure and they are required to be effectively grounded by other parts of the Code, they must be bonded together and effectively grounded by the connection methods listed in this rule. The methods listed are a combination of the messenger and guy connection requirements provided in Rules 092C1 and 092C2.

092D. Current in Grounding Conductor. This rule recognizes that multigrounded systems, for example, a 12.47/7.2-kV, three-phase, four-wire circuit that has four or more grounds in each mile, may develop objectionable current flow on the grounding conductor (pole ground). This rule provides methods to alleviate the objectionable current flow.

Objectionable current flow may exist due to stray earth currents or other reasons. Fault currents and lightning discharge currents are not considered objectionable current flows when applying this rule and some amount of current will always be present on the grounding conductor (pole ground) during normal operation. Separating primary and secondary neutrals on multigrounded systems to address stray voltage concerns is addressed in Rule 097D2.

092E. Fences. When conductive electric supply station fences are required to be effectively grounded by Part 1 of the Code (primarily in Rule 110A1), they must be connected to a grounding conductor as shown in Fig. 092-11.

This rule provides both specific requirements for conductive electric supply station fence grounding (Rules 092E1 through 092E6) and general requirements by noting IEEE Standard 80, which is the industry standard for substation grounding. Rule 093C6 also applies to fences. Fence mesh strands are only required to be bonded if the fence posts are nonconducting. For conducting (metal) fence posts, the fence mesh must be under tension and electrically connected to the post for the mesh to be grounded. A grounding conductor feed up to barbwire strands at the top of a fence can be woven through the chain-link mesh for added grounding continuity. A ground grid which is typically buried under an electric supply station and connected to the station fence is discussed in Rule 096B. An example of electric supply station fence grounding is shown in Fig. 092-12.

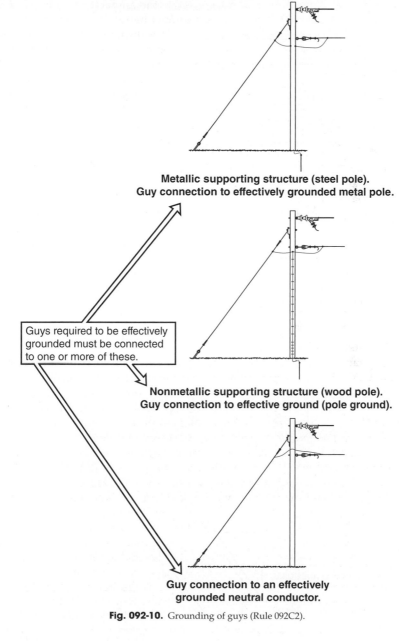

**Metallic supporting structure (steel pole).
Guy connection to effectively grounded metal pole.**

Guys required to be effectively
grounded must be connected
to one or more of these.

**Nonmetallic supporting structure (wood pole).
Guy connection to effective ground (pole ground).**

**Guy connection to an effectively
grounded neutral conductor.**

Fig. 092-10. Grounding of guys (Rule 092C2).

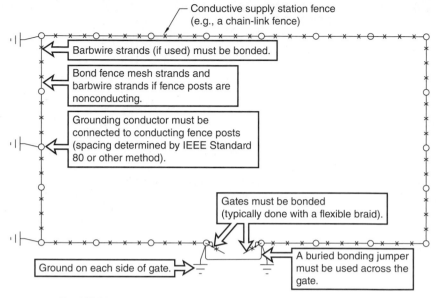

Conductive supply station fence
(e.g., a chain-link fence)

Barbwire strands (if used) must be bonded.

Bond fence mesh strands and
barbwire strands if fence posts are
nonconducting.

Grounding conductor must be
connected to conducting fence posts
(spacing determined by IEEE Standard
80 or other method).

Gates must be bonded
(typically done with a flexible braid).

Ground on each side of gate.

A buried bonding jumper
must be used across the
gate.

Fig. 092-11. Conductive electric supply station fence grounding (Rule 092E).

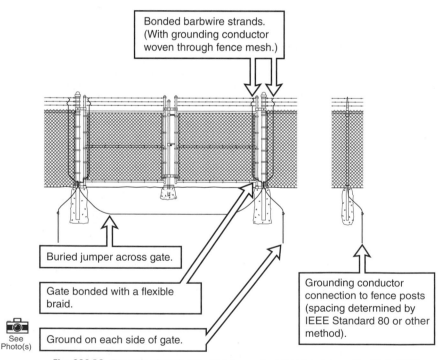

Bonded barbwire strands.
(With grounding conductor
woven through fence mesh.)

Buried jumper across gate.

Gate bonded with a flexible
braid.

See
Photo(s)

Ground on each side of gate.

Grounding conductor
connection to fence posts
(spacing determined by
IEEE Standard 80 or other
method).

Fig. 092-12. Example of conductive electric supply station fence grounding (Rule 092E).

093. GROUNDING CONDUCTOR
AND MEANS OF CONNECTION

093A. Composition of Grounding Conductors. Grounding conductors can be copper or other metals or combinations of metals that will not corrode during their expected service life under the existing conditions. Surge arrester connections must be short, straight, and free from sharp bends. Metallic electrical equipment cases or the structural metal frame of a building or structure can also be used as a grounding conductor. An example of a copper grounding conductor (pole ground) on a wood pole and an example of a structural metal grounding conductor (steel pole) are shown in Fig. 093-1.

Many utilities use copper for the entire length of the grounding conductor (pole ground). The size of the grounding conductor (pole ground) is covered in Rule 093C. Some utilities use aluminum or ACSR. If aluminum or ACSR is used above grade, it is typically spliced to copper, which then runs below grade (see Rule 093E5). Some utilities utilize copper-coated steel. Copper-coated steel has become popular due to copper theft. Rule 093A states that the grounding conductor

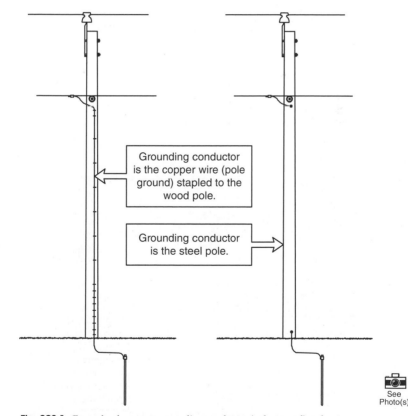

Grounding conductor is the copper wire (pole ground) stapled to the wood pole.

Grounding conductor is the steel pole.

See Photo(s)

Fig. 093-1. Example of a copper grounding conductor (pole ground) and a structural metal grounding conductor (steel pole) (Rule 093A).

should be without a joint or splice. See Rule 093E for an example of an unavoidable splice used in a pole ground.

The grounding conductor must not have a switching device connected to it. Some exceptions apply, including high-voltage DC systems, testing under competent supervision, and surge arrester operation. This rule provides an important note stating that the normally grounded base of the surge arrester may be at line potential (fully energized) following the operation of the disconnector.

093B. Connection of Grounding Conductors. The connection between the grounding conductor (pole ground) and grounded conductor (neutral) must be made considering the metals involved and exposure to the environment. The connector must not corrode and must be rated for the type of metals it is connecting. Dissimilar metals connected together with an improper connector will set up a battery action which will accelerate corrosion. Soldering is not acceptable, except on lead sheath cable, as fault currents will produce enough heat to melt the solder. Suitable connection methods and clarification of the terms grounded and grounding are shown in Fig. 093-2.

093C. Ampacity and Strength. This rule defines short-time ampacity requirements for bare and insulated grounding conductors. A bare conductor can carry a larger fault current than an insulated conductor of the same size because the bare conductor is only limited by melting or damaging the conductor material. The insulated grounding conductor has the additional constraint of not damaging the insulation. See Fig. 093-3.

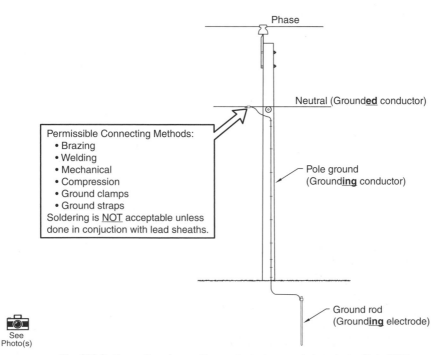

Phase

Neutral (Ground**ed** conductor)

Permissible Connecting Methods:
- Brazing
- Welding
- Mechanical
- Compression
- Ground clamps
- Ground straps

Soldering is <u>NOT</u> acceptable unless done in conjuction with lead sheaths.

Pole ground
(Ground**ing** conductor)

Ground rod
(Ground**ing** electrode)

See Photo(s)

Fig. 093-2. Connection of grounding conductor to grounded conductor (Rule 093B).

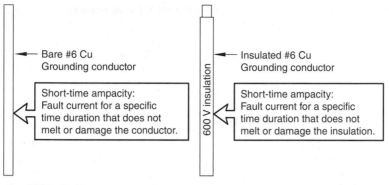

Fig. 093-3. Short-time ampacity of bare and insulated grounding conductors (Rule 093C).

Short-time ampacity of both bare and insulated conductors can be obtained from conductor manufacturers. This information is typically referred to as a conductor short-circuit withstand chart or a conductor damage curve.

Short-time ampacity for a single-grounded system is shown in Fig. 093-4.

Short-time ampacity for a multigrounded AC system is shown in Fig. 093-5.

Rule 093C2 references Rule 093C8, which also specifies ampacity limits based on the ampacity of phase conductors and grounding electrode resistance. The one-fifth ampacity requirement applies to the normal operating current, not to the short-time fault ampacity. An example of pole ground sizing is shown in Fig. 093-6.

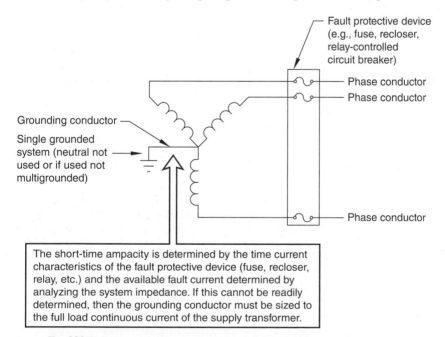

Fig. 093-4. System grounding conductor for single-grounded systems (Rule 093C1).

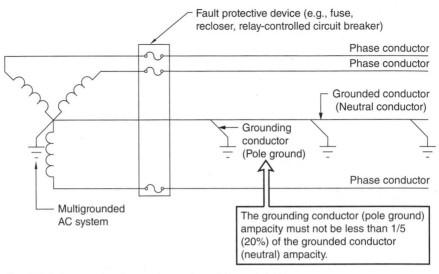

Fig. 093-5. System grounding conductors for multigrounded AC systems (Rule 093C2).

In addition to checking the pole ground ampacity relative to the primary neutral, the service transformer neutral should also be considered. A bare AWG No. 6 copper pole ground connected to the neutral of a large secondary service may not have the required one-fifth ampacity of the secondary neutral. Large secondary services require careful application of Rules 093C2 and 093C8.

In addition to single-grounded and multigrounded system requirements, Rule 093C requires AWG No. 12 copper or larger conductors to ground instrument transformer cases and instrument transformer secondary circuits, and AWG No. 6 copper or AWG No. 4 aluminum or larger conductors to ground primary surge arresters. The primary surge arrester rule, 093C4, has an exception permitting use of copper-clad or aluminum-clad steel wires.

Per Rule 093C5, grounding conductors for equipment, messenger wires, and guys must have a short-time ampacity based on the available fault current and operating time of the circuit protective device. If the circuit does not have an overcurrent or fault protection device (e.g., fuse, recloser, relay-controlled circuit breaker, etc.), then the design and operating conditions of the circuit must be analyzed and the grounding conductor cannot be smaller than AWG No. 8 copper. If a conductor enclosure (e.g., rigid steel conduit) is connected to a metal equipment enclosure with suitable lugs, bushings, etc., the metallic conduit and metallic equipment path can be used as an equipment-grounding conductor. Grounding conductors for equipment, messenger wires, and guys must be connected to a suitable lug, terminal, or other device without disruption during normal inspection, maintenance, or operation. The Code does not specifically address using the messenger or guy hardware as the bond between the messenger or guy and the grounding conductor. For example, a

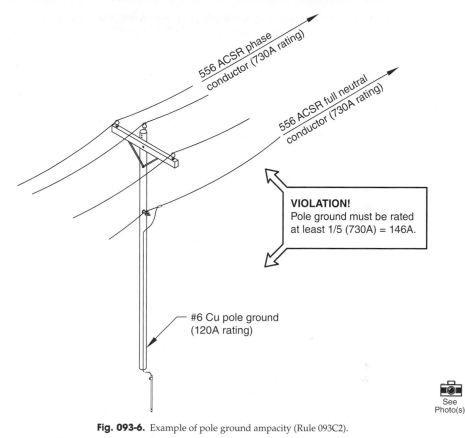

Fig. 093-6. Example of pole ground ampacity (Rule 093C2).

transmission tower static wire may be bonded to the steel tower through the static wire hardware or a separate bonding jumper may be used between the static wire and the transmission tower. Another example is an anchor guy may be bonded to a grounding conductor through the guy wire hardware or a separate bonding jumper may be used between the grounding conductor and the guy wire. Rule 012C, which requires accepted good practice, must be used for these examples. Use of a separate bonding jumper provides a connection that meets this rule, meets the definition of "bonding" in Sec. 09, and does not rely on hardware that can become loose or slack during the inspection, maintenance, or operation of the line.

The ampacity and strength of the grounding conductor used for grounding fences must also have adequate short-time ampacities or must be Stl WG No. 5 or larger.

Bonding of equipment frames and enclosures must consist of a metallic path back to the grounded terminal of the local supply. If the supply is remote, metallic parts within reach must be bonded and connected to ground.

Rule 093C8 specifies an ampacity limit such that no grounding conductor needs to have an ampacity greater than either:

- The phase conductor that would supply the ground fault, or
- The maximum current in the grounding conductor calculated by dividing the supply voltage by the electrode resistance

Consider an example related to Rule 093C8b. Assuming a 7200-V phase to ground circuit and assuming a 25-Ω ground rod resistance, 7200 V divided by 25 Ω = 288 A. For a 120/240-V secondary, 120 V to ground divided by a 25-Ω ground rod resistance would be 4.8 A. Rule 093C8 may limit the size of the ground wire specified in other parts of Rule 093C based on required ampacity. Secondary services may have large grounded (neutral) conductors; however, the grounding (pole ground) conductor size may be limited by applying Rule 093C8. In this example, the assumption of a 25-Ω ground rod resistance is just that, an assumption. Ground rod resistance will vary by type of soil, moisture in the soil, length of rod, etc. Field measurements must be taken to determine actual ground rod resistance.

The mechanical strength of grounding conductors must be suitable to the conditions they are exposed to (i.e., lawn mowers, weed eaters, car bumpers, etc.). Unguarded grounding conductors must have a tensile strength equal to or greater than AWG No. 8 soft-drawn copper except for conductors noted in Rule 093C3 (i.e., AWG No. 12 copper for instrument transformers).

093D. Guarding and Protection.   Guards over grounding conductors (i.e., pole grounds) are only required for single-grounded systems that are exposed to the public. If the grounding electrode is on a single-grounded system that is not exposed to the public (e.g., in a fenced substation), it does not have to be guarded. Grounding conductors on multigrounded systems are not required to be guarded even if they are exposed to mechanical damages. A multigrounded system requires at least four grounds in each mile per Rule 096C, and Rule 214 requires inspection of overhead lines. These two requirements provide a method to assure safe grounding on multigrounded systems; therefore, guards on multigrounded systems are not required. If guards are not required but they are installed, they should be installed in a manner as if they were required.

Rules 239D and 360A provide additional guarding requirements for various types of conductors. If guarding of the grounding conductor is required, guards must be suitable for the damage to which they will be exposed and be at least 8 ft above the ground or other surface. If guarding of the grounding conductor is not required, a typical installation method is stapling the grounding conductor to a wood pole. If the grounding conductor that is not required to be guarded is exposed to mechanical damage, in addition to stapling, where practical, the grounding conductor is to be located on the side of the pole with least exposure to damage (e.g., away from car bumpers in a parking lot). The requirements for grounding conductors with or without guards are outlined in Fig. 093-7.

Rule 093D4 recognizes that an inductive choke is created when a conductor is run through a metallic raceway. This can create a hazardous voltage during a lightning strike (or even during a fault condition). The Code requires a nonmetallic guard to avoid this condition. The strength of nonmetallic materials (i.e., plastics) has increased to the point where they can be used for protection without cracking or breaking. A U-shaped metallic raceway is acceptable, as it does not completely enclose the grounding

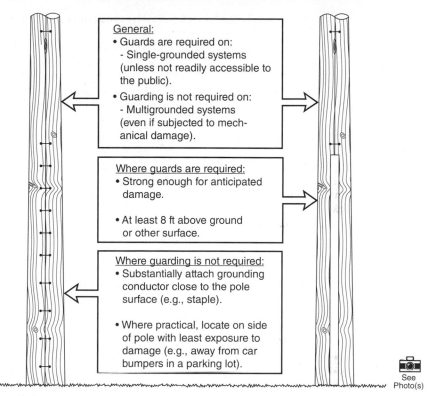

Fig. 093-7. Requirements for grounding conductors with or without guards (Rules 093D1, 093D2, and 093D3).

conductor. If a metallic guard similar to a steel pipe or rigid metal conduit is used, it must be bonded to the grounding conductor at both ends, as shown in Fig. 093-8.

093E. Underground.   Grounding conductors laid underground require slack due to the settling of the earth. Direct-buried joints or splices must be made with corrosion resistance in mind. Splices should be kept to a practical minimum. A cable insulation shield (e.g., concentric neutral, metallic foil, braid, etc.) must be connected to other grounded supply equipment in manholes, handholes, and vaults. Exceptions exist for cathodic protection and cross bonding. An example of grounding interconnection in a manhole is shown in Fig. 093-9.

Looped magnetic elements must not be positioned between the grounding conductor and the phase conductors. The metals used for grounding in earth, concrete, or masonry must not corrode. This rule specifically notes that aluminum is not generally acceptable when used underground. An example of an aluminum ground wire that transitions to copper for underground burial is shown in Fig. 093-10.

Sheath transposition connections, also termed cross bonding, are sometimes used to neutralize induced voltages and therefore eliminate or minimize

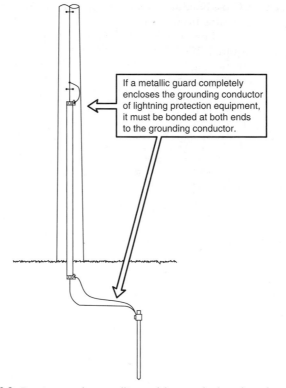

If a metallic guard completely encloses the grounding conductor of lightning protection equipment, it must be bonded at both ends to the grounding conductor.

Fig. 093-8. Requirements for a metallic guard that completely encloses the grounding conductor of lightning protection equipment (Rule 093D4).

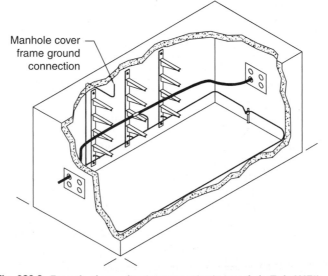

Manhole cover frame ground connection

Fig. 093-9. Example of grounding interconnection in a manhole (Rule 093E3).

circulating currents. Cross bonding of cable shields or sheaths involves insulating the cable shields or sheaths from ground at sectionalized points along the cable route. The insulation level for cross bonding the cable shields or sheaths must be 600 V or greater if required. The cross bonding jumpers must be sized to carry the available fault current. See Fig. 093-11.

093F. Common Grounding Conductor for Circuits, Metal Raceways, and Equipment. This rule allows one common grounding conductor for both the supply system (neutral) and equipment (e.g., a recloser) where the ampacity of the grounding conductor is adequate for both. Ampacity for the system grounding conductor and equipment grounding conductor is discussed in Rule 093C. Rule 097 addresses a common grounding conductor for primary and secondary neutrals at transformer locations. An example of one common grounding conductor for the circuit and equipment is shown in Fig. 093-12.

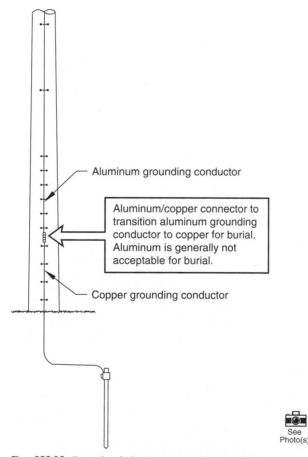

Aluminum grounding conductor

Aluminum/copper connector to transition aluminum grounding conductor to copper for burial. Aluminum is generally not acceptable for burial.

Copper grounding conductor

See Photo(s)

Fig. 093-10. Example of aluminum grounding conductor transitioning to copper for burial (Rule 093E5).

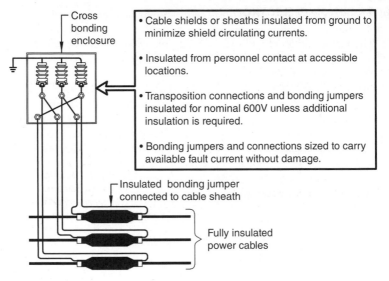

Cross bonding enclosure

• Cable shields or sheaths insulated from ground to minimize shield circulating currents.

• Insulated from personnel contact at accessible locations.

• Transposition connections and bonding jumpers insulated for nominal 600V unless additional insulation is required.

• Bonding jumpers and connections sized to carry available fault current without damage.

Insulated bonding jumper connected to cable sheath

Fully insulated power cables

Fig. 093-11. Sheath transposition connections (cross bonding) (Rule 093E6).

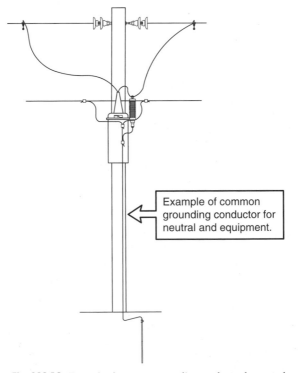

Example of common grounding conductor for neutral and equipment.

See Photo(s)

Fig. 093-12. Example of common grounding conductor for neutral and equipment (Rule 093F).

094. GROUNDING ELECTRODES

Grounding electrodes can be existing electrodes or made electrodes. Existing electrodes are existing conductive items buried in the earth for a purpose other than grounding but can also serve as a grounding electrode. Most utilities use made electrodes, which are purposely constructed and buried to serve as grounding electrodes. Requirements for existing electrodes are outlined in Figs. 094-1 through 094-3.

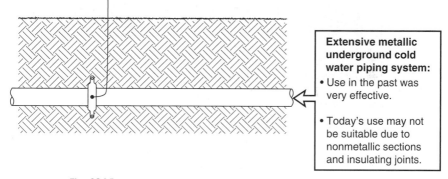

Extensive metallic underground cold water piping system:
• Use in the past was very effective.

• Today's use may not be suitable due to nonmetallic sections and insulating joints.

Fig. 094-1. Existing electrode—metallic water piping system (Rule 094B1).

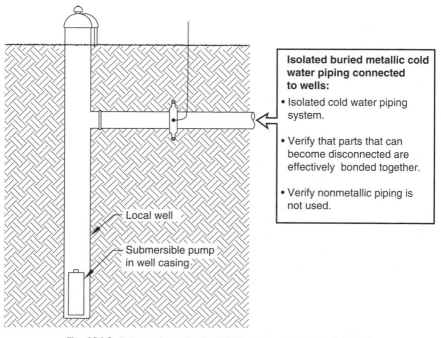

Isolated buried metallic cold water piping connected to wells:
• Isolated cold water piping system.

• Verify that parts that can become disconnected are effectively bonded together.

• Verify nonmetallic piping is not used.

Local well

Submersible pump in well casing

Fig. 094-2. Existing electrode—local (water piping) system (Rule 094B2).

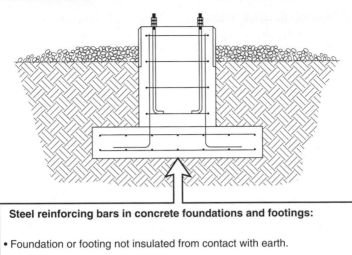

Steel reinforcing bars in concrete foundations and footings:

• Foundation or footing not insulated from contact with earth.

• Extends at least 3' below grade.

• Steel structure on top of foundation can be used as a grounding conductor when it is interconnected by bonding between the anchor bolts and reinforcing bars or by cable from the reinforcing bars to the structure above.

• Steel ties are considered to provide adequate bonding between bars of the reinforcing cage.

• A note warns about fault currents damaging the concrete foundation/footing.

Fig. 094-3. Existing electrode—steel reinforcing bars in concrete foundations and footings (Rule 094B3).

Made electrodes should penetrate the moisture level and be below the frost line. They must be metal or combined metals that do not corrode and they must not be painted, enameled, or covered in any way with an insulating material. For the purposes of this rule (primarily for strips, plates or sheets, and pole butt plates), stainless steel with appropriate non-corrosive properties is considered to be nonferrous metal. The driven ground rod is the most commonly used made electrode. However, buried wire, strips, and plates are considered equivalent if they meet the **Code** requirements. Other made electrodes may be used if supported by a qualified engineering study. Many utilities require the ground rod to be located in undisturbed earth a fixed distance away from the pole hole, although no such requirement is provided in the **Code**. The rules for ground rods use the term "driven rods," which implies driven into the earth, not dropped in the pole hole or laid in a trench. Throughout Rule 094, the terms resistance and resistivity are used. Ground resistance of an electrode such as a ground rod is measured in ohms (Ω). Soil resistivity, which is a measure of how much the soil resists the flow of electricity, is measured in ohm-centimeters ($\Omega \cdot$ cm). The requirements for various types of made electrodes listed in the **Code** are outlined in Figs. 094-4 through 094-12.

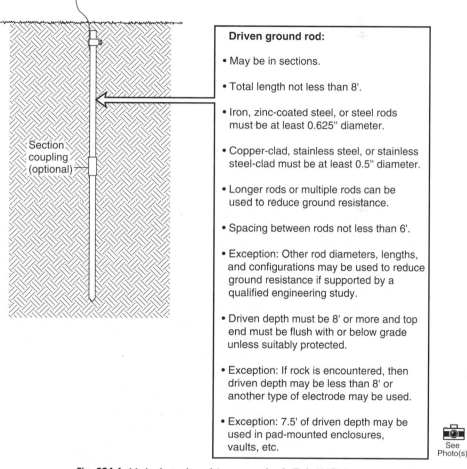

Driven ground rod:

• May be in sections.

• Total length not less than 8'.

• Iron, zinc-coated steel, or steel rods must be at least 0.625" diameter.

• Copper-clad, stainless steel, or stainless steel-clad must be at least 0.5" diameter.

• Longer rods or multiple rods can be used to reduce ground resistance.

• Spacing between rods not less than 6'.

• Exception: Other rod diameters, lengths, and configurations may be used to reduce ground resistance if supported by a qualified engineering study.

• Driven depth must be 8' or more and top end must be flush with or below grade unless suitably protected.

• Exception: If rock is encountered, then driven depth may be less than 8' or another type of electrode may be used.

• Exception: 7.5' of driven depth may be used in pad-mounted enclosures, vaults, etc.

See Photo(s)

Fig. 094-4. Made electrodes—driven ground rods (Rule 094C2a).

Buried wire (counterpoise):

- Material must be suitable for direct burial.
- Must be at least 0.162" in diameter and at least 100' long.
- Must be buried at least 18" deep.
- Must be laid approximately straight.
- May be arranged in a grid.
- Exception: 18" burial depth may be reduced for rock.
- May be useful in areas of high soil resistivity, or shallow bedrock, or where lower resistance is required than obtainable with rods.

Fig. 094-5. Made electrodes—buried wire (counterpoise) (Rule 094C2b(1)).

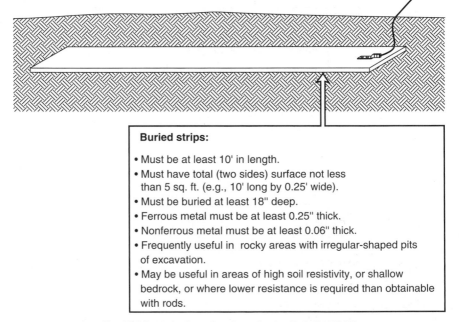

Buried strips:

- Must be at least 10' in length.
- Must have total (two sides) surface not less than 5 sq. ft. (e.g., 10' long by 0.25' wide).
- Must be buried at least 18" deep.
- Ferrous metal must be at least 0.25" thick.
- Nonferrous metal must be at least 0.06" thick.
- Frequently useful in rocky areas with irregular-shaped pits of excavation.
- May be useful in areas of high soil resistivity, or shallow bedrock, or where lower resistance is required than obtainable with rods.

Fig. 094-6. Made electrodes—buried strips (Rule 094C2b(2)).

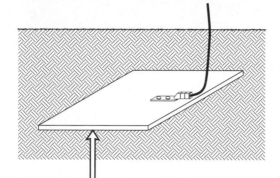

Buried plates or sheets:

• Must have at least 2 sq. ft. of surface exposed to soil.
  Therefore, 1' × 1' if both top and bottom are exposed to soil.
  If the top was exposed to soil and the bottom was exposed
  to rock or if the bottom was exposed to soil and the top was
  exposed to the bottom and sides of a pole, then 1' × 2' would
  be required.
• Must be buried at least 5' deep.
• Ferrous metal must be at least 0.25" thick.
• Nonferrous metal must be at least 0.06" thick.
• May be useful in areas of high soil resistivity, or shallow
  bedrock, or where lower resistance is required than obtainable
  with rods.

Fig. 094-7. Made electrodes—buried plates or sheets (Rule 094C2b(3)).

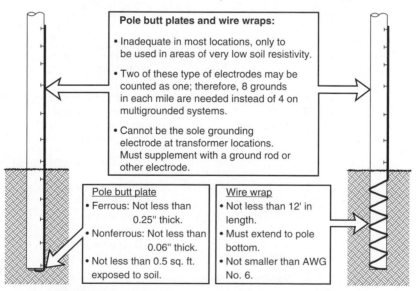

Pole butt plates and wire wraps:

• Inadequate in most locations, only to
  be used in areas of very low soil resistivity.

• Two of these type of electrodes may be
  counted as one; therefore, 8 grounds
  in each mile are needed instead of 4 on
  multigrounded systems.

• Cannot be the sole grounding
  electrode at transformer locations.
  Must supplement with a ground rod or
  other electrode.

Pole butt plate
• Ferrous: Not less than
  0.25" thick.
• Nonferrous: Not less than
  0.06" thick.
• Not less than 0.5 sq. ft.
  exposed to soil.

Wire wrap
• Not less than 12' in
  length.
• Must extend to pole
  bottom.
• Not smaller than AWG
  No. 6.

Fig. 094-8. Made electrodes—butt plates and wire wraps (Rule 094C3).

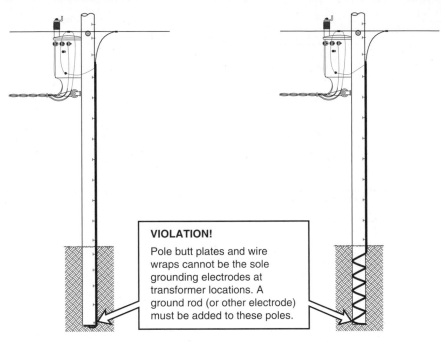

VIOLATION!

Pole butt plates and wire wraps cannot be the sole grounding electrodes at transformer locations. A ground rod (or other electrode) must be added to these poles.

Fig. 094-9. Made electrodes—butt plates and wire wraps at transformer locations (Rule 094C3a).

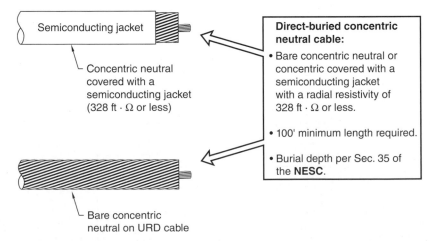

Semiconducting jacket

Concentric neutral covered with a semiconducting jacket (328 ft · Ω or less)

Direct-buried concentric neutral cable:

• Bare concentric neutral or concentric covered with a semiconducting jacket with a radial resistivity of 328 ft · Ω or less.

• 100' minimum length required.

• Burial depth per Sec. 35 of the **NESC**.

Bare concentric neutral on URD cable

Fig. 094-10. Made electrodes—direct-buried concentric neutral cable (Rule 094C4).

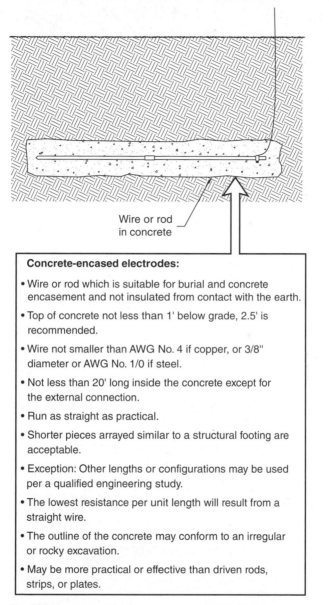

Wire or rod
in concrete

Concrete-encased electrodes:

• Wire or rod which is suitable for burial and concrete encasement and not insulated from contact with the earth.

• Top of concrete not less than 1' below grade, 2.5' is recommended.

• Wire not smaller than AWG No. 4 if copper, or 3/8" diameter or AWG No. 1/0 if steel.

• Not less than 20' long inside the concrete except for the external connection.

• Run as straight as practical.

• Shorter pieces arrayed similar to a structural footing are acceptable.

• Exception: Other lengths or configurations may be used per a qualified engineering study.

• The lowest resistance per unit length will result from a straight wire.

• The outline of the concrete may conform to an irregular or rocky excavation.

• May be more practical or effective than driven rods, strips, or plates.

Fig. 094-11. Made electrodes—concrete-encased electrodes (Rule 094C5).

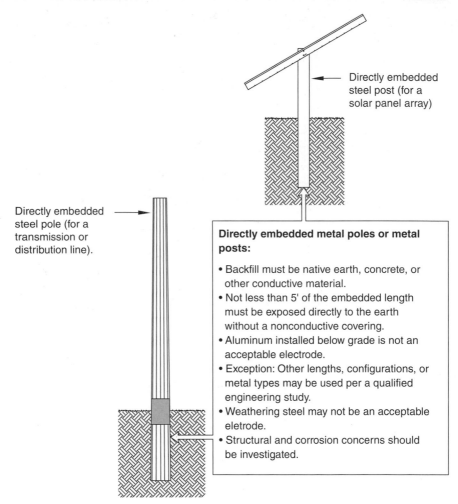

Fig. 094-12. Made electrodes—directly embedded metal poles or metal posts (Rule 094C6).

095. METHOD OF CONNECTION TO ELECTRODE

The connection to the grounding electrode must be permanent (except for removal due to inspection or maintenance) and be mechanically sound, corrosion-resistant, and have the required ampacity for the fault current to which it will be subjected. Suitable connection methods are shown in Fig. 095-1.

The **Code** also has specific rules for connecting to steel framed and non-steel-framed structures. The connection to water piping systems is also outlined. When water piping is used as the grounding electrode, bonds must be made around meters or other removable fittings.

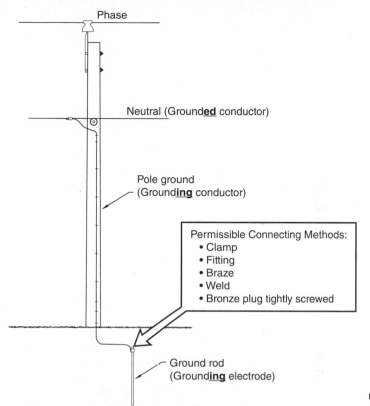

Fig. 095-1. Connection of grounding conductor to grounding electrode (Rule 095A).

The **Code** (in Sec. 094, "Grounding Electrodes") does not list gas piping as an acceptable electrode. Made electrodes or grounded structures should be separated from high-pressure (150 lb/in² or greater) pipelines containing flammable liquids or gases by a distance of 10 ft or more. No distance is specified for separating grounding electrodes from low-pressure gas lines. High-pressure pipelines are used as transmission facilities. Low-pressure gas lines are most commonly used to supply natural gas to homes. The requirements for separating grounding electrodes from high-pressure pipelines are shown in Fig. 095-2.

Rule 095C requires that the connection to the grounding electrode be free from rust, enamel, or scale. This may be done by cleaning or using fittings that penetrate such coatings.

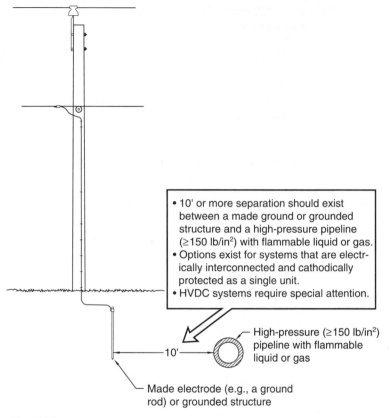

• 10' or more separation should exist between a made ground or grounded structure and a high-pressure pipeline ($\geq$150 lb/in^2) with flammable liquid or gas.
• Options exist for systems that are electrically interconnected and cathodically protected as a single unit.
• HVDC systems require special attention.

High-pressure ($\geq$150 lb/in^2) pipeline with flammable liquid or gas

10'

Made electrode (e.g., a ground rod) or grounded structure

Fig. 095-2. Grounding electrode separation from high-pressure pipelines (Rule 095B2).

096. GROUND RESISTANCE REQUIREMENTS

096A. General. The main intent of Rule 096 is to minimize hazards to persons by limiting step potential and assuring a grounding resistance low enough to permit prompt operation of circuit protective devices (e.g., fuses, reclosers, relay-controlled circuit breakers, etc.).

096B. Supply Stations. Supply stations typically require extensive grounding systems consisting of a ground grid or mat combined with grounding electrodes. They are designed to limit touch, step, mesh, and transferred potentials. The **Code** notes IEEE Standard 80 as a reference for substation grounding. Typically, the design of a substation ground grid starts with taking earth (soil) resistivity measurements. Earth resistivity is a measure of how much the soil resists the flow of electricity and is commonly expressed in ohm-centimeters ($\Omega \cdot$ cm). The final ground resistance (not resistivity) of the substation ground grid is measured in ohms (Ω) or sometimes a fraction of one ohm (Ω). Rules 092E and 093C6 apply to grounding the fence enclosing the electric supply station. The requirements of Rule 096B are outlined in Fig. 096-1.

096C. Multigrounded Systems. Multigrounded systems are the most common type of distribution system. A typical 12.47/7.2-kV, three-phase, four-wire grounded-wye distribution system is multigrounded. For a system to be multigrounded, the following must occur:

- The circuit must have a neutral of sufficient size and ampacity.
- The neutral must be connected to a grounding electrode at each transformer location.
- The neutral should be connected to a grounding electrode not less than four times in each mile of line. The grounds at transformers can be counted in the four grounds in each mile, but the grounds at individual services (i.e., meters) cannot be counted.

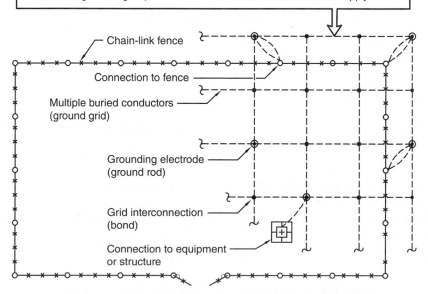

- Supply stations may require extensive grounding systems consisting of buried conductors, grounding electrodes, or interconnected combinations of both.

- The grounding system must be designed to limit touch, step, mesh, and transferred potentials. (No specifics are provided for conductor sizes, conductor spacing, or number of ground rods. IEEE Standard 80 is noted as a reference.)

- The fence grounding requirements of Rules 092E and 093C6 also apply.

Chain-link fence

Connection to fence

Multiple buried conductors
(ground grid)

Grounding electrode
(ground rod)

Grid interconnection
(bond)

Connection to equipment
or structure

Fig. 096-1. Ground resistance requirements for electric supply stations (Rule 096B).

The intent of a multigrounded system is to always carry a neutral and to have not less than four grounds in each mile of line. Typically overhead lines in urban areas or underground lines in subdivision areas have lots of service transformers and therefore lots of ground connections. Lines in rural areas or express feeders without many services typically need a review for not less than four grounds in each mile. To check the four grounds in each mile requirement, a "one-mile window" can be used. Examples are shown in Fig. 096-2.

For underground installations where the supply cable has an insulating jacket over the concentric neutral or the supply cable is in conduit, the cable must be terminated and grounded so that there are not less than four grounds in each mile along the line. If an express direct-buried underground feeder is

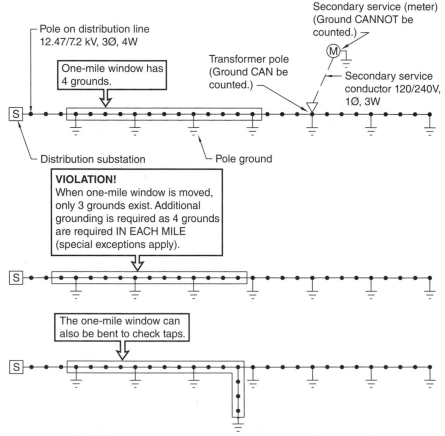

Fig. 096-2. Example of checking "four grounds in each mile" (Rule 096C).

constructed with an insulating jacket but without frequent termination points, the cable jacket can be stripped back and a suitable grounding electrode must be connected not less than four times in each mile along the line. If a supply cable has a semiconducting jacket, the cable can be treated similar to a bare concentric neutral cable. The semiconducting jacket must not exceed 328 ft·Ω radial resistivity per Rule 094C4. Use of semiconducting jacketed cable is not very common due to the fact that these cables are higher in cost than insulated jacketed cable.

Rule 096C provides two exceptions to the four grounds in every mile requirement. The exceptions are outlined below:

- Underwater crossings and underground installations (special conditions apply).
- Overhead for special terrain areas such as river crossings or mountainous areas.

For the underwater and underground exceptions, the neutral conductor must be effectively grounded at locations that are accessible to personnel. For the overhead special terrain exception, all available structures should be grounded. Grounding on each side of the exceptions should be given special attention to make up for any lack of grounding in the area of the exceptions.

Rule 096C provides a note that discusses using this rule for shield wires (also referred to as overhead ground wires, static wires, and surge-protection wires) which are typically located at the top of transmission lines. See Fig. 096-3.

Rule 096C provides a note indicating that the **Code** does not specify a ground resistance of an individual ground for multigrounded systems. A 25 Ω requirement does exist for single-grounded systems in Rule 096D, but no such requirement exists in Rule 096C for multigrounded systems. The **Code** notes that multigrounded systems are dependent on the multiplicity of grounding electrodes, not the ground resistance of any individual electrode. See Sec. 02, Definitions, for additional information on multigrounded system, effectively grounded, and effectively grounded neutral conductor.

There are instances in the **Code** where eight grounds in each mile of line are required instead of four grounds in each mile. Rule 092C1 requires eight grounds in each mile for a communications messenger that is not an adequate grounding conductor. Rule 094B4 requires eight grounds in each mile for pole butt plates and wire wraps. Rule 354D3 requires eight grounds in each mile for direct-buried power and communications conductors in random separation (less than 12 in apart).

096D. Single-Grounded (Unigrounded or Delta) Systems.   Single-grounded systems, typically grounded wye transmission systems, that do not carry a neutral and are grounded only at the source transformer must have a ground resistance not exceeding 25 Ω. This rule states that if a single electrode exceeds 25 Ω, then other grounding methods must be used. If the single-grounded system originates in a substation, Rule 096B also applies. A delta primary system with a grounded wye secondary system also results in a single grounded system and the not-to-exceed 25 Ω requirement applies.

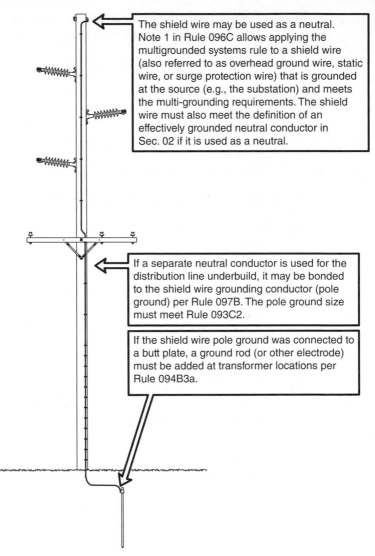

The shield wire may be used as a neutral. Note 1 in Rule 096C allows applying the multigrounded systems rule to a shield wire (also referred to as overhead ground wire, static wire, or surge protection wire) that is grounded at the source (e.g., the substation) and meets the multi-grounding requirements. The shield wire must also meet the definition of an effectively grounded neutral conductor in Sec. 02 if it is used as a neutral.

If a separate neutral conductor is used for the distribution line underbuild, it may be bonded to the shield wire grounding conductor (pole ground) per Rule 097B. The pole ground size must meet Rule 093C2.

If the shield wire pole ground was connected to a butt plate, a ground rod (or other electrode) must be added at transformer locations per Rule 094B3a.

Fig. 096-3. Note permitting applying the multigrounded systems rule to shield wires (Rule 096C).

097. SEPARATION OF GROUNDING CONDUCTORS

Rule 097A requires that separate grounding conductors be run for primary surge arresters over 750 V, secondary circuits under 750 V, and shield wires. Rule 097B allows a single grounding conductor and single grounding electrode if a ground connection exists at each surge arrester location and the primary neutral or shield wire and secondary neutral are connected together. When the primary and secondary neutrals are connected, Rule 097C requires the common neutral to be multigrounded (see Rule 096C). Rule 097A can be applied to a sufficiently heavy ground bus or system ground cable or Rule 097A can be applied in conjunction with Rule 097D1. Examples of these applications for a delta–grounded-wye transformer bank are shown in Figs. 097-1 and 097-2.

Rules 097B and 097C are typically applied to grounded-wye–grounded-wye three-phase systems and grounded-wye single-phase systems fed from a multigrounded primary system as shown in Fig. 097-3.

On multigrounded systems the primary and secondary neutrals should be interconnected. The **NESC** uses the word "should" in this case, not "shall," as there are times when separation of primary and secondary neutrals on a multigrounded system is applicable. The most common reason for separating primary and secondary neutrals on a multigrounded system is to minimize stray voltage on the secondary neutral imposed by the primary neutral. Normal and objectionable current in the grounding conductor (pole ground) is addressed in Rule 092D. The requirements separating primary and secondary neutrals for stray voltage or other valid reasons are outlined in Fig. 097-4.

If a made electrode is used to ground surge arresters on an ungrounded system exceeding 15 kV phase to phase, the **NESC** requires that the ground rod(s) be at least 20 ft from buried communication cables.

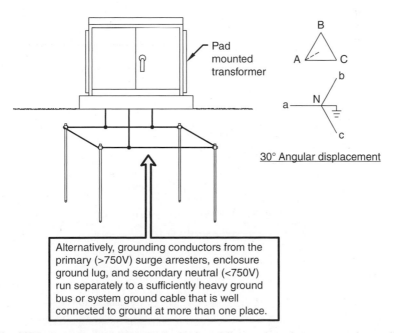

Fig. 097-1. Example of grounding conductors from different voltage classes connected to a sufficiently heavy ground bus (Rule 097A).

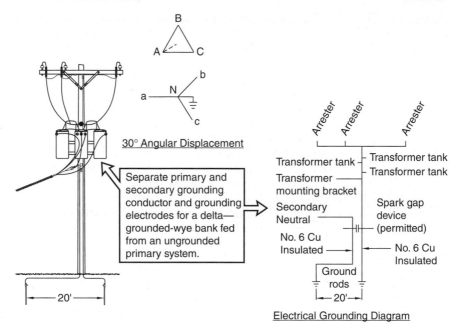

30° Angular Displacement

Separate primary and secondary grounding conductor and grounding electrodes for a delta— grounded-wye bank fed from an ungrounded primary system.

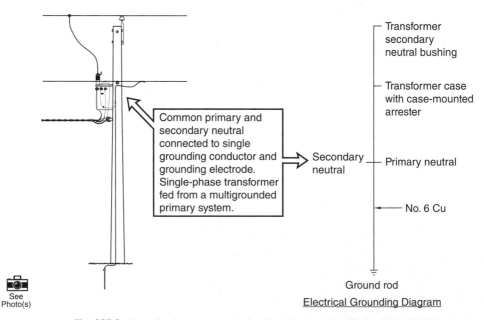

Arrester   Arrester   Arrester

Transformer tank
Transformer tank

Transformer tank
Transformer mounting bracket

Secondary Neutral

No. 6 Cu Insulated

Spark gap device (permitted)

No. 6 Cu Insulated

Ground rods

20'

20'

Electrical Grounding Diagram

Fig. 097-2. Example of separate primary and secondary grounding (Rules 097A and 097D1).

Common primary and secondary neutral connected to single grounding conductor and grounding electrode. Single-phase transformer fed from a multigrounded primary system.

Secondary neutral

Transformer secondary neutral bushing

Transformer case with case-mounted arrester

Primary neutral

No. 6 Cu

Ground rod

See Photo(s)

Electrical Grounding Diagram

Fig. 097-3. Example of a common neutral with single grounding (Rules 097B and 097C).

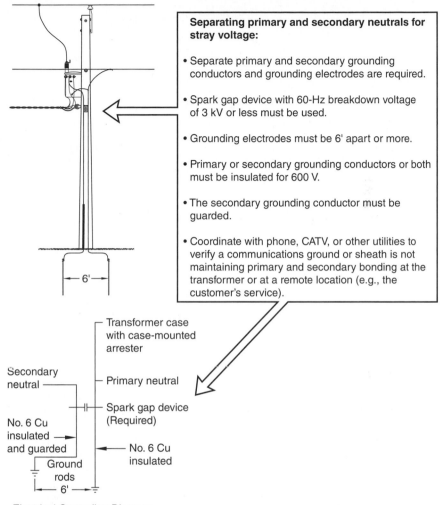

Separating primary and secondary neutrals for stray voltage:

- Separate primary and secondary grounding conductors and grounding electrodes are required.

- Spark gap device with 60-Hz breakdown voltage of 3 kV or less must be used.

- Grounding electrodes must be 6' apart or more.

- Primary or secondary grounding conductors or both must be insulated for 600 V.

- The secondary grounding conductor must be guarded.

- Coordinate with phone, CATV, or other utilities to verify a communications ground or sheath is not maintaining primary and secondary bonding at the transformer or at a remote location (e.g., the customer's service).

Transformer case with case-mounted arrester

Secondary neutral

Primary neutral

Spark gap device (Required)

No. 6 Cu insulated and guarded

No. 6 Cu insulated

Ground rods
6'

Electrical Grounding Diagram

Fig. 097-4. Separating primary and secondary neutrals for stray voltage (Rule 097D2).

Rule 097G focuses on bonding requirements for joint-use (power and communication) poles. Where both electric supply systems and communication systems are grounded on a joint-use structure and a single grounding conductor (pole ground) is present, it must be connected to both systems (i.e., the supply neutral and the communications messenger). If separate grounding conductors (pole grounds) are run to the supply neutral and the communications messenger, a bond between the pole grounds must exist. Most utilities use a single-pole ground for grounding both power and communications. The single-pole ground method

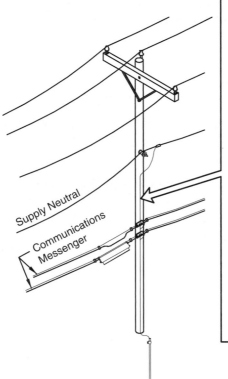

- Where both electric supply systems and communication systems are grounded on a joint-use structure, and a single grounding conductor (pole ground) is present, it must be connected to both the supply neutral and the communications messenger.

- If separate grounding conductors (pole grounds) are run to the supply neutral and the communications messenger, they must be bonded together.

- Exceptions apply when separate grounding conductors are required by other rules (e.g., delta primary systems).

- Exceptions apply when isolation is being maintained between primary and secondary neutrals (e.g., for stray voltage).

See
Photo(s)

Fig. 097-5. Bonding of communication systems to electric supply systems on a joint-use (power and communication) structure (Rule 097G).

will require a review for special cases like a delta to grounded-wye transformation or a stray voltage application. See Fig. 097-5.

098. NUMBER 098 NOT USED IN THIS EDITION

099. ADDITIONAL REQUIREMENTS FOR GROUNDING AND BONDING OF COMMUNICATION APPARATUS

This rule outlines how to ground communication apparatus when they are required to be effectively grounded in other parts of the **Code**. This rule references Note 2 of Rule 097D2, which discusses cooperation between supply and communications employees to isolate primary and secondary neutrals (typically for resolving stray-voltage problems).

A communications grounding conductor shall preferably be made of copper or other material that will not corrode and shall not be less than AWG No. 6. The communications grounding conductor must be connected as shown in Fig. 099-1.

A separate communications ground rod is not required per Rule 099A. If a communications ground rod is used because a supply service does not exist, the communications ground rod may be smaller in diameter and length per the exception to Rule 099A3. However, if a supply service does exist and a communications ground rod is used to supplement the supply grounding system, the exception to Rule 099A3 permitting smaller rods does not apply. Rule 099A does not prohibit a supplemental communications ground rod, but only if the supply service does not exist can the smaller communications-size ground rod be used. If a standard-size ground rod (per Rule 094B2) is used for communications grounding to supplement the supply ground rod, an AWG No. 6 copper or equivalent jumper must bond the two ground rods together. The No. 6 copper jumper is commonly used when the supply and communications service are adjacent to each other on a building. An equivalent jumper is typically used

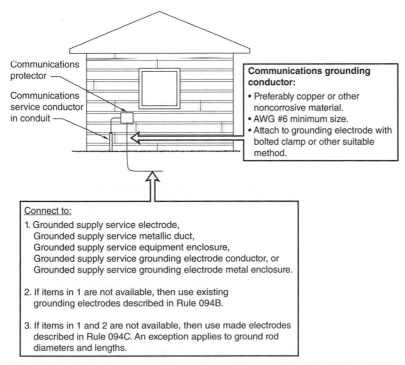

Fig. 099-1. Additional requirements for communications grounding (Rules 099A and 099B).

on a large building that has the supply service and communications service on opposite ends of the building. The requirements in this rule overlap the requirements in the National Electrical Code (NEC). The NEC should be reviewed to resolve any service entrance issues with the local building inspection authority. See Fig. 099-2.

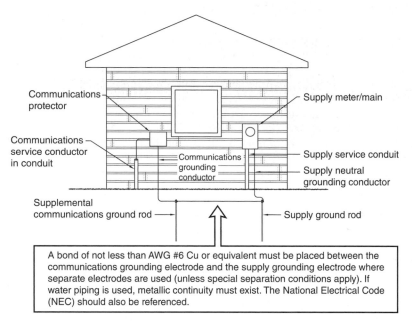

Fig. 099-2. Bonding of communications and supply electrodes (Rule 099C).

Safety Rules for the Installation and Maintenance of Electric Supply Stations and Equipment

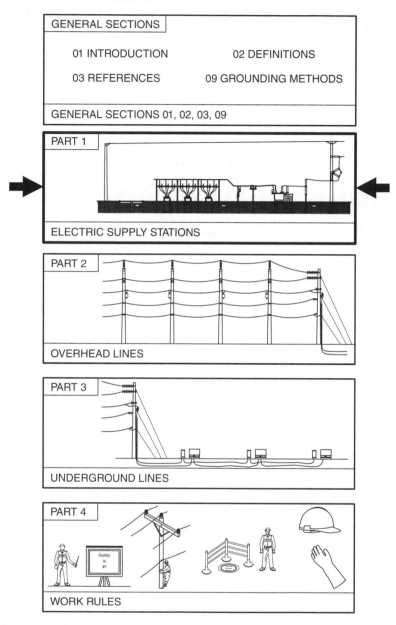

GENERAL SECTIONS

01 INTRODUCTION 02 DEFINITIONS

03 REFERENCES 09 GROUNDING METHODS

GENERAL SECTIONS 01, 02, 03, 09

PART 1

ELECTRIC SUPPLY STATIONS

PART 2

OVERHEAD LINES

PART 3

UNDERGROUND LINES

PART 4

WORK RULES

Section 10

Purpose and Scope of Rules

100. PURPOSE

The purpose of Part 1, Electric Supply Stations, is similar to the purpose of the entire **NESC** outlined in Rule 010, except Rule 100 is specific to electric supply stations and equipment. Part 1 of the **NESC** focuses on the practical safeguarding of persons during the installation, operation, and maintenance of electric supply stations and equipment.

101. SCOPE

The scope of Part 1, Electric Supply Stations, includes electric supply conductors and equipment (in electric supply stations), and associated structural arrangements (in electric supply stations). Electric supply stations can consist of generating stations, substations, and switching stations. The term arrangements is important, as Part 1, Electric Supply Stations, provides rules for arranging items in electric supply stations for clearance purposes, but Part 1 does not provide strength and loading requirements in the form of ice and wind loads for the structural components inside the electric supply station. Part 1, Rule 162A, does have a reference to the strength and loading requirements in Part 2, Overhead Lines, for conductors that extend outside the electric supply station. See the discussion and a figure in Rule 162.

A key phrase in this rule is, "accessible only to qualified personnel." The rules of Part 1, Electric Supply Stations, assume that the general public is not exposed to the conductors and equipment located in the electric supply stations. For example, in Part 2, Overhead Lines, **NESC** Table 232-1 specifies a vertical clearance of 18.5 ft for a 12.47/7.2-kV conductor above a roadway. When this same 12.47/7.2-kV rigid live part is located inside a substation fence (accessible only to qualified personnel), **NESC** Table 124-1 specifies a vertical clearance of 9.0 ft for a 15-kV phase to phase, 110-kV BIL rigid live part above the permanent supporting surface for workers. This example shows that Part 1, Electric Supply Stations, is applicable when the supply facilities are accessible to qualified personnel only, with access limited by a fence, locked room, or other method (see Rule 110A), and that the clearance values are lower in electric supply stations than in Part 2, Overhead Lines.

The last sentence of Rule 101 clarifies the application of the **NESC** versus the National Electrical Code (NEC) to electric supply stations. Rule 011 discusses the scope of the **NESC** and the NEC. The **NESC** covers conductors and equipment in an electric supply station when they are serving a utility function (not an office building wiring function). The **NESC** electric supply station rules cover utility functions. Generation stations in particular and even the control buildings of substations and switching stations involve utilization wiring for lighting, ventilation, and controls. The **NESC** does not provide specific rules for utilization wiring. Rule 012C, which requires accepted good practice, must be applied when specific conditions are not covered. The NEC is an excellent reference for accepted good practice in this case.

As evidenced by reading all the rules in Part 1, Electric Supply Stations, the scope of Part 1 applies to both indoor and outdoor substations.

Communications utility personnel can skip over Part 1, Electric Supply Stations, as it does not apply to them. The definition of electric supply station in Sec. 02 includes generating stations and substations, but not communications central offices. A communications utility may provide service to a substation (e.g., a phone line, cable TV line, or fiber line), but the Part 1, Electric Supply Stations, rules do not address this installation. Rule 012C, which requires accepted good practice, must be applied. The communications protective requirements of Rules 223 and 315 provide guidance for overhead and underground lines near substations.

The definition of Electric Supply Station in Sec. 02 of the **NESC** includes generating stations, substations, and switching stations. Most of the discussions and many of the examples presented in this Handbook refer to substations, but it is important to remember that the discussions and examples may also apply to generating stations and switching stations. Examples of various types of electric supply stations including generating stations and substations are shown in Fig. 101-1.

There is some overlap between Electric Supply Stations (Part 1) and Overhead Lines (Part 2). The overlap occurs when an overhead conductor extends outside the electric supply station fence. Conductors that extend outside the substation and their supports (e.g., poles, deadend towers, etc.) must meet the Part 2 requirements for loading and strength. Conductors and supports within the substation must meet the Part 1 requirements for loading and strength. There are distinct differences between overhead line clearances. Overhead lines and equipment inside the electric supply station fence must meet the Part 1, Electric Supply Station, rules for clearances above the substation gravel or equipment pads. Overhead lines and

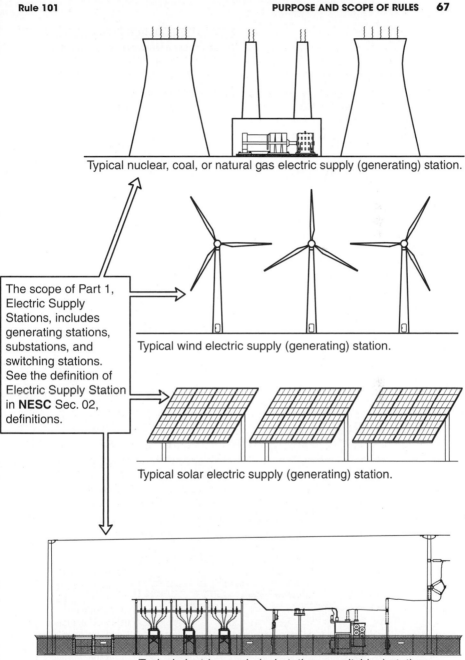

Typical nuclear, coal, or natural gas electric supply (generating) station.

The scope of Part 1, Electric Supply Stations, includes generating stations, substations, and switching stations. See the definition of Electric Supply Station in **NESC** Sec. 02, definitions.

Typical wind electric supply (generating) station.

Typical solar electric supply (generating) station.

Typical electric supply (substation or switching) station.

Fig. 101-1. Examples of electric supply stations (Rule 101).

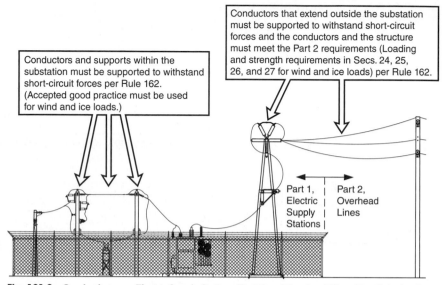

Conductors that extend outside the substation must be supported to withstand short-circuit forces and the conductors and the structure must meet the Part 2 requirements (Loading and strength requirements in Secs. 24, 25, 26, and 27 for wind and ice loads) per Rule 162.

Conductors and supports within the substation must be supported to withstand short-circuit forces per Rule 162. (Accepted good practice must be used for wind and ice loads.)

Part 1, | Part 2,
Electric | Overhead
Supply | Lines
Stations |

Fig. 101-2. Overlap between Electric Supply Stations (Part 1) and Overhead Lines (Part 2) for loading and strength (Rule 101).

equipment outside the electric supply station fence must meet the Part 2, Overhead Line, clearance rules for clearances above the ground or other surface. See Figs. 101-2 and 101-3.

There are distinct differences between Electric Supply Stations (Part 1) and Underground Lines (Part 3). Underground lines outside the electric supply station are covered in Part 3. Conductors and conduit inside the electric supply station are covered in Sec. 16 of Part 1. See Fig. 101-4.

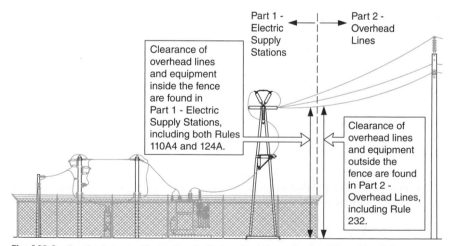

Part 1 -
Electric
Supply
Stations

Part 2 -
Overhead
Lines

Clearance of overhead lines and equipment inside the fence are found in Part 1 - Electric Supply Stations, including both Rules 110A4 and 124A.

Clearance of overhead lines and equipment outside the fence are found in Part 2 - Overhead Lines, including Rule 232.

Fig. 101-3. Overlap between Electric Supply Stations (Part 1) and Overhead Lines (Part 2) for clearances (Rule 101).

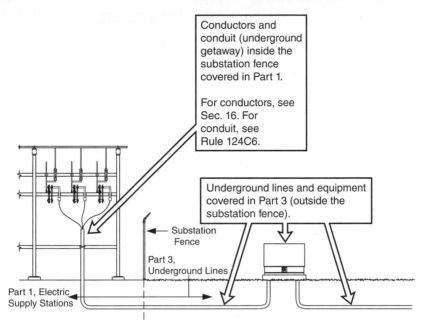

Fig. 101-4. Differences between Electric Supply Stations (Part 1) and Underground Lines (Part 3) (Rule 101).

102. APPLICATION OF RULES

Rule 102 references Rule 013 for the general application of **Code** rules; see Rule 013 for a discussion.

103. REFERENCED SECTIONS

This rule references four sections related to Part 1, Electric Supply Stations, so that rules do not have to be duplicated and the reader of the **Code** realizes that other sections are related to the information provided in Part 1. The related sections are:
- Introduction—Sec. 01
- Definitions—Sec. 02
- References—Sec. 03
- Grounding Methods—Sec. 09

The rules in Part 1, predominantly Rules 110A and 123, will provide the requirements for grounding electric supply stations. The grounding methods are provided in Sec. 09.

Section 11

Protective Arrangements in Electric Supply Stations

110. GENERAL REQUIREMENTS

110A. Enclosure of Equipment. Rule 110A is the defining rule of Part 1, Electric Supply Stations. If Rule 110A is met, the rules in Part 1, Electric Supply Stations, can be used. If Rule 110A is not met, then the rules in Part 2, Overhead Lines, or Part 3, Underground Lines, apply instead of Part 1, Electric Supply Stations. An example of how Rule 110A applies to Part 1, Electric Supply Stations, or Part 2, Overhead Lines, is shown in Fig. 110-1.

An example of how Rule 110A applies to Part 1, Electric Supply Stations, or Part 3, Underground Lines, is shown in Fig. 110-2.

NESC Rule 110A1 discusses enclosures of rooms (for indoor applications) and spaces (for outdoor applications). The following are examples of barriers required to enclose the room or space:

- Fences,
- Screens,
- Partitions, or
- Walls.

The enclosure formed is required to "limit the likelihood" of entrance by unauthorized people (i.e., the general public) or unauthorized workers. Even the best prison system cannot avoid an escape, therefore, "limit the likelihood" is used rather than "prevent entry." The entrance to the room or space must be locked or under observation by an authorized attendant. This wording can become critical

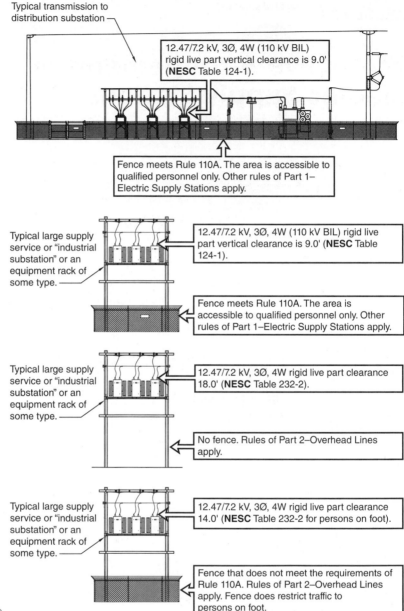

Typical transmission to distribution substation

12.47/7.2 kV, 3Ø, 4W (110 kV BIL) rigid live part vertical clearance is 9.0' (**NESC** Table 124-1).

Fence meets Rule 110A. The area is accessible to qualified personnel only. Other rules of Part 1– Electric Supply Stations apply.

Typical large supply service or "industrial substation" or an equipment rack of some type.

12.47/7.2 kV, 3Ø, 4W (110 kV BIL) rigid live part vertical clearance is 9.0' (**NESC** Table 124-1).

Fence meets Rule 110A. The area is accessible to qualified personnel only. Other rules of Part 1–Electric Supply Stations apply.

Typical large supply service or "industrial substation" or an equipment rack of some type.

12.47/7.2 kV, 3Ø, 4W rigid live part clearance 18.0' (**NESC** Table 232-2).

No fence. Rules of Part 2–Overhead Lines apply.

Typical large supply service or "industrial substation" or an equipment rack of some type.

12.47/7.2 kV, 3Ø, 4W rigid live part clearance 14.0' (**NESC** Table 232-2 for persons on foot).

Fence that does not meet the requirements of Rule 110A. Rules of Part 2–Overhead Lines apply. Fence does restrict traffic to persons on foot.

See Photo(s)

Fig. 110-1. Example of how Rule 110A applies to Part 1, Electric Supply Stations, or Part 2, Overhead Lines (Rule 110A).

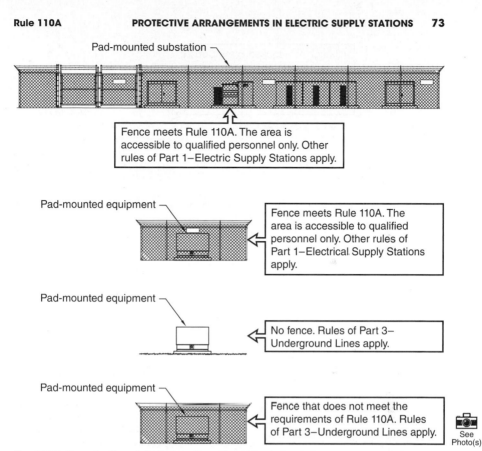

Pad-mounted substation

Fence meets Rule 110A. The area is accessible to qualified personnel only. Other rules of Part 1–Electric Supply Stations apply.

Pad-mounted equipment

Fence meets Rule 110A. The area is accessible to qualified personnel only. Other rules of Part 1–Electrical Supply Stations apply.

Pad-mounted equipment

No fence. Rules of Part 3– Underground Lines apply.

Pad-mounted equipment

Fence that does not meet the requirements of Rule 110A. Rules of Part 3–Underground Lines apply.

See Photo(s)

Fig. 110-2. Example of how Rule 110A applies to Part 1, Electric Supply Stations, or Part 3, Underground Lines (Rule 110A).

when utility employees are working inside a substation with the gate open. At least one employee, designated as an "authorized attendant," must observe the unlocked gate, or the gate must be locked after the employees enter the substation.

Rule 110A2 addresses safety signs. The **Code** requires a safety sign on or beside the gate or door at each entrance of the electric supply station. Fenced or walled electric supply stations without roofs must have a safety sign located on each side of the fenced or walled enclosure. If the electric supply station is entirely enclosed by walls and a roof, a safety sign is only required at ground level entrances. Where entrance is gained through sequential doors, the safety sign should be located at the inner door position. Nothing in the **Code** prevents an additional sign at the outer door position if desired. The electric supply station may be a generating station, substation, or switching station. All of these installations may be enclosed in a building or by a fence. The requirements for safety signs for electric supply stations with and without roofs are shown in Fig. 110-3.

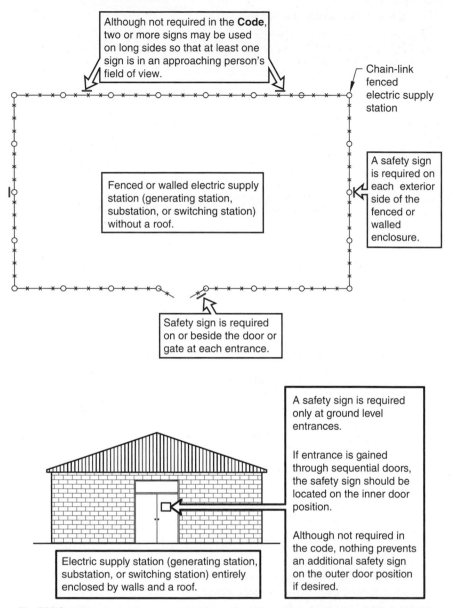

Although not required in the **Code**, two or more signs may be used on long sides so that at least one sign is in an approaching person's field of view.

Chain-link fenced electric supply station

Fenced or walled electric supply station (generating station, substation, or switching station) without a roof.

A safety sign is required on each exterior side of the fenced or walled enclosure.

Safety sign is required on or beside the door or gate at each entrance.

A safety sign is required only at ground level entrances.

If entrance is gained through sequential doors, the safety sign should be located on the inner door position.

Although not required in the code, nothing prevents an additional safety sign on the outer door position if desired.

Electric supply station (generating station, substation, or switching station) entirely enclosed by walls and a roof.

Fig. 110-3. Safety sign locations on an electric supply station with and without a roof (Rule 110A2).

The **NESC** notes ANSI Z535 series documents for sign applications. Substation fences and pad-mounted transformers and enclosures are two of the most common signage applications for electric supply utilities. The ANSI Z535 approach to signage uses the philosophy that a "warning" sign is appropriate on the outer barrier (i.e., fence or enclosure), and if that barrier is breached, a "danger" sign is then appropriate. Traditionally, the "danger" sign was the most common choice for sub-station fence applications with little or no signage used inside the substation. Using the ANSI Z535 signage philosophy, "warning" signs would be placed on the substation fence and "danger" signs would be placed inside the substation on structures that support energized parts. This same philosophy can be applied to a pad-mounted transformer. A "warning" sign is placed on the outside of the enclosure and a "danger" sign on the inside. Utilities should consult the ANSI Z535 documents, federal or state regulatory agencies, and the utility's insurance company for signage applications. Neither the **NESC** nor the ANSI Z535 signage documents are specific as to what words or pictorials are required for individual signage applications. Rule 012C, which requires accepted good practice, must be used. It is important to note that **NESC** Rule 110A2 does not require safety signs inside the substation among the energized parts. Safety signs are required for special cases on switchgear and motor control centers covered in Sec. 18. Examples of safety signs are shown in Fig. 110-4.

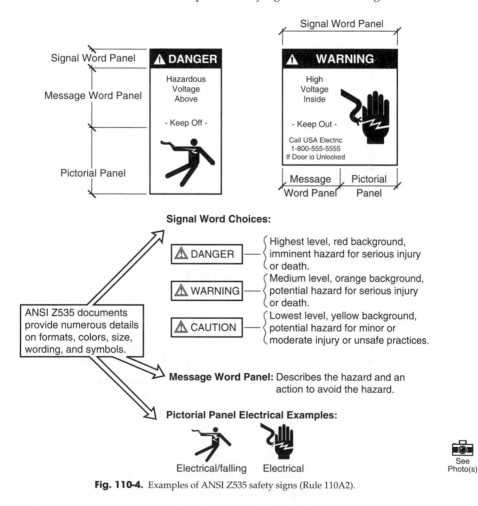

Fig. 110-4. Examples of ANSI Z535 safety signs (Rule 110A2).

Rule 110A1 addresses fences in general, and Rule 110A3 addresses metal fences. When a chain-link fence is used to enclose a substation, the **NESC** details very specific height requirements. The construction requirements discuss the fence fabric and barbed wire strands but do not provide details on the gauge of the fence mesh or diameter of the fence posts or rails. These details are left to the designer. Rule 110A1 starts out with the general statement that the enclosure must limit the likelihood of entrance of unauthorized persons. Other types of construction must present an equivalent barrier to climbing and unauthorized entry as the chain-link fence. To provide an equivalent barrier to climbing, fences should not have handholes or footholes more predominant than the mesh on a chain-link fence. Although not stated in the **Code**, pad-mounted equipment, park benches, parked vehicles, etc., should not be placed near a substation fence, as they can create "steps" for climbing the fence, reducing the effectiveness of the fence height. The requirements for barriers are outlined in Figs. 110-5 and 110-6.

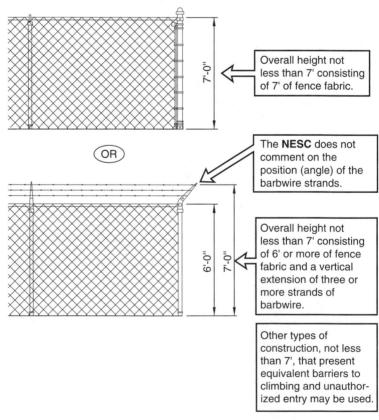

7'-0"

Overall height not less than 7' consisting of 7' of fence fabric.

OR

The **NESC** does not comment on the position (angle) of the barbwire strands.

6'-0"   7'-0"

Overall height not less than 7' consisting of 6' or more of fence fabric and a vertical extension of three or more strands of barbwire.

Other types of construction, not less than 7', that present equivalent barriers to climbing and unauthorized entry may be used.

See Photo(s)

Fig. 110-5. Barrier requirements (Rule 110A1).

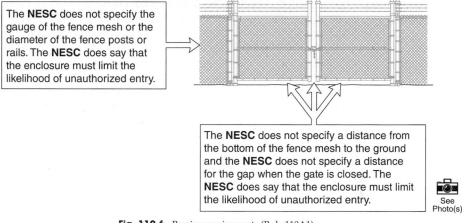

The **NESC** does not specify the gauge of the fence mesh or the diameter of the fence posts or rails. The **NESC** does say that the enclosure must limit the likelihood of unauthorized entry.

The **NESC** does not specify a distance from the bottom of the fence mesh to the ground and the **NESC** does not specify a distance for the gap when the gate is closed. The **NESC** does say that the enclosure must limit the likelihood of unauthorized entry.

See Photo(s)

Fig. 110-6. Barrier requirements (Rule 110A1).

The **Code** does address neighboring fences or similar structures. Neighboring fences or similar structures must not be connected to or located within 6 ft of an electric supply station fence without concurrence of the substation owner. This requirement is addressing two primary issues related to neighboring fences. The first is that a neighboring fence of a different height can create a "step" for climbing the electric supply station fence. The second is that the electric supply station fence that consists of metal chain-link fence or other metallic barrier must be grounded in accordance with the grounding methods in Sec. 09 (see Rule 092E). The question becomes what grounding methods are needed for the neighboring metal fence that is attached to the electric supply station metal fence. Keeping the neighboring fence 6 ft away from the electric supply station fence mitigates these concerns. Locating the electric supply station fence 6 ft inside the electric supply station property line may be one solution to meeting this rule. If the electric supply station fence is on the property line, working with the neighboring fence owner to use a 6 ft long non-metallic fence section the same height as the substation fence may be another solution. This is just one example of applying some accepted good practice (Rule 012C) by the substation owner before concurrence is given by the substation owner. If the substation owner does not see any issue with the neighboring fence being located within 6 ft of the electric supply station fence, a simple concurrence by the substation owner is all that is needed. The requirements for neighboring fences within 6 ft of an electric supply station fence are outlined in Fig. 110-7.

An example of accepted good practice for connecting a neighboring fence to an electric supply station fence is shown in Fig. 110-8.

Many utilities establish substation fence height values by using the **Code** requirement plus an adder. The adder (1 ft, for example) can be thought of as a design or construction tolerance adder to maintain the required fence height over time. There are several factors that can jeopardize the fence height. Factors could include the addition of gravel or some other type of fill outside the substation. Installing a substation fence with an overall height of 8 ft can help maintain the **Code**-required 7-ft height over the life of the installation. For additional information on the concept of **Code** plus an adder, see Rule 010.

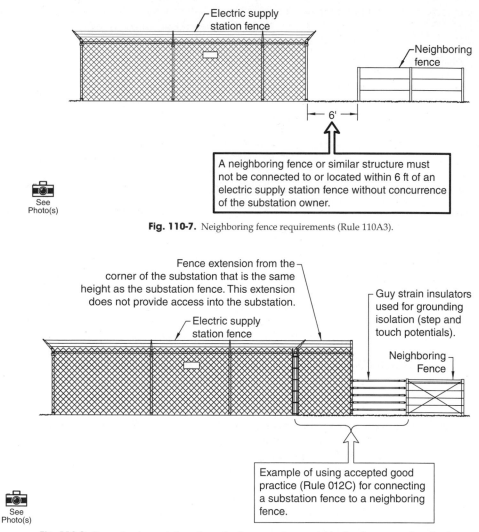

Fig. 110-7. Neighboring fence requirements (Rule 110A3).

Fig. 110-8. Example of accepted good practice for connecting a neighboring fence to a substation fence (Rule 110A3).

In addition to the fence height requirements in Rule 110A1, Rule 110A4 specifies a safety clearance zone from the substation fence to exposed live (energized) parts inside the substation. The method used depends on the type of barrier around the supply station that keeps the public out of the substation.

If a metal chain-link fence is used, Rule 110A4a applies. Rule 110A4a requires the use of **NESC** Fig. 110-1 and **NESC** Table 110-1 to determine the setback of exposed live parts from the fence. The distances in **NESC** Table 110-1 are required to place live parts far enough back from the fence so that a person poking an object through the fence or swinging an object over the fence will not contact energized parts.

Per **NESC** Rule 124A1 and the note at the bottom of **NESC** Table 110-1, the values in **NESC** Table 110-1 are for altitudes of 3300 ft or less. See Rule 124A for a discussion of altitude adjustments to clearance values. An example of meeting Rule 110A4a by applying **NESC** Fig. 110-1 and **NESC** Table 110-1 to a chain-link fence is shown in Fig. 110-9.

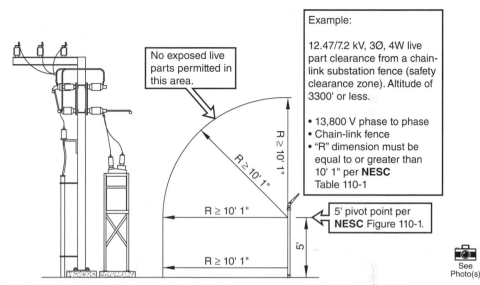

Fig. 110-9. Example of how to apply the safety clearance zone to a chain-link fence per NESC Table 110-1 and NESC Fig. 110-1 (Rule 110A4a).

If an impenetrable barrier is used such as a solid fence (e.g., an outdoor concrete block fence) or wall (e.g., inside a building), Rule 110A4b applies. Rule 110A4b requires the use of **NESC** Fig. 110-2 and **NESC** Table 110-1 to determine the setback of exposed live parts from the fence. The distances in **NESC** Table 110-1 are combined with a formula to determine the setback distance. The calculated setback distance will vary depending on the voltage of the exposed live parts and the height of the impenetrable portion of the fence. The impenetrable portion of the fence does not have to cover the entire fence. It is acceptable to have the fence consisting of penetrable and impenetrable portions. If the impenetrable portion of the fence does cover the entire fence and the fence height is equal to or greater than the "R" dimension in **NESC** Table 110-1 plus 5 ft, a safety clearance zone will not exist. If the safety clearance zone is less than the required working space in Rule 125, then Rule 125 applies to the space between the impenetrable fence and the energized parts. See Rule 125 for more information. Sometimes the impenetrable fence is only located on one side or one portion of a substation; in this case the width of the impenetrable barrier must be such that the distance from the outer edge of the impenetrable barrier to the nearest live part is equal to or greater than the "R" dimension in **NESC** Table 110-1 for the voltage involved. If there are openings below the impenetrable portion of the fence, the "R" dimension of the **NESC** Table 110-1 for the voltage involved must be applied from the lowest impenetrable point to the closest energized part. An example of this application is using plywood sheets bolted onto a chain-link fence to shorten the safety clearance zone between the fence and exposed live (energized) parts inside the substation. There are times when a concern arises that the plywood sheets will greatly increase the wind load on the chain-link fence. Some utilities will not cover the fence near the bottom of the fence to minimize the wind load on the fence, and in this case the "R" dimension of the **NESC** Table 110-1 must be applied from the lowest impenetrable point to the closest energized part. Examples of meeting Rule 110A4b by applying **NESC** Fig. 110-2 and **NESC** Table 110-1 are shown in Figs. 110-10 and 110-11.

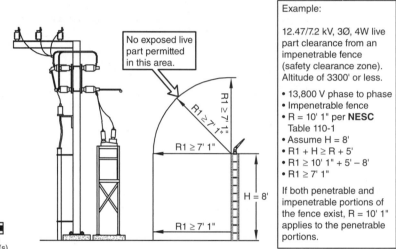

Example:

12.47/7.2 kV, 3Ø, 4W live part clearance from an impenetrable fence (safety clearance zone). Altitude of 3300' or less.

- 13,800 V phase to phase
- Impenetrable fence
- R = 10' 1" per **NESC** Table 110-1
- Assume H = 8'
- R1 + H ≥ R + 5'
- R1 ≥ 10' 1" + 5' − 8'
- R1 ≥ 7' 1"

If both penetrable and impenetrable portions of the fence exist, R = 10' 1" applies to the penetrable portions.

Fig. 110-10. Example of how to apply the safety clearance zone to an impenetrable fence per **NESC** Table 110-1 and **NESC** Fig. 110-2 (Rule 110A4b).

The exception to the safety clearance zone involves internal electric supply station fences. An internal fence does not need to comply with the safety clearance zone as an interior fence inside a substation fence is accessible to qualified employees only. The exception is outlined in Fig. 110-12.

The North American Electric Reliability Corporation (NERC) standards apply to protecting transmission substations from a physical attack. NERC Standard CIP-014 provides information for transmission substation owners. The NERC physical security standards were developed to address cascading outages.

110B. Rooms and Spaces.   The rooms (i.e., interior) or spaces (i.e., exterior) that comprise an electric supply station must be noncombustible. The **Code** uses the phrase, "as much as practical noncombustible." This wording recognizes the oil-filled equipment may be combustible; however, the electric supply station structure should not be. Rules 152 and 173 provide additional requirements related to oil-filled equipment located in electric supply stations. See Rule 152B for additional information related to locating an electric supply station indoors. Steel is the most common choice for modern outdoor substation construction, but the **Code** recognizes that wood poles are still commonly located within the substation fence.

For an impenetrable fence height (H) of 15' 1", the safety clearance zone (R1) is equal to or greater than 0 ft. However, the working space requirements of Rule 125 must still be met. Per Rule 125, using **NESC** Table 124-1 the horizontal working space for 15 kV (110 kV BIL) is 3 ft-6 in. The minimum approach distances in **NESC** Rule 441 should be considered. Additional working room may be needed for tools (e.g., hot sticks).

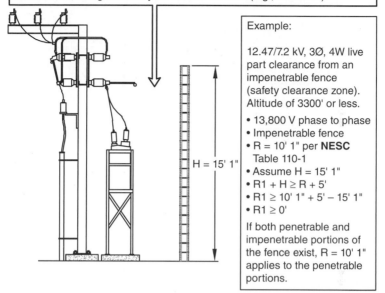

Example:

12.47/7.2 kV, 3Ø, 4W live part clearance from an impenetrable fence (safety clearance zone). Altitude of 3300' or less.

• 13,800 V phase to phase
• Impenetrable fence
• R = 10' 1" per **NESC** Table 110-1
• Assume H = 15' 1"
• R1 + H ≥ R + 5'
• R1 ≥ 10' 1" + 5' − 15' 1"
• R1 ≥ 0'

If both penetrable and impenetrable portions of the fence exist, R = 10' 1" applies to the penetrable portions.

H = 15' 1"

Fig. 110-11. Example of how to apply the safety clearance zone to an impenetrable fence per NESC Table 110-1 and NESC Fig. 110-2 (Rule 110A4b).

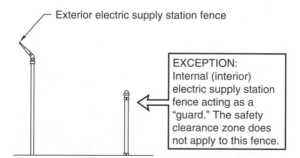

Exterior electric supply station fence

EXCEPTION:
Internal (interior) electric supply station fence acting as a "guard." The safety clearance zone does not apply to this fence.

Fig. 110-12. Exception to the safety clearance zone requirements (Rule 110A4).

Dry grass or weeds should be removed from an outdoor substation to maintain the noncombustible requirement.

The substation room or space must not contain combustible materials or fumes and must not be used for manufacturing or storage. Three exceptions to Rule 110B2 apply to storage of materials in an electric supply station.

The first exception permits storage of material, equipment, and vehicles that are essential for maintenance of the electric supply station, for example, spare fuses, a spare substation transformer, or a bucket truck used for maintenance of the supply station equipment. The material, equipment, or vehicle must be guarded (e.g., stored in a shed) or separated from live parts per Rule 124. See Fig. 110-13.

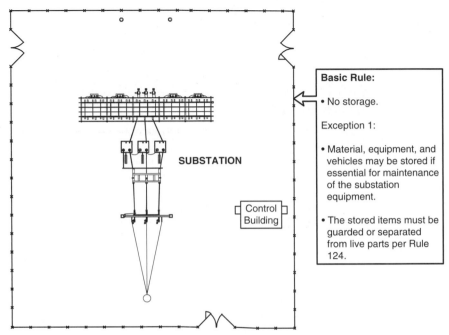

Fig. 110-13. Electric supply station use (storage) (Rule 110B2).

   The second exception permits storage of material, equipment, and vehicles for construction, operations, or maintenance work of station, transmission, and distribution facilities (not just supply station facilities). This exception requires the stored items to be fenced separately from the electric supply substation equipment. The fence separating the electric supply equipment and the storage materials must meet the requirements of Rule 110A. If this exception is applied, the storage effectively ends up not being in the electric supply station per se, but in a separate space adjacent to it. See Fig. 110-14.

   The third exception permits storage of material, equipment, and vehicles on a temporary basis (no time period is specified) for material, equipment, and vehicles for construction, operations, or maintenance of station, transmission, and distribution facilities (not just supply station facilities) work in progress. For example, if a new distribution line is being built near a substation, the substation may temporarily be used to store items for the project. This exception requires the stored items to be associated with work in progress. In other words, the substation site cannot be used as a storage yard for permanent warehousing. To apply the third exception, the **Code** lists five conditions that must be met to maintain a safe working area. See Fig. 110-15.

   Rules 110B3 and 110B4 require that the ventilation in the room or space must be adequate and the room or space, if indoors, should be dry. If the electric supply

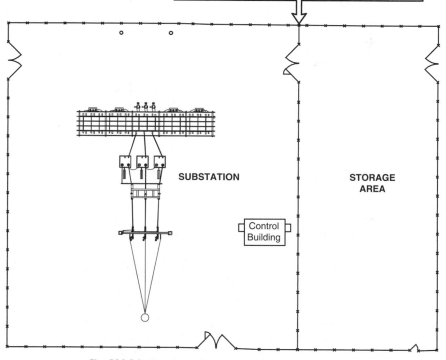

Basic Rule:

• No storage.

Exception 2:

• Material, equipment, and vehicles may be stored for construction, operations, or maintenance work of station, transmission, and distribution facilities (not just supply station facilities).

• The stored items must be separated from the electric supply station equipment by a fence meeting Rule 110A.

SUBSTATION

STORAGE AREA

Control Building

Fig. 110-14. Electric supply station use (storage) (Rule 110B2).

station space is located outdoors, the equipment in the space must be designed for the atmospheric conditions.

110C. Electric Equipment and Supporting Structures. Electric equipment and supporting structures in the supply station must withstand the anticipated conditions of service. The **Code** does not specifically address any seismic (earthquake) construction requirement. Rule 012C, which requires accepted good practice, must be applied in this case. IEEE Std. C57.114, *IEEE Seismic Guide for Power Transformers and Reactors*, is one reference for accepted good practice. The note in this rule references ASCE-113 *(Substation Structure Design Guide)* and IEEE Std. 605 *(IEEE Guide for Bus Design in*

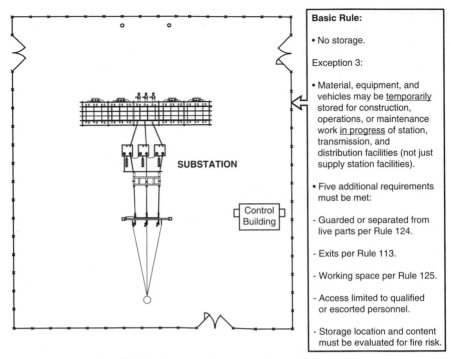

Basic Rule:

• No storage.

Exception 3:

• Material, equipment, and vehicles may be <u>temporarily</u> stored for construction, operations, or maintenance work <u>in progress</u> of station, transmission, and distribution facilities (not just supply station facilities).

• Five additional requirements must be met:

- Guarded or separated from live parts per Rule 124.

- Exits per Rule 113.

- Working space per Rule 125.

- Access limited to qualified or escorted personnel.

- Storage location and content must be evaluated for fire risk.

Fig. 110-15. Electric supply station use (storage) (Rule 110B2).

Air Insulated Substations) for additional information. With proper consideration, heavy equipment such as a substation transformer can be secured in place by its own weight. Heavy equipment such as an electric generator that has dynamic (rotating movement) forces will require anchoring measures in addition to its own weight. See Fig. 110-16.

111. ILLUMINATION

111A. Under Normal Conditions. This rule provides illumination (lighting) levels for electric supply station rooms and spaces. **NESC** Table 111-1 provides illumination values in lux (metric) and foot-candles (English) for generating station areas, both interior and exterior, although outdoor lighting is not required at *unattended* stations. **NESC** Table 111-1 also contains illumination values for specific generating station and substation areas. The specific areas include control building interior, general exterior horizontal (e.g., the substation gravel) and equipment vertical (e.g., a row of disconnect switches), and remote areas. Multiple footnotes to the table provide additional information. Various light fixture manufacturers publish calculation aids for determining lighting levels. The Illuminating Engineering Society (IES) publishes books on lighting design applications. Rule 111A does not require that the lighting be permanently installed; therefore, portable lighting can be used to meet the rule. Rules 111A, 111B, and 111C discuss

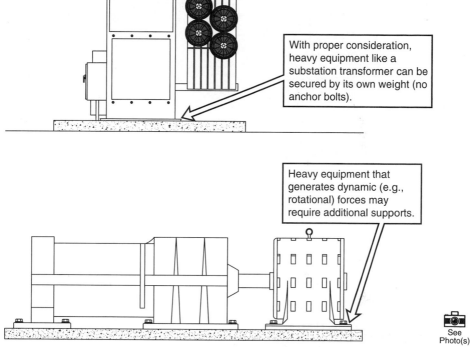

With proper consideration, heavy equipment like a substation transformer can be secured by its own weight (no anchor bolts).

Heavy equipment that generates dynamic (e.g., rotational) forces may require additional supports.

See Photo(s)

Fig. 110-16. Supporting and securing heavy equipment (Rule 110C).

receptacles and portable cords for cord and plug light fixtures. The rules for illumination under normal conditions are outlined in Fig. 111-1.

111B. Emergency Lighting.   Attended electric supply stations must have automatically initiated emergency lighting for power failure. The exit paths in attended stations must have 1 foot-candle of lighting at all times. This can be done using an emergency generator or storage batteries. The duration of backup lighting should be evaluated, but in no case should the duration be less than 90 min (1.5 h). It is recommended that the wiring for the emergency lighting fixtures be kept independent from the normal wiring.

111C. Fixtures.   Portable cords must not be brought dangerously close to live parts (e.g., in a substation) or moving parts (e.g., in a generating station). Consideration needs to be given to the location of permanent fixtures and plug receptacles to avoid this hazard. Switches for lighting must be in a safely accessible location.

111D. Attachment Plugs and Receptacles for General Use.   Plugs and receptacles used in electric supply stations must disconnect all poles by one operation and must be of the grounding type. Special voltages, amperages, or frequencies must have plugs and receptacles that are not interchangeable. Manufacturers of wiring

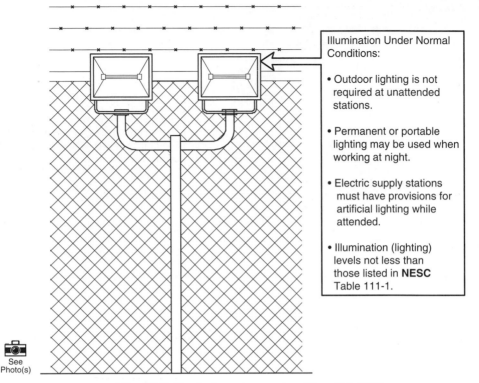

Illumination Under Normal Conditions:

• Outdoor lighting is not required at unattended stations.

• Permanent or portable lighting may be used when working at night.

• Electric supply stations must have provisions for artificial lighting while attended.

• Illumination (lighting) levels not less than those listed in **NESC** Table 111-1.

See Photo(s)

Fig. 111-1. Illumination under normal conditions (Rule 111A).

devices (e.g., receptacles, plugs, switches, etc.) use National Electrical Manufacturers Association (NEMA) standard configurations for various voltage, phase, and current ratings.

111E. Receptacles in Damp or Wet Locations. If the receptacle is in a damp or wet location, it must have ground-fault circuit interrupter (GFCI) protection as part of either the receptacle or the circuit breaker feeding the receptacle. As an alternative or in addition to using GFCI protection, the **NESC** allows testing of a grounded circuit as often as experience has shown necessary. A GFCI device may also be applied between the receptacle and the load. Many times, the National Electrical Code (NEC) is used for accepted good practice for wiring related to lighting and receptacles in the control building of a substation.

112. FLOORS, FLOOR OPENINGS, PASSAGEWAYS, AND STAIRS

This rule gives special attention to floors, floor openings, passageways, and stairs, as they are accident-prone areas. The requirements for floors and passageways are outlined in Fig. 112-1.

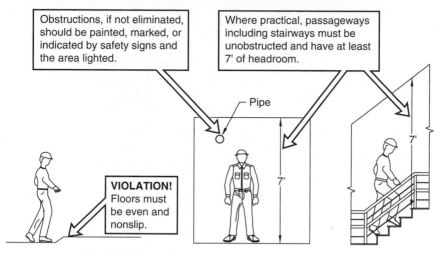

Fig. 112-1. Requirements for floors and passageways (Rule 112).

Railings are required for floor openings and raised platforms or walkways in excess of 1 ft in height. Handrails are required for stairways with four or more risers. A 3-in unobstructed clearance is required around handrails to assure an adequate grip. A note in the rule references **OSHA** for additional information related to handrails. The **NESC** rule requiring handrails for stairways consisting of four or more risers is outlined in Fig. 112-2.

Fig. 112-2. Requirements for stair handrails (Rule 112).

113. EXITS

Exits in spaces and rooms must be kept clear of obstructions. Double exits must be provided if the arrangement of equipment and an accident can make a single exit inaccessible. The exit doors must swing out and have some type of panic hardware (e.g., a push bar, not a door knob) except for fence gates in outdoor substations and doors in rooms containing only low-voltage nonexplosive equipment. See Fig. 113-1.

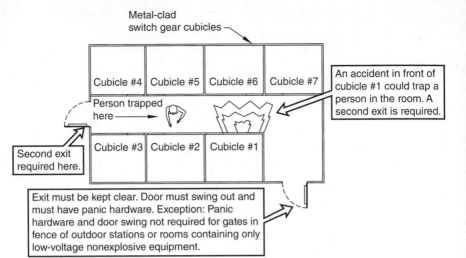

Metal-clad switch gear cubicles

Cubicle #4 | Cubicle #5 | Cubicle #6 | Cubicle #7

An accident in front of cubicle #1 could trap a person in the room. A second exit is required.

Person trapped here

Second exit required here.

Cubicle #3 | Cubicle #2 | Cubicle #1

Exit must be kept clear. Door must swing out and must have panic hardware. Exception: Panic hardware and door swing not required for gates in fence of outdoor stations or rooms containing only low-voltage nonexplosive equipment.

Fig. 113-1. Electric supply station exit requirements (Rule 113).

Section 12

Installation and Maintenance of Equipment

120. GENERAL REQUIREMENTS

This rule discusses installation and maintenance of electric supply station equipment. Safeguarding personnel during installation, construction, and maintenance is the primary concern of Sec. 12. The rules of Sec. 12 apply to both alternating-current (AC) and direct-current (DC) electric supply stations.

121. INSPECTIONS

The inspections discussed in this rule and the inspections discussed in Rule 214, Part 2, Overhead Lines, and in Rule 313, Part 3, Underground Lines, form the basic requirements for inspecting electric supply stations and supply and communication lines. The inspections required in this rule for electric supply stations are outlined in Fig. 121-1.

Any equipment or wiring found defective must be permanently disconnected or promptly corrected. New equipment must be tested in accordance with industry practice. The **Code** does not specify the details of the inspection program; it simply states that it be regular and scheduled. Accepted good practice, which is discussed in Rule 012C, is commonly used to help develop an electric supply station inspection program.

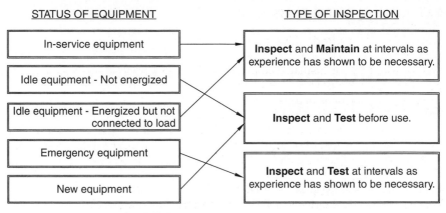

Fig. 121-1. Inspection requirements (Rule 121).

122. GUARDING SHAFT ENDS, PULLEYS, BELTS, AND SUDDENLY MOVING PARTS

Mechanical parts located in an electric supply station must be safeguarded. Mechanical transmission machinery is abundant in generating stations. Substations typically do not have much mechanical machinery. This rule requires the use of ANSI/ASME B15.1. Many times the **Code** makes note of a standard, but in this case, the standard is required, as it is part of the **Code** text. Suddenly moving parts must also be guarded or isolated.

123. PROTECTIVE GROUNDING

This rule states the requirements for electric supply station grounding. The methods of protective grounding are found in Sec. 09, "Grounding Methods for Electric Supply and Communications Facilities."

Non-current-carrying metal parts in the electric supply station must be grounded or isolated. Metallic fences, rails, and other guards around electric equipment must be grounded. IEEE Standard 80 is noted in this rule and in Sec. 09, as it is the standard for electric supply station grounding.

Provisions must also exist for grounding during maintenance. When a conductor, bus section, or piece of equipment is disconnected for maintenance, it must be grounded. The grounding can be done with permanent grounding switches or a readily accessible means for connecting portable grounding jumpers. The Part 4 Work Rules are referenced for proper procedures.

Direct-current (DC) systems have unique rules for grounding, which are discussed in Sec. 09.

124. GUARDING LIVE PARTS

124A. Where Required. Live parts in an electric supply station over 300 V phase-to-phase must be guarded, isolated by location (using vertical and horizontal clearance), or insulated to avoid inadvertent contact by qualified utility personnel in the substation. The basic intent of this rule is that utility personnel who are qualified to be in the substation (but not necessarily working on the substation) can walk around without accidentally contacting energized parts. The clearances in this section do not apply to the general public. Rule 110A, "General Requirements, Enclosure of Equipment," must be met before the clearances of Rule 124 apply. Otherwise Part 2, Overhead Lines, is applicable. See Rule 110A for an example.

Rule 124 constantly uses the word "guard," which implies a physical barrier between the utility employee and the live part. A common guard in a substation is the grounded enclosure of metal-clad switchgear. For substations consisting of open bus construction, physical guards are not nearly as common as the alternative to providing a guard, which is providing adequate clearance.

To determine the adequate clearance of live (i.e., energized) parts in the electric supply station, Rule 124A1 requires the use of **NESC** Table 124-1 and **NESC** Fig. 124-1 and the live parts must meet the safety clearance zone to the fence or wall which is covered in Rule 110A4. The vertical clearances in **NESC** Table 124-1 can use the taut-string method described in Rule 124D. The footnotes to **NESC** Table 124-1 provide information on Basic Impulse Insulation Levels (BIL) selection methods. Appendix D in the **NESC** provides additional information on overvoltage factors. **NESC** Rule 124A1 states that the clearance values in **NESC** Table 124-1 and **NESC** Table 110-1 are for altitudes of 3300 ft or less. A method is provided for increasing the substation clearance values due to higher altitudes. Altitude correction formulas are commonly used in the **NESC** in Part 2, Overhead Lines. See Rule 232C for a discussion and a figure related to clearance adders based on altitude. A note in Rule 124A1 references two IEEE Standards for additional information. Examples of altitude correction methods for increasing substation clearances can also be found in the Rural Utilities Service (RUS) Bulletin 1724E-300, *Design Guide for Rural Substations*.

In addition to clearance to live parts, Rule 124A3 specifies an 8-ft, 6-in (8.5-ft) vertical clearance to parts on indeterminate potential. A bushing or insulator has a surface along it of indeterminate potential. The top of the bushing has a known voltage, for example, 7.2 kV to ground. The bottom of the bushing has a known voltage, 0 V if it is grounded. The surface between the top of the bushing and the bottom of the bushing is of unknown voltage, or as the **Code** calls it, indeterminate potential. It is somewhere between 7.2 kV and 0 V. Per Rule 124A3, the 8-ft, 6-in (8.5-ft) vertical clearance must exist from the bottom of the part of indeterminate potential (bottom of the bushing) to the surface below. The vertical clearance in Rule 124A3 can use the taut-string method described in Rule 124D.

If the substation equipment purchased is not tall enough to meet the required vertical clearances to the top of the equipment bushings per **NESC** Table 124-1 and **NESC** Fig. 124-1 and the 8-ft, 6-in (8.5-ft) clearance to the bottom of a part of indeterminate potential (bottom of the bushing), then a concrete pad and/or equipment stands must be used to provide the required height for both the live part clearance (top of the bushing) and the indeterminate voltage clearance (bottom of the bushing). Both clearances (top and bottom of the bushing) must be met or exceeded to avoid a **Code** violation. An example of how to apply the vertical clearances of **NESC** Table 124-1, **NESC** Fig. 124-1, and **NESC** Rule 124A3 is shown in Fig. 124-1.

Both Rule 124A1 and Rule 124A3 require clearance to any permanent supporting surface for workers inside the substation. If a concrete pad under substation equipment is large enough to stand on, then the measurement needs to be taken from the top of the concrete pad. The interpretation of what size concrete pad is a permanent supporting surface for workers is not well defined in the **Code**. If the concrete pad is oversized for easy maneuverability of a worker, the clearance measurement must be made from the top of the concrete pad. If the pad is used for additional vertical clearance of the equipment but a worker has to "hug the equipment" to stay standing on the concrete pad, the clearance measurement can be made from the substation surface (e.g., gravel) instead of the top of the concrete pad. This rule is outlined in Fig. 124-2.

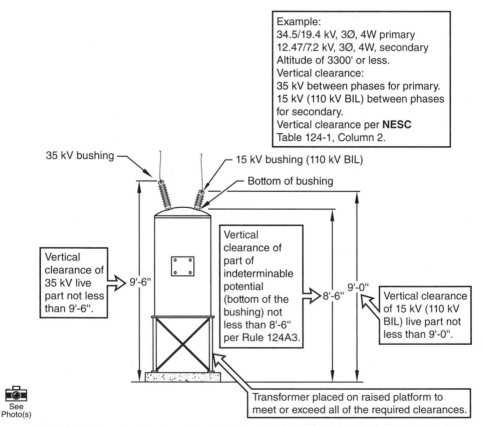

Example:
34.5/19.4 kV, 3Ø, 4W primary
12.47/7.2 kV, 3Ø, 4W, secondary
Altitude of 3300' or less.
Vertical clearance:
35 kV between phases for primary.
15 kV (110 kV BIL) between phases for secondary.
Vertical clearance per **NESC** Table 124-1, Column 2.

35 kV bushing — 15 kV bushing (110 kV BIL)

Bottom of bushing

Vertical clearance of 35 kV live part not less than 9'-6".

9'-6"

Vertical clearance of part of indeterminable potential (bottom of the bushing) not less than 8'-6" per Rule 124A3.

8'-6"   9'-0"

Vertical clearance of 15 kV (110 kV BIL) live part not less than 9'-0".

See Photo(s)

Transformer placed on raised platform to meet or exceed all of the required clearances.

Fig. 124-1. Example of how to apply vertical clearances per **NESC** Table 124-1, **NESC** Fig. 124-1, and Rule 124A3 (Rules 124A1 and 124A3).

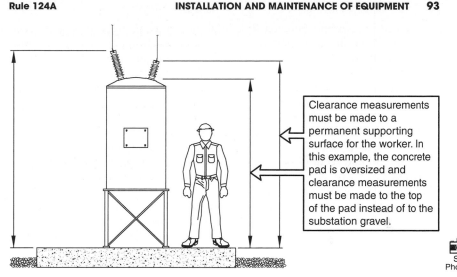

Clearance measurements must be made to a permanent supporting surface for the worker. In this example, the concrete pad is oversized and clearance measurements must be made to the top of the pad instead of to the substation gravel.

See Photo(s)

Fig. 124-2. Clearance measurements made to a permanent supporting surface (Rule 124A1).

Rule 124A2 recognizes that additional clearances or guarding may be needed where material may be carried such as passageways, corridors, storage areas, etc. (primarily indoor areas). Additional clearance values are not specified. If physical guards are used for these areas, they must be removable only with tools or keys.

The **Code** does not specify energized conductor or bus clearances to vehicles in a substation. Many substations are designed for bucket truck access. The **NESC** Part 4 Work Rules and OSHA Standard 1910.269 applies. See Fig. 124-3.

The **Code** does not specify bus-to-bus clearances, conductor-to-bus clearances, conductor-to-conductor clearances, or energized part-to-grounded part clearances

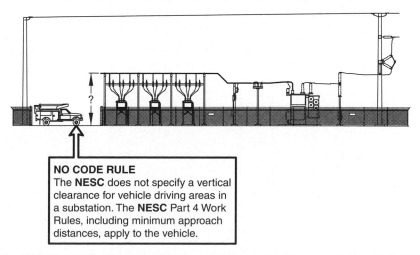

NO CODE RULE
The **NESC** does not specify a vertical clearance for vehicle driving areas in a substation. The **NESC** Part 4 Work Rules, including minimum approach distances, apply to the vehicle.

Fig. 124-3. Absence of a **Code** rule related to vertical clearance to vehicles in a substation (Rule N/A).

in Part 1, Electric Supply Stations. Rule 012C, which requires accepted good practice, must be applied. The conductor-to-conductor clearances given in Part 2, Overhead Lines, Sec. 23, are not required to be used in Part 1, Electric Supply Stations, but they are a reference for accepted good practice. Other common references for accepted good practice for bus-to-bus clearance in outdoor electrical substations are ANSI C37.32, NEMA SG6, and Rural Utilities Service (RUS) Bulletin 1724E-300, *Design Guide for Rural Substations*. The absence of a **Code** rule related to bus clearance is outlined in Fig. 124-4.

124B. Strength of Guards. Physical guards, when used instead of clearance, must be rigid and secure such that a person falling or slipping will not displace or deflect the guard. A common physical guard in a substation is the grounded enclosure of metal-clad switchgear.

124C. Types of Guards. The first sentence of this rule points out that meeting the safety clearance zone for the electric supply station fence in Rule 110A4 and meeting the live part clearances in **NESC** Table 124-1 permit guarding by location.

Providing adequate clearance is the most common form of guarding by location. When guarding by isolation is used, entrances to the guarded space

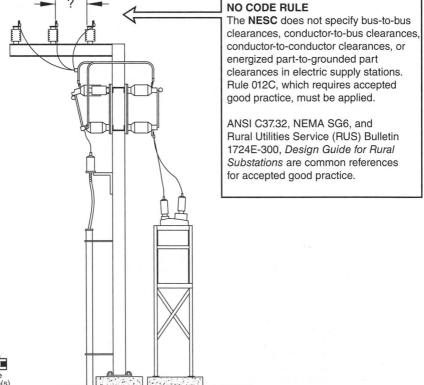

NO CODE RULE
The **NESC** does not specify bus-to-bus clearances, conductor-to-bus clearances, conductor-to-conductor clearances, or energized part-to-grounded part clearances in electric supply stations. Rule 012C, which requires accepted good practice, must be applied.

ANSI C37.32, NEMA SG6, and Rural Utilities Service (RUS) Bulletin 1724E-300, *Design Guide for Rural Substations* are common references for accepted good practice.

See
Photo(s)

Fig. 124-4. Absence of **Code** rule related to bus to bus clearances (Rule N/A).

must be locked, barricaded, or roped off, and safety signs must be posted at entrances.

Rules 124C2 through 124C6 discuss various types of physical guards including shields, enclosures, barriers, mats, supporting surfaces for persons above live parts, and insulating covering. When railings or fences are used as guards, **NESC** Fig. 124-2 applies. The requirement in Rule 124C3 to locate the guard railing or fence "preferably not more than 4 ft" from the nearest point in the guard zone may not be practical in some cases. A NOTE to the rule indicates that additional working space may be required (more than 4 ft) when the working space in Rule 125 and the minimum approach distances in Rule 441 are considered for working with hot sticks. An example of a railing or fence used as a guard is shown in Fig. 124-5.

Rule 124C6 discusses conductor guarding using insulation. Specific requirements apply to various voltage levels and cable types. See Fig. 124-6.

124D. Taut-String Distances.   The taut-string clearance distance is composed of two components, the vertical clearance component which must be not less than 5 ft, and the shortest diagonal or horizontal clearance component. The shortest diagonal or horizontal clearance component allows the setback of a bushing at the top of a piece of equipment to be factored into the overall vertical clearance distance. **NESC** Fig. 124-3 is provided to illustrate the taut-string measurement. The taut-string distance can be applied to the energized (live) part clearances in Rule 124A1 and to

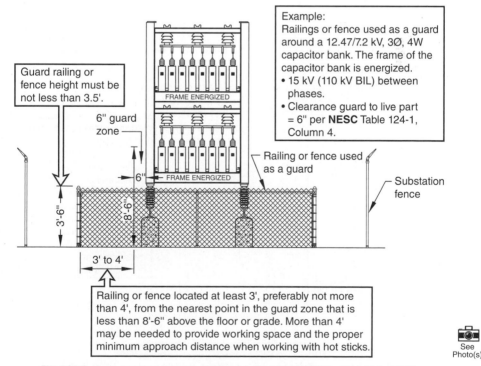

Fig. 124-5. Example of how to apply **NESC** Table 124-1 and **NESC** Fig. 124-2 (Rule 124C3).

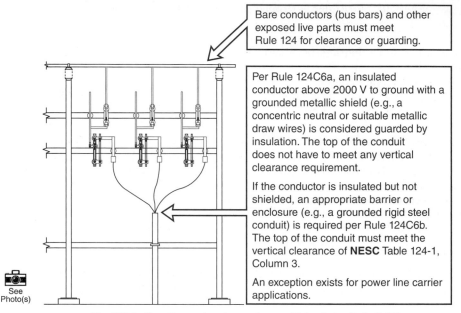

Bare conductors (bus bars) and other exposed live parts must meet Rule 124 for clearance or guarding.

Per Rule 124C6a, an insulated conductor above 2000 V to ground with a grounded metallic shield (e.g., a concentric neutral or suitable metallic draw wires) is considered guarded by insulation. The top of the conduit does not have to meet any vertical clearance requirement.

If the conductor is insulated but not shielded, an appropriate barrier or enclosure (e.g., a grounded rigid steel conduit) is required per Rule 124C6b. The top of the conduit must meet the vertical clearance of **NESC** Table 124-1, Column 3.

An exception exists for power line carrier applications.

See Photo(s)

Fig. 124-6. Guarding a substation conductor with insulation (Rule 124C6).

the part of indeterminate potential clearance in Rule 124A3. The taut-string vertical and diagonal components for an energized part (top of bushing) and the taut-string vertical and horizontal components for a part of indeterminate potential (bottom of the bushing) are shown in Fig. 124-7.

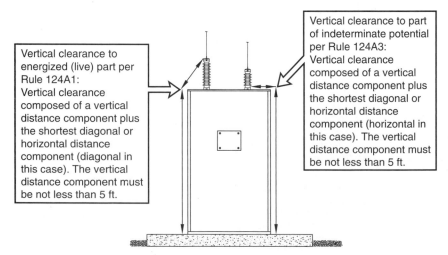

Vertical clearance to energized (live) part per Rule 124A1:
Vertical clearance composed of a vertical distance component plus the shortest diagonal or horizontal distance component (diagonal in this case). The vertical distance component must be not less than 5 ft.

Vertical clearance to part of indeterminate potential per Rule 124A3:
Vertical clearance composed of a vertical distance component plus the shortest diagonal or horizontal distance component (horizontal in this case). The vertical distance component must be not less than 5 ft.

Fig. 124-7. Taut-string distances (Rule 124D).

125. WORKING SPACE ABOUT
ELECTRIC EQUIPMENT

125A. Working Space (600 V or Less). Working space is required around electrical equipment for inspection or servicing. Adequate working space avoids equipment crowding and provides a safe working environment. The working space required for equipment operated at 600 V or less is outlined in **NESC** Table 125-1. In addition to the horizontal distances shown in the table, a minimum of 7 ft of headroom is required and a width of not less than 30 in is required. If the equipment is wider than 30 in, then the working space must be available for the full width of the equipment.

The **Code** specifically states that concrete, brick, or tile walls are considered grounded. A sheetrock wall is not referenced.

The back of a switchboard is assumed to be nonaccessible if all parts replacements, wire connections, etc., can be done from the front. If this were not the case, working space would also be needed behind a switchboard. The distance must be measured from the front of the enclosure if the energized parts are normally enclosed.

The conditions in **NESC** Table 125-1 are for exposed energized parts. If the parts are always de-energized during inspection, servicing, etc., then the **NESC** does not specify a workspace dimension. See Figs. 125-1 and 125-2.

The working space in Fig. 125-1 must be guarded to avoid encroachment by others into the equipment or into the service person when the working space is in an open area or passageway. There must be at least one entrance for access into the working space. See Rule 113 for exit requirements. The working space must not be used for storage. See Fig. 125-3.

125B. Working Space over 600 V. For voltages above 600 V, the working space is provided in accordance with **NESC** Table 124-1. If the horizontal clearance of unguarded parts from **NESC** Table 124-1 is used as the working space around equipment, additional working room may be needed for tools such as hot sticks. A note to Rule 125B states that the minimum approach distances in Rule 441 should be considered. Examples of working space in areas with equipment over 600 V are shown in Figs. 125-4 and 125-5.

126. EQUIPMENT FOR WORK
ON ENERGIZED PARTS

This rule indirectly ties the work rules of Part 4 into the working space of Rule 125. If a worker is within the guard zone of **NESC** Table 124-1, Column 4, then they must utilize protective equipment that is properly tested and rated for the voltage involved. The Work Rules in Part 4, specifically Rule 441, contain minimum approach distances to live parts.

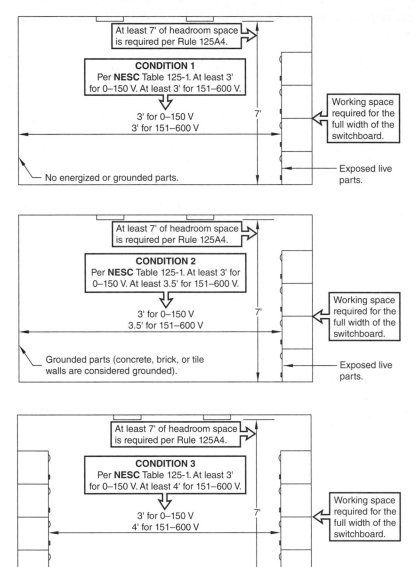

At least 7' of headroom space is required per Rule 125A4.

CONDITION 1
Per **NESC** Table 125-1. At least 3' for 0–150 V. At least 3' for 151–600 V.

3' for 0–150 V
3' for 151–600 V

7'

Working space required for the full width of the switchboard.

No energized or grounded parts.

Exposed live parts.

At least 7' of headroom space is required per Rule 125A4.

CONDITION 2
Per **NESC** Table 125-1. At least 3' for 0–150 V. At least 3.5' for 151–600 V.

3' for 0–150 V
3.5' for 151–600 V

7'

Working space required for the full width of the switchboard.

Grounded parts (concrete, brick, or tile walls are considered grounded).

Exposed live parts.

At least 7' of headroom space is required per Rule 125A4.

CONDITION 3
Per **NESC** Table 125-1. At least 3' for 0–150 V. At least 4' for 151–600 V.

3' for 0–150 V
4' for 151–600 V

7'

Working space required for the full width of the switchboard.

Exposed live parts on both sides with operator between.

Fig. 125-1. Working space for 600 V or less (Rule 125A).

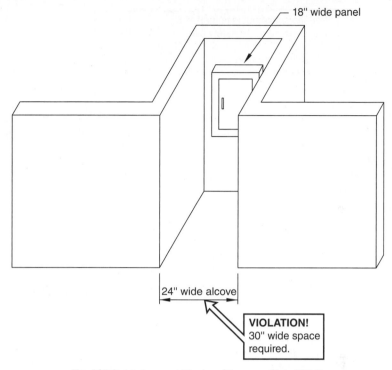

Fig. 125-2. Minimum width of working space (Rule 125A3).

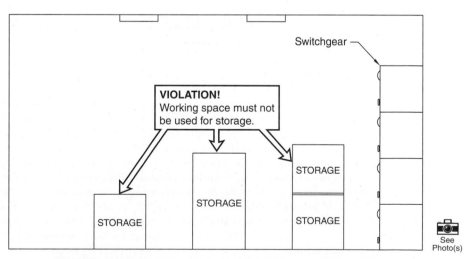

Fig. 125-3. Storage materials must not be in the working space (Rule 125A1).

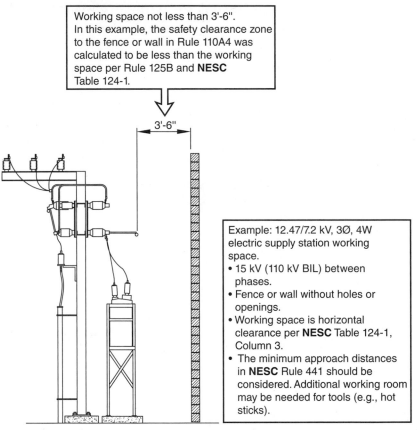

Working space not less than 3'-6".
In this example, the safety clearance zone to the fence or wall in Rule 110A4 was calculated to be less than the working space per Rule 125B and **NESC** Table 124-1.

3'-6"

Example: 12.47/7.2 kV, 3Ø, 4W electric supply station working space.
• 15 kV (110 kV BIL) between phases.
• Fence or wall without holes or openings.
• Working space is horizontal clearance per **NESC** Table 124-1, Column 3.
• The minimum approach distances in **NESC** Rule 441 should be considered. Additional working room may be needed for tools (e.g., hot sticks).

Fig. 125-4. Example of working space over 600 V (Rule 125B).

127. CLASSIFIED LOCATIONS

Classified locations are locations where fire or explosion hazards may exist due to flammable gases, vapors, liquids, dust, or fibers. The **NESC** requires that classified locations in the vicinity of electric supply stations meet the National Electrical **Code** (NEC) hazardous (classified) locations requirements. The NEC has very detailed requirements related to classified locations, and rather than repeating them, the **NESC** requires that the NEC rules be met. In addition to the NEC requirements, the **NESC** provides requirements for coal-handling areas and various other hazardous locations primarily related to electric supply generating stations. The potential for substations associated with or in proximity to refineries, fuel dispensaries, and chemical processing facilities to be classified locations should also be evaluated. Several National Fire Protection Association (NFPA) documents are required to be referenced for specific types of installations.

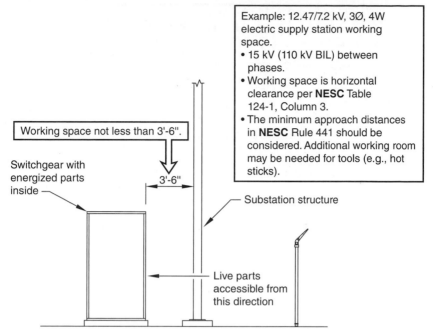

Example: 12.47/7.2 kV, 3Ø, 4W electric supply station working space.
• 15 kV (110 kV BIL) between phases.
• Working space is horizontal clearance per **NESC** Table 124-1, Column 3.
• The minimum approach distances in **NESC** Rule 441 should be considered. Additional working room may be needed for tools (e.g., hot sticks).

Working space not less than 3'-6".

Switchgear with energized parts inside

3'-6"

Substation structure

Live parts accessible from this direction

Fig. 125-5. Example of working space over 600 V (Rule 125B).

Separation from or ventilation of a hazardous area can reduce the classified area requirements. High-voltage facilities are typically located away from classified areas. Low-voltage equipment (i.e., under 600 V) can be installed in explosionproof enclosures and located in a classified location in accordance with the NEC rules.

128. IDENTIFICATION

This rule requires identification or labeling of equipment and devices in the electric supply station. This includes indoor switchboards and outdoor equipment. The identification must be uniform throughout any one station and not be placed on removable covers or doors that could be interchanged, thereby creating a labeling error. The Code is not specific as to the type of identification, but equipment must be sufficiently labeled for safe use and operation. Identification may include voltage levels, nameplate information, phase color coding, feeder numbering, etc. Rule 171 also addresses identification of equipment. An example of identification in a substation is shown in Fig. 128-1.

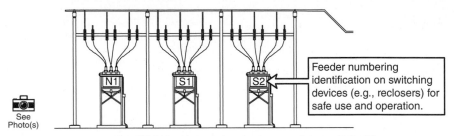

Fig. 128-1. Example of identification in a substation (Rule 128).

129. MOBILE HYDROGEN EQUIPMENT

Hydrogen is primarily used in generating stations. It is highly flammable. Bonding of mobile hydrogen equipment will reduce the chance for a difference in potential, which will reduce the chance of sparks that may cause an explosion.

Section 13

Rotating Equipment

130. SPEED CONTROL AND STOPPING DEVICES

Section 13 deals with generators, motors, motor generators, and rotary converters in electric supply stations. **NESC** Sec. 02, "Definitions of Special Terms," defines an electric supply station as a generating station or substation. Rotating equipment is more applicable to generating stations, as substations typically do not involve very much rotating equipment, although it still applies to substation equipment like transformer fans and climate control machinery, to a limited degree.

Section 13 only provides general rules related to basic safety features. Other IEEE standards are available, but not specifically referenced in the **Code**, for generation station design. IEEE Standards and the National Electrical Code (NEC) are great examples of accepted good practice resources to aid in specific design features and particulars not specified in the **NESC**, as required by Rule 012C.

Rule 130 requires overspeed trip of prime movers (i.e., diesel engines, turbines, etc.) in addition to governors. Manual stopping devices are also required. Speed-limiting devices are required for separately excited DC motors and series motors, as their designs are potentially susceptible to problems with overspeed or runaway. Adjustable-speed motors that are controlled by field regulation must also have a speed-limiting device and must be equipped or connected to avoid weak fields that produce overspeed. It should be noted that this is not applicable to more modern variable frequency drives because their speed is controlled by the frequency of their input power supply. Mechanical protection of the control circuits for stopping and speed limiting is also required. This provision is intended to ensure that a significant malfunction of machinery does not damage the control circuitry in a way to prevent a system shutdown.

131. MOTOR CONTROL

Rule 131 requires that motors do not restart automatically after a power outage if the unexpected starting could create injury to personnel. Two types of basic motor starting circuits are used in the motor control industry, low-voltage protection and low-voltage release.

Low-voltage protection (three-wire control) requires manual restarting by an equipment operator. This rule does provide an option to use low-voltage release (two-wire control with automatic restarting) if warning signals and time-delay features are designed into the control scheme. Examples of simple two-wire and three-wire motor control schemes are shown in Fig. 131-1.

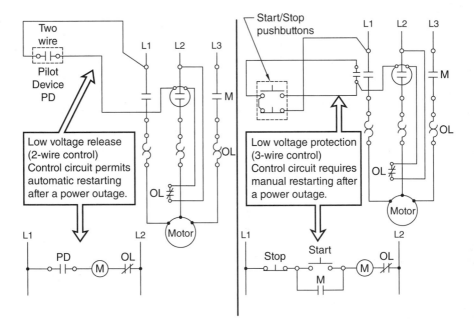

Fig. 131-1. Example of simple two-wire and three-wire motor control schemes (Rule 131).

132. NUMBER 132 NOT USED IN THIS EDITION

133. SHORT-CIRCUIT PROTECTION

This rule requires short-circuit protection for electric motors. Only a simple requirement is stated. Rule 012C, which requires accepted good practice, should be applied
for specific details. The National Electrical Code (NEC) specifies maximum fuse and
circuit breaker sizes for various types and sizes of motors. The NEC is an excellent
reference for accepted good practice in this case.

An example of motor protection components and terminology is shown in
Fig. 133-1.

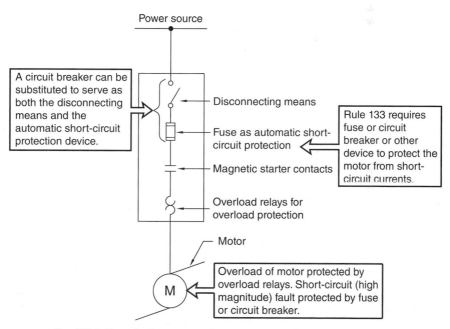

Fig. 133-1. Example of motor protection components and terminology (Rule 133).

Section 14

Storage Batteries

140. GENERAL

This section outlines the requirements for stationary storage batteries. Stationary battery storage systems fall primarily into several categories. Plant batteries provide power and/or backup power for control and operation of generating stations. Switchgear batteries, including most substation battery applications, provide power for safety systems, circuit breakers, circuit switchers, reclosers, etc., enabling continued system management and protection even upon loss of grid power. Grid storage batteries draw energy from and discharge energy into the electric power grid as a normal part of operation. Rule 140 is applicable to all stationary storage battery systems, while Rule 141 addresses switchgear and plant batteries, and Rule 142 addresses grid storage batteries.

Several different battery technologies are used in these types of applications, including lead-acid, nickel-cadmium (NiCd), Lithium Ion (Li-Ion), and Lithium Iron Phosphate (LiFePO4 or LFP). The hazards associated with battery systems include shock and arc flash hazards, similar to most high-energy electrical systems, but also may include chemical hazards, depending on the battery technology, such as corrosive electrolytes. Some types of battery systems also produce combustible gases (e.g., hydrogen), which pose a significant explosion hazard if allowed to accumulate. The requirements of Section 14 are written with sufficient generality to address all of the different hazards, allowing application as needed for the specific types of batteries used. An example of storage batteries used in a substation control building is shown in Fig. 140-1.

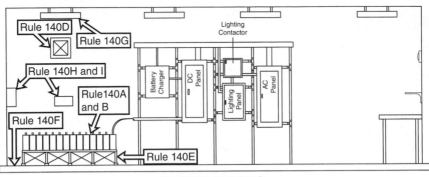

CONTROL BUILDING

Rule 140B. Working space • Adequate space for inspection, maintenance, testing, and replacement of batteries.	**Rule 140F. Floor protection** • Corrosion-resistant floor where batteries contain corrosive liquids. • Provisions to contain spilled electrolyte.
Rule 140C. Enclosure • Batteries located in a protective enclosure or area accessible only to qualified personnel.	**Rule 140G. Illumination** • Guard or isolate lighting fixtures (e.g., lenses or wire guards). • No receptacles or light switches in battery area where batteries may produce combustible gases.
Rule 140D. Ventilation • Natural or powered ventilation. • Annunciate powered ventilation failure.	**Rule 140H. Signs** • Safety sign prohibiting smoking, sparks, or flames where batteries may produce combustible gases.
Rule 140E. Racks • Racks firmly anchored, preferably to the floor. • Metal racks must be grounded.	**Rule 140I. Neutralizing agent** • Water or neutralizing agent for eyes and skin must be available while performing battery installation and maintenance.

See
Photo(s)

Fig. 140-1. Example of storage batteries used in a substation control building (Rule 140).

The requirement in 140B that sufficient space be provided for the purposes of inspection, maintenance, testing, and replacement of batteries is an example of one that will depend on battery chemistry. For example, if a lead-acid battery system were used, sufficient space for testing electrolytes and adding water would be required, while some types of Li-Ion systems require only monitoring of voltages to verify the functionality of battery charge management systems. Infrared inspection of connections to and between batteries will likely be required for all types of systems. Eventual replacement of batteries also needs to be considered in the design of the space in which the batteries are installed.

Another example of a requirement that applies differently depending on the battery chemistry is Rule 140D, Ventilation. Not all batteries are subject to producing combustible gases, but care must be taken in the design of the systems that do, to provide adequate ventilation to prevent the combustible gases from accumulating to explosive levels. For example, a concentration of as little as

4% hydrogen in air will explode, so levels of hydrogen should be held well below that level. Battery manufacturers can provide guidance for calculating hydrogen gas emissions and ventilation recommendations for specific systems. Given the potential catastrophic impact of an explosion, ventilation systems should be monitored, and an alarm generated if there is a ventilation failure, where the combustible gas emissions are a concern. Other rules also address the hazard of a potentially explosive atmosphere including the requirement that receptacles and switches for lighting be located outside of battery areas (Rule 140G) and the mandate for safety signs prohibiting smoking, sparks, and flame (Rule 140H).

Safe work practices pertaining to work on or around batteries are addressed in Rule 420G, but Rule 140I does include a related requirement that water facilities or neutralizing agents be available to rinse eyes and skin in the event of a chemical spill (e.g., electrolyte) while performing maintenance and installation. The required facilities may be stationary or portable but must be present while work is being undertaken.

141. SWITCHGEAR AND PLANT BATTERIES

This rule addresses switchgear and plant batteries, as described under Rule 140. Rule 141B specifically prohibits the usage of switchgear/plant batteries for grid applications. This is presumably because it is critical that the switchgear and plant applications always be available because of their safety and protection functions. Indeed, the only other requirement specific to this application is that the battery system be sized so that it has enough energy that it can continue to operate as needed for as long as it takes to remediate the loss of, or a fault on, the dc charging system (Rule 141C). If the switchgear and plant batteries were integrated with a grid storage application, it would be difficult to ensure that the system always retained sufficient energy to meet the capacity requirement.

142. GRID STORAGE BATTERIES

As discussed under Rule 140, Rule 142 applies to grid storage applications. Rule 142B prohibits the use of grid storage batteries in switchgear and plant battery applications, as discussed under Rule 141B. Grid storage batteries generally have very large energy capacities and power delivery rates. Because of this, there may be a more elevated risk of fire in the event of a battery cell failure. Therefore, Rule 142C includes a specific requirement that the battery installations that pose a fire hazard include protective measures to prevent such a fire from damaging adjacent structures and equipment. Rule 142D is similar to the requirement for containment set forth in Rule 140F, except that it generalizes to include any hazardous liquid (whereas Rule 140F only refers to electrolytes). Similarly, Rule 142E requires safety signage as does Rule 140H, except that Rule 142E more broadly addresses fire, toxic chemicals, and other hazards.

Section 15

Transformers and Regulators

150. CURRENT-TRANSFORMER SECONDARY
CIRCUITS PROTECTION WHEN EXCEEDING 600 V

This rule requires secondary circuits of current transformers (with primary circuits over 600 V) to be mechanically protected from damage by duct, covering, or some other protection. If the duct (conduit) or covering is metal, it must be effectively grounded and consideration must be given to circulating currents. Current transformers (CTs) must also have a provision for shorting the secondary wiring. These requirements are needed, as open or damaged current-transformer secondary circuits may cause a hazardous high voltage and arcing.

An example of protecting current-transformer secondary circuits is shown in Fig. 150-1.

151. GROUNDING SECONDARY CIRCUITS
OF INSTRUMENT TRANSFORMERS

Instrument transformers consist of both voltage transformers (VTs) and current transformers (CTs). Voltage transformers are sometimes referred to as potential transformers (PTs). The secondaries of VTs and CTs must be effectively grounded. An example of basic VT and CT connections is shown in Fig. 151-1.

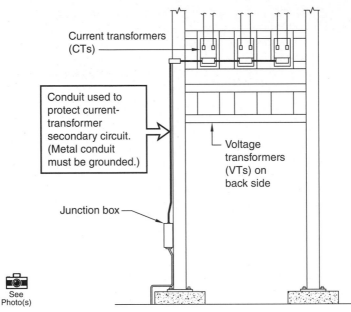

Current transformers (CTs)

Conduit used to protect current-transformer secondary circuit. (Metal conduit must be grounded.)

Voltage transformers (VTs) on back side

Junction box

See Photo(s)

Fig. 150-1. Example of protecting current-transformer secondary circuits (Rule 150).

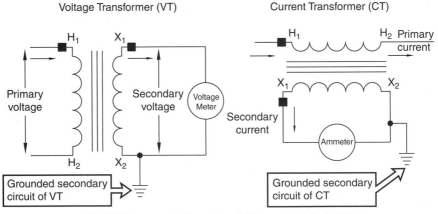

Voltage Transformer (VT)   Current Transformer (CT)

H_1 X_1 H_1 H_2 Primary current

Primary voltage   Secondary voltage   Voltage Meter

Secondary current

H_2 X_2 X_1 X_2

Grounded secondary circuit of VT   Ammeter

Grounded secondary circuit of CT

Fig. 151-1. Example of basic VT and CT connections (Rule 151).

152. LOCATION AND ARRANGEMENT OF POWER TRANSFORMERS AND REGULATORS

152A. Outdoor Installations. This rule requires that energized parts of power transformers be enclosed, guarded, or physically isolated as discussed in Rule 124. The case (enclosure) of a substation transformer and regulator must be effectively grounded or guarded.

Most large substation transformers and generator step up transformers, as well as a significant percentage of regulators and service transformers in electric supply stations, are filled with a highly refined mineral oil as an insulating and cooling medium. This oil is flammable, and is also present in such large quantities that, if ever ignited, the risk of fire spread and the challenge of extinguishment are extreme, particularly when a high-energy internal failure ruptures the transformer tank. To address this, the **Code** requires that fire hazards related to liquid-filled power transformers installed in outdoor electric supply stations be minimized using one or more of the following methods:

- Less flammable liquids
- Space separation
- Fire-resistant barriers
- Automatic extinguishing systems
- Absorption beds
- Enclosures

The **Code** requires that the degree of the fire hazard and the amount and characteristics of the liquid be considered when selecting a method, but no specifics are provided for the amount or type of oil, separation distance, etc. Rule 012C, which requires accepted good practice, must be applied. IEEE Standard 979, *IEEE Guide for Substation Fire Protection,* is an excellent reference for accepted good practice in this case.

The **NESC** does not address Spill Prevention Control & Countermeasure (SPCC) plans. SPCC plans are required by the Environmental Protection Agency (EPA). The amount of oil contained in the substation and the location of the substation relative to water are two of the many considerations that trigger the need for a substation SPCC plan.

There are times when an engineer or designer is trying to determine how close an oil-filled service transformer can be placed to a building. Rule 152 is in Part 1, which only applies to electric supply stations, not to overhead or pad-mounted service transformers outside the substation fence. Part 2, Overhead Lines, and Part 3, Underground Lines, do not provide any additional rule related to locating oil-filled transformers near buildings. Part 2 does provide clearances to buildings for live parts and equipment cases, but these clearances are based on electrical safety, not fire safety. Rule 012C, which requires accepted good practice, must be applied in this instance. The National Electrical Code (NEC), Underwriters Laboratories (UL), and Factory Mutual (FM) are all good references for determining service transformer location standards.

152B. Indoor Installations. Transformers and regulators installed indoors require even greater consideration of fire hazards. Rule 152B is divided into three parts. Rule 152B1 covers traditional oil-filled transformers, 75 kVA and above. It requires ventilation for the room or vault in which the transformer is installed so that smoke generated by a fire involving such equipment may be released outdoors rather than being forced into the building. It also requires containment to prevent the flow of burning oil from spreading the fire outside the room, and fire doors and fire walls between the room or vault and the rest of the building. The second paragraph covers dry-type transformers. The third paragraph covers oil-filled transformers with less flammable oil. Traditional oil-filled transformers use mineral oil, which is considered flammable. Less flammable oil is typically a fluid consisting of fire-resistant hydrocarbon. Dry-type transformers use no fluid oil for cooling. They are vented for air cooling. Dry-type transformers usually do not have the overload capability that oil-filled transformers have. The term "fire walls" is used but no specifics as to the fire wall ratings are provided (i.e., 1 hour rating, 2 hour rating, etc.). In addition to the accepted good practice documents discussed in Rule 152A, the Uniform Building Code (UBC) or other applicable building code document must be referenced for fire wall construction details. When an electric supply station is located indoors, for example, as part of a high-rise building in a large city, both the **NESC** and applicable building codes need to be addressed. NFPA 850, *Recommended Practice for Fire Protection for Electric Generating Plants and High Voltage Direct Current Converter Stations*, can be referenced when designing generating stations.

The requirements for locating traditional oil-filled transformers indoors are outlined in Fig. 152-1.

Rule 152B2 addresses dry-type transformers and those filled with a nonflammable liquid or gas. Dry-type transformers use no fluid oil for cooling. They are

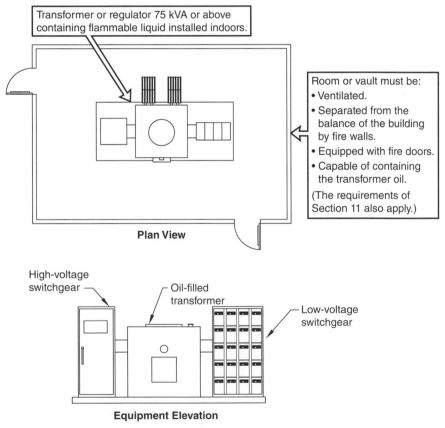

Fig. 152-1. Oil-filled transformer installed indoors (Rule 152B1).

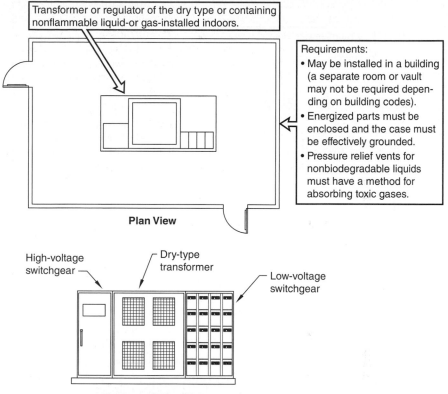

Transformer or regulator of the dry type or containing nonflammable liquid-or gas-installed indoors.

Requirements:
• May be installed in a building (a separate room or vault may not be required depending on building codes).
• Energized parts must be enclosed and the case must be effectively grounded.
• Pressure relief vents for nonbiodegradable liquids must have a method for absorbing toxic gases.

Plan View

High-voltage switchgear

Dry-type transformer

Low-voltage switchgear

Equipment Elevation

Fig. 152-2. Dry-type transformer installed indoors (Rule 152B2).

vented for air cooling. Dry-type transformers usually do not have the overload capability that oil-filled transformers have, and are generally larger than liquid-filled transformers for the same capacity. The requirements for locating dry-type transformers indoors are outlined in Fig. 152-2.

Rule 152B3 covers oil-filled transformers with less flammable liquids than typical mineral insulating oil. Less flammable oil is typically a fluid consisting of fire-resistant hydrocarbon. The requirements for locating less flammable liquid transformers indoors are outlined in Fig. 152-3.

153. SHORT-CIRCUIT PROTECTION OF POWER TRANSFORMERS

Electric supply station power transformers are required to have short-circuit protection. A short circuit within the transformer provides a high-magnitude fault current and may lead to catastrophic transformer explosions and/or fires. The short-circuit protection device typically will not be sized to protect transformer overload although the same circuit breaker or other protective device may

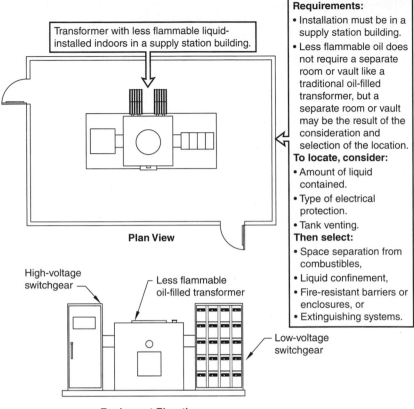

Requirements:
- Installation must be in a supply station building.
- Less flammable oil does not require a separate room or vault like a traditional oil-filled transformer, but a separate room or vault may be the result of the consideration and selection of the location.

To locate, consider:
- Amount of liquid contained.
- Type of electrical protection.
- Tank venting.

Then select:
- Space separation from combustibles,
- Liquid confinement,
- Fire-resistant barriers or enclosures, or
- Extinguishing systems.

Transformer with less flammable liquid-installed indoors in a supply station building.

Plan View

High-voltage switchgear

Less flammable oil-filled transformer

Low-voltage switchgear

Equipment Elevation

Fig. 152-3. Less flammable oil-filled transformer installed indoors (Rule 152B3).

be used, controlled by separate short circuit and overload relaying functions. Transformer overload is commonly monitored by metering the load and comparing the metered load to the transformer nameplate capacity. It should be noted that all power sources need to be disconnected, so if the transformer is not radially connected, both primary and secondary devices will be required.

This rule provides a list of automatically disconnecting short-circuit protection devices. Circuit breakers typically open all three phases of a three-phase line simultaneously. Fuses, on the other hand, open just the faulted phase. Removing just the faulted phase is acceptable. In the case of large generators where placing a circuit breaker between the generator and its step-up transformer is impractical, it is acceptable to remove the power source by interrupting the field source and disconnecting or otherwise releasing the mechanical energy of the prime mover. This rule does not apply to every transformer in the station, only to power transformers. For example, current transformers (CTs) are not fused for reasons discussed under Rule 150. Voltage transformers (VTs) may be fused, but the **Code** does not require them to be fused. The choices for short-circuit protection of power transformers are outlined in Fig. 153-1.

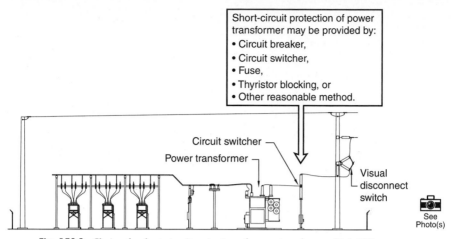

Short-circuit protection of power transformer may be provided by:
• Circuit breaker,
• Circuit switcher,
• Fuse,
• Thyristor blocking, or
• Other reasonable method.

Circuit switcher

Power transformer

Visual disconnect switch

See Photo(s)

Fig. 153-1. Choices for short-circuit protection of power transformers (Rule 153).

Section 16

Conductors

160. APPLICATION

Selecting conductors for use in an electric supply station may appear to be a simple task, but there are literally thousands of choices including material, cross sectional area and geometry, coatings/plating, etc. This rule requires that the following conductor and termination hardware features must be considered when choosing a conductor.

- Suitable for the installation
- Suitable for the ampacity (conductor and termination size)
- Suitable for the voltage (insulation rating)
- Suitable for other applicable ratings

This section only applies to conductors and termination hardware in an electric supply station. Part 2 applies to overhead line conductors outside the substation, and Part 3 applies to underground conductors outside the substation. Conductors in this rule, in Part 2, Overhead Lines, and in Part 3, Underground Lines, are discussed in general terms. The NESC does not specify conductor ampacity for various conductor sizes and types. To determine a conductor's ampacity, Rule 012C, which requires accepted good practice, must be applied. Ampacity tables can be found in the National Electrical Code (NEC) and in conductor manufacturers' literature. Both of these sources are excellent applications of Rule 012C (accepted good practice). The NEC also specifies fuse and circuit breaker sizes for protecting 600 V conductors.

Conductors, by the definition in Sec. 02, include substation bus bars, as well as wires, which are typically bare, and cables, which are typically insulated.

Typical examples of conductors found in substations are shown in Fig. 160-1.

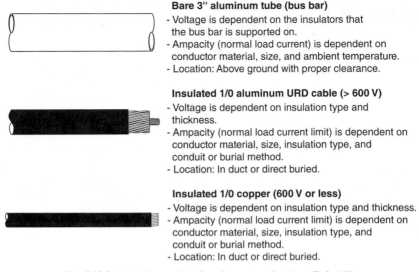

Bare 3" aluminum tube (bus bar)
- Voltage is dependent on the insulators that the bus bar is supported on.
- Ampacity (normal load current) is dependent on conductor material, size, and ambient temperature.
- Location: Above ground with proper clearance.

Insulated 1/0 aluminum URD cable (> 600 V)
- Voltage is dependent on insulation type and thickness.
- Ampacity (normal load current limit) is dependent on conductor material, size, insulation type, and conduit or burial method.
- Location: In duct or direct buried.

Insulated 1/0 copper (600 V or less)
- Voltage is dependent on insulation type and thickness.
- Ampacity (normal load current limit) is dependent on conductor material, size, insulation type, and conduit or burial method.
- Location: In duct or direct buried.

Fig. 160-1. Typical examples of conductors in substations (Rule 160).

161. ELECTRICAL PROTECTION

After the conductors are properly selected and applied, electrical protection of conductors is required. Overcurrent protection of conductors consists of two components, overload and short-circuit protection. Overload protection keeps the load current from exceeding the conductor's normal load current ampacity. Short-circuit protection keeps the fault current from exceeding the conductor's short-time fault current ampacity.

This rule requires overcurrent (both overload and short-circuit) protection by the following methods:
- Design of the system *and*
- Overcurrent devices *or*
- Alarm devices *or*
- Indication devices *or*
- Trip devices

A typical 120-V, 20-A, single-pole, thermal-magnetic circuit breaker protecting a No. 12 AWG copper, 600-V conductor provides both overload protection via the circuit breaker thermal element and short-circuit protection via the circuit breaker magnetic element. A typical 15-kV recloser protecting an overhead or underground distribution feeder may be set to only provide short-circuit protection. The overload protection of the feeder may be monitored by checking percent conductor loading of the conductor during a system load study.

The electrical protection device may protect the conductor and a piece of equipment. For example, the high-side fuse of a substation transformer can provide short-circuit protection for the substation transformer and the bus bars. Overload protection of the transformer and bus bars may be provided by monitoring the load and comparing the load to the transformer and bus bar ratings. An exception permits short lengths of insulated power cables operating at voltages less than 1,000 V to not have short-circuit protection in situations where the cable leads from the power source (e.g., transformer or battery) to the first distribution point (e.g., distribution panel or fused disconnect), as long at the cable is protected to minimize the likelihood of a short circuit. This is a practical exception as it may not be possible to place the first protective device right at the power source.

Grounded conductors must be configured without overcurrent protection, so there is not an interruption in the grounding protection. When a single-phase or three-phase circuit is fused or connected to a circuit breaker or recloser, the grounded conductor (neutral) must not be fused or connected to the circuit breaker or recloser contacts so that the continuity to ground is maintained.

Rule 161 applies to conductors and cables inside the electric supply station. There is not a corresponding rule in Part 2 (Overhead Lines) or Part 3 (Underground Lines) requiring overcurrent protection for conductors outside the substation; therefore, accepted good practice (Rule 012C) must be applied.

162. MECHANICAL PROTECTION AND SUPPORT

Conductors contained within the electric supply station must be adequately supported to withstand forces caused by the maximum short-circuit current that the conductor may experience. During short circuits, forces due to magnetic fields can occur on the substation conductors, including rigid bus bars. A rigid bus design must be adequately supported to withstand the short-circuit forces. The bus center-to-center spacing and the magnitude of short-circuit current both have an effect on the short-circuit forces. The strength of the substation structure and the strength of the insulators are typically based on the short-circuit forces and wind and ice loading; however, for the conductors contained inside the substation fence, the **Code** does not specify wind and ice conditions. For wind and ice loads inside the substation, Rule 012C, which requires accepted good practice, must be applied. For overhead conductors that extend outside the electric supply station, the wind and ice loading and strength requirements in Part 2, Overhead Lines, Secs. 24, 25, 26, and 27 must be applied. For example, a transmission-deadend tower inside the substation that has overhead conductors that leave the substation must have the same loading and strength rules applied to it as the transmission pole outside the substation. Some utilities use a reduced tension span between the substation deadend tower and the first transmission pole outside the substation to reduce the loading and strength requirements on the substation deadend tower. This method typically requires guying on the first transmission pole outside the substation.

Commonly referenced design standards for substation conductor supports are IEEE Standard 605, *IEEE Guide for Design of Substation Rigid Bus Structures*, ASCE 113, *Substation Structure Design Guide*, and ASCE 7, *Minimum Design Loads for Buildings and Other Structures*. The Uniform Building Code (UBC) is commonly referenced for seismic (earthquake) zones. The Rural Utilities Service (RUS) publishes Bulletin 1724E-300, *Design Guide for Rural Substations*, which contains detailed information on substation design calculations including calculations for short-circuit forces.

If a conductor, its insulation, or support is subject to mechanical damage, the following are required to limit the likelihood of damage or disturbance:
- Casing,
- Armor, or
- Other means

Typically in a substation, galvanized rigid steel conduit is used to protect and support insulated conductors subject to mechanical damage.

The mechanical protection and support requirements of this rule are outlined in Fig. 162-1.

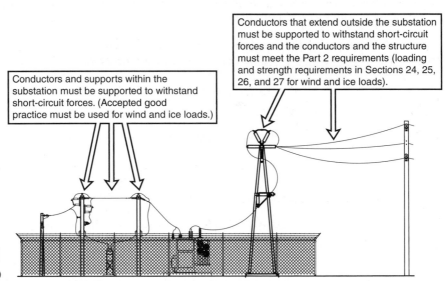

Conductors that extend outside the substation must be supported to withstand short-circuit forces and the conductors and the structure must meet the Part 2 requirements (loading and strength requirements in Sections 24, 25, 26, and 27 for wind and ice loads).

Conductors and supports within the substation must be supported to withstand short-circuit forces. (Accepted good practice must be used for wind and ice loads.)

See Photo(s)

Fig. 162-1. Substation conductor mechanical protection and support (Rule 162).

163. ISOLATION

Conductors inside the substation fence must be treated like the live parts in Rule 124. Nonshielded, insulated, and jacketed conductors may be installed in accordance with Rule 124C6, which discusses conductor guarding using insulation. See Rule 124 for additional information and a figure.

164. CONDUCTOR TERMINATIONS

The rules for conductor terminations in an electric supply substation are outlined in Fig. 164-1.

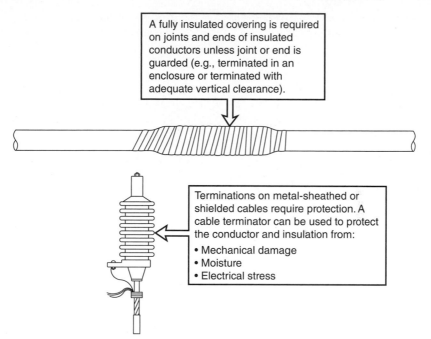

A fully insulated covering is required on joints and ends of insulated conductors unless joint or end is guarded (e.g., terminated in an enclosure or terminated with adequate vertical clearance).

Terminations on metal-sheathed or shielded cables require protection. A cable terminator can be used to protect the conductor and insulation from:

• Mechanical damage
• Moisture
• Electrical stress

Fig. 164-1. Conductor terminations (Rule 164).

Section 17

Circuit Breakers, Reclosers, Switches, Fuses, and Other Equipment

170. GENERAL

Rule 170 defines the term "equipment" for the purpose of defining equipment as it applies to Sec. 17. Since the definition is specific to Sec. 17, it is provided in Rule 170 instead of Sec. 02, "Definitions of Special Terms."

171. ARRANGEMENT

This rule requires equipment to be accessible only to qualified persons. Section 17 is part of Part 1, Electric Supply Stations; therefore, the rules of this section apply only to equipment located in electric supply stations. Equipment in electric supply stations are accessible only to qualified persons when Rule 110A is met. "Equipment" is defined in Rule 170. For the purposes of Rule 171, Equipment includes circuit breakers, reclosers, switches, fuses, and arresters.

To protect persons from energized parts or arcing, Rule 170 requires the following:
- Walls,
- Barriers,
- Latched doors,

- Location,
- Isolation, or
- Other means

The requirements of Rule 124, Guarding Live Parts, also apply.

Conspicuous and unique markings must be provided at the switching device or at any remote operating point to identify the equipment that is being controlled. To avoid confusion, device identification must not be duplicated within the same electric supply station. The identification is for qualified personnel who are authorized to operate switching devices. See the discussion of identification and a related figure in Rule 128.

When switch contacts are not normally visible (e.g., under oil, contained in a vacuum bottle, etc.), the switching device must be equipped with an operating position indicator. The work rules in Part 4 of the **NESC** provide switching control procedures and rules for de-energizing equipment or lines to protect employees. An example of a switch position indicator is shown in Fig. 171-1.

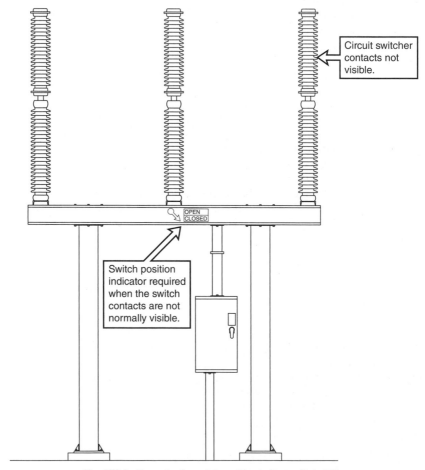

See
Photo(s)

Fig. 171-1. Example of a switch position indicator (Rule 171).

172. APPLICATION

The following ratings must be considered when using equipment in the electric supply station:

- Voltage
- Continuous current
- Momentary current
- Short-circuit current interrupt rating

Per the definition of "equipment" in Rule 170, equipment refers to circuit breakers, reclosers, switches, fuses, and arresters. The maximum short-circuit current interrupt rating must be considered if the device is used to interrupt fault current.

When to apply ratings for momentary currents and interrupt currents is dependent on how the switch is used. If a switch is used as a disconnect only (i.e., it does not open under a fault condition), then it must be able to withstand a fault current flowing through it but it does not need to be rated to interrupt the fault current. The fault current it must withstand is termed the momentary fault current rating. If a switch is used to interrupt a fault, it must be rated to withstand the momentary fault current and interrupt the fault current without damage to the switch itself. A device that interrupts fault current must have a fault current interrupt rating also referred to as amps interrupting capacity (AIC).

It is important to note that the **NESC** does not address the selection of fuse, relay, or recloser settings such as time–current curves, delay intervals, number of operations to lockout, etc. These ratings are typically addressed in a system sectionalizing or coordination study. Careful selection of these and other settings is required for the safe and reliable operation of the electric system, and accepted good practice must be applied in this case. The work rules in Part 4 of the **NESC** do address disabling a reclosing feature during work activities.

The interrupting capacity should be reviewed prior to each significant system change. For example, if a substation recloser is rated to interrupt 1250 A of short-circuit current and the substation transformer is replaced, the available fault current may increase to a value larger than 1250 A, which could cause damage to the recloser and personnel when the recloser operates to interrupt a fault. An example is given in Fig. 172-1.

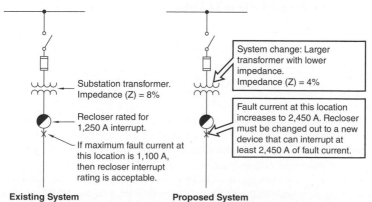

Substation transformer.
Impedance (Z) = 8%

Recloser rated for
1,250 A interrupt.

If maximum fault current at
this location is 1,100 A,
then recloser interrupt
rating is acceptable.

System change: Larger
transformer with lower
impedance.
Impedance (Z) = 4%

Fault current at this location
increases to 2,450 A. Recloser
must be changed out to a new
device that can interrupt at
least 2,450 A of fault current.

Existing System   **Proposed System**

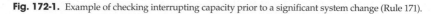

Fig. 172-1. Example of checking interrupting capacity prior to a significant system change (Rule 171).

173. CIRCUIT BREAKERS, RECLOSERS, AND SWITCHES CONTAINING OIL

Circuit breakers, reclosers, and switches containing oil receive special attention due to the flammability of the oil. Similar requirements are outlined in Rule 152 for oil-filled power transformers and regulators.

Segregation of oil-filled circuit-interrupting devices is required in the electric supply station. Segregation of oil-filled circuit-interrupting devices minimizes fire damage to adjacent equipment or buildings. Segregation may be provided by the following methods:
- Spacing,
- Fire-resistant barrier walls, or
- Metal cubicles

Gas-release vents require oil-separating devices or piping to a safe location. Means to control oil discharges from vents or tank rupture are required. Methods to control oil are:
- Absorption beds,
- Pits,
- Drains, or
- Any combination of the above

This list is slightly different from the requirements for oil-filled transformers and regulators in Rule 152, but the general intent is the same. IEEE Standard 979, *IEEE Guide for Substation Fire Protection,* is an excellent reference for fire protection requirements.

Buildings or rooms housing circuit breakers, reclosers, and switches containing flammable oil must be of fire-resistant construction (see Rule 152 for additional information). Not all circuit breakers, reclosers, and switches contain oil. Some are air-break, some are vaccuum-break, and some are SF_6 gas-insulated. This rule applies only to circuit breakers, reclosers, and switches that contain oil and are located in an electric supply station.

174. SWITCHES AND DISCONNECTING DEVICES

Switches and disconnecting devices must have capacity for the following system ratings:
- Voltage
- Current
- Load break current (if required)

Switches can be used to break load currents or open under no-load conditions. If required to break load current, the load current they are rated to interrupt must be marked on the switch. This value should not be confused with the short-circuit (fault) current-interrupt rating discussed in Rule 171.

Switches and disconnectors must be able to be locked open and locked closed, or plainly tagged where locks are not practical. Part 4 of the **NESC** specifies the work rules applicable to operating, locking, and tagging switches. Switches that are operated remotely and automatically must have a disconnecting means for the control circuit near the disconnecting apparatus to limit the likelihood of accidental operation of the switch, such as when maintenance work is underway.

175. DISCONNECTION OF FUSES

Disconnecting an energized fuse can be dangerous at any voltage. This rule requires fuses in circuits of more than 150 V to ground or more than 60 A (at any voltage) to be classified as disconnecting, or arranged to be disconnected from the source of power, or removed with insulating handles.

In most cases, fused cutout-type switches are configured such that the switch-blade or fuse is "dead" when the switch is in the open position. Loop feeds can create energized blades when the switch is in the open position. The NESC does not have a requirement for how the switch blades or fuses are energized but does require the proper handling of the fuse. Proper consideration must be given to clearance requirements when fuse and switchblades are in the open position.

176. OPEN GAPS AND BREAKS

When a switch is opened to de-energize a line or equipment, the open gap or break must have the appropriate dielectric withstand for the system conditions. One main concern is that the open gap must be sufficient for the voltage involved. An open gap under oil will have an insulating rating based on the oil. An open gap in air must be long enough based on the insulating value of air. Rule 444C2 references NESC Table 444-1 which specifies minimum clearances for open air gaps (cut or open jumpers). Rule 176 also contains a note referencing two IEEE standards that contain gap information.

177. ARRESTERS

177A. General Requirements. Surge arresters are used to protect equipment from overvoltage primarily due to switching surges and lighting. They are used throughout the electric utility system. Rule 177 applies to surge arresters installed in an electric supply station. Critical equipment is located in the electric supply station. The careful application of surge arresters is used to limit damage to expensive equipment such as the substation power transformer. Locating a surge arrester as close as practical to the equipment it protects limits voltage buildup across the equipment insulation. IEEE Standard C62.1 and IEEE Standard C62.11 are noted in the NESC. They are excellent references for surge arrester applications.

177B. Indoor Locations.   Arresters can discharge hot gases and produce electric arcs. To avoid damage to its surroundings, if an arrester is located inside a building, it must be enclosed or located far away from passageways and combustible parts.

177C. Grounding Conductors.   Short conductive, low-impedance grounding leads are required to keep the voltage buildup across equipment insulation as low as possible. Suppression of the voltage surge can develop significant current that flows through the arrester and through the ground lead. Excess resistance or impedance in the ground lead would result in a significant reduction in the effectiveness of the arrester because of the voltage developed on the ground lead when subjected to high current. The grounding methods of Sec. 09 are required to be met.

177D. Installation.   Surge arresters can be installed in various locations throughout the electric supply station. The rules regarding surge arrester installation in electric supply stations are outlined in Fig. 177-1.

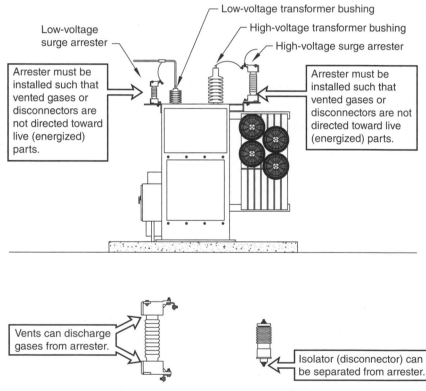

See
Photo(s)

Fig. 177-1.   Arrester installation (Rule 177D).

Section 18

Switchgear and Metal-Enclosed Bus

180. SWITCHGEAR ASSEMBLIES

180A. General Requirements for All Switchgear. This rule covers general requirements for all switchgear. Examples of switchgear found in electric supply stations are shown in Fig. 180-1.

The general requirements for all switchgear are outlined below:

- Secure to minimize movement.
- Support cables to minimize force on terminals.
- Locate away from liquid or hazardous gas piping unless switchgear is protected against pipe failure.
- Locate away from flammable gases or liquids.
- Install after general construction or provide temporary protection.
- Protect when doing maintenance in the area.
- Do not use as a physical support unless so designed.
- Interiors shall not be used for storage unless so designed.
- Metal instrument cases are to be effectively grounded, enclosed in effectively grounded metal covers, or made of insulating materials.

180B. Metal-Enclosed Power Switchgear. In addition to the general requirements for all switchgear, requirements exist for metal-enclosed power switchgear. The rules for metal-enclosed power switchgear are outlined in Fig. 180-2.

180C. Dead-Front Power Switchboards. Dead-front power switchboards with uninsulated rear connections must be installed in rooms capable of being locked

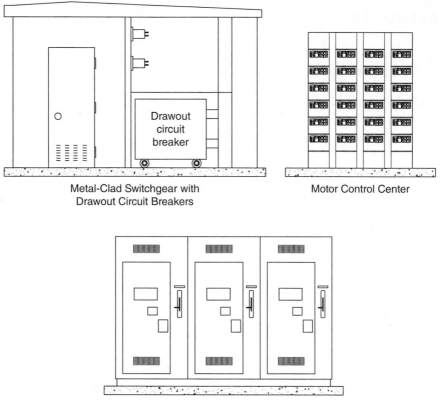

Metal-Clad Switchgear with
Drawout Circuit Breakers

Motor Control Center

Fused Switch Cubicles

Fig. 180-1. Examples of switchgear assemblies (Rule 180A).

with access limited to qualified personnel only. The uninsulated rear connections typically will not meet the vertical and horizontal clearances of Rule 124; therefore, a separate locked room is required for these energized parts.

180D. Motor Control Centers. The fault current within the electric supply station can be very high due to low source impedance and little or no distribution-line impedance. The bus withstand ratings of motor control centers must therefore be given proper consideration. Bus bracing in low-voltage (e.g., 480-V) motor control centers commonly come in 50,000-, 100,000-, and 200,000-A values. This is a withstand rating, not an interrupt rating, as the bus must withstand the force of a momentary fault but not interrupt it. A fuse or circuit breaker in the motor control center will interrupt the fault. Current-limiting fuses can reduce the available fault current to which the motor control center bus and motor control center circuit breakers are exposed. Peak let-through currents for current-limiting fuses can be obtained from fuse manufacturers' literature. This rule also requires a safety sign

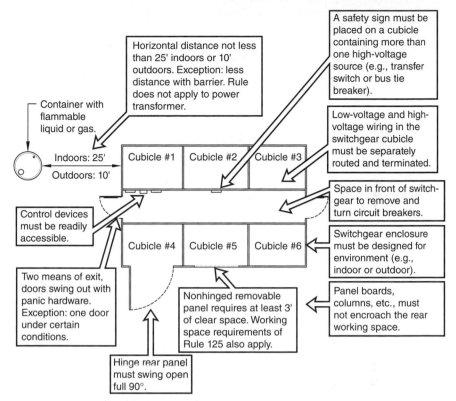

Fig. 180-2. Metal-enclosed power switchgear (Rule 180B).

for a motor control center cubicle having more than one voltage source. The rules regarding motor control center short-circuit ratings are outlined in Fig. 180-3.

180E. Control Switchboards. The requirements for control switchboards are outlined below:

- Control switchboards include cabinets with meters, relays, annunciators, computers, etc., for substation and generating station controls.
- Carpeting in control switchboard rooms must be antistatic and minimize toxic gas emissions under any condition (e.g., water damage, fire damage, etc.).
- Adequate clearance in front and rear to read meters without stools.
- Personnel openings must be covered when not in use and must not be used for cable routing.

Reading control switchboard meters from stools produces a tripping or falling hazard. The rule related to adequate clearance for reading meters without stools is outlined in Fig. 180-4.

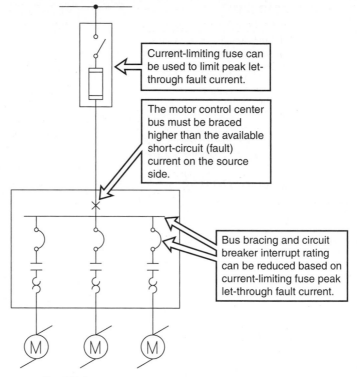

Current-limiting fuse can be used to limit peak let-through fault current.

The motor control center bus must be braced higher than the available short-circuit (fault) current on the source side.

Bus bracing and circuit breaker interrupt rating can be reduced based on current-limiting fuse peak let-through fault current.

Fig. 180-3. Motor control center short-circuit ratings (Rule 180D).

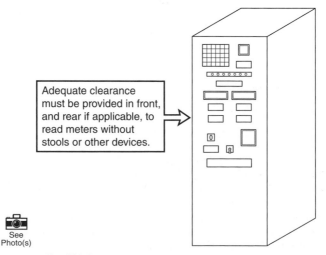

Adequate clearance must be provided in front, and rear if applicable, to read meters without stools or other devices.

See Photo(s)

Fig. 180-4. Adequate clearance for reading meters on control switchboards (Rule 180E3).

181. METAL-ENCLOSED BUS

181A. General Requirements for All Types of Bus.   The rules for metal-enclosed bus are outlined in Fig. 181-1.

181B. Isolated-Phase Bus.   Metal-enclosed bus is available in three common designs: segregated (isolated) phase bus bar, nonsegregated phase bus bar, and cable bus.

The use of isolated-phase bus requires the following special conditions:
- Clearance to magnetic material per manufacturer's recommendations
- Nonmagnetic duct or conduit for alarm circuits
- Piping if enclosure drains are used
- Nonmagnetic wall plates
- Non-ferrous duct or conduit for grounding conductors

Examples of metal-enclosed bus are shown in Fig. 181-2.

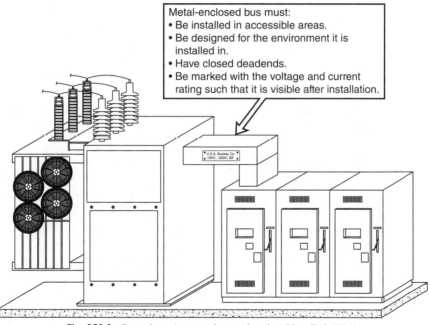

Metal-enclosed bus must:
- Be installed in accessible areas.
- Be designed for the environment it is installed in.
- Have closed deadends.
- Be marked with the voltage and current rating such that it is visible after installation.

U.S.A. Busway Co.
15KV, 1200A, 3Ø

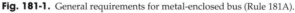

Fig. 181-1. General requirements for metal-enclosed bus (Rule 181A).

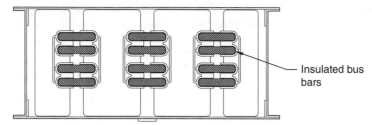

Nonsegregated phase metal-enclosed bus

Insulated bus bars

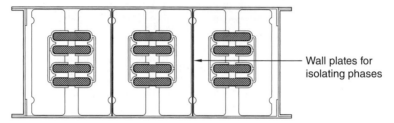

Segregated (isolated) phase metal-enclosed bus

Wall plates for isolating phases

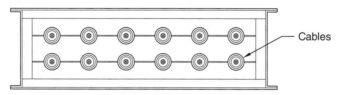

Cables

Cable type metal-enclosed bus

Fig. 181-2. Examples of metal-enclosed bus (Rule 181).

Section 19

Photovoltaic Generating Stations

190. GENERAL

Section 19 applies only to utility scale photovoltaic (PV), but does not apply to smaller scale, non-utility, systems such as those that might be mounted on occupied commercial and residential structures. For systems that may not fall directly into either category, Rule 190 also includes a note referring the user to Rule 011 which defines the scope of the **Code**, breaking down with greater detail what is and is not covered by the **NESC**. Examples of systems that are and are not covered by this section are illustrated in Fig. 190-1.

191. LOCATION

Because this is part of Part 1 which defines the rules that apply to generating stations, the specific enclosure and access control requirements of Rule 110 apply to PV generating stations as well. It is important to emphasize that utility scale PV systems can generate substantial voltages and deliver significant levels of current, so the restriction of access to only authorized personnel has a safety element as well as protection of equipment from damage by unauthorized persons.

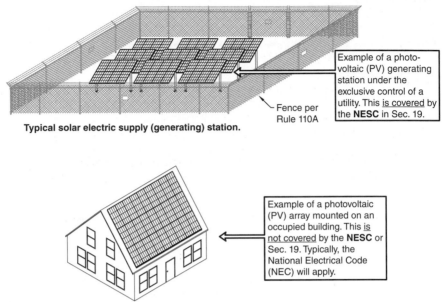

Example of a photo-
voltaic (PV) generating
station under the
exclusive control of a
utility. This <u>is covered</u> by
the **NESC** in Sec. 19.

Fence per
Rule 110A

Typical solar electric supply (generating) station.

Example of a photovoltaic
(PV) array mounted on an
occupied building. This <u>is</u>
<u>not covered</u> by the **NESC** or
Sec. 19. Typically, the
National Electrical Code
(NEC) will apply.

See
Photo(s)   **Typical solar panels attached to a residence.**

Fig. 190-1. Examples of photovoltaic (PV) generation installations covered and not covered by
Sec. 19 (Rule 190).

192. GROUNDING CONFIGURATIONS

Photovoltaic (PV) cells convert the energy of the sun to direct current and voltage.
The large-scale systems discussed in this section also include inverter systems that
convert the direct current (DC) to alternating current (AC) for interfacing with the
interconnected power grid. Rule 192 addresses the grounding configurations for
both the DC and AC portions of the PV generating stations.

For the DC portion of the system, two options are allowed. The first option
allows the use of ungrounded photovoltaic array circuits. The second option
allows the use of DC conductors that are of the functional grounded type. A definition of "functional grounded PV DC conductor" is given under the "conductor"
heading of "Section 2: Definitions of Special Terms." The requirements of grounding for other equipment also apply to grounding of the DC system, as referenced
in Rules 123A through 123C. The requirement 123D, which requires that DC systems operating in excess of 750 V be grounded, is not referenced in Rule 192A,
because that would contradict the allowance that the PV array DC circuits are
permitted to be ungrounded, although there is not a specific exception stated. If
the functional grounded approach is used, then best practices would recommend
consideration of Rule 123D for systems of greater than 750 VDC. For the AC portion of the system, no specific instruction is given, except to require compliance
with the extensive grounding rules of Sec. 09.

193. VEGETATION MANAGEMENT

One might be surprised to see a section on vegetation management in the section on PV systems as the phrase is usually applied to the issue of overhanging trees, and such would cast shadows that would undermine the PV power production. It is notable, however, that where PV arrays are generally installed over dirt or gravel as opposed to pavement, there is a real potential for groundcover below the PV panels, if not properly managed, to grow sufficiently to interfere with supply and communication conductors. For example, climbing vines may become entwined with the PV cables and deliver sufficient weight to pull connectors free. Of course, to the extent that trees are present, such as near the boundaries of the PV generating station, or for any other suspended communication or supply line, the same principles of vegetation management for overhead lines as outlined in Rule 218 should also be considered.

194. DC OVERCURRENT PROTECTION

Unlike typical power generating systems, PV energy conversion, by its nature, can generally sustain a short circuit without catastrophic damage. Consequently, Rule 194 allows the system to be implemented without overcurrent protection on DC cables that are rated to carry the available continuous short-circuit current, which, for PV systems, is typically only fractionally greater than the full load current, except on collector portions of the system where the power from multiple individual strings of PV panels is combined. See Rule 161 for additional discussion regarding overcurrent protection.

195. DC CONDUCTORS

This rule gives specific guidance with regard to the conductors in the DC portion of the PV generating station. In particular, all DC conductors, whether ungrounded or functional grounded, are required to be insulated unless otherwise guarded.

Utility scale PV systems are usually mounted on racks, often with motorized tilting mechanisms for tracking the sun for more consistent solar exposure and maximum energy production. Rule 195 requires that where the DC cables are secured to those types of structures, they must be protected from physical damage. This is particularly important at locations where inadequate protection would allow a cable to become caught in a pinch point of a tracking system or subject to frictional wear as the cable and/or structure move due to intentional operation of the tracking control system or due to the long-term vibrations of the structure due to wind and/or thermal expansion.

Safety Rules for the Installation and Maintenance of Overhead Electric Supply and Communication Lines

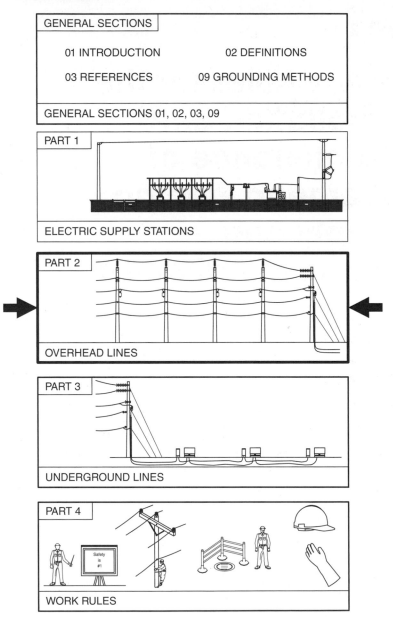

GENERAL SECTIONS

01 INTRODUCTION   02 DEFINITIONS

03 REFERENCES   09 GROUNDING METHODS

GENERAL SECTIONS 01, 02, 03, 09

PART 1

ELECTRIC SUPPLY STATIONS

PART 2

OVERHEAD LINES

PART 3

UNDERGROUND LINES

PART 4

WORK RULES

Purpose, Scope, and Application of Rules

200. PURPOSE

The purpose of Part 2, Overhead Lines, is similar to the purpose of the entire NESC outlined in Rule 010, except Rule 200 is specific to overhead supply and communication lines and equipment. Part 2 of the NESC focuses on the practical safeguarding of persons during the installation, operation, and maintenance of overhead supply (power) and communication lines and equipment.

201. SCOPE

The scope of Part 2, Overhead Lines, includes supply (power) and communication (phone, cable TV, fiber, etc.) conductors, cables, and equipment in overhead line applications. Separate rules apply to Electric Supply Stations (Part 1) and Underground Lines (Part 3). Part 2 covers overhead structural arrangements (i.e., poles, towers, guys, etc.) and extensions into buildings. Requirements are provided in Part 2 for:

- Spacing (center-to-center measurement)
- Clearances (surface-to-surface measurement)
- Strength of construction

There is some overlap between Overhead Lines (Part 2) and Underground Lines (Part 3). The overlap is at the underground riser on the overhead pole. Risers are covered in Sec. 23, Rule 239 of Part 2, and in Sec. 36 of Part 3. See Fig. 201-1.

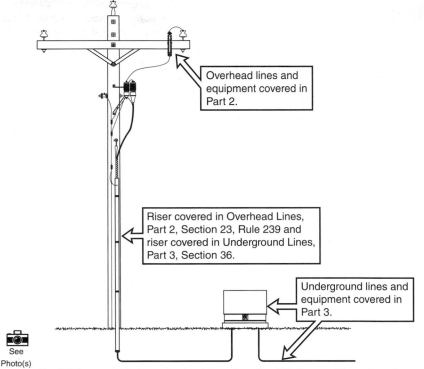

Fig. 201-1. Overlap between Overhead Lines (Part 2) and Underground Lines (Part 3) (Rule 201).

There is some overlap between Overhead Lines (Part 2) and Electric Supply Stations (Part 1). The overlap occurs when an overhead conductor extends outside the electric supply station fence. Conductors that extend outside the substation and their supports (e.g., poles, deadend towers, etc.) must meet the Part 2 requirements for loading and strength. Conductors and supports within the substation must meet the Part 1 requirements for loading and strength. There are more distinct differences between overhead line clearances. Overhead lines and equipment inside the electric supply station fence must meet the Part 1, Electric Supply Station, rules for clearances above the substation gravel or equipment pads. Overhead lines and equipment outside the electric supply station fence must meet the Part 2, Overhead Line, clearance rules for clearances above the ground or other surface. See Figs. 201-2 and 201-3.

When overhead supply equipment is fenced, the rules of Part 1, Electric Supply Stations, may apply. See Rule 110A for a discussion.

The notes at the end of Rule 201 remind the reader that Part 4 of the **NESC** and the Occupational Health and Safety Administration (OSHA) regulations cover the work rules related to overhead lines. See Part 4, Work Rules, for a discussion.

See Rule 011, Scope, for additional information on the scope of the **NESC**.

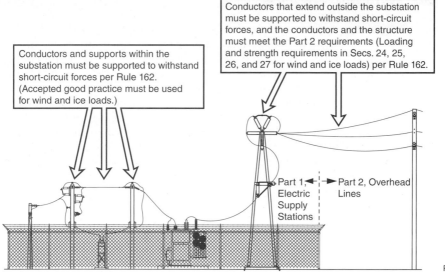

Fig. 201-2. Overlap between Overhead Lines (Part 2) and Electric Supply Stations for loading and strength (Part 1) (Rule 201).

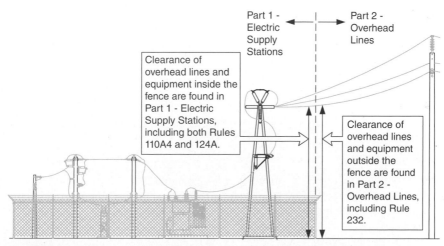

Fig. 201-3. Overlap between Overhead Lines (Part 2) and Electric Supply Stations (Part 1) for clearances (Rule 201).

202. APPLICATION OF RULES

Rule 202 references Rule 013 for the general application of **Code** rules. See Rule 013 for a discussion. Rule 202 has a "however" statement that is essentially an exception to Rule 013. If a supporting structure (e.g., pole) is being replaced, the arrangement of equipment must conform to the current **NESC** requirements for vertical clearance between span wires or brackets carrying luminaires, traffic signals, or trolley conductors, and communications equipment covered in Rule 238C. No note or other explanation is provided in the **Code** as to why Rule 238C was singled out from all the overhead line rules, but it may be an acknowledgment of the significant technological developments over the years in lighting, transportation, and communication that might not have been accounted for in earlier versions of the **Code** that would govern the grandfathering of replacement structures.

Section 21

General Requirements

210. REFERENCED SECTIONS

This rule references four sections related to Part 2, Overhead Lines, so that rules do not need to be duplicated and the reader of the **Code** realizes that other sections are related to the information provided in Part 2. The related sections are:

- Introduction—Sec. 01
- Definitions—Sec. 02
- References—Sec. 03
- Grounding Methods—Sec. 09

The rules in Part 2, predominantly Rule 215, will provide the requirements for grounding overhead lines and equipment. The grounding methods are provided in Sec. 09.

211. NUMBER 211 NOT USED IN THIS EDITION

212. INDUCED VOLTAGES

Rule 212 recognizes that voltages induced by the supply line may affect the communication line in some manner. Noise on the communication line due to the supply line is one example of induced voltage problems. Specific requirements to deal with induced voltage are not specified other than cooperation and advance notice. The clearances provided in Part 2 between supply and communication lines and equipment are for safety purposes. They were not developed to mitigate induced voltage.

Note 1 in Rule 212 provides a pair of references for more information on induced voltage effects on communication lines (IEEE Standards 776 and 1137). Note 2 in Rule 212 provides a statement that other types of induced voltages may exist including induced voltages on pipelines. Another example that is not referenced in Rule 212 is induced voltage from a supply line, such as a high-voltage transmission line, on to an adjacent fence line. Rule 012C, which requires accepted good practice, must be applied in these situations.

Rule 232C1c does require limiting electrostatic effects between a supply line exceeding 98 kV to ground (exceeding 170 kV phase to phase) and a truck, vehicle, or equipment under the line. See Rule 232C for more information and a figure. Rules for induced voltages that occur when working on or near supply lines are covered in the Part 4 Work Rules.

213. ACCESSIBILITY

This rule generally states the requirement that overhead parts that need to be examined or adjusted during operation must be accessible to authorized supply and communication workers. This rule requires the following, all of which are discussed in more detail throughout Part 2:
- Climbing space
- Working space
- Working facilities
- Clearances between conductors

Climbing space is specifically discussed in Rule 236. Working space is covered in Rule 237. Clearances between conductors and between conductors and equipment are covered in various rules throughout Sec. 23.

214. INSPECTION AND TESTS
OF LINES AND EQUIPMENT

The rules for inspection and testing on lines and equipment are broken down into two parts, when in-service (Rule 214A) and when out-of-service (Rule 214B). The open or closed status of a switch, fused cutout, recloser, etc., will determine if a line is in- or out-of-service. A line that is permanently out of service must be removed or maintained in a safe condition. Opening the cutouts on a line and grounding the line may make it electrically safe but not structurally safe. As an example, if the permanently out-of-service line does not meet the strength requirements in Sec. 26, it must be removed or repaired to a safe condition. The rules for inspecting and testing overhead lines and equipment are outlined in Fig. 214-1.

Rule 214 applies to overhead supply and communications lines. Similar inspection rules are outlined in Rule 121 for electric supply stations and Rule 313 for underground lines.

Rule 214 does not provide a specific requirement as to how often to inspect overhead lines. The inspection frequency must be determined by the utility

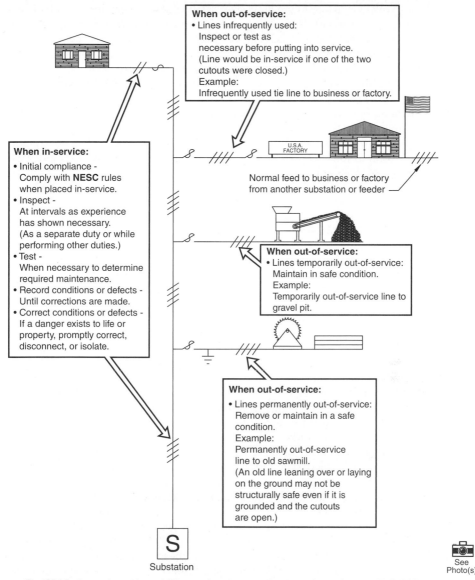

When out-of-service:
• Lines infrequently used:
  Inspect or test as
  necessary before putting into service.
  (Line would be in-service if one of the two
  cutouts were closed.)
  Example:
  Infrequently used tie line to business or factory.

Normal feed to business or factory
from another substation or feeder

When in-service:
• Initial compliance -
  Comply with **NESC** rules
  when placed in-service.
• Inspect -
  At intervals as experience
  has shown necessary.
  (As a separate duty or while
  performing other duties.)
• Test -
  When necessary to determine
  required maintenance.
• Record conditions or defects -
  Until corrections are made.
• Correct conditions or defects -
  If a danger exists to life or
  property, promptly correct,
  disconnect, or isolate.

When out-of-service:
• Lines temporarily out-of-service:
  Maintain in safe condition.
  Example:
  Temporarily out-of-service line to
  gravel pit.

When out-of-service:
• Lines permanently out-of-service:
  Remove or maintain in a safe
  condition.
  Example:
  Permanently out-of-service
  line to old sawmill.
  (An old line leaning over or laying
  on the ground may not be
  structurally safe even if it is
  grounded and the cutouts
  are open.)

S

Substation

See
Photo(s)

Fig. 214-1. Inspection and tests of lines and equipment when in- and out-of-service (Rule 214).

"at such intervals as experience has shown to be necessary." This statement should consider what the local conditions are. For example: "Do poles rot faster in some areas due to acidic soil conditions?" "Do insulators fail more in some areas due to salt fog, pollution from a nearby factory, lightning, etc.?" Even factors such as age, repair history, exposure to weather events, damage induced by wildlife, etc., are well to consider in the process of establishing inspection intervals. These considerations are part of applying accepted good practice for the given local conditions. See **NESC** Rule 012C. The phrase "as experience has shown to be necessary," is not restricted to the individual utility's experience but includes experience and published information from the industry as a whole.

Most electric power utilities have an inspection program that has some type of record keeping and inspection frequency; however, the frequency of the inspection is not specified in the **NESC**. The authority having jurisdiction over utilities (i.e., a State Public Service Commission or State Public Utility Commission) sometimes sets a fixed inspection frequency for utilities in a particular state. It is important for utilities to check the authority having jurisdiction to determine if fixed inspection intervals apply.

A note in Rule 214A2 recognizes that inspections may be performed as a separate duty or while performing other duties. An example of an inspection performed as a separate operation is an electric utility sending an inspector to poles in a geographic area and testing them for rot. An example of an inspection performed while performing other duties is a communications utility technician looking for damaged hardware and low hanging wires while adding, changing, or removing a service drop. If a supply or communications employee notices a defect while performing another task, the defect must be addressed, not ignored.

Typically, supply (power) utility employees inspect power lines, power equipment, and associated power facilities and communications utility employees inspect communication lines, communication equipment, and associated communication facilities. Inspection of the pole is typically done by the pole owner. Inspection responsibilities on joint-use (power and communication) poles are commonly outlined in the joint-use agreement between the power and communication companies. Nevertheless, the contractual arrangements of a joint-use-agreement do not displace the requirement for each utility with facilities on the pole to comply with the requirements of the **Code**.

Rule 214 also does not provide any type of inspection checklist. The specifics as to what items are inspected and to what detail is not specified. Rule 012C, which requires accepted good practice, must be used. Sometimes electric utilities use both detailed and patrol inspections. These inspection terms are not found in the **NESC**. A State Public Service Commission or State Public Utility Commission may define detailed and patrol inspection tasks, or accepted good practice (**NESC** Rule 012C) may be used. It is also important to recognize the need to evaluate all aspects of compliance with the **Code** for all of the utility's overhead infrastructure. For example, if a utility uses separate contractors for pole inspections and line inspections, the responsibility for the integrity of the crossarms must be clearly designated to the contractor(s) so that defects in the crossarms are not overlooked.

Rule 214A4 requires conditions or defects found during inspections or tests to be recorded if the conditions or defects are not promptly corrected. The records of the conditions or defects must be maintained until the conditions or defects are corrected. The **NESC** does not provide a time frame for correcting conditions or defects found during inspections. For example, the correction could be done immediately or the correction could be done in the future during a line rebuild project.

Per Rule 214A5, if the condition or defect can reasonably be expected to endanger persons or property, the condition or defect must be promptly corrected, disconnected, or isolated, but again the NESC does not provide a time frame for the prompt correction. Rule 012C, which requires accepted good practice, must be used to determine the correction time frame for both Rules 214A4 and 214A5. The direct hazard to persons and property, together with the the dictionary definition of the term "prompt," strongly suggests that the corrective measures should be implemented as soon as the utility can assemble the equipment and personnel to remediate the life-safety condition. If any delay in the implementation of corrective measure increases the hazard to the public, the requirements of 214A5a require that the line or equipment be disconnected or isolated until it can be corrected.

215. GROUNDING AND BONDING

Rule 215 is broken into three main paragraphs. Rule 215A reminds us that the methods of grounding are specified in Sec. 09 and the requirements of grounding are provided in this rule for circuits and non-current-carrying parts. Rule 215B focuses on the requirements for grounding circuits. See Sec. 02, Definitions, for a discussion of the term "effectively grounded." The circuit grounding requirements of Rule 215B are outlined in Fig. 215-1.

Rule 215C focuses on the requirements for grounding non-current-carrying parts. When a communication messenger is grounded and bonded to the supply neutral conductor, it can carry current. When a transformer enclosure is grounded and bonded to the supply neutral conductor, it can carry current. But, by definition of a current-carrying part in Sec. 02 of the NESC, messengers and transformer enclosures are not considered current-carrying parts as they are not connected to a source of voltage. Rule 314, which discusses grounding of underground circuits and equipment, uses the term "conductive parts to be grounded" instead of "non-current-carrying parts."

Per Rule 215C2, anchor guys and span guys must be effectively grounded or insulated (per the exceptions). This requirement applies to supply (power) guys supporting any voltage and communication guys. If guys are grounded, Rule 092C2 provides the methods for grounding of guys. If guys are insulated, the exception in Rule 215C2a (for anchor guys) and the exception in Rule 215C2b (for span guys) provide requirements for the location of the guy insulator or insulators. In addition, the guy and span insulator requirements specified in Rule 279A must also be met and the clearance requirements for guys in NESC Table 232-1 must be met. It is difficult to determine a "one size fits all" guy insulator standard as guyed structures vary in voltage, location, configuration, and complexity.

The non-current-carrying parts grounding requirements of Rule 215C are outlined in Figs. 215-2 through 215-4.

Examples of the use of insulator links in guy wires are shown in Figs. 215-5 through 215-8.

Rule 215D applies to bonding of supply and communication systems on the same supporting structure. See Fig. 215-9.

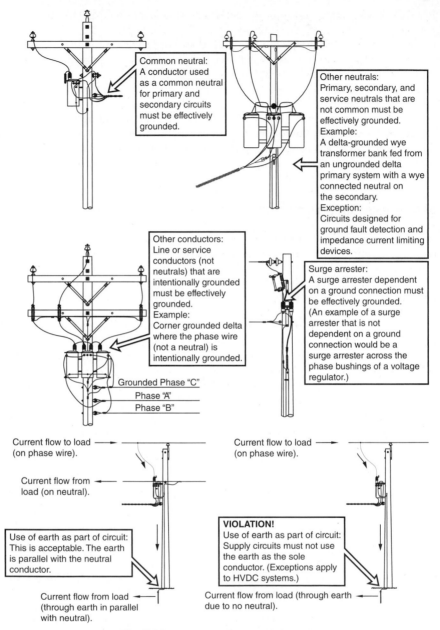

Common neutral:
A conductor used as a common neutral for primary and secondary circuits must be effectively grounded.

Other neutrals:
Primary, secondary, and service neutrals that are not common must be effectively grounded.
Example:
A delta-grounded wye transformer bank fed from an ungrounded delta primary system with a wye connected neutral on the secondary.
Exception:
Circuits designed for ground fault detection and impedance current limiting devices.

Other conductors:
Line or service conductors (not neutrals) that are intentionally grounded must be effectively grounded.
Example:
Corner grounded delta where the phase wire (not a neutral) is intentionally grounded.

Surge arrester:
A surge arrester dependent on a ground connection must be effectively grounded. (An example of a surge arrester that is not dependent on a ground connection would be a surge arrester across the phase bushings of a voltage regulator.)

Grounded Phase "C"
Phase "A"
Phase "B"

Current flow to load
(on phase wire).

Current flow from load (on neutral).

Use of earth as part of circuit:
This is acceptable. The earth is parallel with the neutral conductor.

Current flow from load
(through earth in parallel with neutral).

Current flow to load
(on phase wire).

VIOLATION!
Use of earth as part of circuit:
Supply circuits must not use the earth as the sole conductor. (Exceptions apply to HVDC systems.)

Current flow from load (through earth due to no neutral).

Fig. 215-1. Grounding of circuits (Rule 215B).

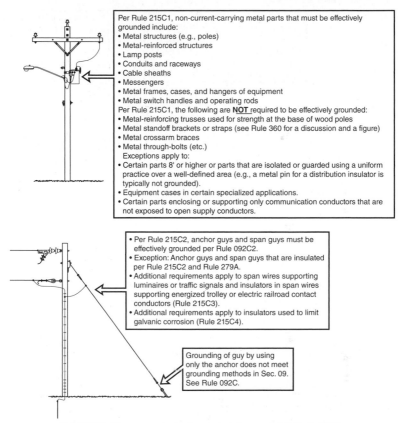

Per Rule 215C1, non-current-carrying metal parts that must be effectively grounded include:
• Metal structures (e.g., poles)
• Metal-reinforced structures
• Lamp posts
• Conduits and raceways
• Cable sheaths
• Messengers
• Metal frames, cases, and hangers of equipment
• Metal switch handles and operating rods
Per Rule 215C1, the following are **NOT** required to be effectively grounded:
• Metal-reinforcing trusses used for strength at the base of wood poles
• Metal standoff brackets or straps (see Rule 360 for a discussion and a figure)
• Metal crossarm braces
• Metal through-bolts (etc.)
Exceptions apply to:
• Certain parts 8' or higher or parts that are isolated or guarded using a uniform practice over a well-defined area (e.g., a metal pin for a distribution insulator is typically not grounded).
• Equipment cases in certain specialized applications.
• Certain parts enclosing or supporting only communication conductors that are not exposed to open supply conductors.

• Per Rule 215C2, anchor guys and span guys must be effectively grounded per Rule 092C2.
• Exception: Anchor guys and span guys that are insulated per Rule 215C2 and Rule 279A.
• Additional requirements apply to span wires supporting luminaires or traffic signals and insulators in span wires supporting energized trolley or electric railroad contact conductors (Rule 215C3).
• Additional requirements apply to insulators used to limit galvanic corrosion (Rule 215C4).

Grounding of guy by using only the anchor does not meet grounding methods in Sec. 09. See Rule 092C.

Fig. 215-2. Non-current-carrying parts to be grounded (Rule 215C).

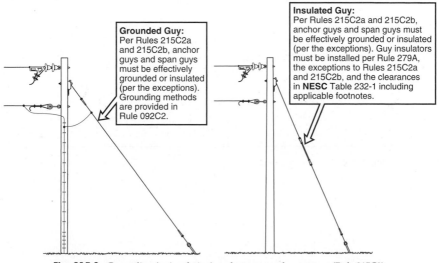

Grounded Guy:
Per Rules 215C2a and 215C2b, anchor guys and span guys must be effectively grounded or insulated (per the exceptions). Grounding methods are provided in Rule 092C2.

Insulated Guy:
Per Rules 215C2a and 215C2b, anchor guys and span guys must be effectively grounded or insulated (per the exceptions). Guy insulators must be installed per Rule 279A, the exceptions to Rules 215C2a and 215C2b, and the clearances in **NESC** Table 232-1 including applicable footnotes.

Fig. 215-3. Grounding (or insulating) anchor guys and span guys (Rule 215C2).

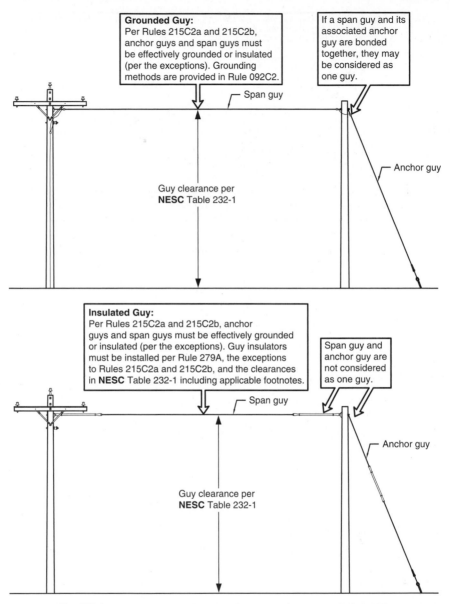

Grounded Guy:
Per Rules 215C2a and 215C2b, anchor guys and span guys must be effectively grounded or insulated (per the exceptions). Grounding methods are provided in Rule 092C2.

If a span guy and its associated anchor guy are bonded together, they may be considered as one guy.

Span guy

Anchor guy

Guy clearance per **NESC** Table 232-1

Insulated Guy:
Per Rules 215C2a and 215C2b, anchor guys and span guys must be effectively grounded or insulated (per the exceptions). Guy insulators must be installed per Rule 279A, the exceptions to Rules 215C2a and 215C2b, and the clearances in **NESC** Table 232-1 including applicable footnotes.

Span guy and anchor guy are not considered as one guy.

Span guy

Anchor guy

Guy clearance per **NESC** Table 232-1

Fig. 215-4. Grounding (or insulating) anchor guys and span guys (Rule 215C2).

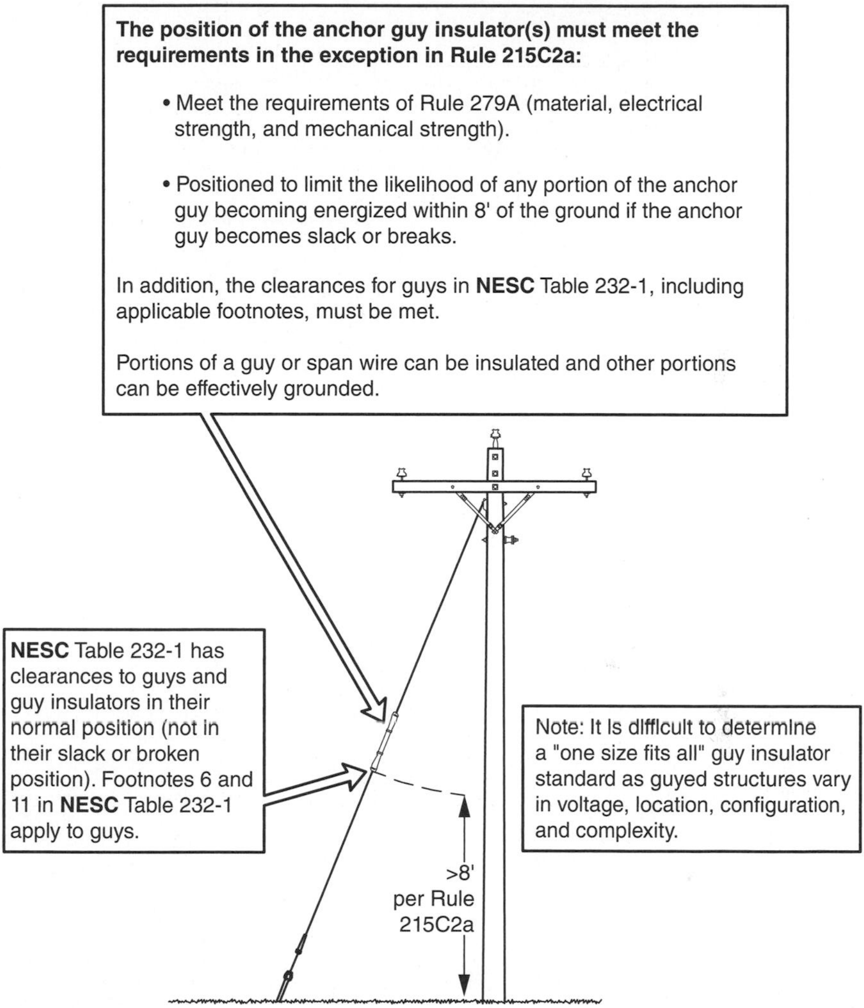

The position of the anchor guy insulator(s) must meet the requirements in the exception in Rule 215C2a:

- Meet the requirements of Rule 279A (material, electrical strength, and mechanical strength).

- Positioned to limit the likelihood of any portion of the anchor guy becoming energized within 8' of the ground if the anchor guy becomes slack or breaks.

In addition, the clearances for guys in **NESC** Table 232-1, including applicable footnotes, must be met.

Portions of a guy or span wire can be insulated and other portions can be effectively grounded.

NESC Table 232-1 has clearances to guys and guy insulators in their normal position (not in their slack or broken position). Footnotes 6 and 11 in **NESC** Table 232-1 apply to guys.

Note: It is difficult to determine a "one size fits all" guy insulator standard as guyed structures vary in voltage, location, configuration, and complexity.

>8'
per Rule
215C2a

Fig. 215-5. Example of the use of insulators in anchor guys (Rules 215C2a, 232, and 279A).

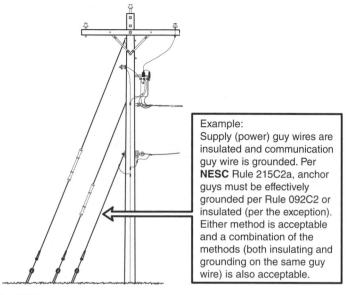

Example:
Supply (power) guy wires are insulated and communication guy wire is grounded. Per **NESC** Rule 215C2a, anchor guys must be effectively grounded per Rule 092C2 or insulated (per the exception). Either method is acceptable and a combination of the methods (both insulating and grounding on the same guy wire) is also acceptable.

Fig. 215-6. Example of the use of insulators in anchor guys (Rules 215C2a and 279A).

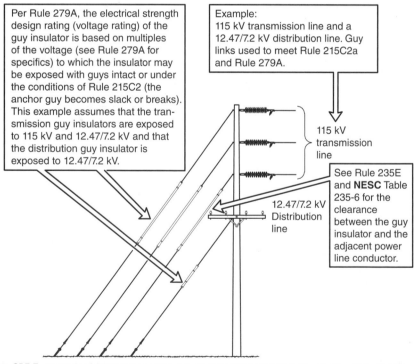

Per Rule 279A, the electrical strength design rating (voltage rating) of the guy insulator is based on multiples of the voltage (see Rule 279A for specifics) to which the insulator may be exposed with guys intact or under the conditions of Rule 215C2 (the anchor guy becomes slack or breaks). This example assumes that the transmission guy insulators are exposed to 115 kV and 12.47/7.2 kV and that the distribution guy insulator is exposed to 12.47/7.2 kV.

Example:
115 kV transmission line and a 12.47/7.2 kV distribution line. Guy links used to meet Rule 215C2a and Rule 279A.

115 kV transmission line

See Rule 235E and **NESC** Table 235-6 for the clearance between the guy insulator and the adjacent power line conductor.

12.47/7.2 kV Distribution line

Fig. 215-7. Example of determining guy insulator link electrical strength (voltage) rating (Rules 215C2a and 279A).

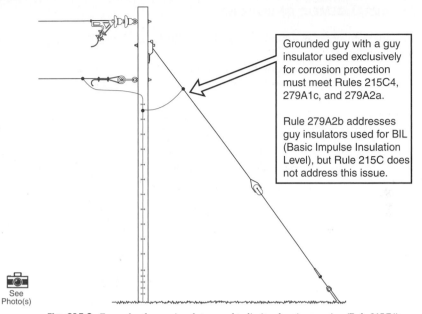

Grounded guy with a guy insulator used exclusively for corrosion protection must meet Rules 215C4, 279A1c, and 279A2a.

Rule 279A2b addresses guy insulators used for BIL (Basic Impulse Insulation Level), but Rule 215C does not address this issue.

See Photo(s)

Fig. 215-8. Example of a guy insulator used to limit galvanic corrosion (Rule 215C4).

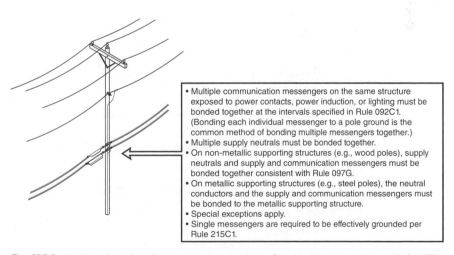

• Multiple communication messengers on the same structure exposed to power contacts, power induction, or lighting must be bonded together at the intervals specified in Rule 092C1. (Bonding each individual messenger to a pole ground is the common method of bonding multiple messengers together.)
• Multiple supply neutrals must be bonded together.
• On non-metallic supporting structures (e.g., wood poles), supply neutrals and supply and communication messengers must be bonded together consistent with Rule 097G.
• On metallic supporting structures (e.g., steel poles), the neutral conductors and the supply and communication messengers must be bonded to the metallic supporting structure.
• Special exceptions apply.
• Single messengers are required to be effectively grounded per Rule 215C1.

Fig. 215-9. Bonding of supply and communication systems on the same supporting structure (Rule 215D).

216. ARRANGEMENT OF SWITCHES

This rule addresses overhead line switches. Rule 381C covers underground or pad-mounted switches, and Sec. 17 covers switches in electric supply stations. Transmission and distribution line switches are the most common application of Rule 216. The **Code** rules related to the arrangement of overhead line switches are outlined in Fig. 216-1.

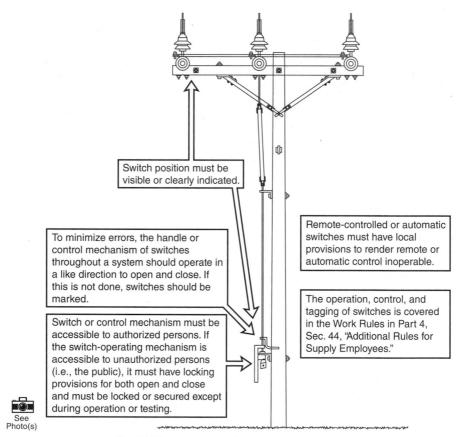

Switch position must be visible or clearly indicated.

Remote-controlled or automatic switches must have local provisions to render remote or automatic control inoperable.

To minimize errors, the handle or control mechanism of switches throughout a system should operate in a like direction to open and close. If this is not done, switches should be marked.

The operation, control, and tagging of switches is covered in the Work Rules in Part 4, Sec. 44, "Additional Rules for Supply Employees."

Switch or control mechanism must be accessible to authorized persons. If the switch-operating mechanism is accessible to unauthorized persons (i.e., the public), it must have locking provisions for both open and close and must be locked or secured except during operation or testing.

See Photo(s)

Fig. 216-1. Arrangement of overhead line switches (Rule 216).

The intent of the requirements in Rule 216 is to minimize switching errors. Switching is sometimes done during poor weather conditions or low visibility. Accessibility, marking, uniform positioning, etc., help minimize switching errors. The operation, control, and tagging of switches is covered in the Work Rules in Part 4, Sec. 44, "Additional Rules for Supply Employees."

217. GENERAL

Rule 217 provides general information regarding supporting structures (e.g., poles and towers) except for clearance to other objects. Clearance of a supporting structure to other objects is covered in the overhead line clearance rules in Sec. 23, Rule 231. Supporting structures includes poles, lattice towers, etc. Rule 217A1a addresses vehicular damage to supporting structures in established parking areas, alleys, and driveway areas subject to vehicular traffic abrasion. This Rule has a note explaining that it is not practical to protect supporting structures from contact by out-of-control vehicles or vehicles operating outside of established areas. The terms roadway, shoulder, and traveled way are all defined in Sec. 02, Definitions of Special Terms, to help apply this Rule. Rule 231B addresses clearances of structures (e.g., poles and anchor guys) from roadways. The rules related to protection of supporting structures covered in Rule 217A1 are outlined in Fig. 217-1.

Rule 217A2 focuses on unauthorized climbing of supporting structures. The Code uses the term "readily climbable" in Rule 217A2a when discussing lattice towers. A definition of "supporting structure" (readily climbable and not readily climbable) is provided in Sec. 02 of the NESC. The Code uses 8-ft spacing in Rules 217A2b, 217A2c, and 217A2d between pole steps, between riser standoff brackets, and between equipment and associated support brackets to limit climbing. The concern of readily climbable structures, steps, riser standoff brackets, and equipment is to keep unauthorized people (e.g., the public) from climbing the pole or structure. See Rule 362 for additional information and a figure related to riser standoff brackets.

The rules related to climbing of supporting structures covered in Rule 217A2 are outlined in Figs. 217-2 through 217-5.

Rule 217A3 provides requirements for identifying supporting structures (e.g., poles and towers) on overhead lines. Structure identification is required but not specifically limited to numbering. Location, construction, or marking can also be used to identify poles and towers for employees. Identification of switch poles (or the switches themselves) is especially critical so that switching operations discussed in Part 4, Work Rules, are done correctly. Some utilities number every pole and some only number corner, tap, or other selected poles. A small rural utility can use construction and location for identification (e.g., the transformer pole at the Smith Ranch). The method used to identify poles should consider the size and complexity of the utility's system. Rule 411A2 in Part 4, Work Rules, requires that diagrams (i.e., maps) be maintained on file for employees to use. Rule 220D applies to identification of the conductors on the supporting structures.

The rules related to the identification of supporting structures covered in Rule 217A3 are outlined in Fig. 217-6.

Rule 217A4 covers attachments, decorations, and obstructions on supporting structures (e.g., poles). The supporting structure owner must agree (provide concurrence) to attachments made on structures of utility lines. This requirement is for any kind of attachment (utility or non-utility). The supporting structure

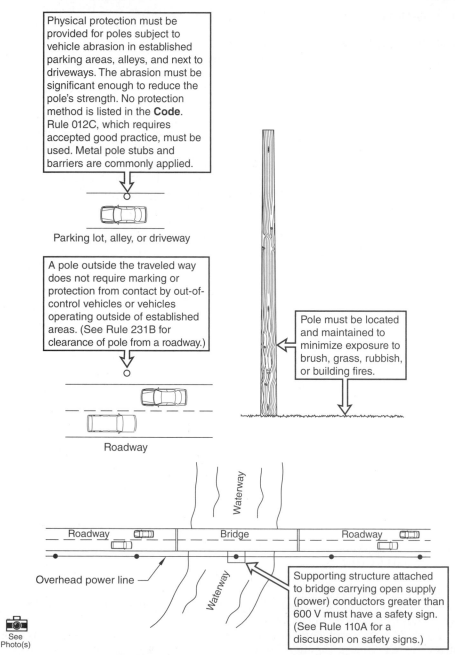

Physical protection must be provided for poles subject to vehicle abrasion in established parking areas, alleys, and next to driveways. The abrasion must be significant enough to reduce the pole's strength. No protection method is listed in the **Code**. Rule 012C, which requires accepted good practice, must be used. Metal pole stubs and barriers are commonly applied.

Parking lot, alley, or driveway

A pole outside the traveled way does not require marking or protection from contact by out-of-control vehicles or vehicles operating outside of established areas. (See Rule 231B for clearance of pole from a roadway.)

Roadway

Pole must be located and maintained to minimize exposure to brush, grass, rubbish, or building fires.

Waterway

Roadway   Bridge   Roadway

Overhead power line

Waterway

See
Photo(s)

Supporting structure attached to bridge carrying open supply (power) conductors greater than 600 V must have a safety sign. (See Rule 110A for a discussion on safety signs.)

Fig. 217-1. Protection of supporting structures (Rule 217A1).

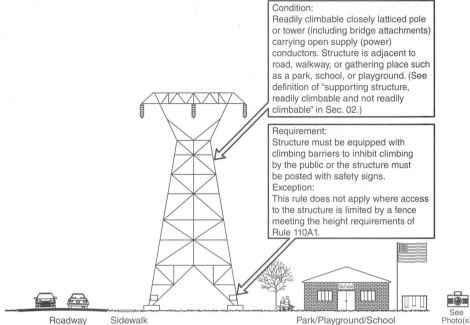

Condition:
Readily climbable closely latticed pole or tower (including bridge attachments) carrying open supply (power) conductors. Structure is adjacent to road, walkway, or gathering place such as a park, school, or playground. (See definition of "supporting structure, readily climbable and not readily climbable" in Sec. 02.)

Requirement:
Structure must be equipped with climbing barriers to inhibit climbing by the public or the structure must be posted with safety signs.
Exception:
This rule does not apply where access to the structure is limited by a fence meeting the height requirements of Rule 110A1.

See Photo(s)

Roadway   Sidewalk   Park/Playground/School

Fig. 217-2. Readily climbable supporting structures (Rule 217A2a).

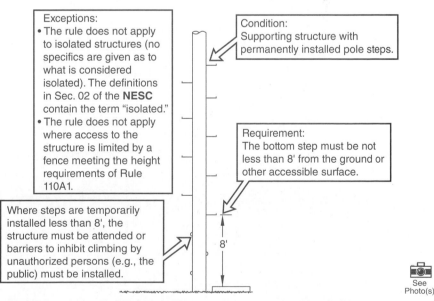

Exceptions:
• The rule does not apply to isolated structures (no specifics are given as to what is considered isolated). The definitions in Sec. 02 of the **NESC** contain the term "isolated."
• The rule does not apply where access to the structure is limited by a fence meeting the height requirements of Rule 110A1.

Where steps are temporarily installed less than 8', the structure must be attended or barriers to inhibit climbing by unauthorized persons (e.g., the public) must be installed.

Condition:
Supporting structure with permanently installed pole steps.

Requirement:
The bottom step must be not less than 8' from the ground or other accessible surface.

8'

See Photo(s)

Fig. 217-3. Permanently mounted and temporary steps on supporting structures (Rule 217A2b).

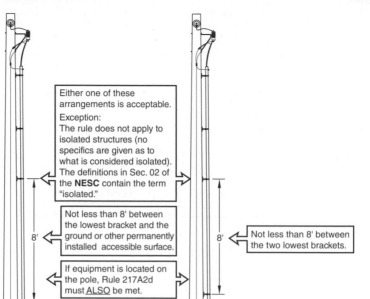

Either one of these arrangements is acceptable.

Exception:
The rule does not apply to isolated structures (no specifics are given as to what is considered isolated). The definitions in Sec. 02 of the **NESC** contain the term "isolated."

Not less than 8' between the lowest bracket and the ground or other permanently installed accessible surface.

Not less than 8' between the two lowest brackets.

If equipment is located on the pole, Rule 217A2d must <u>ALSO</u> be met.

See Photo(s)

Fig. 217-4. Arrangement of riser standoff brackets (Rule 217A2c).

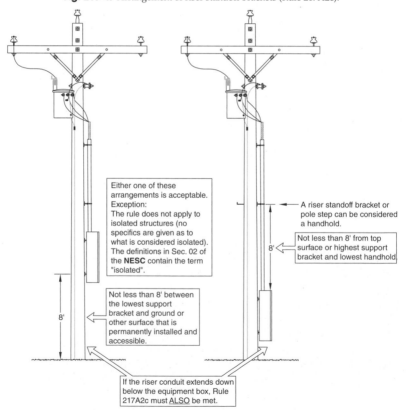

Either one of these arrangements is acceptable.
Exception:
The rule does not apply to isolated structures (no specifics are given as to what is considered isolated). The definitions in Sec. 02 of the **NESC** contain the term "isolated".

A riser standoff bracket or pole step can be considered a handhold.

Not less than 8' from top surface or highest support bracket and lowest handhold.

Not less than 8' between the lowest support bracket and ground or other surface that is permanently installed and accessible.

If the riser conduit extends down below the equipment box, Rule 217A2c must <u>ALSO</u> be met.

Fig. 217-5. Equipment and associated support brackets on supporting structures (Rule 217A2d).

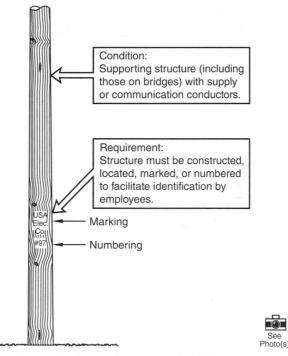

Fig. 217-6. Identification of supporting structures (Rule 217A3).

owner is typically the power company or the traditional telephone company. The structure can also be owned by multiple parties. Structure ownership is typically defined in joint use agreements (contracts) between power and communication utilities. See Rule 222 for a discussion of joint use agreements. Typically, traditional Cable TV companies do not own poles; they only rent space on poles. The word typically is used in this paragraph as other ownership situations can exist. The word traditional is used in this paragraph to refer to telephone and cable utilities before the present-day mixing of communication services. Non-utility attachments (e.g., a stop sign mounted on a wood utility pole) must have agreement (concurrence) of the occupant or occupants of the space in which the attachment is made (the supply or communications space) as well as the pole owner. For example, assume a wood pole is owned by the power company and has power, phone, and Cable TV lines attached to it. If a stop sign is mounted on the pole, both the pole owner (the power company) and any occupants of the communication space (the phone and Cable TV company) must provide concurrence if the stop sign is in the communication space. In this example, the stop sign would have to be above the lowest communication cable or conductor. See the definition of communication space in Sec. 02 of the **NESC**. Attachments must not result in violations (noncompliance) of the **NESC** rules. Rule 012C, which requires accepted good practice, may need to be applied as the NESC may not specifically address Code issues related to the attachment. For example, no clearance requirement exists between a stop sign and a communications cable above it and no grounding requirement exists for stop signs. The wind load on the stop sign should be considered when the strength and loading rules are applied. Attachments must not obstruct the climbing space or present a climbing hazard to utility workers (lineworkers). See Rule 236 for a discussion of climbing space. Through-bolts must be properly trimmed, but no dimension is provided. Rule 012C, which requires accepted good practice, must be used. A through-bolt that is too long can be trimmed or replaced with a shorter bolt. Vines, nails, tacks, or other items that may interfere with climbing should be removed before climbing to avoid a slip or fall. The rules related to attachments, decorations, and obstructions on supporting structures are outlined in Fig. 217-7.

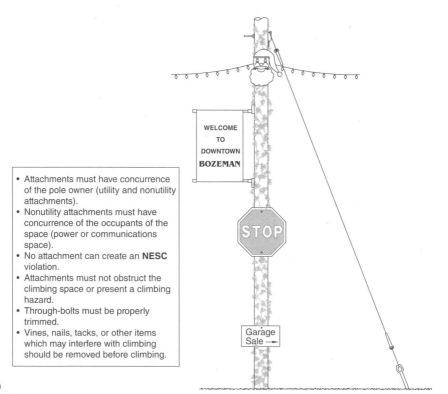

- Attachments must have concurrence of the pole owner (utility and nonutility attachments).
- Nonutility attachments must have concurrence of the occupants of the space (power or communications space).
- No attachment can create an **NESC** violation.
- Attachments must not obstruct the climbing space or present a climbing hazard.
- Through-bolts must be properly trimmed.
- Vines, nails, tacks, or other items which may interfere with climbing should be removed before climbing.

WELCOME
TO
DOWNTOWN
BOZEMAN

Garage
Sale →

See
Photo(s)

Fig. 217-7. Attachments, decorations, and obstructions on supporting structures (Rule 217A4).

Rule 217B discusses unusual conductor supports. The NESC does not prohibit unusual conductor supports. It does require applying Code rules and additional precautions deemed necessary by the administrative authority to assure safety. For example, a power line supported on a bridge requires complying with the **NESC** rules for clearance to the bridge and the **NESC** rules for loading and strength of the power line structure and any additional rules required by the highway department or bridge authority. The rules required by the highway department or bridge authority may involve additional loading and strength requirements related to the power line being attached to the bridge and additional clearance requirements for persons using or maintaining the bridge. One of the most recognizable unusual conductor supports in the United Sates is the "Mickey Mouse" tower outside of Disney World in Orlando, FL. See Fig. 217-8.

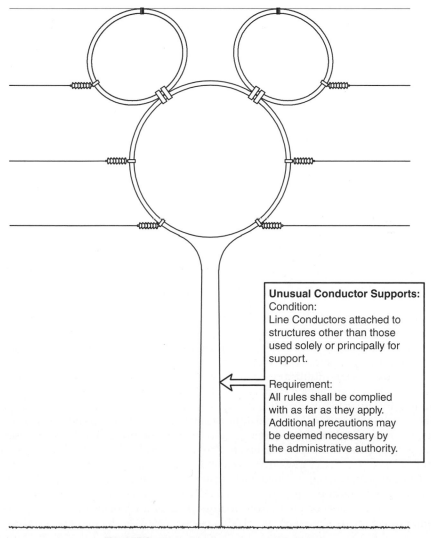

Unusual Conductor Supports:
Condition:
Line Conductors attached to structures other than those used solely or principally for support.

Requirement:
All rules shall be complied with as far as they apply. Additional precautions may be deemed necessary by the administrative authority.

Fig. 217-8. Unusual conductor supports (Rule 217B).

Rule 217B also addresses supporting conductors on trees and roofs. The **Code** states that supporting (attaching) overhead conductors to trees or roofs "should be avoided." The **Code** does not use stronger language like "shall not be attached." Even though "should" is used instead of "shall," everything practical must be done to avoid attaching conductors to trees or roofs. See Rule 015 for a discussion of the intent of the words "should" and "shall." The requirement to avoid supporting conductors on trees and roofs is shown in Fig. 217-9.

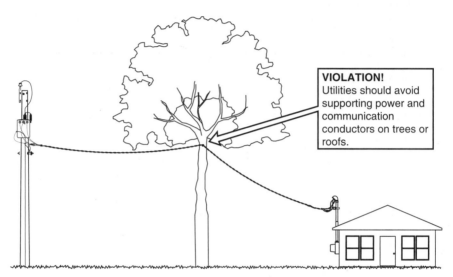

VIOLATION!
Utilities should avoid supporting power and communication conductors on trees or roofs.

Fig. 217-9. Supporting conductors on trees and roofs (Rule 217B).

Rule 217C applies to the protection and marking of guys. Utilities are required to use guy markers on each guy in pedestrian areas. To be more specific, this rule requires a guy marker on each guy "adjacent to regularly traveled pedestrian thoroughfares, or places where persons are normally encountered or reasonably anticipated." Some utilities use guy markers at every guy location, but this is not a **Code** requirement. The note in Rule 217C1 clarifies that there is no intent to require guy markers at all guy locations, only those locations required by the rule. Some utilities use bold fluorescent multicolored markers; others use conspicuous single color markers. A guy marker or protection must be used for guys in established parking areas. The note in Rule 217C2 clarifies that it is not practical to protect guys from contact by out-of-control vehicles or vehicles operating outside of established parking areas. Similar wording appears in Rule 217A1a for supporting structures. Rule 231B addresses clearances of structures (e.g., poles and anchor guys) from roadways. Examples of using guy markers and protection are shown in Fig. 217-10.

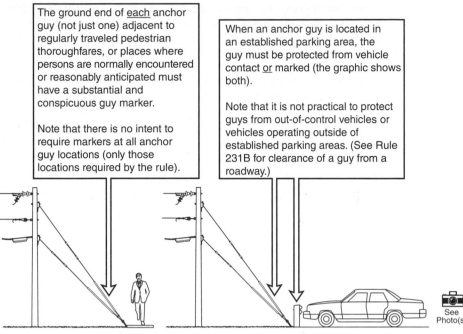

The ground end of <u>each</u> anchor guy (not just one) adjacent to regularly traveled pedestrian thoroughfares, or places where persons are normally encountered or reasonably anticipated must have a substantial and conspicuous guy marker.

Note that there is no intent to require markers at all anchor guy locations (only those locations required by the rule).

When an anchor guy is located in an established parking area, the guy must be protected from vehicle contact <u>or</u> marked (the graphic shows both).

Note that it is not practical to protect guys from out-of-control vehicles or vehicles operating outside of established parking areas. (See Rule 231B for clearance of a guy from a roadway.)

See Photo(s)

Fig. 217-10. Examples of protection and marking of guys (Rule 217C).

218. VEGETATION MANAGEMENT

The NESC does not provide a specific clearance from a supply or communication line (of any voltage) to vegetation (e.g., trees) nor does it provide a specific frequency (time period) for performing vegetation management (e.g., tree trimming). The first sentence in Rule 218A1 addresses both supply and communication lines and is very general. Note 1 in Rule 218A1 lists factors to consider in determining the extent of vegetation management. The factors to consider apply to both selecting the distance (clearance) between the supply or communication line and the vegetation and the frequency of the pruning or removal (e.g., tree trimming cycle). The first factor to consider in Note 1 is the line voltage class. Communication lines have little or no voltage, and therefore the resulting vegetation management needs are less. High-voltage primary lines have greater vegetation management needs. Low-voltage secondary lines are somewhere in the middle of communication lines and high-voltage primary lines. The second sentence in Rule 218A1 is slightly less general than the first sentence and states that vegetation that may damage ungrounded supply conductors should be pruned or removed. Some examples of ungrounded supply conductors are the bare phase wires of a 12.47/7.2 kV distribution line or the insulated line conductors of a 120/240 V secondary triplex circuit. Some examples that are not ungrounded supply conductors are the effectively grounded neutral conductor associated with a 12.47/7.2 kV distribution line, the effectively grounded bare messenger of a 120/240 V secondary triplex circuit, and a communication cable. Since Rule 218 is not specific, Rule 012C, which requires accepted good practice, must be applied. It is common in the electric power utility industry to have some accepted good practice distance between a 12.47/7.2 kV power line and a tree and is common for there to be some tree trimming cycle for a 12.47/7.2 kV power line. It is common in the electric power utility industry to see a 120/240 V secondary triplex circuit going through a tree but to use some method (trimming or tree guard over the triplex cable) to prevent abrasion of the 120/240 V secondary triplex circuit. It is common in the communication utility industry to see a communications cable going through a tree. There is little or no electrical hazard in this case; however, tree trimming for communication cables can be important to avoid damage to lashing wires which can cause the communication cable to separate from the messenger and to provide a clear working space for the communication lineworker or technician during installation, operation, or maintenance of the communication line and equipment. Note 2 in Rule 218A1 provides a statement that it is not practical to prevent all tree-conductor contacts on overhead lines. It is possible for new vegetation growth to contact an ungrounded supply conductor but not damage the ungrounded supply conductor. To determine what amount of tree–conductor contact is practical to prevent and what amount of tree-conductor contact is not practical to prevent requires accepted good practice (Rule 012C). Note 2 should not be applied as an excuse for a poor vegetation management program, a utility needs to exercise due diligence when performing vegetation management, and then if applicable, apply Note 2 on a case-by-case basis. Workers who trim trees in the vicinity of energized conductors must be qualified to do so.

The **Code** recognizes that at times tree trimming or removal may not be practical. If tree trimming or removal is not practical, the **Code** requires using methods to separate the conductor from the tree to avoid damage by abrasion and grounding. This requirement is very general. No specifics are provided as to how to accomplish the separation. One common example is using a factory-made PVC tree guard over a 120/240-V, single-phase, three-wire triplex service drop in contact with a tree. The general **NESC** rules for vegetation management are outlined in Fig. 218-1.

Although not addressed in the **Code**, pruning or removal of vegetation near supply conductors mitigates the chance of starting forest fires or wildfires and mitigates the chance of killing or injuring a person climbing a tree. The general **Code** rules for vegetation management are outlined in Fig. 218-1. As discussed in Rule 010, the State of California does not adopt the **NESC**. The State of California writes its own code titled, General Order 95 for Overhead Lines. This document does contain rules with specific distances between power lines and trees and may be referenced as an example of accepted good practice.

The North American Electric Reliability Corporation (NERC) standards apply to transmission line vegetation management. NERC Standard FAC-003 provides specific values for clearance between transmission line conductors and vegetation (trees). The NERC vegetation management standards were developed to address vegetation-related outages of transmission lines which can lead to cascading outages.

Line crossings, railroad crossings, limited-access highway crossings, and navigable waterways receive special considerations to minimize downed lines. Tree trimming for these areas should consider decaying trees or limbs and trees that are overhanging the line (sometimes referred to as danger trees). See Fig. 218-2.

• Vegetation management should be performed around supply and communication lines as experience has shown necessary.

• Vegetation that may damage ungrounded supply conductors should be pruned or removed.

• Where pruning or removal is not practical, the conductor should be separated from the tree with suitable materials or devices to avoid abrasion and grounding of the circuit through the tree.

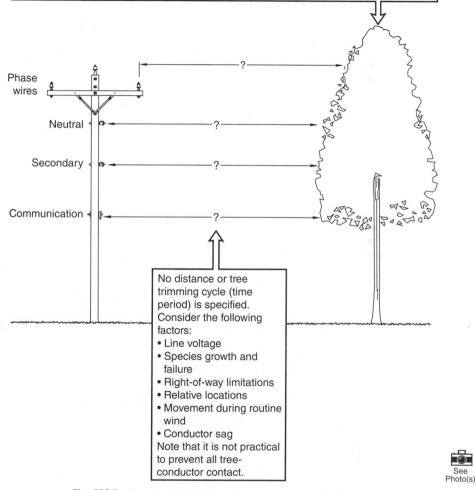

Phase wires

Neutral   ——————?——————→

Secondary   ——————?——————→

Communication   ——————?——————→

No distance or tree trimming cycle (time period) is specified. Consider the following factors:
• Line voltage
• Species growth and failure
• Right-of-way limitations
• Relative locations
• Movement during routine wind
• Conductor sag
Note that it is not practical to prevent all tree-conductor contact.

See Photo(s)

Fig. 218-1. General vegetation management requirements (Rule 218A).

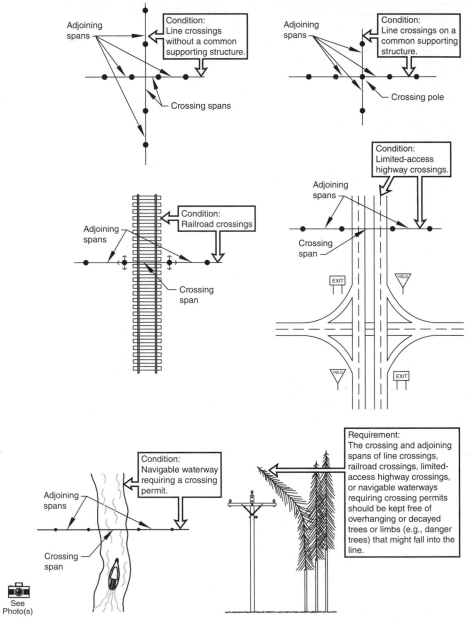

Fig. 218-2. Vegetation management at line, railroad, limited-access highway, and navigable waterway crossings (Rule 218B).

Section 22

Relations between Various Classes of Lines and Equipment

220. RELATIVE LEVELS

Rule 220A requires that the levels of different classes of conductors should be standardized by agreement of the utilities concerned. The **NESC** does not provide relative levels or specific locations for multiple communication circuits in the communications space. Years ago, telephone was the primary joint-use communications attachment on a power pole. Then came community antenna television (CATV) systems; these systems were typically installed above telephone. Today, pole owners are receiving numerous requests for attachments by fiber utilities. Both **NESC** Rule 220A and Rule 235H (Vertical Clearance and Spacing between Communication Conductors, Wires and Cables in the Communication Space) must be referenced to determine the relative levels of communication cables in the communication space. Rules 220A and 235H require agreement between the utilities concerned or the parties involved. See Rule 235H for more information. Once standardization of levels has been agreed upon, there are several other **NESC** Rules that must be met including, but not limited to, clearance rules and strength and loading rules. A joint-use agreement between supply and communications utilities is one type of agreement that can be used to standardize utility locations on a pole. Rule 220B1 states that it is preferred that supply (power) conductors and equipment be located above communications conductors and equipment. Four exceptions apply to Rule 220B1. See Fig. 220-1.

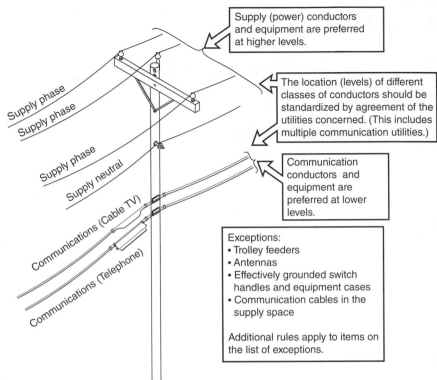

Supply (power) conductors and equipment are preferred at higher levels.

The location (levels) of different classes of conductors should be standardized by agreement of the utilities concerned. (This includes multiple communication utilities.)

Communication conductors and equipment are preferred at lower levels.

Exceptions:
• Trolley feeders
• Antennas
• Effectively grounded switch handles and equipment cases
• Communication cables in the supply space

Additional rules apply to items on the list of exceptions.

Supply phase
Supply phase
Supply phase
Supply neutral
Communications (Cable TV)
Communications (Telephone)

See Photo(s)

Fig. 220-1. Standardization of levels of supply and communication conductors (Rules 220A and 220B1).

Rule 220B2, Special Construction for Railroad Supply Circuits, of 600 V or Less and Carrying Power Not in Excess of 5 kW Associated with Railroad Communication Circuits, is a rule for special construction related to railroad signaling. There are seven conditions (paragraphs a through g) that must be met to apply this rule. Rule 220B2 does not apply to modern cable television and telephone circuits.

Rule 220C discusses where to position supply (power) lines of different voltages on overhead structures. The terms "crossings" and "conflicts" are used in Rule 220C1. Crossings are discussed in detail in Rule 241C. Conflicts are discussed in Rule 221 and a definition of "structure conflict" is provided in Sec. 02 of the **NESC**. Relative levels of supply lines at crossings and conflicts are shown in Figs. 220-2 and 220-3.

Rule 220C2 has requirements for structures used only for supply (power) conductors. Rule 220C2a covers structures with circuits owned by one utility, and Rule 220C2b covers structures with circuits owned by separate supply utilities. See Figs. 220-4 and 220-5.

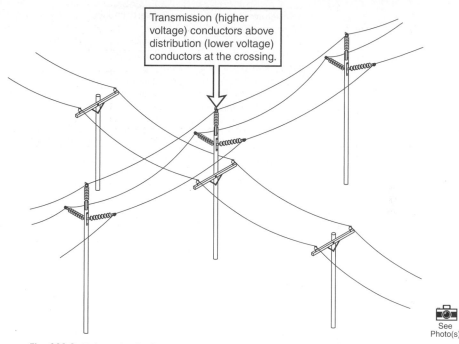

Transmission (higher voltage) conductors above distribution (lower voltage) conductors at the crossing.

See Photo(s)

Fig. 220-2. Relative levels of supply lines of different voltages at crossings (Rule 220C1).

Positioning a higher voltage line above a lower voltage line is a common construction practice. Clearance above ground is greater for higher voltage circuits; therefore, having higher voltage circuits at higher positions permits greater clearance. Rule 220C states the relative levels of conductors, not the clearance between them. **NESC** Rule 235 is referenced in Rule 220C as this rule provides tables for horizontal and vertical clearances between conductors on a common supporting structure. Lower voltage circuits are typically worked on more than higher voltage circuits, so having lower voltage circuits at lower positions provides easier access. Having the higher voltage circuits in the more elevated position also makes sense when considering vertical clearances as discussed in Rule 235.

Rules 220D and 220E require uniform positions of supply and communication conductors and cables (Rule 220D) and equipment (Rule 220E) or constructing, locating, marking, or numbering to facilitate identification by authorized employees who have to work on them. This rule is similar to Rule 217A3, Identification of Supporting Structures. Identifying overhead conductors by attachment to distinctive insulators or crossarms is an acceptable method. Using uniform positions of conductors does not prohibit changing locations systematically.

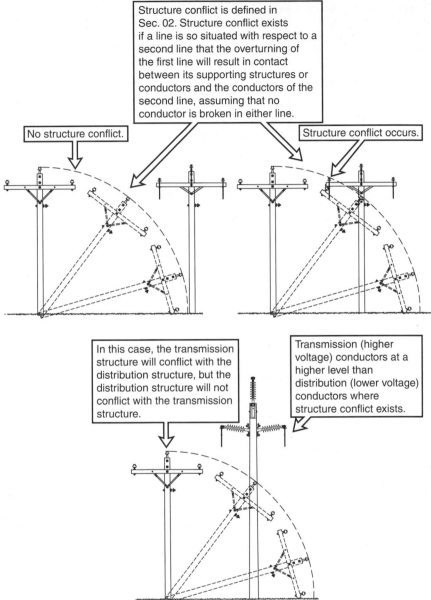

Structure conflict is defined in Sec. 02. Structure conflict exists if a line is so situated with respect to a second line that the overturning of the first line will result in contact between its supporting structures or conductors and the conductors of the second line, assuming that no conductor is broken in either line.

No structure conflict.

Structure conflict occurs.

In this case, the transmission structure will conflict with the distribution structure, but the distribution structure will not conflict with the transmission structure.

Transmission (higher voltage) conductors at a higher level than distribution (lower voltage) conductors where structure conflict exists.

See Photo(s)

Fig. 220-3. Relative levels of supply lines of different voltages at structure conflict locations (Rule 220C1).

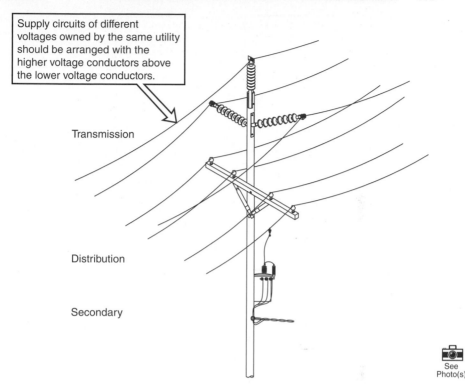

Supply circuits of different voltages owned by the same utility should be arranged with the higher voltage conductors above the lower voltage conductors.

Transmission

Distribution

Secondary

See Photo(s)

Fig. 220-4. Relative levels of supply circuits of different voltages owned by one utility (Rule 220C2a).

A neutral conductor, when on a crossarm with the phase conductors, is sometimes identified with a different color or style insulator or by labeling the crossarm with a letter "N" (Neutral) or letters "CN" (Common Neutral) below the neutral insulator. A pole with crossarms owned by multiple utilities can have the crossarms labeled (i.e., marked or numbered) with the utility name, or the crossarms can be constructed or located such that the employees authorized to work on them can recognize them as their own. Similarly, a pole with multiple communication cable attachments can have the communication cables labeled (i.e., marked or numbered) with color coding or a numbering system or the utility name, or the cables can be constructed or located such that the employees authorized to work on them can recognize them as their own. An example of identification of overhead conductors is shown in Fig. 220-6.

Supply circuits of different voltages owned by separate supply utilities can use the basic highest voltage to lowest voltage rule.

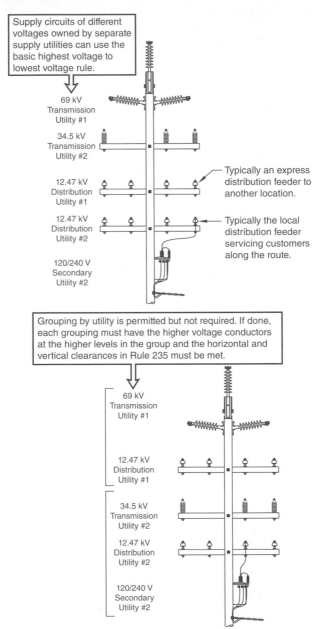

69 kV
Transmission
Utility #1

34.5 kV
Transmission
Utility #2

12.47 kV
Distribution
Utility #1

Typically an express distribution feeder to another location.

12.47 kV
Distribution
Utility #2

Typically the local distribution feeder servicing customers along the route.

120/240 V
Secondary
Utility #2

Grouping by utility is permitted but not required. If done, each grouping must have the higher voltage conductors at the higher levels in the group and the horizontal and vertical clearances in Rule 235 must be met.

69 kV
Transmission
Utility #1

12.47 kV
Distribution
Utility #1

34.5 kV
Transmission
Utility #2

12.47 kV
Distribution
Utility #2

120/240 V
Secondary
Utility #2

Fig. 220-5. Relative levels of supply circuits of different voltages owned by separate utilities (Rule 220C2b).

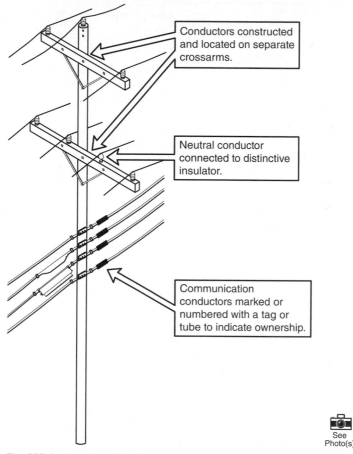

Conductors constructed and located on separate crossarms.

Neutral conductor connected to distinctive insulator.

Communication conductors marked or numbered with a tag or tube to indicate ownership.

See Photo(s)

Fig. 220-6. Example of identification of overhead conductors (Rule 220D).

221. AVOIDANCE OF CONFLICT

The term "conflict" was introduced in Rule 220C. See Rule 220 for a discussion and a figure related to structure conflict. Avoidance of conflict can be accomplished by sufficient separation of lines, or by structure strength, or by combining the lines on the same structure. See Fig. 221-1.

Many times, right-of-way constraints will prohibit two separate lines and collinear or joint-use construction will be required.

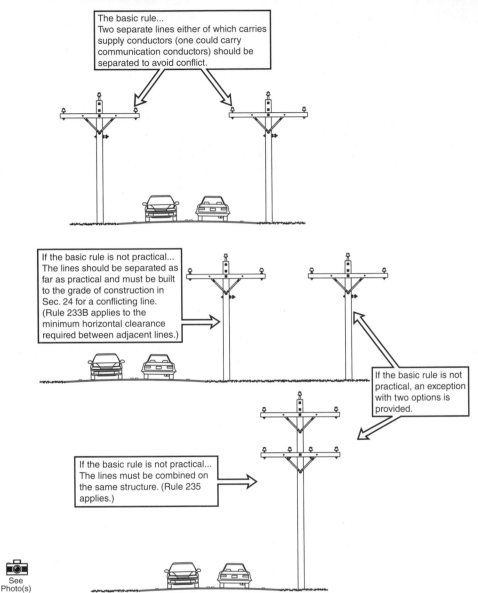

The basic rule...
Two separate lines either of which carries supply conductors (one could carry communication conductors) should be separated to avoid conflict.

If the basic rule is not practical...
The lines should be separated as far as practical and must be built to the grade of construction in Sec. 24 for a conflicting line. (Rule 233B applies to the minimum horizontal clearance required between adjacent lines.)

If the basic rule is not practical, an exception with two options is provided.

If the basic rule is not practical...
The lines must be combined on the same structure. (Rule 235 applies.)

See Photo(s)

Fig. 221-1. Avoiding conflict between two separate lines (Rule 221).

222. JOINT USE OF STRUCTURES

Per the definition of "joint use" in Sec. 02 of the **NESC**, joint use refers to simultaneous use by two or more utilities, which can be two or more of the same kind of utility (e.g., two power utilities) or two or more of different kinds of utilities (e.g., power and communications). The **NESC** encourages considering the use of

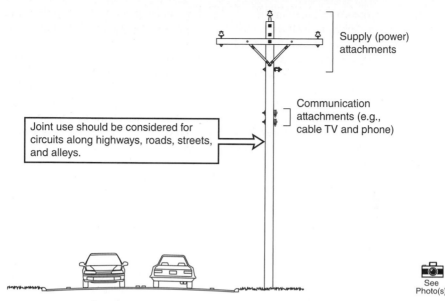

Fig. 222-1. Example of joint use of structures (Rule 222).

joint-use construction along highways, roads, streets, and alleys. An example of a joint use structure along a roadway is shown in Fig. 222-1.

To decide between joint use and separate lines along highways, roads, streets, and alleys, the following must be considered by the utilities considering the joint occupancy:

- Character of circuits
- Worker safety
- Total number and weight of conductors
- Tree conditions
- Number of branches (taps)
- Number of service drops
- Structure conflict
- Availability of right of way, etc.

When joint use is used, it must meet the grade of construction specified in Sec. 24. In addition to applying the appropriate grade of construction for loading and strength issues, numerous other rules apply to joint-use construction including the overhead line clearance rules in Part 2 and the underground separation rules in Part 3. Although not addressed in the **NESC**, a joint-use agreement is appropriate when joint-use construction is used. Joint-use agreements typically include wording regarding an attachment fee that is paid by the utility that is attaching to a pole owned by a different utility. Joint-use agreements also commonly include references to **NESC** requirements, methods to address payment for structures that must be replaced to meet joint-use requirements, and attachment locations. Some supply utilities reserve the top 10 or 12 ft of a distribution pole for supply attachments. This method assures supply space will be available for future distribution service transformers, primary taps, secondary services, etc. Other supply utilities permit communication attachments as long as space is available. This method may require that the communications utility vacate the pole or pay for a pole upgrade if distribution service transformers, primary taps, secondary services, etc., need to be added in the future and space is not available for these items. Similar approaches may be applied to underground ducts. Federal Communications Commission (FCC) or state regulations may specify pole attachment or underground duct procedures and rental fee methodologies. A joint-use agreement is appropriate for collecting pole rental fees, resolving make-ready issues, and assuring joint-use attachments are in accordance with the **NESC**.

223. COMMUNICATIONS PROTECTIVE REQUIREMENTS

Rule 223 requires that a communication apparatus that is subjected to lightning, contact with supply conductors exceeding 300 V to ground, a ground potential rise greater than 300 V, or a steady-state induced voltage of a hazardous level be protected by insulation and, where necessary, surge arresters in conjunction with fusible elements. Additional communication protective devices are also listed for severe conditions.

A typical joint-use (power and communication) overhead distribution line is subjected to the conditions listed in Rule 223A. The most common method used for the means of protection required in Rule 223B is insulating the communication conductors (in the form of a communication cable) and bonding the grounded communication messenger to the grounded supply neutral. This measure must be taken to satisfy Footnote 7 of **NESC** Table 242-1 and to meet the messenger grounding requirements in Rules 215C1 and 215D and the grounding and bonding method in Rule 097G.

It is common to find communication protective devices at building service entrance points and at the head end or central office of a communications system. The additional communication protective devices for severe conditions are typically applied to a communication line entering an electric supply station. The electric supply station typically has large fault current duties, which can severely damage a metallic communication cable. Commonly, isolation equipment or a fiber-optic communication cable is used to serve substations to mitigate this concern.

224. COMMUNICATION CIRCUITS LOCATED WITHIN THE SUPPLY SPACE AND SUPPLY CIRCUITS LOCATED WITHIN THE COMMUNICATION SPACE

Rule 224A applies to communication circuits located in the supply space. Examples of communication circuits located in the supply space are a communication circuit operated by a supply utility and used for communicating between supply stations or a communication circuit operated by a supply utility as a line of business to provide Internet, data, television, or telephone services. Another example is a communication utility locating a communications line in the supply space (as opposed to the communication space) using qualified supply line employees. The terms "supply space" and "communications space" are defined in Sec. 02 of the **NESC**. The "communication worker safety zone" is described in **NESC** Rules 235C4 and 238E.

If a communication circuit is located in the supply space (not the communication space), it must be installed and maintained by an employee qualified to work in the supply space per the work rules (Secs. 42 and 44) of the **NESC**. The employee in this case is typically a power lineworker. If a communication circuit is located in the communication space (not in the supply space), it can be installed and maintained by an employee qualified to work in the communication space per the work rules (Secs. 42 and 43) of the **NESC**. The employee in this case is typically a communications lineworker or communications technician.

Rule 224A2 states that an insulated communication cable supported by an effectively grounded messenger and located in the supply space must have the same clearance as neutrals meeting Rule 230E1 from communication cables in the communication space and from supply conductors in the supply space. This requirement is also conveyed in **NESC** Table 235-5, Footnote 5 and Footnote 9. See Rules 235C and 238 for examples and additional information. Fiber-optic cables located in the supply space must meet Rule 230F1. Rule 224 addresses communication circuits in the supply space. See Rule 238F for antennas in the supply space.

Examples of communication cables located in the supply space and the communication space are shown in Fig. 224-1.

Rule 224A3 discusses the voltage requirements of communication circuits located in the supply space. The voltage requirements in Rule 224A3 for communication circuits in the supply space are different than the voltage requirements of a communication circuit in the communications space. See the definition of lines/communication lines/located in the communication space in Sec. 02, Definitions of Special Terms.

Rule 224A4 provides conditions for locating a communication circuit in the supply space in one portion of the system and in the communication space in another portion of the system. The transition of the communication cable from the supply space to the communication space must occur on a single structure. See Fig. 224-2.

Rule 224B applies to special supply circuits used exclusively in the operation of communication circuits. Rule 224B1 applies to open wire (e.g., bare noninsulated) circuits. Rule 224B2 applies to a communication cable with a supply circuit embedded in it. Rule 224 addresses communication circuits in the supply space. See Rule 238F for antennas in the supply space.

225. ELECTRIC RAILWAY CONSTRUCTION

Electric railways can be in the form of electric locomotives on railroad tracks or electric trolleys on streetcar tracks. Rule 225 provides general information related to these types of installations. It uses the term "electric railway," but there are

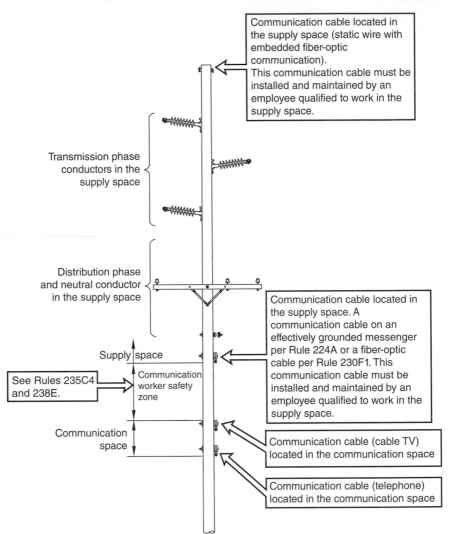

Communication cable located in the supply space (static wire with embedded fiber-optic communication).
This communication cable must be installed and maintained by an employee qualified to work in the supply space.

Transmission phase conductors in the supply space

Distribution phase and neutral conductor in the supply space

Supply space

See Rules 235C4 and 238E.

Communication worker safety zone

Communication space

Communication cable located in the supply space. A communication cable on an effectively grounded messenger per Rule 224A or a fiber-optic cable per Rule 230F1. This communication cable must be installed and maintained by an employee qualified to work in the supply space.

Communication cable (cable TV) located in the communication space

Communication cable (telephone) located in the communication space

Fig. 224-1. Examples of communication cables located in the supply space and the communication space (Rule 224A).

times where trolley contact conductors are used for electric trolley buses that have tires on a roadway. The **NESC** addresses clearances of electric railway and trolley contact conductors in Rule 225 and throughout the clearance rules in Sec. 23. Size, strength, and loading issues related to electric railway and trolley contact conductors are addressed in Rule 261H3 and throughout Secs. 24, 25, 26, and 27.

Rule 225, as well as Rule 215C and other rules in Secs. 23, 24, 25, 26, and 27, addresses the span wires that support electric railway or trolley contact conductors.

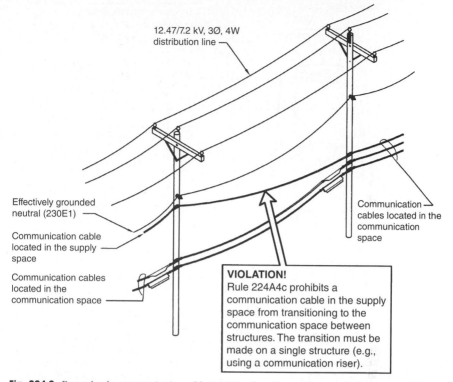

12.47/7.2 kV, 3Ø, 4W
distribution line

Effectively grounded
neutral (230E1)

Communication cable
located in the supply
space

Communication cables
located in the
communication space

Communication
cables located in the
communication
space

VIOLATION!
Rule 224A4c prohibits a
communication cable in the supply
space from transitioning to the
communication space between
structures. The transition must be
made on a single structure (e.g.,
using a communication riser).

Fig. 224-2. Example of a communication cable transition from the supply space to the communication space (Rule 224A4c).

Span wires or brackets carrying trolley conductors may be located in the communication worker safety zone. See Rule 238 for more information.

Many times electric railway or electric trolley systems are operated at DC voltages. See Rule 230G for a discussion of DC vs. AC voltages and how to apply DC voltages to the clearance tables in the **NESC**. Part 4 of the **NESC** contains minimum approach distances for DC voltages in Sec. 44.

Another reference for electric railway and electric trolley design is the American Railway Engineering and Maintenance-of-Way Association (AREMA). AREMA publishes a document titled *Manual for Railway Engineering*.

Additional terms specific to electric railway and trolley systems are pantograph, trolley pole, and third-rail. Examples of common terms used in electric railway and trolley systems are shown in Fig. 225-1.

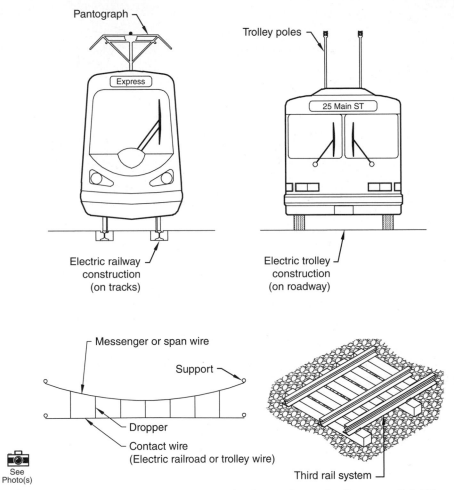

Fig. 225-1. Examples of common terms used in electric railway and trolley systems (Rule 225).

Section 23

Clearances

Section 23 is probably the most referenced section in the **NESC**. Rule 232, Vertical Clearance of Wires, Conductors, Cables, and Equipment Aboveground, Roadway, Rail, or Water Surfaces, is probably the most referenced rule. Chances are that if a person only uses the **NESC** once a year, he or she will be using it to find a clearance value somewhere in this section. When an overhead line clearance question comes up, one of the first problems a person has is determining which rule applies to the **Code** clearance question at hand. An outline of each rule in Sec. 23 with an icon providing a graphical representation of what the rule covers and a column indicating if a sag chart is needed to determine clearance values is shown in Fig. 23-1.

Many of the clearances found in the **NESC** tables in this section are based on conductor sags. The **NESC** specifies maximum sags, minimum sags, sags with wind, sags with ice, sags at specified temperatures, initial sags, and final sags. It is important that the person using the **Code** to determine overhead line clearances understands how to apply a sag and tension chart. Important concepts that apply to sag and tension charts are outlined below:

- Sag and tension charts are created for a conductor type (e.g., 1/0 ACSR, 6/1 stranding, code word: Raven) at a ruling span (e.g., 300 ft) in a loading district (e.g., Medium Loading) and clearance zone (e.g., Zone 2). This information typically appears at the heading of the sag chart.
- A sag and tension chart describes what position a conductor will be in at various temperature and physical loading conditions. Examples of physical loading conditions are ice and wind.
- A ruling span is the span that "rules" or "governs" the behavior of all the spans between two conductor deadends.

INFO	RULE 230	General	Sag chart needed to check clearance? ☐ YES ☒ NO
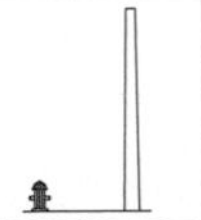	RULE 231	Clearances of supporting structures from other objects	☐ YES ☒ NO
	RULE 232	Vertical clearances of wires, conductors, cables, and equipment aboveground, roadway, rail, or water surfaces	☒ YES ☐ NO
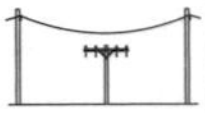	RULE 233	Clearances between wires, conductors, and cables carried on different supporting structures	☒ YES ☐ NO
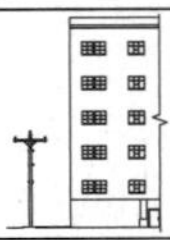	RULE 234	Clearances of wires, conductors, cables, and equipment from buildings, bridges, rail cars, swimming pools, supporting structures and other installations	☒ YES ☐ NO
	RULE 235	Clearance for wires, conductors, or cables carried on the same supporting structure	☒ YES ☐ NO
	RULE 236	Climbing space	☐ YES ☒ NO
	RULE 237	Working space	☐ YES ☒ NO
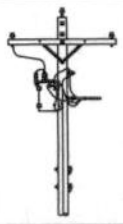	RULE 238	Clearances at the support between specified communications and supply facilities located on the same structure	☐ YES ☒ NO
	RULE 239	Clearance of vertical and lateral facilities from other facilities and surfaces on the same supporting structure	☐ YES ☒ NO

Fig. 23-1. Summary of overhead line clearance rules (Sec. 23).

- A ruling span is calculated between deadends by using the following formula:

$$RS = \sqrt{\frac{S_1^3 + S_2^3 + S_3^3 + \cdots + S_n^3}{S_1 + S_2 + S_3 + \cdots + S_n}}$$

where RS = Ruling Span
$S_1, S_2, S_3, \cdots, S_n$ = 1st, 2nd, 3rd, $\cdots$, nth span lengths between the conductor deadends.

- The sags on a sag and tension chart are for a span length equal to the ruling span. To calculate sags for spans longer or shorter than the ruling span, the following formula can be used:

$$\text{Sag in feet} = \left(\frac{\text{span length}}{\text{ruling span length}}\right)^2 \times \text{ruling span sag in feet}$$

- The curved shape of a conductor suspended between two rigid supports is described as a catenary curve. The catenary equation is a complex equation involving hyperbolic terms and is typically solved using computer programs. Manual sag and tension calculations are typically solved using a simple parabolic equation to approximate the more complex catenary equation.
- The temperature on a sag chart is the conductor temperature, not the ambient air temperature. The conductor temperature is based on the ambient air temperature, the cooling effect of the wind, the radiant heating effect of the sun, and the amount of electrical current flowing through the conductor. IEEE Standard 738 titled, IEEE Standard for Calculating the Current-Temperature Relationships of Bare Overhead Conductors, provides methods to calculate the current–temperature relationship. Below are approximate examples for a 1/0 ACSR conductor:
  ✓ 1/0 ACSR bare conductor has an approximate current-carrying capacity (ampacity) of 230 A. The 230 A is based on a conductor temperature of 167°F.
  ✓ 1/0 ACSR bare conductor:
  Assume the conductor carries a small electrical load (e.g., 11.5 A, which is approximately 5 percent of the 230-A ampacity).
  During summer conditions: Assume a 104°F ambient temperature combined with the heating effect of the electrical current yields a 110°F conductor temperature.
  During winter conditions: Assume a −20°F ambient temperature combined with the heating effect of the electrical current yields a −15°F conductor temperature.
  The maximum sag conditions per Rule 232A would be:
    ▪ 32°F with ice (e.g., 0.25 in of ice for Zone 2)
    ▪ 120°F (since the maximum operating temperature from above is less than 120°F)
  ✓ 1/0 ACSR bare conductor:
  Assume the conductor carries a large electrical load (e.g., 175 A, which is approximately 75 percent of the 230-A ampacity).

During summer conditions: Assume a 104°F ambient temperature combined with the heating effect of the electrical current yields a 167°F conductor temperature.

During winter conditions: Assume a −20°F ambient temperature combined with the heating effect of the electrical current yields a 32°F conductor temperature.

The maximum sag conditions per Rule 232A would be:

- 32°F with ice (e.g., 0.25 in of ice for Zone 2)
- 120°F
- 167°F (since the maximum operating temperature from above is greater than 120°F)

✓ 1/0 ACSR bare conductor:

Assume the conductor carries an emergency electrical load (e.g., 230 A, which is 100 percent of the 230-A ampacity).

During summer conditions: Assume a 104°F ambient temperature combined with the heating effect of the electrical current yields a 212°F conductor temperature.

During winter conditions: Assume a −20°F ambient temperature combined with the heating effect of the electrical current yields a 60°F conductor temperature.

The maximum sag conditions per Rule 232A would be:

- 32°F with ice (e.g., 0.25 in of ice for Zone 2)
- 120°F
- 212°F (since the maximum operating temperature from above is greater than 120°F)

- It is common to see conductor temperatures of 104°F, 167°F, and 212°F on a sag chart as they represent 40°C, 75°C, and 100°C, respectively.
- Electrical loads can peak in the summer or winter or both. Areas with a large amount of electric heating load tend to peak in the winter. Areas with a large amount of air conditioning load tend to peak in the summer.
- Sag and tension charts provide both initial and final sags and tensions. Initial values apply to the day a new conductor is strung up. Final values can occur anywhere in time after that point. For example, if the conductor is exposed to an ice storm the first week it is installed, the conductor could be at final sag. If the conductor was never exposed to ice and was exposed to very little wind, it would take many years of hanging under its own weight for the conductor to reach final sag.
- The additive constant for the heavy, medium, light, and warm islands loading districts is sometimes referred to as a K-factor. The additive constant is also used for clearance zones 1, 2, 3, and 4.
- The maximum sag on a sag chart typically occurs at one of the following conditions:
  ✓ 32°F with ice (e.g., 0.25 in of ice for Zone 2), final
  ✓ 120°F, final
  ✓ Greater than 120°F, final (e.g., 167°F, 212°F, etc.)

Larger sags may be observed at the following conditions, but these conditions are not required for checking vertical clearance:

  ✓ Heavy, medium, light, or warm islands loading (e.g., 0.25 in of ice, 4-lb/ft² wind, 0.20 lb/ft additive constant for medium loading)

- The heavy, medium, light, and warm island conditions (Rule 250B for loading) are also referred to as the Zone 1, 2, 3, and 4 conditions (Rule 230B for clearance).
- One of the Zone 1, 2, 3, and 4 conditions (e.g., Zone 2, 15°F, 0.25 in of ice, 4 lb/ft² wind, 0.20 lb/ft additive constant) must be applied to the conductor for physical loading purposes before the final sag is checked at 32°F with ice (e.g., 0.25 in of ice for Zone 2), 120°F, and greater than 120°F if so designed. Note that sag is not required to be checked at 15°F, 0.25 in of ice, 4 lb/ft² wind, 0.20 lb/ft additive constant. The heavy, medium, light, or warm islands (Zone 1, 2, 3, or 4) conditions must be applied to the conductor for physical loading purposes, but sag is checked at other conditions.
  - ✓ 60°F with a 6-lb/ft² wind (this condition is required for conductor horizontal blowout, not vertical sag)
  - ✓ 60°F with extreme wind (e.g., 17.76 lb/ft²) (this condition is required for conductor tension limits, not vertical sag)
  - ✓ 15°F with extreme ice with concurrent wind (e.g., 0.25 in of ice with 2.30 lb/ft²) (this condition is required for conductor tension limits, not vertical sag)
- The minimum sag on a sag chart usually occurs at the initial sag of the coldest temperature without ice or wind loading (e.g., −20°F, initial).
- The tension must be checked at the heavy, medium, light, or warm islands loading condition (Rule 250B), the extreme wind condition (Rule 250C), if applicable, and the extreme ice with concurrent wind condition (Rule 250D), if applicable, to determine the maximum tension which is commonly referred to as the "design tension." It is possible for the maximum tension to occur at the minimum temperature condition.
- The minimum tension usually occurs at the maximum sag condition due to high temperatures at final tension.
- The sag and tensions on a sag chart can be varied by increasing tension which will decrease sag or by decreasing tension which will increase sag. Too much tension is not good for guying and structure strength and loading issues. Too much sag is not good for clearance issues. A balance must be reached between the two.
- The sag values on a sag and tension chart are used to calculate clearance. The tensions on a sag and tension chart are used to calculate strength and loading of structures and conductor supports.
- The following headings and abbreviations are used on a typical sag and tension chart:
  - ✓ Conductor: Raven (this is the nickname for the conductor).
  - ✓ 1/0 AWG, 6/1 stranding, ACSR: 1/0 AWG (American Wire Gauge) is the conductor size. ACSR stands for Aluminum Conductor, Steel Reinforced. The 6/1 represents six strands of aluminum twisted around one strand of steel.
  - ✓ Area = cross-sectional area of the conductor
  - ✓ Dia = diameter of the conductor
  - ✓ Wt = weight of the conductor
  - ✓ RTS = rated tensile strength of the conductor
  - ✓ Span = ruling span

✓ The design condition is the limiting design condition entered into the program. It may be an **NESC** tension limit, a conductor manufacturer tension limit, or a user-defined sag or tension limit.

- Final sag values on the sag and tension chart include the effects of conductor inelastic deformation due to ice and wind loads and long-term material deformation (sometimes referred to as creep). **NESC** Appendix B provides additional information on conductor creep.

- **NESC**-defined tension limits (percentages) must be entered into the sag and tension program.

- User-defined sag and tension limits can be entered into a sag and tension program in addition to **NESC** design conditions. An example of a user-defined tension limit is 20 percent tension at the initial condition at the average annual minimum temperature (e.g., 20 percent tension at 0°F, initial). This limit is used to control aeolian vibration. Another example of a user-defined limit is specifying a conductor sag for a lower circuit to match the sag of a higher circuit. The sags of two different conductor sizes or types cannot perfectly match throughout the entire temperature range. Matching sags at 60°F at the final tension typically produces the closest matching sags throughout the entire temperature range. Other examples of user-defined sag and tension limits are 3 ft of sag at 60°F final, a sag of 1 percent of the ruling span at 60°F final, a 2000-lb maximum tension limit, etc. The conductor or cable manufacturer may also specify a sag or tension limit. When a user-defined tension is entered into a sag and tension program, the **NESC** design conditions must also be entered to verify that none of the **NESC** design conditions is exceeded.

- The phase and neutral conductor of the same circuit can carry different amounts of current and therefore operate at different temperatures. The neutral conductor may carry less current due to phase balance and neutral current cancellation or due to the fact that the earth and a communication messenger on a joint-use pole are in parallel with a multigrounded neutral. It is possible for the neutral and communication messenger on a joint-use (power and communications) line to operate at 120°F, while the phase wires on the same line operate at temperatures above 120°F.

- If a line is existing, a sag chart can be created for the existing line by measuring the sag at a known conductor temperature and span length. This information is then entered into a program to create a sag chart. Another method is to find the third or fifth return wave for a known conductor temperature and span length. The return wave time can be converted to sag using the following formula:

$$\text{Sag in feet} = 4.025 \left(\frac{\text{time in seconds}}{2 \times \text{number of return waves}} \right)^2$$

In both the methods described above, the line must be assumed to be in its initial or final sag condition. The return wave method is not accurate when used on deadend spans as the deadend insulator string absorbs the conductor wave and distorts the wave timing. Splices in the span can also distort the wave timing.

- Sag and tension charts must be run for secondary and communication conductors and cables to accurately check clearance and strength issues for these circuits.
- Sag and tension charts are used to create stringing charts for installing conductors in the field. Stringing charts, combined with sag and tension charts, typically provide sag in inches and third and fifth return wave time in seconds for various installation temperatures and span lengths contained by the ruling span.
- Methods of installing conductors in the field include the sighting method (using a target temporarily nailed to each pole), the tension method (using a dynamometer), and the return wave method (using a stopwatch).

Required sag and tension values in Rules 230, 232, 233, 234, 235, 250B, 250C, 250D, 251, and 261H are noted in the sample sag and tension chart provided in Fig. 23-2.

The sag and tension chart provides a sag value at the midspan or center of the span. There are times when sag needs to be checked somewhere else in the span, for example, if a rise in the height of the surface below the line occurs around the quarter span instead of the midspan. Another example is if two lines cross in the span, one at 10% of the span distance and the other at 33% of the span distance. A graph can be used to estimate the percentage of midspan sag at various distances along the span length. See Fig. 23-3.

Sample Sag and Tension Chart

Conductor RAVEN   #1/0 AWG 6/ 1 Stranding ACSR

Area= .0968 Sq. In   Dia= .398 In   Wt= .145 Lb/F RTS= 4380 Lb
Span= 300.0 Feet (NESC Medium Load District /NESC Clearance Zone 2)

		Design Points					Final			Initial	
	Temp	Ice	Wind	K	Weight	Sag	Tension	RTS	Sag	Tension	RTS
	F	In	Psf	Lb/F	Lb/F	Ft	Lb	%	Ft	Lb	%
Note 1	15.	.25	4.00	.20	.658	5.43	1366.	31.2	5.43	1366.	31.2
Note 2	15.	.25	2.30	.00	.387	4.31	1012.	23.1	4.07	1070.	24.4
Note 3	32.	.25	.00	.00	.346	4.49	868.	19.8	4.13	945.	21.6
Note 4	60.	.00	17.53	.00	.599	6.11	1104.	25.2	5.89	1146.	26.2
Note 5	60.	.00	6.00	.00	.246	4.66	595.	13.6	4.03	688.	15.7
Note 6	-20.	.00	.00	.00	.145	1.78	916.	20.9	1.61	1015.	23.2
	-10.	.00	.00	.00	.145	1.98	825.	18.8	1.73	945.	21.6
Note 12	0.	.00	.00	.00	.145	2.20	740.	16.9	1.86	876.	20.0*
	10.	.00	.00	.00	.145	2.46	663.	15.1	2.02	808.	18.5
Note 7	15.	.00	.00	.00	.145	2.60	628.	14.3	2.10	775	17.7
	20.	.00	.00	.00	.145	2.74	595.	13.6	2.20	743.	17.0
	30.	.00	.00	.00	.145	3.04	536.	12.2	2.40	680.	15.5
	32.	.00	.00	.00	.145	3.11	526.	12.0	2.44	668.	15.3
	40.	.00	.00	.00	.145	3.36	486.	11.1	2.62	622.	14.2
	50.	.00	.00	.00	.145	3.68	444.	10.1	2.87	569.	13.0
Note 8	60.	.00	.00	.00	.145	4.00	408.	9.3	3.14	520.	11.9
	70.	.00	.00	.00	.145	4.32	378.	8.6	3.42	478.	10.9
	80.	.00	.00	.00	.145	4.63	353.	8.1	3.70	441.	10.1
	90.	.00	.00	.00	.145	4.93	331.	7.6	4.00	408.	9.3
	100.	.00	.00	.00	.145	5.22	313.	7.1	4.29	380.	8.7
	104.	.00	.00	.00	.145	5.34	306.	7.0	4.41	370.	8.5
	110.	.00	.00	.00	.145	5.47	299.	6.8	4.58	356.	8.1
Note 9	120.	.00	.00	.00	.145	5.61	291.	6.7	4.84	335.	7.7
Note 10	167.	.00	.00	.00	.145	6.25	262.	6.0	6.14	266.	6.1
Note 11	212.	.00	.00	.00	.145	6.84	239.	5.5	6.76	242.	5.5

*Design Condition

Fig. 23-2. Sample sag and tension chart (Sec. 23).

Note 1:

15°F, 0.25" ice, 4 lb/ft² wind, 0.20 lb/ft K-factor (additive constant).
This condition (Medium Loading) is specified in Rules 250B and 251.
This condition (Zone 2) is specified in Rule 230B.
The tension from this line,1366 lb initial, is used for
guying and other structure strength and loading calculations.
This tension must be compared to the tension for extreme wind loading (Rule 250C), if
applicable, and the tension for extreme ice with concurrent wind loading (Rule 250D), if
applicable, to determine the maximum design tension. The maximum tension on the chart
is commonly referred to as the "design tension."
The percent rated tensile strength from this line, 31.2%, is used in Rule 261H1a.
Rule 261H1a specifies a 60% limit.

Note 2:

15°F, 0.25" ice, 2.30 lb/ft² wind.
This condition is specified in Rule 250D.
A 2.30 lb/ft² wind correspondents to a 30 mph wind in western Washington
State. 0.25"of ice is also required at this location.
The tension from this line, 1070 lb initial, is used for guying and
other structure strength and loading calculations when Rule 250D applies
(for structures and supported facilities over 60' above ground).
The percent rated tensile strength from this line, 24.4%, is used in Rule 261H1a.
Rule 261H1a specifies an 80% limit. This tension must be compared
to the tension for heavy, medium, light, and warm islands loading (Rule 250B) and the tension
for extreme wind loading (Rule 250C), if applicable, to determine the maximum design tension.

Note 3: (Zone 2)

32°F, 0.25" ice.
This condition is specified in Rules 232A, 233A, 234A, and 235C2b(1)(c)(ii).
The sag from this line, 4.49' final, is used for clearance calculations.
This sag must be compared to the sag at 120°F, final, and greater than 120°F,
final, if so designed, to determine the largest final sag.

Note 4:

60°F, 17.53 lb/ft² wind.
This condition is specified in Rule 250C. A 17.53 lb/ft² wind corresponds to an 85 mph
wind in central Washington State. See Fig. 250-3 for the 17.53 lb/ft² calculation.
The tension from this line, 1146 lb initial, is used for guying calculations and other structure
strength and loading calculations when Rule 250C applies (for structures and supported
facilities over 60' above ground). The percent rated tensile strength from this line, 26.2%,
is used in Rule 261H1a. Rule 261H1a specifies an 80% limit. This tension must be
compared to the tension for heavy, medium, light, and warm islands loading (Rule 250B)
and the tension for extreme ice with concurrent wind loading (Rule 250D), if applicable, to
determine the maximum design tension.

Note 5:

60°F, 6 lb/ft² wind.
This condition is specified in Rules 233A and 234A.
The sags from this line, 4.66' final and 4.03' initial, are used for horizontal
clearance calculations. The sags from this line are "resultant" sags in a diagonal
position. Options typically exist in sag and tension programs to show both the
horizontal and vertical components of the resultant sag value.
The data shown below were produced with the horizontal and vertical sag option enabled:

Design points					Final			Initial		
Cond.	Ice	Wind	K	Weight	H-Sag	V-Sag	Tension	H-Sag	V-Sag	Tension
°F	in	lb/ft²	lb/ft	lb/ft	ft	ft	lb	ft	ft	lb
60.0	0.00	6.00	0.00	0.246	3.77	2.74	595	3.26	2.37	688

This line only accounts for the conductor movement. Insulator swing and structure deflection
may also be applicable.

Fig. 23-2. Sample sag and tension chart (Sec. 23). (*Continued*)

Note 6:
−20°F.
This condition is required in Rule 234A1d (A minimum conductor temperature for a specific area.)
The sag from this line, 1.61' initial, is used for vertical clearance under signs or buildings.

Note 7:
15°F (Medium Loading).
This condition is used in Rule 261H1c.
The percent rated tensile strengths from this line, 14.3% final, and 17.7% initial, are used in Rule 261H1c. Rule 261H1c specifies a 35% initial tension limit and a 25% final tension limit.

Note 8:
60°F.
This condition is used in Rules 233A, 235B, and 235C2b(3).
The sags from this line, 4.00' final and 3.14' initial, are used for vertical and horizontal clearance calculations.

Note 9:
120°F.
This condition is specified in Rules 232A, 233A, 234A, and 235C2b(1)(c)(i).
The sag from this line, 5.61' final, is used for clearance calculations.
The sag must be compared to the sag at 32°F, 0.25" ice, final, and greater than 120°F, final, if so designed, to determine the largest final sag.

Note 10:
167°F.
This condition is specified in Rules 232A, 233A, 234A, and 235C2b(1)(c)(i).
The sag from this line, 6.25' final, is used for clearance calculations.
This condition is only required if the line is designed to operate at a temperature greater than 120°F.

Note 11:
212°F.
This condition is specified in Rules 232A, 233A, 234A, and 235C2b(1)(c)(i).
The sag from this line, 6.82' final, is used for clearance calculations.
This condition is only required if the line is designed to operate at a temperature greater than 120°F.

Note 12:
0°F initial, 20% tension limit.
This condition is not specified in the **Code**. It is a user-defined condition. For example, a 20% tension limit at the average minimum cold temperature may help prevent aeolian vibration. (See Note 3 in Rule 261H1c.) Other user-defined conditions can also be entered into a sag and tension program. The program will determine the most limiting condition of all the conditions entered. In this case the 20% tension limit at 0°F initial was the most limiting condition and the software flagged this value as the "*Design Condition."

Fig. 23-2. Sample sag and tension chart (Sec. 23). (*Continued*)

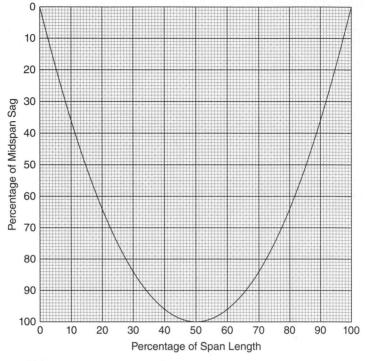

Fig. 23-3. Estimate of the percentage of midspan sag at various distances along the span length (Sec. 23).

230. GENERAL

230A. Application. Rule 230A provides an introduction to Sec. 23. Section 23 covers clearances for overhead supply and communication lines. Burial depths for underground lines are covered in Part 3 of the NESC, and clearances in electric supply stations are covered in Part 1 of the NESC.

Rule 230A provides a note regarding the development of clearance values and makes reference to Appendix A of the NESC, which outlines how NESC clearances are calculated starting with the 1990 edition of the NESC. Prior to 1990, clearance was calculated using a 60°F sag condition with a clearance adder for long spans instead of using a maximum sag condition. Therefore, clearance values prior to 1990 cannot be directly compared to today's clearances. The examples and discussion presented in this Handbook will all focus on using today's Code and today's clearance calculations. Reading and understanding prior methods may be needed if a person is trying to apply the "grandfather" requirements of Rule 013.

Temporary clearances are required to always be the same as permanent clearances. In other words, no clearance reduction is permitted for a temporary situation. This rule is conveyed in Fig. 230-1.

TEMPORARY   PERMANENT
CLEARANCE   ＝   CLEARANCE

Fig. 230-1. Relationship between temporary and permanent clearance (Rule 230A1).

Temporary installations are permitted to have lower grades of construction and an exception with special conditions applies to underground lines. See Rule 014C.

Emergency installations do permit some slight decreases in clearance if certain conditions are met. Rule 014B is referenced as it provides general waiver information. Rule 230A2 provides the specifics, the most common of which are outlined in Fig. 230-2.

> No exact value is specified. Horizontal clearances may also be reduced using accepted good practice.

CONDITION	TYPE OF CONDUCTOR		
	NEUTRAL (230E1) OR COMMUNICATIONS	SECONDARY DUPLEX, TRIPLEX, OR QUADRUPLEX (230C3)	PRIMARY VOLTAGE OPEN (NONINSULATED) SUPPLY CONDUCTORS 750 V TO 22 KV
Roads, streets, and other areas subject to truck traffic	Normal: 15.5' Emergency: 15.5'	Normal: 16.0' Emergency: 15.5'	Normal: 18.5' Emergency: No exact value is specified. Adjust for voltage and local conditions. Value must be greater than 15.5'.
Spaces and ways subject to persons on foot (pedestrians) or restricted traffic only.	Normal: 9.5' Emergency: 9.0'	Normal: 12.0' Emergency: 9.0'	Normal: 14.5' Emergency: No exact value is specified. Adjust for voltage and local conditions. Value must be greater than 9.0'.

> Normal values from **NESC** Table 232-1 (footnotes may apply). Emergency values from Rule 230A2.Caution: Use is very limited. See Rule 230A2a for a description of what qualifies as a pedestrian (person on foot) only area.

> Normal values from **NESC** Table 232-1 (footnotes may apply). Emergency values from Rule 230A2. For the purposes of this rule, a truck is defined as a vehicle exceeding 8' in height. An ambulance or fire rescue unit responding to an emergency could be 8' high or higher.

Fig. 230-2. Reductions of overhead clearances for emergency conditions (Rule 230A2).

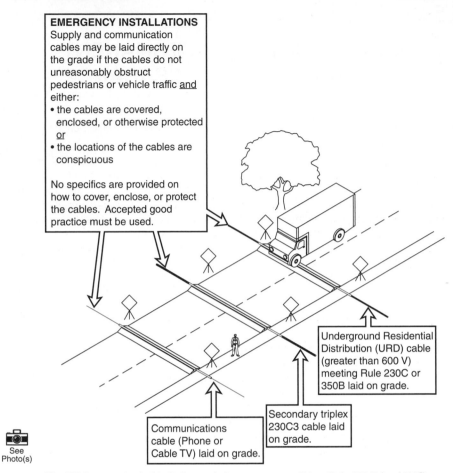

EMERGENCY INSTALLATIONS
Supply and communication cables may be laid directly on the grade if the cables do not unreasonably obstruct pedestrians or vehicle traffic <u>and</u> either:
• the cables are covered, enclosed, or otherwise protected <u>or</u>
• the locations of the cables are conspicuous

No specifics are provided on how to cover, enclose, or protect the cables. Accepted good practice must be used.

Underground Residential Distribution (URD) cable (greater than 600 V) meeting Rule 230C or 350B laid on grade.

Secondary triplex 230C3 cable laid on grade.

Communications cable (Phone or Cable TV) laid on grade.

See Photo(s)

Fig. 230-3. Example of cables laid on grade for emergency conditions (Rules 230A2d and 311C).

Emergency installations permit laying certain supply (power) and communication cables directly on the ground. This same permission can be found in Rule 311C. An example of supply and communication cables laid on the ground is shown in Fig. 230-3.

Where access is limited to qualified personnel only, as inside a properly fenced electric substation, no emergency clearance is specified.

Rule 014B4 requires emergency installations to be removed, replaced, or relocated as soon as practical.

Rule 230A3 describes how clearance and spacing are measured. Rule 230A3 also addresses energized metallic hardware used to secure or support supply line conductors and communication equipment used to secure or support communication line conductors. The basic method for measuring clearance and spacing is shown in Fig. 230-4.

Throughout Sec. 23, the term clearance is used more often than spacing. Part 3, Underground Lines, uses the term separation which is measured similar to clearance.

Rule 230A4 provides rounding requirements specific to the clearance calculations in Sec. 23. In general, Rule 018 permits rounding "off" to the nearest

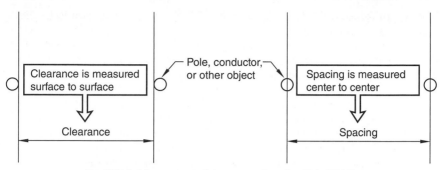

Fig. 230-4. Measurement of clearance and spacing (Rule 230A3).

significant digit unless otherwise specified in applicable rules. One rule where other rounding is specified is Rule 230A4 which requires rounding "up" as Sec. 23 deals with overhead line clearances which are typically specified as "not less than" clearances. Rounding "off" follows the rules of traditional rounding learned in math class. An example of rounding "off" is rounding 20.02 down to 20.0 or rounding 20.66 up to 20.7. An example of rounding "up" for "not less than" clearance is rounding 20.02 to 20.1 because rounding "off" to 20.0 would not meet a clearance required to be "not less than" 20.02. The number of decimal places for the final rounded value depends on the number of decimal places for the original starting value from the **Code** table or rule. An exception is provided for millimeters which must be rounded up to the next multiple of 25 mm. For example, the clearance values in **NESC** Table 232-1 have one decimal place (e.g., 18.5 ft). Therefore, a calculated clearance for a voltage greater than 22 kV per Rule 232C would be rounded "up" to the tenths place (e.g., 20.02 would be rounded up to 20.1). An exception and an example are provided in Rule 230A4 requiring rounding down when determining clearance at specified conditions based on field measurements. The example pertains to vertical clearance required in Rule 232 and **NESC** Table 232-1. The example in **NESC** Rule 230A4 not only explains when rounding down is appropriate but also explains the difference between field-measured sag and the maximum sag conditions that are required per Rule 232A.

230B. Ice and Wind Loading for Clearances.   Rule 230B specifies ice and wind loading for clearance purposes. **NESC** Fig. 230-1 and **NESC** Tables 230-1 and 230-2 are the three basic references to determine ice and wind loading for Zone 1, Zone 2, Zone 3, and Zone 4. Zone 4 has values for above and below 9,000 ft of elevation as some warm islands can have high mountains with ice storms. **NESC** Fig. 230-1 is nearly identical to **NESC** Fig. 250-1. The only difference is Zone 1 in **NESC** Fig. 230-1 is labeled "Heavy" in **NESC** Fig. 250-1, Zone 2 is labeled "Medium," Zone 3 is labeled "Light," and Zone 4 is labeled "Warm Islands." The values specified in **NESC** Tables 230-1 and 230-2 are comparable to the values specified in **NESC** Tables 250-1 and 251-1. The values are identical; however, the presentation format varies slightly. The requirements in Sec. 25 (**NESC** Fig. 250-1, and **NESC** Tables 250-1 and 251-1) are for determining conductor tension and calculating strength and loading issues. The requirements in Sec. 23 (**NESC** Fig. 230-1, and **NESC** Tables 230-1 and 230-2) are for determining conductor sag and therefore calculating clearance. **NESC** Appendix B provides additional information on applying ice and wind loads for clearance and strength. The requirements in **NESC** Table 230-2 (e.g., Zone 2, 15°F, 0.25 in of ice, 4 lb/ft² wind, 0.20 lb/ft additive constant) must be applied to the conductor for physical loading purposes before the final sag is checked at 32°F with ice (e.g., 0.25

in of ice for Zone 2), 120°F, and greater than 120°F if so designed (see Rule 232A). Note that sag is not required to be checked at 15°F, 0.25 in of ice, 4 lb/ft² wind, 0.20 lb/ft additive constant. This condition must be on the sag and tension chart (for clearance Zone 2 and Medium Loading), but clearance is checked at the temperature and loading conditions in Rule 232A (or other applicable rule). The Zone 1, 2, 3, or 4 conditions (whichever is applicable) from Table 230-2 must be applied to the conductor for physical loading purposes, but sag is checked at the ice condition (if applicable) in **NESC** Table 230-1 and high temperature conditions defined in the individual clearance rules (e.g., Rule 232A). More specifically, the values specified in **NESC** Table 230-2 for Zones 1, 2, 3, and 4 are for "clearance loading" in Sec. 23. Clearance measurement conditions in Section 23 are covered in Rules 232A, 233A, 234A, 235B, and 235C2b(1)(c). Rule 250C (extreme wind) and 250D (extreme ice with concurrent wind) can impose greater loads (and therefore more sag) than the loads in Zones 1, 2, 3, and 4. However, the loads specified in Rules 250C and 250D are for strength calculations, only the loads of Zones 1, 2, 3, and 4 are required for clearance purposes (see **NESC** Appendix B). Rule 230I, Maintenance of Clearances and Spacings, requires that conductors be resagged if an excessive ice or wind storm stretched conductors to the point of a clearance violation. Rules 230B3 and 230B4 are similar to Rules 251A and 251B. See Rules 251A and 251B for a discussion and related figures. The requirements for Zones 1, 2, 3, and 4 are outlined in Fig. 230-5.

230C. Supply Cables.   The terms 230C1, 230C2, and 230C3 cables will be used over and over throughout the rules of Sec. 23 and throughout the clearance tables in Sec. 23. The most common of these three cables used in construction today is the 230C3 cable, which is an overhead secondary duplex, triplex, or quadruplex cable. The rules defining the construction of 230C1, 230C2, and 230C3 supply cables are outlined in Fig. 230-6.

A bare messenger or neutral is required for 230C1, 230C2, and 230C3 cables. An insulated neutral is more common for underground secondary duplex, triplex, and quadruplex construction. The 230C1, 230C2, and 230C3 cables are relying on the effectively grounded bare messenger or neutral to carry fault current if an insulation failure occurs.

230D. Covered Conductors.   Covered conductors addressed in Rule 230D are commonly called tree wire. Tree wire cables are not fully insulated like underground residential distribution (URD) cables or 230C cables. They are covered to limit the likelihood of a short circuit in case of momentary contact with a tree limb. Covered conductors can be attached to insulators on crossarms or used as part of a spacer cable system. Since 230D cables are not fully insulated, they must be considered bare conductors for clearance purposes (2 exceptions apply). See Fig. 230-7.

230E. Neutral Conductors.   The phrase "neutral conductors meeting 230E1" will be used over and over throughout the rules of Sec. 23 and throughout the clearance tables in Sec. 23. See Sec. 02, Definitions, for a discussion of the term "effectively grounded." The rules for a 230E1 neutral are outlined in Fig. 230-8.

230F. Fiber-Optic Cable.   Fiber-optic supply cable and fiber-optic communication cable are **NESC** terms to categorize where a fiber-optic cable is located. Examples of communication cables located in the supply space and in the communication space are discussed in Rule 224. Rule 224A2a applies to insulated communication circuits supported on an effectively grounded messenger. Rule 224A2b references Rule 230F1 for fiber-optic cable requirements.

Clearance Zone	Ice and Wind Condition at a Specified Temperature (Clearance Loading)	Location
Zone 1	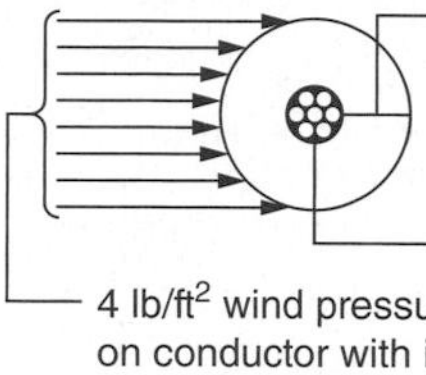 1/2" of radial ice Conductor at 0°F 4 lb/ft² wind pressure on conductor with ice Ice weight: 56 lb/ft³ Additive constant: 0.30 lb/ft	See a map of the USA in **NESC** Fig. 230-1.
Zone 2	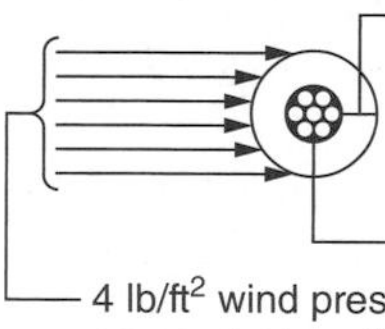 1/4" of radial ice Conductor at 15°F 4 lb/ft² wind pressure on conductor with ice Ice weight: 56 lb/ft³ Additive constant: 0.20 lb/ft	See a map of the USA in **NESC** Fig. 230-1.
Zone 3	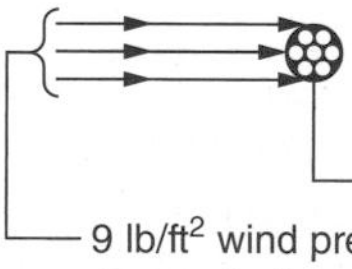 Conductor at 30°F 9 lb/ft² wind pressure on conductor without ice no ice Additive constant: 0.05 lb/ft	See a map of the USA in **NESC** Fig. 230-1.
Zone 4 (for altitudes of sea level to 9000 ft)	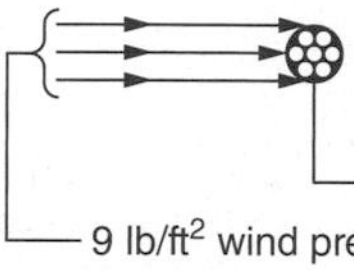 Conductor at 50°F 9 lb/ft² wind pressure on conductor without ice no ice Additive constant: 0.05 lb/ft	See a map of the USA in **NESC** Fig. 230-1 (American Samoa, Guam, Hawaii, Puerto Rico, Virgin Islands, and other islands located from latitude 25 degrees south through 25 degrees north).
Zone 4 (for altitudes above 9000 ft)	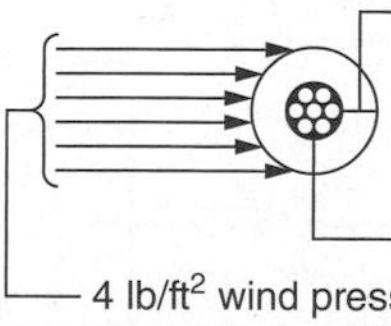 1/4" of radial ice Conductor at 15°F 4 lb/ft² wind pressure on conductor with ice Ice weight: 56 lb/ft³ Additive constant: 0.20 lb/ft	See a map of the USA in **NESC** Fig. 230-1 (American Samoa, Guam, Hawaii, Puerto Rico, Virgin Islands, and other islands located from latitude 25 degrees south through 25 degrees north).

Values shown above are for "clearance loading" in Sec. 23. Clearance measurement conditions in Sec. 23 are covered in Rules 232A, 233A, 234A, 235B, and 235C2b(1)(c).

Fig. 230-5. Zone 1, 2, 3, and 4 ice and wind loading for clearances (Rule 230B).

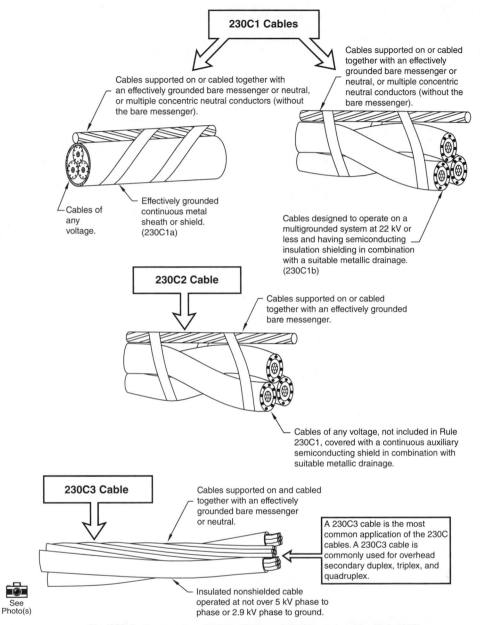

Fig. 230-6. Construction of 230C1, 230C2, and 230C3 supply cables (Rule 230C).

Fiber-optic cables that are strung on overhead pole lines are supported on a messenger or they are contained in an All-Dielectric Self-Supporting (ADSS) cable. They can also be embedded in a metallic messenger or static wire. An example of an ADSS cable is shown in Fig. 230-9.

Fiber-optic cables that are located in the supply space and embedded in a messenger or conductor must have the same clearance from communication facilities

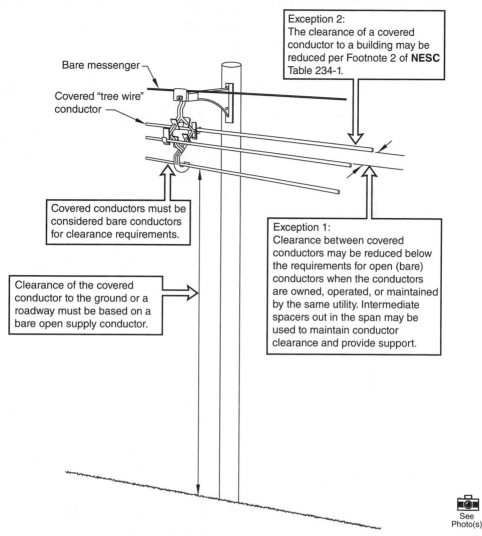

Bare messenger

Covered "tree wire" conductor

Exception 2:
The clearance of a covered conductor to a building may be reduced per Footnote 2 of **NESC** Table 234-1.

Covered conductors must be considered bare conductors for clearance requirements.

Exception 1:
Clearance between covered conductors may be reduced below the requirements for open (bare) conductors when the conductors are owned, operated, or maintained by the same utility. Intermediate spacers out in the span may be used to maintain conductor clearance and provide support.

Clearance of the covered conductor to the ground or a roadway must be based on a bare open supply conductor.

See Photo(s)

Fig. 230-7. Covered conductors (Rule 230D).

as the messenger or conductor the fiber-optic cables are embedded in. Fiber-optic cables that are supported on an effectively grounded messenger or contained in an ADSS cable are required to have the same clearance as neutrals meeting Rule 230E1 from communication facilities. This requirement can be applied to **NESC** Table 235-5, Footnotes 5 and 9. See Rules 235C and 238 for examples, figures, and additional information.

Fiber-optic cables located in the communication space, whether or not supported by a messenger, must have the clearance from supply facilities as required for a communication messenger.

230G. Alternating- and Direct-Current Circuits. The clearances in Sec. 23 apply to AC and DC circuits. The voltage used in a typical American home is 120-V alternating current at 60 Hz (60 cycles per second). The 120-V value is a root mean square (rms)

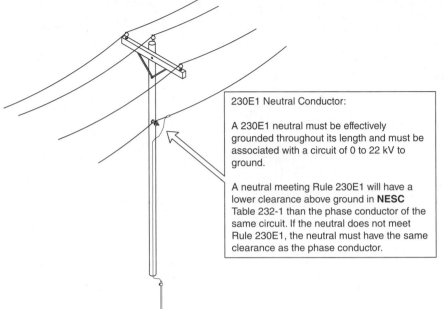

230E1 Neutral Conductor:

A 230E1 neutral must be effectively grounded throughout its length and must be associated with a circuit of 0 to 22 kV to ground.

A neutral meeting Rule 230E1 will have a lower clearance above ground in **NESC** Table 232-1 than the phase conductor of the same circuit. If the neutral does not meet Rule 230E1, the neutral must have the same clearance as the phase conductor.

See
Photo(s)

Fig. 230-8. Neutral conductors (Rule 230E).

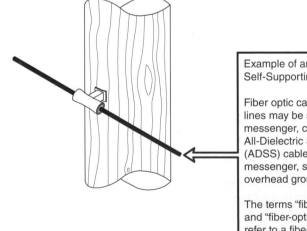

Example of an All-Dielectric Self-Supporting (ADSS) cable.

Fiber optic cables on overhead lines may be supported on a messenger, contained in an All-Dielectric Self-Supporting (ADSS) cable, or impeded in a messenger, static wire, or overhead ground wire.

The terms "fiber-optic–supply" and "fiber-optic–communication" refer to a fiber optic cable in the supply space or the communication space, respectively.

See
Photo(s)

Fig. 230-9. Example of an All-Dielectric Self-Supporting (ADSS) cable (Rule 230F).

measurement. Another term used for the rms voltage is the effective voltage. If an oscilloscope were plugged into a home receptacle, the scope would show that the top of the sine wave for the 120-V circuit actually crests at $120 \times \sqrt{2}$ or 169.7 V. This same discussion is true for 7.2 kV, 115 kV, 500 kV, etc. Per the definition of voltage in Sec. 02, the clearance tables in Sec. 23 apply to the rms voltage (e.g., 120 V) not the crest voltage (e.g., 169.7 V). The **NESC** does use the crest value when applying the formula for alternate clearances in Rule 232D and in other alternate clearance calculations in Sec. 23.

Values for DC, AC crest, and AC rms are outlined in Fig. 230-10.

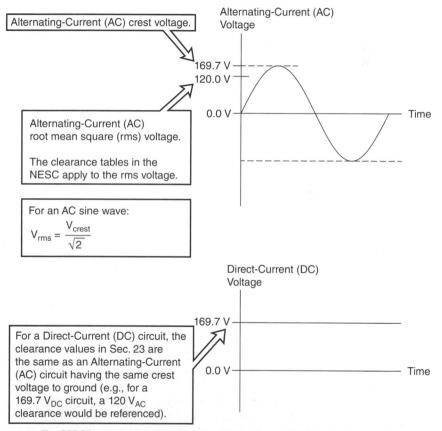

Fig. 230-10. Alternating-Current (AC) and Direct-Current (DC) circuits (Rule 230G).

For three-phase circuits, it is important to note the difference between phase-to-phase and phase-to-ground voltage. The phase-to-ground voltage is equal to the phase-to-phase voltage divided by the square root of 3. An example of phase-to-phase and phase-to-ground voltages is shown in Fig. 230-11.

Many of the clearance tables in Sec. 23 use phase-to-ground values; however, some tables that involve voltage between conductors will use phase-to-phase values.

230H. Constant-Current Circuits. Normal utility circuits are constant voltage and the current varies with the load connected to the circuit. Constant-current

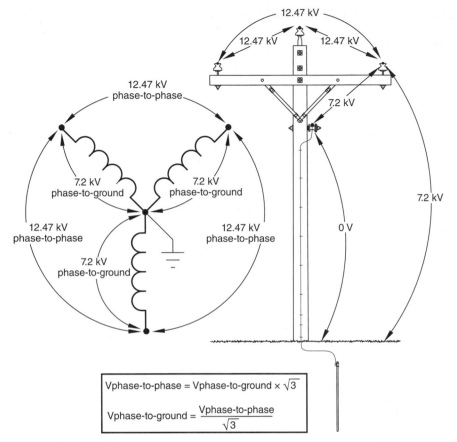

Fig. 230-11. Example of phase-to-phase and phase-to-ground voltages (Rule 230G).

circuits are just the opposite. They operate with a constant current, and the voltage varies with the load connected to the circuit. Constant-current circuits were once very common for street lighting and can still be found in use today. The voltage used to determine the clearance of constant-current circuit is the normal full-load voltage of the circuit. See Fig. 230-12.

230I. Maintenance of Clearances and Spacings.   This is a small rule with a big implication. Clearances and spacing must be maintained. Forever. Per Rule 010, the **NESC** applies during installation, operation, and maintenance of supply and communication lines, not just during the initial installation. Several changes can occur that force a utility to be active in maintaining clearance. One is a change in land use under the supply or communications line. Another is a change in structures (buildings, signs, etc.) under or adjacent to the supply or communications line. Another is excessive sag due to a major ice- or wind-storm. In each of these cases, it is the responsibility of the utility to maintain clearance. The note in Rule 230I that references Rule 013 is a reminder that when maintaining a line, the "grandfather clause" of Rule 013B may be applied and the inspection rules and work rules in the current **Code** edition must be applied per Rule 013C. Applying Rule 013B can help determine which edition of the **Code** is applicable. Applying Rule 013B does not excuse a utility from correcting a violation once the applicable

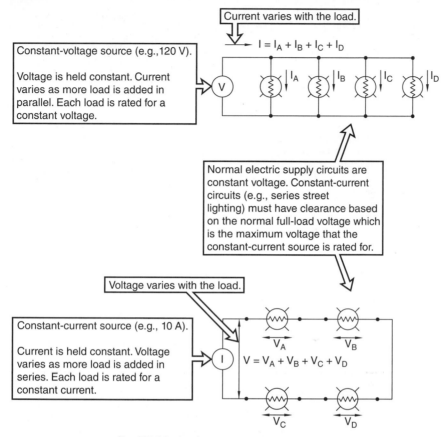

Current varies with the load.

$I = I_A + I_B + I_C + I_D$

Constant-voltage source (e.g.,120 V).

Voltage is held constant. Current varies as more load is added in parallel. Each load is rated for a constant voltage.

Normal electric supply circuits are constant voltage. Constant-current circuits (e.g., series street lighting) must have clearance based on the normal full-load voltage which is the maximum voltage that the constant-current source is rated for.

Voltage varies with the load.

Constant-current source (e.g., 10 A).

Current is held constant. Voltage varies as more load is added in series. Each load is rated for a constant current.

$V = V_A + V_B + V_C + V_D$

Fig. 230-12. Constant-current circuits (Rule 230H).

edition is determined. See Rule 013 for additional information. Examples of maintenance of clearances and spacings are shown in Fig. 230-13.

In the case of the excessive sag due to a major ice or wind storm, Rule 230I recognizes that clearance cannot be maintained during or after an abnormal event of this nature. There is some reasonable time period for storm damage work to be repaired; however, the **NESC** does not address the time period. This same discussion applies to an abnormal event such as a car hitting a power pole and knocking down the wires or reducing the clearance of the wires. In this instance, clearance cannot be maintained during or for some time after the accident. There is some reasonable time period for repairing damage to the line caused by the car hitting the pole; however, the **NESC** does not address the time period. Notably, the wording regarding abnormal events as described in Rule 230I does not include failures due to inadequate inspections or maintenance on the part of the utility. It is important that utilities be proactive to correct deteriorated conditions prior to failures that significantly increase the safety hazards (e.g., line drop).

A utility must correct a **Code** clearance violation, even when the line was built with the proper **Code** clearance. The **NESC** does not address who pays for the correction. The utility may consider billing a third party for the correction, but the utility is responsible for maintaining clearance or, in other words, fixing the

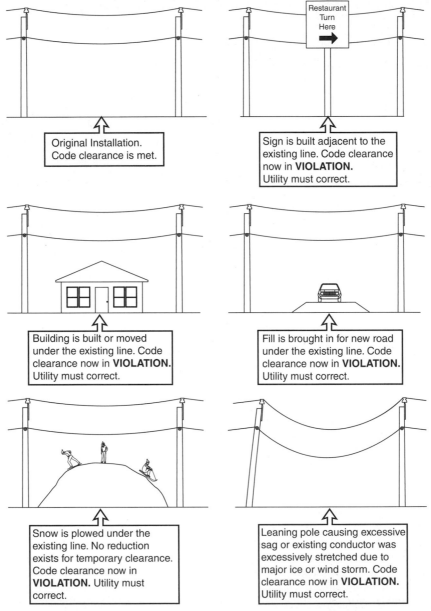

Fig. 230-13. Examples of maintenance of clearances and spacings (Rule 230I).

problem. Some utilities establish clearance values by using the **Code** clearance plus an adder. See Rule 010 for a discussion of clearance adders. Rule 230I states that utilities are responsible for correcting known noncompliant conditions. The line inspection requirements of Rule 214 provide a means to bring clearance issues to the utility's attention and known noncompliant conditions (violations) must be addressed per Rules 214A4 and 214A5, whichever is applicable.

231. CLEARANCES OF SUPPORTING STRUCTURES FROM OTHER OBJECTS

Rule 231 applies to supporting structures. The most common supporting structure is a pole, but since other structures exist (e.g., lattice towers) the **NESC** uses the term supporting structure, not pole. Rule 231 does not apply to conductors, only to the supporting structure. A simple title for this section could be "Where can I set my pole?"

Depending on which Rule is referenced (Rule 231A, 231B, or 231C), included with the supporting structure are the support arms, anchor guys, braces, and equipment. The clearance requirements of this section are between the nearest parts of the objects concerned.

231A. Fire Hydrants. Rule 231A is needed to provide the fire department crews adequate space to connect wrenches, hoses, and equipment to the fire hydrant. The rule for clearance of a supporting structure (including anchor guys and attached equipment) from a fire hydrant is outlined in Fig. 231-1.

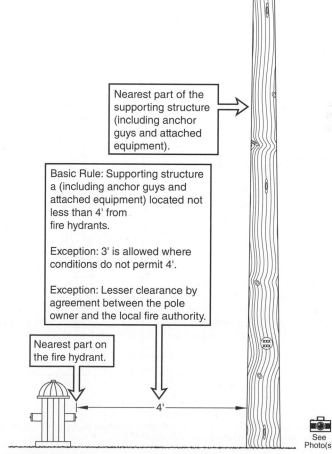

Nearest part of the supporting structure (including anchor guys and attached equipment).

Basic Rule: Supporting structure a (including anchor guys and attached equipment) located not less than 4' from fire hydrants.

Exception: 3' is allowed where conditions do not permit 4'.

Exception: Lesser clearance by agreement between the pole owner and the local fire authority.

Nearest part on the fire hydrant.

4'

See Photo(s)

Fig. 231-1. Clearance of a supporting structure (including anchor guys and attached equipment) from a fire hydrant (Rule 231A).

231B. Streets, Roads, and Highways. Rule 231B provides clearances between supporting structures (poles) and roads. Avoiding vehicle contact with poles can be just as important as maintaining clearance to energized conductors. Although not referenced in Rule 231B, Rules 217A and 217C relate to Rule 231B. The notes in Rules 217A1a, 217C2, and 231B regarding out-of-control vehicles help distinguish between an ordinary vehicle using the road and an out-of-control vehicle or a vehicle operating outside of established traveled ways that cannot be planned for.

Rule 231B1 applies to roads with curbs. No specific dimension is specified for the distance that a supporting structure (including support arms, anchor guys, braces, and attached equipment) must be behind the curb. The rules for clearance of supporting structures (including support arms, anchor guys, braces, and attached equipment) from roads that have curbs are outlined in Fig. 231-2.

Rule 231B2 applies to roads without curbs. No specific dimension is specified, just the requirement to locate the supporting structure (including support arms, anchor guys, braces, and attached equipment) "a sufficient distance from the roadway to avoid contact by ordinary vehicles using and located on the traveled way." The **NESC** has definitions in Sec. 02 for the terms shoulder, roadway, and traveled way to clarify the requirements in this rule. The rules for clearance of supporting structures from roads that do not have a curb are outlined in Fig. 231-3.

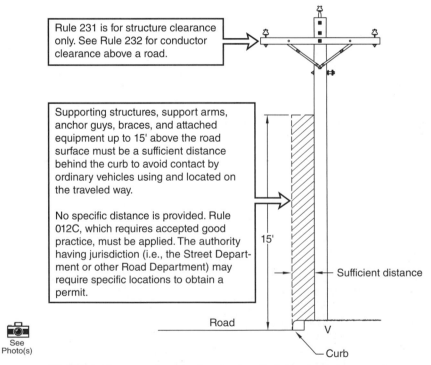

Rule 231 is for structure clearance only. See Rule 232 for conductor clearance above a road.

Supporting structures, support arms, anchor guys, braces, and attached equipment up to 15' above the road surface must be a sufficient distance behind the curb to avoid contact by ordinary vehicles using and located on the traveled way.

No specific distance is provided. Rule 012C, which requires accepted good practice, must be applied. The authority having jurisdiction (i.e., the Street Department or other Road Department) may require specific locations to obtain a permit.

15'

Sufficient distance

Road

V

Curb

See Photo(s)

Fig. 231-2. Clearance of a supporting structure (including support arms, anchor guys, braces, and attached equipment) from a curb (Rule 231B1).

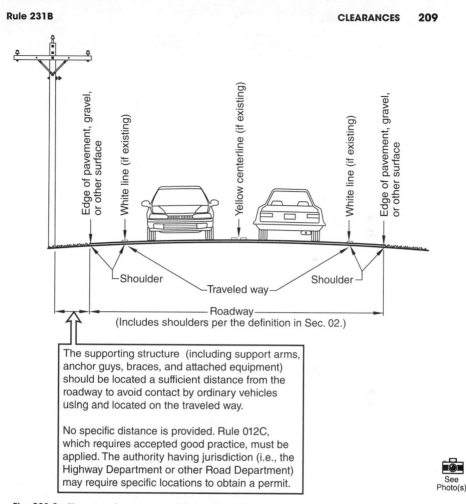

The supporting structure (including support arms, anchor guys, braces, and attached equipment) should be located a sufficient distance from the roadway to avoid contact by ordinary vehicles using and located on the traveled way.

No specific distance is provided. Rule 012C, which requires accepted good practice, must be applied. The authority having jurisdiction (i.e., the Highway Department or other Road Department) may require specific locations to obtain a permit.

See Photo(s)

Fig. 231-3. Clearance of a supporting structure (including support arms, anchor guys, braces, and attached equipment) from a roadway without a curb (Rule 231B2).

Rule 231B3 recognizes various utilities compete for narrow rights-of-way along roads, streets, and highways. Special cases may exist and must be resolved using accepted good practice for the conditions at hand. Protection of supporting structures in parking lots, alleys, and next to driveways is addressed in Rule 217A. Protection and marking of guys is addressed in Rule 217C.

Rule 231B4 recognizes that in many cases a state highway department, county road department, city, or other governmental authority may require a permit to place poles in a right of way. The permit may have special requirements for pole placement. The **NESC** does not reference the Americans with Disabilities Act (ADA). However, this act may apply to utilities when setting a pole in a sidewalk. It is important to work with the sidewalk permitting authority when considering placing poles, guys and anchors, etc., in a sidewalk. See Fig. 231-4.

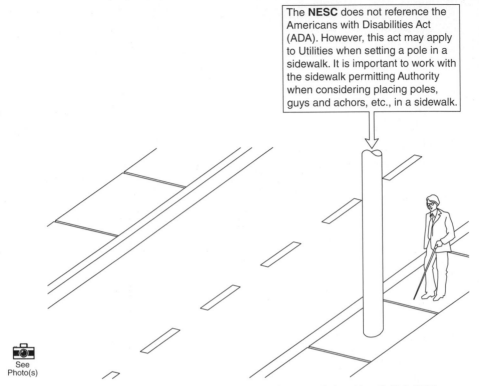

The **NESC** does not reference the Americans with Disabilities Act (ADA). However, this act may apply to Utilities when setting a pole in a sidewalk. It is important to work with the sidewalk permitting Authority when considering placing poles, guys and achors, etc., in a sidewalk.

See Photo(s)

Fig. 231-4. Absence of a Code Rule related to placing a pole in a sidewalk (Rule N/A).

231C. Railroad Tracks. Rule 231C applies to supporting structures (including support arms, anchor guys, braces, and attached equipment) on lines paralleling or crossing railroad tracks. Rule 231C2 and four exceptions to Rule 231C1 permit reductions to the basic clearance provided in Rule 231C1. Many times when applying for a permit to parallel or cross a railroad right-of-way, the railroad company will require greater clearances than listed in this rule. Rule 231C has a note that references Rule 234I, which covers clearance of wires, conductors, and cables to rail cars. Rule 232, which applies to conductor clearance over railroad tracks, should also be checked. Rule 231C only applies to structures, not conductors. The rules for clearance of supporting structures (including support arms, anchor guys, braces, and attached equipment) from railroad tracks are outlined in Fig. 231-5.

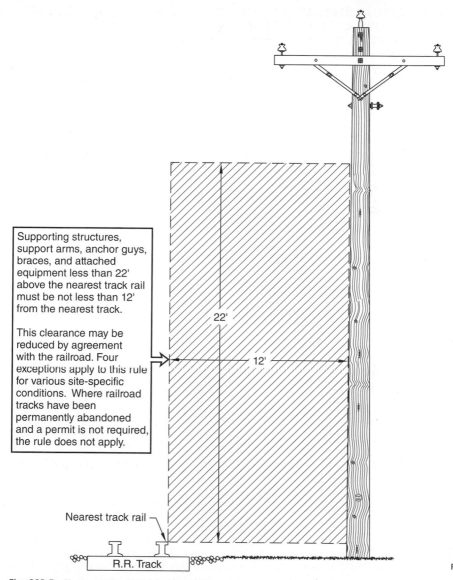

Supporting structures, support arms, anchor guys, braces, and attached equipment less than 22' above the nearest track rail must be not less than 12' from the nearest track.

This clearance may be reduced by agreement with the railroad. Four exceptions apply to this rule for various site-specific conditions. Where railroad tracks have been permanently abandoned and a permit is not required, the rule does not apply.

22'

12'

Nearest track rail

R.R. Track

See Photo(s)

Fig. 231-5. Clearance of a supporting structure (including support arms, anchor guys, braces, and attached equipment) from railroad tracks (Rule 231C).

232. VERTICAL CLEARANCES OF WIRES, CONDUCTORS, CABLES, AND EQUIPMENT ABOVEGROUND, ROADWAY, RAIL, OR WATER SURFACES

Rule 232 is probably referenced more than any other rule in the NESC. NESC Table 232-1 is probably referenced more than any other table. One problem that can occur is referencing NESC Table 232-1 without reading and understanding all the related rules that apply to the table. For example, Rules 230A through 230I define general clearance requirements and types of conductors used in NESC Table 232-1. Rule 232A describes the conductor temperature and loading (sag) conditions that apply to NESC Table 232-1. Just as important as the related rules is the use and understanding of a sag and tension chart. The basics of how to use a sag and tension chart and a sample sag and tension chart are provided at the beginning of Sec. 23 of this Handbook.

232A. Application.   Rule 232A states the temperature and loading conditions that apply to the clearances in NESC Table 232-1. Three conditions are specified, and the condition that produces the largest final sag must be used. The conditions are:

- 120°F, no wind displacement, final sag.
- Greater than 120°F, if so designed, no wind displacement, final sag.
- 32°F, no wind displacement, final sag, with ice from the Zone specified in Rule 230B (e.g., Zone 2 requires 0.25 in of ice).

The conditions above are outlined on the sample sag and tension chart at the beginning of Sec. 23.

The conductor conditions specified in Rule 232A are used to check clearance, not conductor tension. Conductor tension limits for open supply conductors are provided in Rule 261H.

The temperatures listed in Rule 232A are the conductor temperatures, not the ambient air temperature. See the beginning of Sec. 23 for additional information on conductor versus air temperature. The term loading in Rule 232A refers to the physical loads on the conductors in the form of ice and wind, not electrical loads in kilowatts or amperes. The conductor temperature and loading conditions for measuring vertical clearance are outlined in Fig. 232-1.

Using the sample sag and tension chart at the beginning of Sec. 23, 212°F, no wind, no ice, final tension, produces the largest sag (6.84 ft) of the conditions that need to be checked in Rule 232A. If the line were not designed to operate at temperatures above 120°F, then the 120°F, no wind, no ice, final value (5.61 ft) would be used as it is larger than the 32°F, no wind, 0.25 in ice, final value (4.49 ft). This example is not true for every case. Some conductors will have larger sags with ice, others will have larger sags with high temperatures. The wire size, span length, amount of ice, and design tension will all affect which condition produces the largest final sag. Examples that reinforce the fact that Code clearance for overhead lines is based on the largest final sag condition are shown in Figs. 232-2 and 232-3.

The majority of the discussion and examples in Sec. 23 revolve around the sample 1/0 ACSR sag and tension chart at the beginning of Sec. 23. Sag and tension charts can also be produced for secondary duplex, triplex, and quadruplex, communication cables on messengers (including phone, cable TV, and

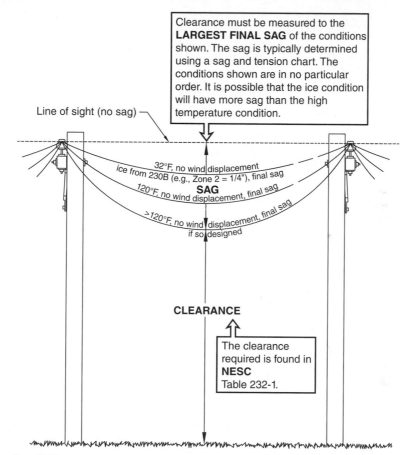

Line of sight (no sag) ⎯

Clearance must be measured to the
LARGEST FINAL SAG of the conditions
shown. The sag is typically determined
using a sag and tension chart. The
conditions shown are in no particular
order. It is possible that the ice condition
will have more sag than the high
temperature condition.

ice from 230B (e.g., Zone 2 = 1/4"), final sag

32°F, no wind displacement

SAG

120°F, no wind displacement, final sag

>120°F, no wind displacement, final sag
if so designed

CLEARANCE

The clearance
required is found in
NESC
Table 232-1.

Fig. 232-1. Conductor temperature and loading conditions for measuring vertical clearance
(Rule 232A).

fiber-optic), self-supporting all-dielectric fiber-optic cables, overhead ground wires
(static wires), and virtually any other type of overhead power and communication
conductor or cable.

The exception to Rule 232A recognizes that electric railroad and trolley car conductors experience different conditions than typical power and communication
conductors. Finally, a note to Rule 232A reminds us that the phase and neutral
conductor may not operate at the same temperatures. The neutral conductor may
carry less current due to phase balance and neutral current cancellation or due to
the fact that the earth and a communication messenger on a joint use pole are in
parallel with a multigrounded neutral.

Prior to 1990, a 60°F sag was used to determine clearance plus additional clearance was required for long spans. In 1990, the **Code** clearance rules experienced
major revisions and the method to determine **Code** clearance was changed from
using the 60°F sag condition to using the maximum sag conditions in Rule 232A.
This change partly came about due to the fact that computer programs became

Example:
• 300' span of 1/0 ACSR conductor.
• Assume the calculated ruling span is 300'.
• The clearance measurement is taken on a 50°F day; the conductor has a light electrical load all year long.
• Assume the conductor temperature is 60°F.
• Assume the maximum temperature the line is designed to operate at is 120°F.
• The midspan clearance measurement to a phase wire of 12.47/7.2 kV, 3Ø 4W effectively grounded circuit is 20'.
• Assume the line was strung using the sag and tension chart at the beginning of Sec. 23.
• Assume the line is at the final sag condition.

Largest final sag the line is designed to operate. In this example, the sag at 120°F, final.

Sag for a 60° conductor temperature.

Line of site (no sag)

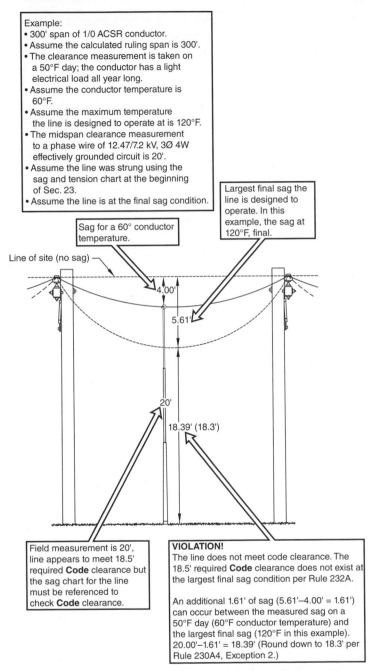

4.00'

5.61'

20'

18.39' (18.3')

Field measurement is 20', line appears to meet 18.5' required **Code** clearance but the sag chart for the line must be referenced to check **Code** clearance.

VIOLATION!
The line does not meet code clearance. The 18.5' required **Code** clearance does not exist at the largest final sag condition per Rule 232A.

An additional 1.61' of sag (5.61'–4.00' = 1.61') can occur between the measured sag on a 50°F day (60°F conductor temperature) and the largest final sag (120°F in this example). 20.00'–1.61' = 18.39' (Round down to 18.3' per Rule 230A4, Exception 2.)

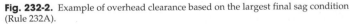

Fig. 232-2. Example of overhead clearance based on the largest final sag condition (Rule 232A).

Example:
• 400' span of 1/0 ACSR conductor.
• The 400' span is one span in a series of five spans from deadend to deadend.
• Assume the calculated ruling span is 300'.
• The clearance measurement is taken on a 50°F day; the conductor has a light electrical load all year long.
• Assume conductor temperature is 60°F.
• Assume the maximum temperature the line is designed to operate at is 120°F.
• The quarter span clearance measurement to phase wire of a 12.47/7.2 kV, 3Ø 4W effectively grounded circuit is 20'.
• The quarter span rise is 5'.
• Assume the line was strung using the sag and tension chart at the beginning of Sec. 23.
• Assume the line is at the final sag conditions.

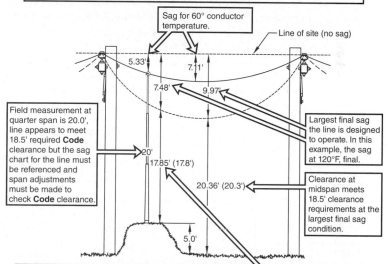

Sag for 60° conductor temperature.

Line of site (no sag)

5.33'

7.11'

7.48'

9.97'

Field measurement at quarter span is 20.0', line appears to meet 18.5' required **Code** clearance but the sag chart for the line must be referenced and span adjustments must be made to check **Code** clearance.

20'

17.85' (17.8')

Largest final sag the line is designed to operate. In this example, the sag at 120°F, final.

Clearance at midspan meets 18.5' clearance requirements at the largest final sag condition.

20.36' (20.3')

5.0'

VIOLATION!
The line does not meet **Code** clearance. The 18.5' required **Code** clearance does not exist at the quarter span location at the largest final condition per Rule 232A.

The sag on a 50° day (60° conductor temperature) for a 400' span at midspan is:

$$\left(\frac{400}{300}\right)^2 \times 4' = 7.11'$$

The sag at the quarter span is 75% of the midspan sag per Fig. 23-3.
$7.11' \times 0.75 = 5.33'$

The largest final sag (at 120°F) in this example for a 400' span at midspan is:

$$\left(\frac{400}{300}\right)^2 \times 5.61' = 9.97'$$

The sag at the quarter span is 75% of the midspan sag per Fig. 23-3.
$9.97 \times 0.75 = 7.48'$

An additional 2.15' of sag (7.48'–5.33') can occur between the measured sag on a 50°F day (60°F conductor temperature) and the largest final sag (120°F in this example) at the quarter span location. 20.00'–2.15' = 17.85' (Round down to 17.8' per Rule 230A4, Exception 1.)

Fig. 232-3. Example of overhead clearance based on the largest final sag condition (Rule 232A).

available to generate detailed sag and tension charts like the sample at the beginning of Sec. 23 of this Handbook. **NESC** Appendix A provides additional information on the uniform system of clearances adopted in the 1990 **NESC**.

232B. Clearance of Wires, Conductors, Cables, Equipment, and Support Arms Mounted on Supporting Structures.   Rule 232B1 references the most commonly used table in the **NESC**, Table 232-1. Table 232-1 cannot be used without first reading and understanding the temperature and loading conditions in Rule 232A and without understanding the application of a sag and tension chart.

NESC Table 232-1 covers vertical clearance of wires, conductors, and cables above ground, roadway, rail, and water surfaces. **NESC** Table 232-2 covers unguarded rigid live parts of equipment. The basic difference between these two tables is that Table 232-1 addresses items that vary due to sag or may be displaced by wind or other changing conditions, and Table 232-2 addresses items that are fixed or rigid and do not vary with sag. Equipment cases fall under Table 232-2. Secondary drip loops are included in Table 232-1 even though they do not vary much with sag.

A note in Rule 232B1 provides additional information for clearances to irrigation equipment. See Rule 234C for a discussion and a figure related to irrigation equipment.

An example of how clearance values are determined in **NESC** Tables 232-1 and 232-2 is shown in Fig. 232-4.

One of the main differences between **NESC** Table 232-1 (conductors) and **NESC** Table 232-2 (rigid parts) is that the mechanical and electrical component of

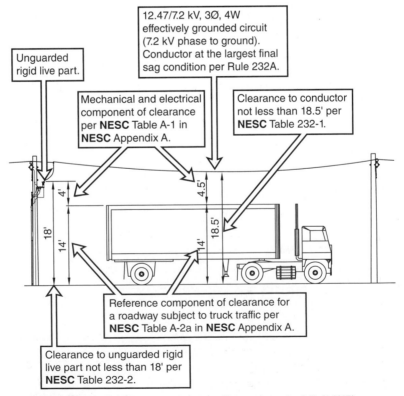

Fig. 232-4. Example of how clearance values are determined (Rule 232B).

clearance is typically reduced by 0.5 ft for rigid parts as they are not subject to sag variations. This is evident when comparing similar clearance conditions in **NESC** Tables 232-1 and 232-2.

NESC Tables 232-1 and 232-2 cover multiple conditions under the line, but the tables cannot cover every specific condition. For conditions not covered, Rule 012C, which requires accepted good practice, must be applied. **NESC** Appendix A explains how the clearances on the table were determined. Footnote 26 of **NESC** Table 232-1 and Footnote 3 of **NESC** Table 232-2 explain how to increase Code clearance for vehicles in excess of 14 ft in height. For example, if the truck in Fig. 232-4 were a 22-ft-high mining truck on an industrial site, the engineer or designer of the line would be required to use the 4.5-ft (conductor) and 4.0-ft (rigid part) mechanical and electrical component of clearance for 12.47/7.2 kV (Rules 232C and 232D discuss when additional clearance is required for higher voltages). The 22-ft height must be substituted for the 14-ft reference component. See Fig. 232-5.

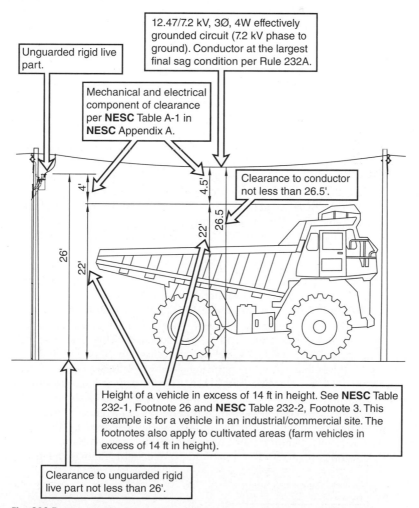

Fig. 232-5. Example of how to calculate clearance for vehicles in excess of 14 ft in height (Rule 232).

Footnote 26 of **NESC** Table 232-1 and Footnote 3 of **NESC** Table 232-2 are also applicable to farming vehicles in excess of 14 ft in height. Footnote 26 of **NESC** Table 232-1 and Footnote 3 of **NESC** Table 232-2 can also be used to calculate the clearance for house moves or other oversized vehicles that exceed 14 ft in height but are traveling on roadways or highways that normally have trucks not exceeding 14 ft. Footnote 26 of **NESC** Table 232-1 and Footnote 3 of **NESC** Table 232-2 apply to communication lines and supply lines. The mechanical and electrical component of the clearance from **NESC** Table A-1 in **NESC** Appendix A is less for a communication line compared to a supply line. The **NESC** does not address who pays the construction costs for increasing power line clearance due to vehicles in excess of 14 ft in height. OSHA Standard 1910.333 has a 4-ft clearance requirement (up to 50 kV) for vehicular and mechanical equipment in transit with its structure lowered. This OSHA requirement, which addresses safe behavior, is in addition to the **NESC** requirements, which addresses the system dimensions. Mobile cranes used in the vicinity of power lines are addressed in the OSHA regulations in OSHA 1926.1408 which is part of OSHA 1926, Subpart CC. While mobile cranes are not subject to the **NESC**, it is good for utilities to be aware of the issues associated with cranes for their lines in construction areas.

The **NESC** clearance tables provide a wealth of information. They must be carefully reviewed before selecting the proper value from the table. Many of the clearances in **NESC** Table 232-1 are the same except that different footnotes apply to each. An explanation of how **NESC** Table 232-1 is formatted is shown in Fig. 232-6.

All of the clearances in **NESC** Table 232-1 are "not less than" values. See Rule 010 for a discussion of using **Code** clearance values plus an adder. Using the **Code** clearance plus an adder can help meet and maintain clearance over the life of an installation as required in Rule 230I. If the authority issuing an occupancy permit (e.g., a railroad company, highway department, etc.) requires greater clearance than the **NESC**, the greater clearance required in the occupancy permit will take precedence over the **NESC** clearance values. The phase conductor, neutral conductor, secondary cable (0 to 750 V), and insulated communication cable clearances for the 10 land use categories in **NESC** Table 232-1 are shown in Figs. 232-7 through 232-16.

NESC Table 232-1 also addresses trolley and electrified railroad contact conductors and associated span or messenger wires. See Rule 225 for a discussion and a figure related to electric railway construction and trolley lines.

Rules 232B2 and 232B3 both reference **NESC** Table 232-2. **NESC** Table 232-2 covers vertical clearance to equipment cases and unguarded rigid live parts above ground, roadway, and water surfaces. The land use categories are similar to **NESC** Table 232-1. The clearances in **NESC** Table 232-2 are commonly $1/2$ ft less than the corresponding clearance in **NESC** Table 232-1 as the rigid parts are not subject to variations in sag. Examples of clearance to equipment cases and ungrounded rigid live parts are shown in Figs. 232-17 and 232-18.

Rule 232B4 includes vertical clearance requirements for effectively grounded and ungrounded conductive parts of luminaires (light fixtures). Modern luminaires are effectively grounded and operate on a constant voltage system. An example of clearance to an effectively grounded street light is shown in Fig. 232-19.

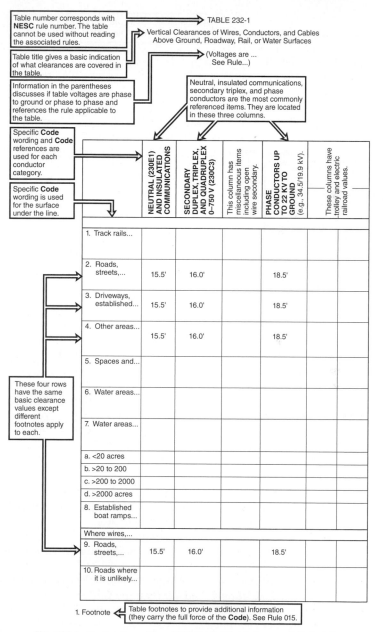

Table number corresponds with **NESC** rule number. The table cannot be used without reading the associated rules. → TABLE 232-1

→ Vertical Clearances of Wires, Conductors, and Cables Above Ground, Roadway, Rail, or Water Surfaces

Table title gives a basic indication of what clearances are covered in the table. → (Voltages are ... See Rule...)

Information in the parentheses discusses if table voltages are phase to ground or phase to phase and references the rule applicable to the table.

Neutral, insulated communications, secondary triplex, and phase conductors are the most commonly referenced items. They are located in these three columns.

Specific **Code** wording and **Code** references are used for each conductor category.

Specific **Code** wording is used for the surface under the line.

		NEUTRAL (230E1) AND INSULATED COMMUNICATIONS	SECONDARY DUPLEX, TRIPLEX, AND QUADRUPLEX 0–750 V (230C3)	This column has miscellaneous items including open wire secondary.	PHASE CONDUCTORS UP TO 22 KV TO GROUND (e.g., 34.5/19.9 kV).	These columns have trolley and electric railroad values.	
1.	Track rails...						
2.	Roads, streets,...	15.5'	16.0'		18.5'		
3.	Driveways, established...	15.5'	16.0'		18.5'		
4.	Other areas...	15.5'	16.0'		18.5'		
5.	Spaces and...						
6.	Water areas...						
7.	Water areas...						
a.	<20 acres						
b.	>20 to 200						
c.	>200 to 2000						
d.	>2000 acres						
8.	Established boat ramps...						
	Where wires,...						
9.	Roads, streets,...	15.5'	16.0'		18.5'		
10.	Roads where it is unlikely...						

These four rows have the same basic clearance values except different footnotes apply to each.

1. Footnote ← Table footnotes to provide additional information (they carry the full force of the **Code**). See Rule 015.

Fig. 232-6. Explanation of how **NESC** Table 231-1 is formatted (Rule 232).

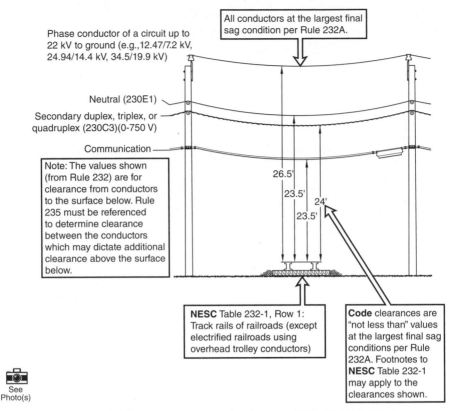

Phase conductor of a circuit up to 22 kV to ground (e.g.,12.47/7.2 kV, 24.94/14.4 kV, 34.5/19.9 kV)

All conductors at the largest final sag condition per Rule 232A.

Neutral (230E1)

Secondary duplex, triplex, or quadruplex (230C3)(0-750 V)

Communication

Note: The values shown (from Rule 232) are for clearance from conductors to the surface below. Rule 235 must be referenced to determine clearance between the conductors which may dictate additional clearance above the surface below.

26.5'

23.5'

24'

23.5'

NESC Table 232-1, Row 1: Track rails of railroads (except electrified railroads using overhead trolley conductors)

Code clearances are "not less than" values at the largest final sag conditions per Rule 232A. Footnotes to NESC Table 232-1 may apply to the clearances shown.

See Photo(s)

Fig. 232-7. Common clearance values from NESC Table 232-1 (Rule 232B1).

Per the title of Rule 232 and the titles of **NESC** Tables 232-1 and 232-2, the clearances specified in Rule 232 are vertical clearances. Per the note at the beginning of Rule 232B, Rule 232B does not provide horizontal or diagonal clearances to land surfaces. Rule 234 provides horizontal clearances to buildings and other structures but not to a land surface. To determine the horizontal or diagonal clearance to a land surface, Rule 012C, which requires accepted good practice, must be applied. See Fig. 232-20.

The **Code** clearances in Rule 232 and clearance values elsewhere in the **Code** are "not less than" values. Clearance greater than the **Code** requires is certainly acceptable and may be needed to maintain the "not less than" clearance over the life of the line, but greater clearance is not required by the **NESC**. Certain agencies like the State Highway Department, railroad company, U.S. Army Corps of Engineers (water areas), or other agency issuing a crossing or occupancy permit may require greater clearance than the **NESC**. These agencies are considered the authority having jurisdiction. See Fig. 232-21.

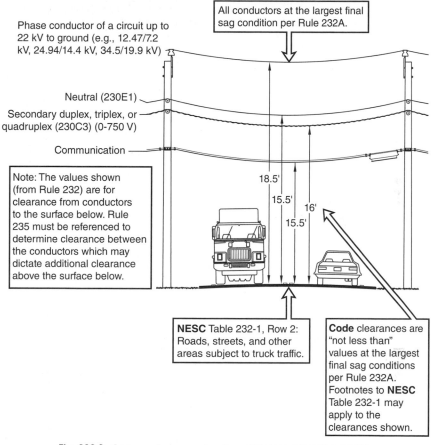

Phase conductor of a circuit up to 22 kV to ground (e.g., 12.47/7.2 kV, 24.94/14.4 kV, 34.5/19.9 kV)

All conductors at the largest final sag condition per Rule 232A.

Neutral (230E1)

Secondary duplex, triplex, or quadruplex (230C3) (0-750 V)

Communication

Note: The values shown (from Rule 232) are for clearance from conductors to the surface below. Rule 235 must be referenced to determine clearance between the conductors which may dictate additional clearance above the surface below.

18.5'
15.5'
16'
15.5'

NESC Table 232-1, Row 2: Roads, streets, and other areas subject to truck traffic.

Code clearances are "not less than" values at the largest final sag conditions per Rule 232A. Footnotes to NESC Table 232-1 may apply to the clearances shown.

Fig. 232-8. Common clearance values from NESC Table 232-1 (Rule 232B1).

See Photo(s)

The figures in this Handbook typically show the phase wire at the pole top position and the neutral wire at a lower position on the pole or on the same cross-sarm as the phase wire. It is not a Code clearance violation to position the neutral wire at the top of the pole and the phase wire below the neutral on a crossarm or on the pole as long as the appropriate "not less than" clearances for each conductor in Rule 232 is met. This construction method is sometimes done in high lightning areas and is sometimes referred to as "neutral high" construction. Communication workers installing communication cables under the bottom phase wire need to be able to recognize the differences between a neutral wire and a phase wire. See Fig. 232-22.

There is one agency that gets involved if too much clearance exists above-ground or near an airport. That agency is the Federal Aviation Administration (FAA). The NESC does not address placing marker balls on conductors or

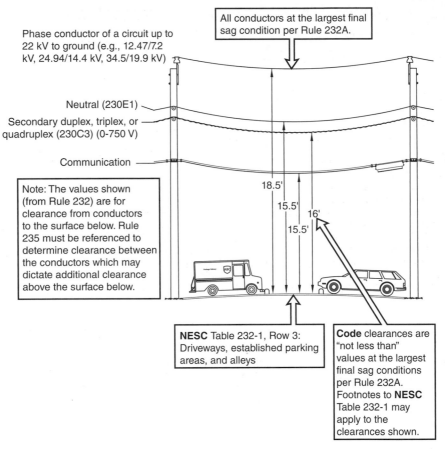

Fig. 232-9. Common clearance values from NESC Table 232-1 (Rule 232B1).

marking of tall structures. FAA Advisory Circular AC 70/7460-1K and AC 70/7460-2K address obstruction marking and lighting for structures including overhead power lines. See Fig. 232-23.

232C. Additional Clearances for Wires, Conductors, Cables, and Unguarded Rigid Live Parts of Equipment. The clearance values in Tables 232-1 and 232-2 must be increased for the following reasons:

- The voltage of the line exceeds 22 kV.
    - ✓ 22 kV to ground for an effectively grounded circuit. The wording in parentheses under the titles of NESC Tables 232-1 and 232-2 permits effectively grounded circuits and those other circuits where all ground faults are cleared by promptly de-energizing the faulted section, both initially and following subsequent breaker operations.

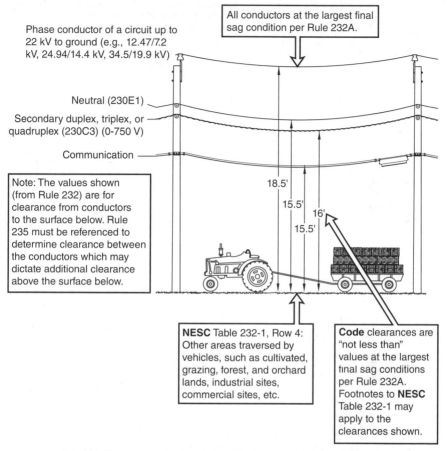

Fig. 232-10. Common clearance values from NESC Table 232-1 (Rule 232B1).

✓ 22 kV phase-to-phase for an ungrounded (e.g., delta) circuit.

✓ The maximum operating voltage, typically 1.05 times the nominal operating voltage, must be used for lines over 50 kV. ANSI C84.1, American National Standard for Electric Power Systems and Equipment—Voltage Ratings (60 Hz), provides a table listing nominal voltage levels and the corresponding maximum voltage level.

✓ The clearance adder is 0.4 in per kilovolt in excess of 22 kV.

• The voltage of the line exceeds 50 kV and the line is above an elevation of 3300 ft.

   ✓ 50 kV to ground for an effectively grounded circuit. The wording in parentheses under the titles of NESC Tables 232-1 and 232-2 permits effectively grounded circuits and those other circuits where all ground faults are

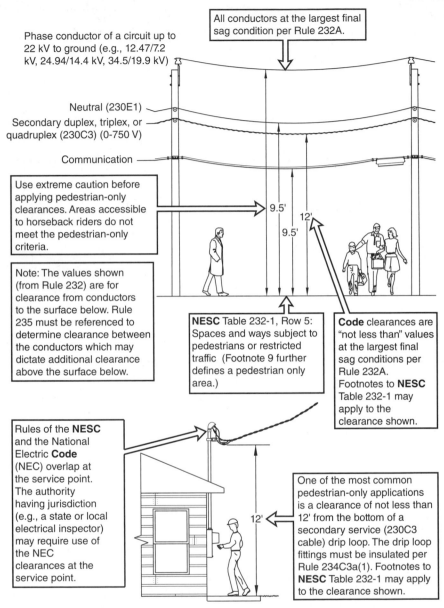

Phase conductor of a circuit up to 22 kV to ground (e.g., 12.47/7.2 kV, 24.94/14.4 kV, 34.5/19.9 kV)

All conductors at the largest final sag condition per Rule 232A.

Neutral (230E1)

Secondary duplex, triplex, or quadruplex (230C3) (0-750 V)

Communication

Use extreme caution before applying pedestrian-only clearances. Areas accessible to horseback riders do not meet the pedestrian-only criteria.

9.5'   12'   9.5'

Note: The values shown (from Rule 232) are for clearance from conductors to the surface below. Rule 235 must be referenced to determine clearance between the conductors which may dictate additional clearance above the surface below.

NESC Table 232-1, Row 5: Spaces and ways subject to pedestrians or restricted traffic (Footnote 9 further defines a pedestrian only area.)

Code clearances are "not less than" values at the largest final sag conditions per Rule 232A. Footnotes to NESC Table 232-1 may apply to the clearance shown.

Rules of the NESC and the National Electric Code (NEC) overlap at the service point. The authority having jurisdiction (e.g., a state or local electrical inspector) may require use of the NEC clearances at the service point.

One of the most common pedestrian-only applications is a clearance of not less than 12' from the bottom of a secondary service (230C3 cable) drip loop. The drip loop fittings must be insulated per Rule 234C3a(1). Footnotes to NESC Table 232-1 may apply to the clearance shown.

12'

See Photo(s)

Fig. 232-11. Common clearance values from NESC Table 232-1 (Rule 232B1).

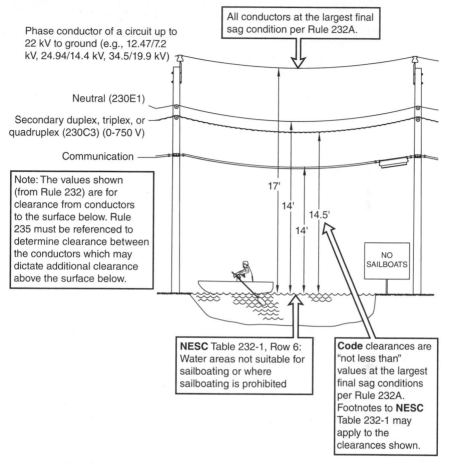

Fig. 232-12. Common clearance values from NESC Table 232-1 (Rule 232B1).

cleared by promptly de-energizing the faulted section, both initially and following subsequent breaker operations.

✓ 50 kV phase-to-phase for an ungrounded (e.g., delta) circuit.

✓ A clearance adder of 3 percent (3%) for each 1000 ft in excess of 3300 ft of elevation. Assuming the 3 percent increase is applied in 1000-ft blocks (or steps), 3 percent is required for altitudes over 3300 ft to 4300 ft, 6 percent is required for altitudes over 4300 ft to 5300 ft, 9 percent is required for altitudes over 5300 ft to 6300 ft, etc. The percent increase is applied to the additional clearance required for lines above 22 kV, not to the total clearance.

✓ No increase in clearance is required for elevations above 3300 ft if the line is rated 50 kV or less.

Phase conductor of a circuit up to 22 kV to ground (e.g., 12.47/7.2 kV, 24.94/14.4 kV, 34.5/19.9 kV)

All conductors at the largest final sag condition per Rule 232A.

Neutral (230E1)

Secondary duplex, triplex, or quadruplex (230C3) (0-750 V)

Communication

Note: The values shown (from Rule 232) are for clearance from conductors to the surface below. Rule 235 must be referenced to determine clearance between the conductors which may dictate additional clearance above the surface below.

Code clearances are "not less than" values at the largest final sag conditions per Rule 232A. Footnotes to **NESC** Table 232-1 may apply to the clearances shown.

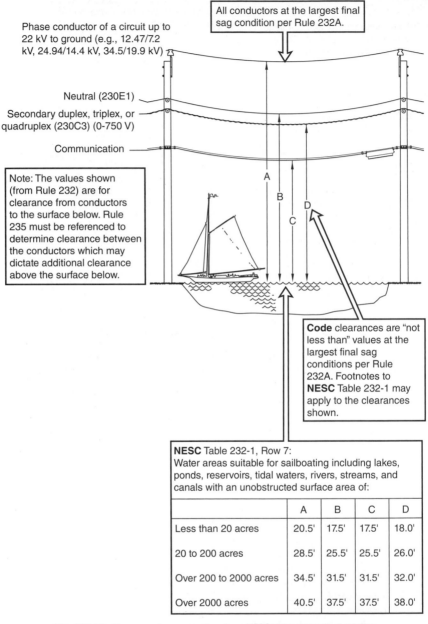

NESC Table 232-1, Row 7:
Water areas suitable for sailboating including lakes, ponds, reservoirs, tidal waters, rivers, streams, and canals with an unobstructed surface area of:

	A	B	C	D
Less than 20 acres	20.5'	17.5'	17.5'	18.0'
20 to 200 acres	28.5'	25.5'	25.5'	26.0'
Over 200 to 2000 acres	34.5'	31.5'	31.5'	32.0'
Over 2000 acres	40.5'	37.5'	37.5'	38.0'

Fig. 232-13. Common clearance values from **NESC** Table 232-1 (Rule 232B1).

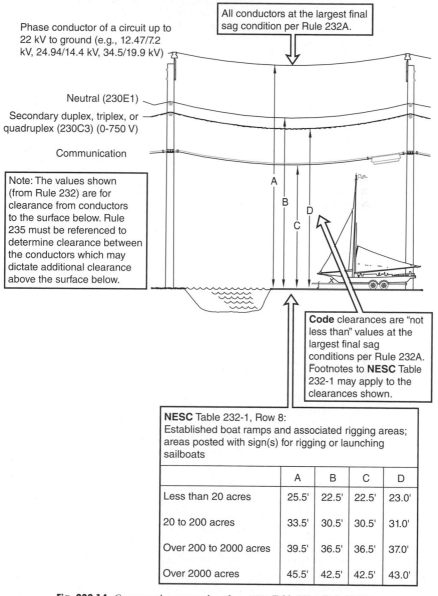

Phase conductor of a circuit up to 22 kV to ground (e.g., 12.47/7.2 kV, 24.94/14.4 kV, 34.5/19.9 kV)

All conductors at the largest final sag condition per Rule 232A.

Neutral (230E1)

Secondary duplex, triplex, or quadruplex (230C3) (0-750 V)

Communication

Note: The values shown (from Rule 232) are for clearance from conductors to the surface below. Rule 235 must be referenced to determine clearance between the conductors which may dictate additional clearance above the surface below.

A
B
C
D

Code clearances are "not less than" values at the largest final sag conditions per Rule 232A. Footnotes to **NESC** Table 232-1 may apply to the clearances shown.

NESC Table 232-1, Row 8:
Established boat ramps and associated rigging areas; areas posted with sign(s) for rigging or launching sailboats

	A	B	C	D
Less than 20 acres	25.5'	22.5'	22.5'	23.0'
20 to 200 acres	33.5'	30.5'	30.5'	31.0'
Over 200 to 2000 acres	39.5'	36.5'	36.5'	37.0'
Over 2000 acres	45.5'	42.5'	42.5'	43.0'

See Photo(s)

Fig. 232-14. Common clearance values from NESC Table 232-1 (Rule 232B1).

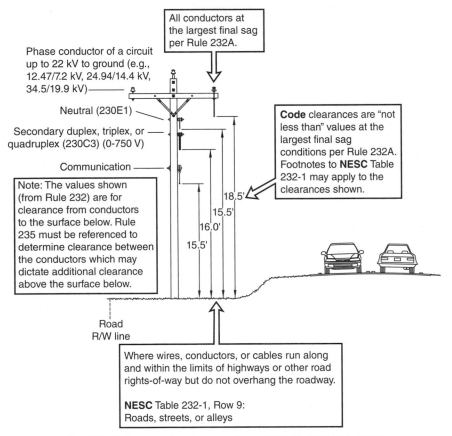

Fig. 232-15. Common clearance values from NESC Table 232-1 (Rule 232B1).

- For voltages exceeding 98 kV AC to ground, additional clearance or other means are required to limit electrostatic field effects to 5 mA or less.
- For voltages over 470 kV, the clearance must be determined using the formulas in Rule 232D. Rule 232D can also be used as an alternate clearance method for lines over 98 kV AC to ground or 189 kV DC to ground.
- The "not less than" rounding requirements of Rule 230A4 apply.

An example of an additional clearance calculation is shown in Fig. 232-24.

Rule 232C1c requires that lines exceeding 98 kV AC to ground have additional clearance or other means to reduce electrostatic field effects to 5 mA or less. High-voltage transmission lines can induce a voltage and therefore induce currents into metal objects like a truck or railroad car under the line. The magnitude of induced currents is strongly related to the configuration of the transmission line phases, the operating voltage, the balance of voltage among phases, and line height.

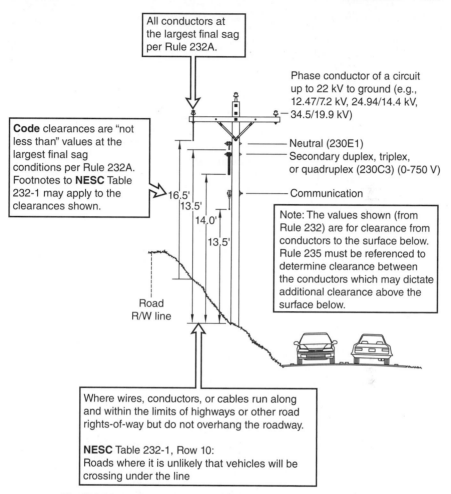

All conductors at the largest final sag per Rule 232A.

Phase conductor of a circuit up to 22 kV to ground (e.g., 12.47/7.2 kV, 24.94/14.4 kV, 34.5/19.9 kV)

Code clearances are "not less than" values at the largest final sag conditions per Rule 232A. Footnotes to **NESC** Table 232-1 may apply to the clearances shown.

Neutral (230E1)

Secondary duplex, triplex, or quadruplex (230C3) (0-750 V)

16.5'
13.5'
14.0'
13.5'

Communication

Note: The values shown (from Rule 232) are for clearance from conductors to the surface below. Rule 235 must be referenced to determine clearance between the conductors which may dictate additional clearance above the surface below.

Road R/W line

Where wires, conductors, or cables run along and within the limits of highways or other road rights-of-way but do not overhang the roadway.

NESC Table 232-1, Row 10:
Roads where it is unlikely that vehicles will be crossing under the line

Fig. 232-16. Common clearance values from **NESC** Table 232-1 (Rule 232B1).

The size of the equipment under the line should also be considered (e.g., tractor trailers near truck stops or recreational vehicles near campgrounds), as well as the orientation of the equipment (i.e., is the roadway parallel to or perpendicular to the line?). The average adult human body can detect an electrical shock at currents around 2 mA. Currents above 5 mA can cause pain and can be harmful to the body. Since no specific clearance adders or other methods are provided, Rule 012C, which requires accepted good practice, must be applied. Accepted good practice in this case may be an engineering analysis or field testing. The rules related to limiting electrostatic field effects to 5 mA are outlined in Fig. 232-25.

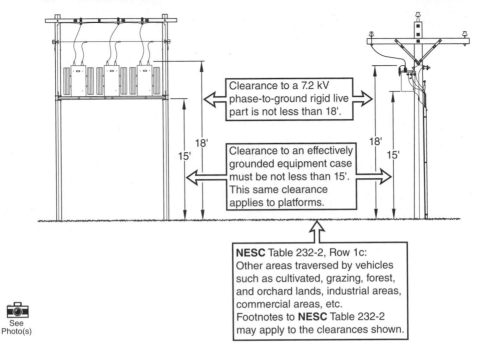

Clearance to a 7.2 kV phase-to-ground rigid live part is not less than 18'.

Clearance to an effectively grounded equipment case must be not less than 15'. This same clearance applies to platforms.

18'
15'
18'
15'

NESC Table 232-2, Row 1c:
Other areas traversed by vehicles such as cultivated, grazing, forest, and orchard lands, industrial areas, commercial areas, etc.
Footnotes to **NESC** Table 232-2 may apply to the clearances shown.

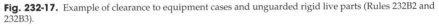

See Photo(s)

Fig. 232-17. Example of clearance to equipment cases and unguarded rigid live parts (Rules 232B2 and 232B3).

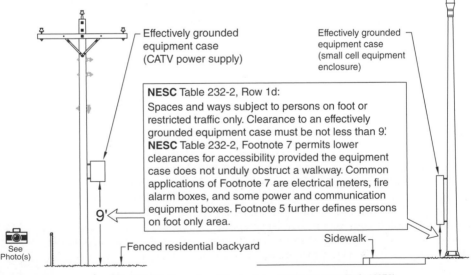

Effectively grounded equipment case (CATV power supply)

Effectively grounded equipment case (small cell equipment enclosure)

NESC Table 232-2, Row 1d:
Spaces and ways subject to persons on foot or restricted traffic only. Clearance to an effectively grounded equipment case must be not less than 9'.
NESC Table 232-2, Footnote 7 permits lower clearances for accessibility provided the equipment case does not unduly obstruct a walkway. Common applications of Footnote 7 are electrical meters, fire alarm boxes, and some power and communication equipment boxes. Footnote 5 further defines persons on foot only area.

9'

Fenced residential backyard

Sidewalk

See Photo(s)

Fig. 232-18. Examples of clearance to equipment cases (Rule 232B3).

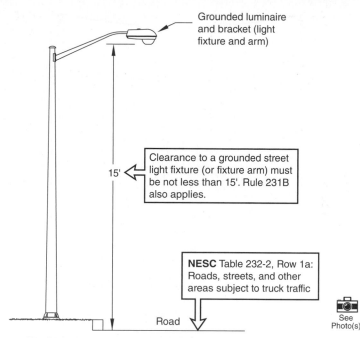

Fig. 232-19. Example of clearance to street lighting (Rule 232B4).

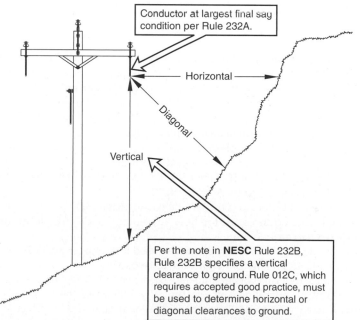

Fig. 232-20. Vertical clearance to land surface (Rule 232).

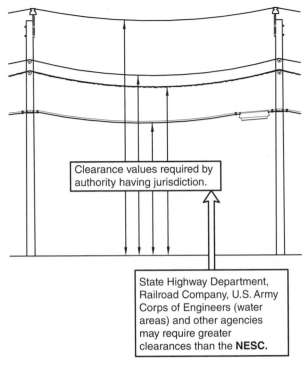

Clearance values required by authority having jurisdiction.

State Highway Department, Railroad Company, U.S. Army Corps of Engineers (water areas) and other agencies may require greater clearances than the **NESC.**

Fig. 232-21. Authority having jurisdiction (Rule 232).

232D. Alternate Clearances for Voltages Exceeding 98 kV AC to Ground or 139 kV DC to Ground.   The use of Rule 232D will permit reduced clearances for some voltages exceeding 98 kV to ground alternating current (AC) and 138 kV to ground direct current (DC). See Rule 230G for a discussion of AC and DC circuits. The best method for checking the application of Rule 232D is to use the formulas in Rule 232D and check the calculated results to **NESC** Table 232-4. If the calculated result does not match the examples provided in Table 232-4 for the same input conditions, a value was incorrectly entered or the formula was misapplied. Rule 232D also requires application of **NESC** Table 232-3, which lists reference heights similar to **NESC** Table A-2a in Appendix A of the **NESC**.

A list of important factors to consider when using Rule 232D is outlined below:

- The alternate clearances are for lines exceeding 98 kV (AC) to ground and 139 kV (DC) to ground only.
- The alternate clearance method must be used for voltages over 470 kV.
- **NESC** Table 232-3 reference heights are to be used.
- Crest voltage (not rms voltage) must be used. See Rule 230G for a discussion.
- The maximum operating voltage, typically 1.05 times the nominal voltage, must be used. ANSI C84.1, American National Standard for Electric Power

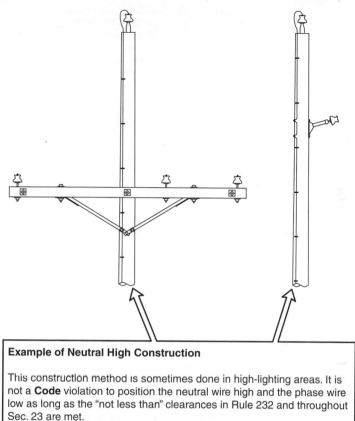

Example of Neutral High Construction

This construction method is sometimes done in high-lighting areas. It is not a **Code** violation to position the neutral wire high and the phase wire low as long as the "not less than" clearances in Rule 232 and throughout Sec. 23 are met.

Communication workers installing communication cables on joint use poles need to be able to recognize the differences between a neutral wire and phase wire.

Fig. 232-22. Examples of neutral high construction (Rule 232).

Systems and Equipment—Voltage Ratings (60 Hertz), provides a table listing nominal voltage levels and the corresponding maximum voltage level.
- Maximum switching surge factors must be known.
- Additional clearances are required for elevation and electrostatic effects similar to Rule 232C. (Increases for elevation are slightly different.)
- There is a lower limit for clearance based on Rule 232C.
- The "not less than" rounding requirements of Rule 230A4 apply.
- **NESC** Appendix D provides additional information on per-unit overvoltage factors.
- **NESC** Table 232-4 can be used to verify that calculations were done correctly.

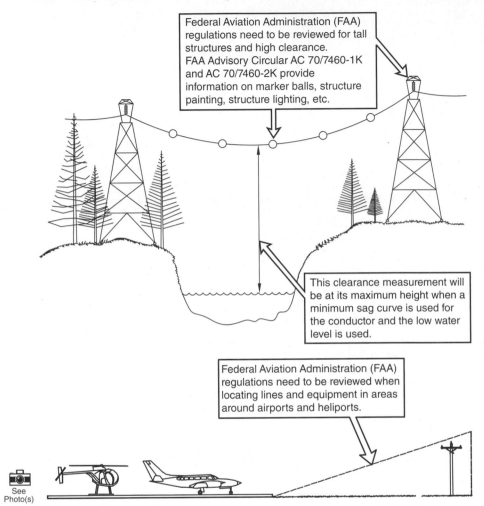

Federal Aviation Administration (FAA) regulations need to be reviewed for tall structures and high clearance.
FAA Advisory Circular AC 70/7460-1K and AC 70/7460-2K provide information on marker balls, structure painting, structure lighting, etc.

This clearance measurement will be at its maximum height when a minimum sag curve is used for the conductor and the low water level is used.

Federal Aviation Administration (FAA) regulations need to be reviewed when locating lines and equipment in areas around airports and heliports.

See Photo(s)

Fig. 232-23. Federal Aviation Administration (FAA) requirements (Rule N/A).

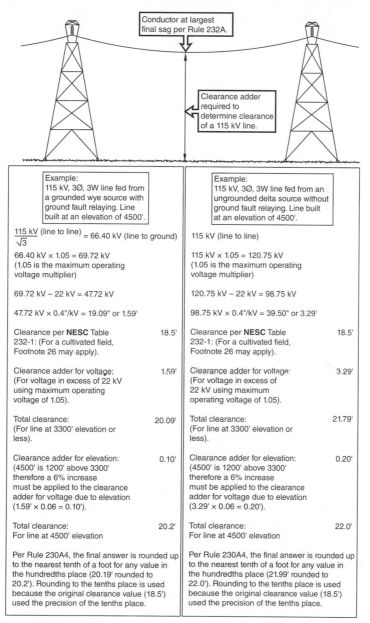

Example:
115 kV, 3Ø, 3W line fed from
a grounded wye source with
ground fault relaying. Line
built at an elevation of 4500'.

$\dfrac{115 \text{ kV}}{\sqrt{3}}$ (line to line) $= 66.40$ kV (line to ground)

66.40 kV × 1.05 = 69.72 kV
(1.05 is the maximum operating
voltage multiplier)

69.72 kV − 22 kV = 47.72 kV

47.72 kV × 0.4"/kV = 19.09" or 1.59'

Clearance per **NESC** Table 232-1: (For a cultivated field, Footnote 26 may apply).	18.5'
Clearance adder for voltage: (For voltage in excess of 22 kV using maximum operating voltage of 1.05).	1.59'
Total clearance: (For line at 3300' elevation or less).	20.09'
Clearance adder for elevation: (4500' is 1200' above 3300' therefore a 6% increase must be applied to the clearance adder for voltage due to elevation (1.59' × 0.06 = 0.10').	0.10'
Total clearance: For line at 4500' elevation	20.2'

Per Rule 230A4, the final answer is rounded up
to the nearest tenth of a foot for any value in
the hundredths place (20.19' rounded to
20.2'). Rounding to the tenths place is used
because the original clearance value (18.5')
used the precision of the tenths place.

Example:
115 kV, 3Ø, 3W line fed from an
ungrounded delta source without
ground fault relaying. Line built
at an elevation of 4500'.

115 kV (line to line)

115 kV × 1.05 = 120.75 kV
(1.05 is the maximum operating
voltage multiplier)

120.75 kV − 22 kV = 98.75 kV

98.75 kV × 0.4"/kV = 39.50" or 3.29'

Clearance per **NESC** Table 232-1: (For a cultivated field, Footnote 26 may apply).	18.5'
Clearance adder for voltage: (For voltage in excess of 22 kV using maximum operating voltage of 1.05).	3.29'
Total clearance: (For line at 3300' elevation or less).	21.79'
Clearance adder for elevation: (4500' is 1200' above 3300' therefore a 6% increase must be applied to the clearance adder for voltage due to elevation (3.29' × 0.06 = 0.20').	0.20'
Total clearance: For line at 4500' elevation	22.0'

Per Rule 230A4, the final answer is rounded up
to the nearest tenth of a foot for any value in
the hundredths place (21.99' rounded to
22.0'). Rounding to the tenths place is used
because the original clearance value (18.5')
used the precision of the tenths place.

Fig. 232-24. Example of an additional clearance calculation (Rules 232C1a and 232C1b).

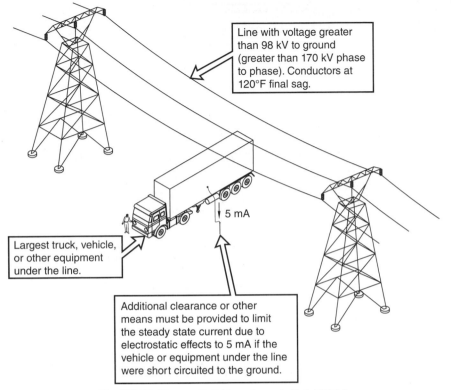

Line with voltage greater than 98 kV to ground (greater than 170 kV phase to phase). Conductors at 120°F final sag.

Largest truck, vehicle, or other equipment under the line.

5 mA

Additional clearance or other means must be provided to limit the steady state current due to electrostatic effects to 5 mA if the vehicle or equipment under the line were short circuited to the ground.

Fig. 232-25. Electrostatic field requirements (Rule 232C1c).

233. CLEARANCES BETWEEN WIRES, CONDUCTORS, AND CABLES CARRIED ON DIFFERENT SUPPORTING STRUCTURES

233A. General.   The first sentence of Rule 233 is very important and very powerful. "Crossings should be made on a common supporting structure, where practical." The wording includes "should" and "where practical." There are times when crossing on a common supporting structure is not practical. Turning lanes at roadway intersections can interfere with the location of a common crossing pole. Transmission lattice towers do not typically provide a practical means of attaching a distribution line crossing. If a line crossing can be attached to a common supporting structure, Rule 235 applies instead of Rule 233. See Rule 233B for a comparison of Rules 233B, 233C, 235B, and 235C. Rule 235, Clearance for Wires, Conductors, or Cables Carried on the Same Supporting Structure, is simpler to apply at a crossing because the conductors are tied onto the structure, not moving around in a span as they are in Rule 233. Examples of line crossings with and without a common supporting structure are shown in Fig. 233-1.

Two "envelopes" are addressed in this rule, the conductor movement envelope in Rule 233A1 that outlines an area of where the conductor can be positioned at various temperature and loading conditions; and the clearance envelope in Rule 233A2 that outlines the horizontal and vertical clearance area that must be maintained between the two conductors.

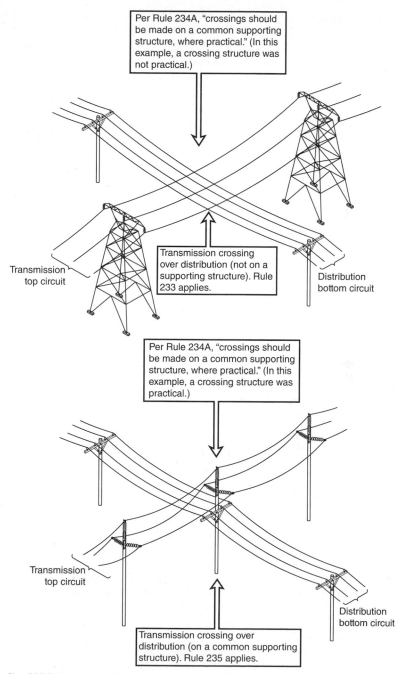

Per Rule 234A, "crossings should be made on a common supporting structure, where practical." (In this example, a crossing structure was not practical.)

Transmission top circuit

Transmission crossing over distribution (not on a supporting structure). Rule 233 applies.

Distribution bottom circuit

Per Rule 234A, "crossings should be made on a common supporting structure, where practical." (In this example, a crossing structure was practical.)

Transmission top circuit

Distribution bottom circuit

Transmission crossing over distribution (on a common supporting structure). Rule 235 applies.

Fig. 233-1. Examples of line crossings with and without a common supporting structure (Rule 233A).

The conductor movement envelope in Rule 233A1a(3) has the same largest final sag conditions described in Rule 232A. See Rule 232A for a discussion. In addition to the largest final sag condition, Rule 233A1a(1) requires a 60°F, no wind displacement, initial and final sag condition and Rule 233A1a(2) requires a 60°F, 6-lb/ft^2 wind, initial and final sag condition. These values are outlined on the sample sag and tension chart at the beginning of Sec. 23.

The temperatures listed in Rule 233A are the conductor temperatures, not the ambient air temperature. See the beginning of Sec. 23 for additional information on conductor versus air temperature. The term loading in Rule 233A refers to the physical loads on the conductors in the form of ice and wind, not electrical loads in kilowatts or amperes. The conductor temperature and loading conditions for measuring the conductor movement envelope are outlined in Fig. 233-2.

Using the sample sag and tension chart at the beginning of Sec. 23, 212°F, no wind, no ice, final condition produces the largest sag (6.84 ft) of the conditions that need to be checked in this rule. If the line were not designed to operate at high temperatures, then the 120°F, no wind, no ice, final value (5.61 ft) would be used as it is larger than the 32°F, no wind, 0.25 in ice, final value (4.49 ft). The 60°F, no wind, no ice, initial condition produces 3.14 ft of sag. The 60°F, no wind, no ice, final condition produces 4.00 ft of sag. The 60°F, 6 lb/ft^2 wind, no ice, initial condition produces 4.03 ft of sag (this value is a resultant sag in a diagonal position). The 60°F, 6 lb/ft^2 wind, no ice, final condition produces 4.66 ft of sag (this value is a resultant sag in a diagonal position). The **NESC** does not provide the details of how to calculate the position of the blownout conductor at 60°F with a 6-lb/ft^2 wind. The conductor "blowout" position can be calculated using a sag and tension software program with horizontal and vertical sag options enabled. See Note 5 in the sample sag and tension chart at the beginning of Sec. 23 for more information on horizontal and vertical sag values. A 6-lb/ft^2 wind pressure on a cylindrical surface like a conductor, cable, or pole is equal to approximately a 50 mile per hour wind. A sag and tension chart will only show the position of the conductor alone. The deflection of suspension insulators and flexible structures must also be considered in the wind blowout calculation. A transmission or distribution line design manual will typically include formulas to account for the deflection of suspension insulators and flexible structures. Some pole loading software programs may also be able to perform these calculations. See Fig. 233-3.

The footnotes to **NESC** Fig. 233-2 provide additional information on how to apply the conductor movement envelope. Footnote 1 requires that different wind directions be considered. The direction that produces the minimum distance between conductors must be used. Footnote 5 applies when one line is above the other. When the largest final sag of the upper conductor is determined, the lower conductor temperature must be equal to the ambient air temperature used to determine the largest final sag of the upper conductor. This requirement is similar to, but worded slightly different than, Rule 235C2b(1)(c). Rule 235C2b(1)(c) requires checking vertical clearances during two conditions, when the top conductor is at a maximum operating temperature (120°F, or greater if so designed) and when the top conductor is at a cold temperature with ice (32°F, with ice from Rule 230B). Footnote 3 of **NESC** Fig. 233-2 only

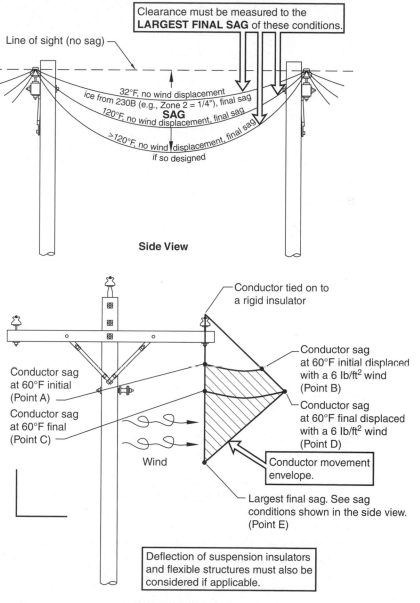

Clearance must be measured to the
LARGEST FINAL SAG of these conditions.

Line of sight (no sag)

32°F, no wind displacement
ice from 230B (e.g., Zone 2 = 1/4"), final sag
SAG
120°F, no wind displacement, final sag
>120°F, no wind displacement, final sag
if so designed

Side View

Conductor tied on to
a rigid insulator

Conductor sag
at 60°F initial
(Point A)

Conductor sag
at 60°F final
(Point C)

Conductor sag
at 60°F initial displaced
with a 6 lb/ft² wind
(Point B)

Conductor sag
at 60°F final displaced
with a 6 lb/ft² wind
(Point D)

Wind

Conductor movement
envelope.

Largest final sag. See sag
conditions shown in the side view.
(Point E)

Deflection of suspension insulators
and flexible structures must also be
considered if applicable.

See
Photo(s)

Front View

Fig. 233-2. Conductor movement envelope (Rule 233A).

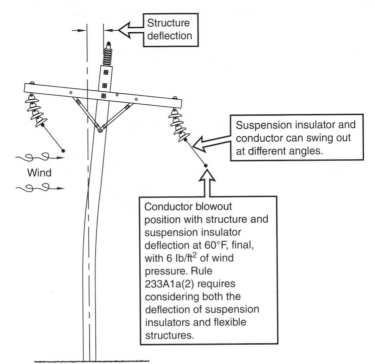

Structure deflection

Suspension insulator and conductor can swing out at different angles.

Conductor blowout position with structure and suspension insulator deflection at 60°F, final, with 6 lb/ft² of wind pressure. Rule 233A1a(2) requires considering both the deflection of suspension insulators and flexible structures.

Wind

Fig. 233-3. Deflection of suspension insulators and flexible structures [Rule 233A1a(2)].

requires checking vertical clearance at one of these two conditions. Both conditions are checked to determine the greatest sag of the top conductor, but then vertical clearance is only required to be checked at the condition that produces the largest sag of the top conductor. The condition that produces the largest sag of the top conductor may not be the condition that produces the greater vertical clearance at the structure. For a thorough review, both conditions can be checked as required in Rule 235C2b(1)(c). See Rule 235C and the beginning of Sec. 23 for discussions related to conductor temperature and ambient temperature.

233B. Horizontal Clearance.   Rule 233B outlines the requirements for the horizontal component of the clearance envelope. The horizontal clearance between conductors on different supporting structures is not less than 5.0 ft. A 0.4-in/kV clearance adder is required for voltages exceeding 22 kV between conductors involved. When comparing line conductors of different circuits, the voltage between the conductors is determined by using the greater of the phasor difference or the phase-to-ground voltage of the higher circuit. See Rule 235A for more information and a figure showing an example of how to calculate the phasor difference between line conductors of different circuits. An example of horizontal clearance between wires carried on different supporting structures is shown in Fig. 233-4.

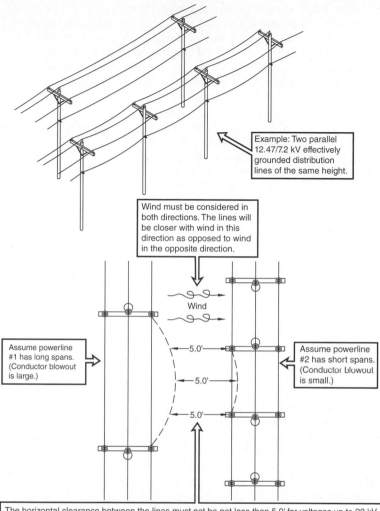

Example: Two parallel 12.47/7.2 kV effectively grounded distribution lines of the same height.

Wind must be considered in both directions. The lines will be closer with wind in this direction as opposed to wind in the opposite direction.

Wind

Assume powerline #1 has long spans. (Conductor blowout is large.)

—5.0'—

—5.0'—

—5.0'—

Assume powerline #2 has short spans. (Conductor blowout is small.)

The horizontal clearance between the lines must not be not less than 5.0' for voltages up to 22 kV between conductors per Rule 233B. A clearance adder of 0.4" per kV must be applied over 22 kV. In this example, the phasor difference between the line conductors is 7.2 kV + 7.2 kV = 14.4 kV (assuming a 180° phasor relationship). Therefore, no additional clearance for voltage applies and the maximum operating voltage and additional clearance for elevation requirements do not apply.

The clearance must be measured after considering the conductor movement envelopes of each conductor.
The horizontal clearance between the conductor of one line and the supporting structure of the other line must also be checked (Rule 234B).

See Photo(s)

Fig. 233-4. Example of horizontal clearance between wires carried on different supporting structures (Rule 233B).

Rule 233B has an exception to the 5.0-ft horizontal clearance for anchor guys.

Alternate clearances may be used for voltages exceeding 98 kV AC to ground or 139 kV DC to ground. The formulas in Rule 235B3 must be used for this calculation. See Rule 232D for a discussion of the alternate clearance method.

Rules 233B, 233C, 235B, and 235C are related to each other but use slightly different methods for determining clearance values. A comparison of the methods used to determine clearance in Rules 233B, 233C, 235B, and 235C is shown below:

- **Rule 233B.** Horizontal Clearance between Wires, Conductors, and Cables Carried on Different Supporting Structures.
    - ✓ Clearance value of 5.0 ft in rule (no table).
    - ✓ The 5.0-ft clearance value applies to supply and communication lines.
    - ✓ Clearance adder for voltage above 22 kV.
    - ✓ Clearance adder for altitude above 3300 ft when the voltage exceeds 50 kV.
    - ✓ The maximum operating voltage is used for lines over 50 kV.
    - ✓ The conductor movement envelope method is used.
    - ✓ The phasor difference voltage calculation method is used.
    - ✓ Alternate clearance calculations are based on Rule 235B3.
- **Rule 233C.** Vertical Clearance between Wires, Conductors, and Cables Carried on Different Supporting Structures.
    - ✓ Various clearance values in **NESC** Table 233-1.
    - ✓ The **NESC** Table 233-1 clearance values address both supply and communication lines.
    - ✓ Clearance adder for voltage above 22 kV.
    - ✓ Clearance adder for altitude above 3300 ft when the voltage exceeds 50 kV.
    - ✓ The maximum operating voltage is used for lines over 50 kV.
    - ✓ The conductor movement envelope method is used.
    - ✓ The phasor difference voltage calculation is not used. Clearance adders are calculated individually for the upper level and lower level conductors.
    - ✓ Alternate clearance calculations are based on Rule 233C3.
- **Rule 235B.** Horizontal Clearance between Line Conductors on the Same Supporting Structure.
    - ✓ Various clearance values in **NESC** Table 235-1 at the support.
    - ✓ Various clearance values in **NESC** Tables 235-2 and 235-3 at the support based on voltage and sag out in the span.
    - ✓ The **NESC** Table 235-1, 235-2, and 235-3 clearance values apply to supply lines and open wire communication lines. Communication line clearance is also covered in Rule 235H.
    - ✓ Clearance adder for voltage shown in the tables (above 8.7 kV and above 50 kV).
    - ✓ Clearance adder for altitude above 3300 ft when the voltage exceeds 50 kV.
    - ✓ The maximum operating voltage is used for lines over 50 kV.

✓ A clearance envelope method is used (wind pressures and sag conditions are specified in individual rules).

✓ The phasor difference voltage calculation method is used for line conductors of different circuits.

✓ Alternate clearance calculations are based on Rule 235B3.

• **Rule 235C.**   Vertical Clearance Between Conductors at the Support on the Same Supporting Structure.

✓ Various clearance values in **NESC** Table 235-5 at the support.

✓ Various clearance values in Rule 235C2b at the support based on voltage and sag out in the span.

✓ The **NESC** Table 235-5 clearance values address both supply and communication lines. Clearance between supply and communication lines is also covered in Rule 235C4. Communication line clearance is also covered in Rule 235H. Antenna clearances are covered in Rule 238F.

✓ Clearance adder for voltage shown in the table and in the rule (above 8.7 kV in the table and above 50 kV in the rule).

✓ Clearance adder for altitude above 3300 ft when the voltage exceeds 50 kV.

✓ The maximum operating voltage is used for lines over 50 kV.

✓ A clearance envelope method is used (wind pressures and sag conditions are specified in individual rules).

✓ The phasor difference voltage calculation method is used for line conductors of different circuits.

✓ Alternate clearance calculations are based on Rule 233C3.

233C. Vertical Clearance.   Rule 233C outlines the requirements for the vertical component of the clearance envelope. **NESC** Table 233-1 is used to find the vertical clearance between conductors on different supporting structures. Rule 233C applies to crossing and adjacent wires, conductors, and cables carried on different supporting structures. See Rule 233B for a comparison of Rules 233B, 233C, 235B, and 235C. Examples of vertical clearance between wires carried on different supporting structures are shown in Figs. 233-5, 233-6, and 233-7.

Rule 233C has an exception that states that no vertical clearance is required between wires, conductors, or cables that are electrically interconnected at the crossing. An example of the exception to vertical clearance between wires that are electrically connected at a crossing is shown in Fig. 233-8.

Alternate clearances may be used for voltages exceeding 98 kV AC to ground or 139 kV DC to ground per Rule 233C3. The formulas in Rule 233C3 must be used for this calculation. The values in **NESC** Table 233-2 can be used to check the proper application of the formulas. See Rule 232D for a discussion of the alternate clearance method.

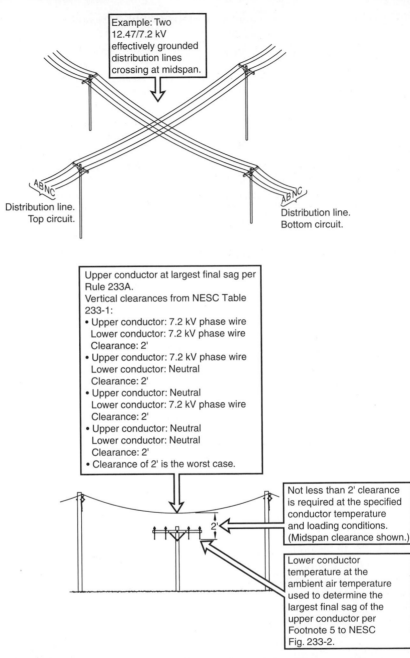

Example: Two 12.47/7.2 kV effectively grounded distribution lines crossing at midspan.

Distribution line. Top circuit.

Distribution line. Bottom circuit.

Upper conductor at largest final sag per Rule 233A.
Vertical clearances from NESC Table 233-1:
• Upper conductor: 7.2 kV phase wire
  Lower conductor: 7.2 kV phase wire
  Clearance: 2'
• Upper conductor: 7.2 kV phase wire
  Lower conductor: Neutral
  Clearance: 2'
• Upper conductor: Neutral
  Lower conductor: 7.2 kV phase wire
  Clearance: 2'
• Upper conductor: Neutral
  Lower conductor: Neutral
  Clearance: 2'
• Clearance of 2' is the worst case.

Not less than 2' clearance is required at the specified conductor temperature and loading conditions. (Midspan clearance shown.)

Lower conductor temperature at the ambient air temperature used to determine the largest final sag of the upper conductor per Footnote 5 to NESC Fig. 233-2.

2'

Fig. 233-5. Example of vertical clearance between wires carried on different supporting structures (Rule 233C).

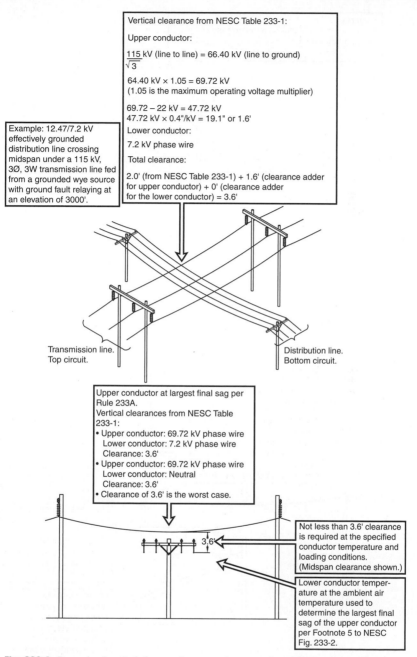

Vertical clearance from NESC Table 233-1:

Upper conductor:

$\dfrac{115}{\sqrt{3}}$ kV (line to line) = 66.40 kV (line to ground)

64.40 kV × 1.05 = 69.72 kV
(1.05 is the maximum operating voltage multiplier)

69.72 – 22 kV = 47.72 kV
47.72 kV × 0.4"/kV = 19.1" or 1.6'

Lower conductor:

7.2 kV phase wire

Total clearance:

2.0' (from NESC Table 233-1) + 1.6' (clearance adder for upper conductor) + 0' (clearance adder for the lower conductor) = 3.6'

Example: 12.47/7.2 kV effectively grounded distribution line crossing midspan under a 115 kV, 3Ø, 3W transmission line fed from a grounded wye source with ground fault relaying at an elevation of 3000'.

Transmission line.
Top circuit.

Distribution line.
Bottom circuit.

Upper conductor at largest final sag per Rule 233A.
Vertical clearances from NESC Table 233-1:
• Upper conductor: 69.72 kV phase wire
  Lower conductor: 7.2 kV phase wire
  Clearance: 3.6'
• Upper conductor: 69.72 kV phase wire
  Lower conductor: Neutral
  Clearance: 3.6'
• Clearance of 3.6' is the worst case.

3.6'

Not less than 3.6' clearance is required at the specified conductor temperature and loading conditions. (Midspan clearance shown.)

Lower conductor temperature at the ambient air temperature used to determine the largest final sag of the upper conductor per Footnote 5 to NESC Fig. 233-2.

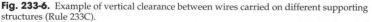

Fig. 233-6. Example of vertical clearance between wires carried on different supporting structures (Rule 233C).

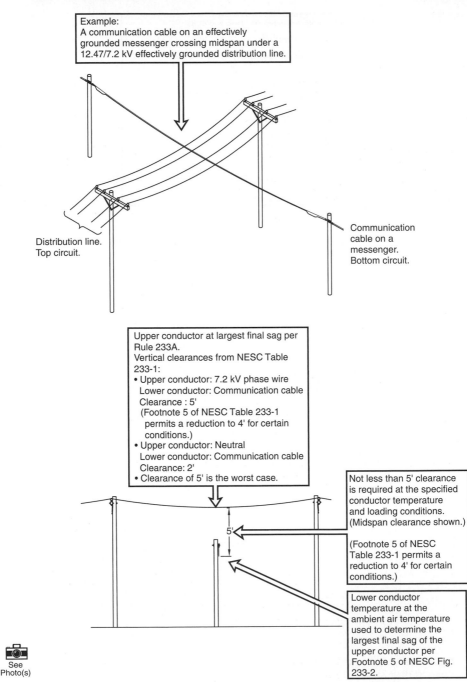

Example:
A communication cable on an effectively grounded messenger crossing midspan under a 12.47/7.2 kV effectively grounded distribution line.

Distribution line.
Top circuit.

Communication cable on a messenger.
Bottom circuit.

Upper conductor at largest final sag per Rule 233A.
Vertical clearances from NESC Table 233-1:
• Upper conductor: 7.2 kV phase wire
  Lower conductor: Communication cable
  Clearance : 5'
  (Footnote 5 of NESC Table 233-1 permits a reduction to 4' for certain conditions.)
• Upper conductor: Neutral
  Lower conductor: Communication cable
  Clearance: 2'
• Clearance of 5' is the worst case.

Not less than 5' clearance is required at the specified conductor temperature and loading conditions. (Midspan clearance shown.)

(Footnote 5 of NESC Table 233-1 permits a reduction to 4' for certain conditions.)

Lower conductor temperature at the ambient air temperature used to determine the largest final sag of the upper conductor per Footnote 5 of NESC Fig. 233-2.

5'

See Photo(s)

Fig. 233-7. Example of vertical clearance between wires carried on different supporting structures (Rule 233C).

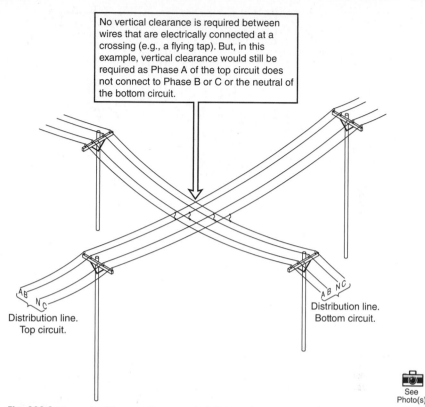

No vertical clearance is required between
wires that are electrically connected at a
crossing (e.g., a flying tap). But, in this
example, vertical clearance would still be
required as Phase A of the top circuit does
not connect to Phase B or C or the neutral of
the bottom circuit.

Distribution line.
Top circuit.

Distribution line.
Bottom circuit.

See
Photo(s)

Fig. 233-8. Example of the exception to vertical clearance between wires that are electrically interconnected at a crossing (Rule 233C).

234. CLEARANCE OF WIRES, CONDUCTORS, CABLES, AND EQUIPMENT FROM BUILDINGS, BRIDGES, RAIL CARS, SWIMMING POOLS, SUPPORTING STRUCTURES, AND OTHER INSTALLATIONS

234A. Application. Rule 234 contains both vertical and horizontal clearance requirements. When horizontal clearance is specified in Rules 234B, 234C, and 234D, the clearance must be checked when the conductor is at rest and when the conductor is displaced by wind. Both conditions must be met, not one or the other. Rule 234A defines the conductor temperature and loading conditions that must be checked before applying the rules and tables in Sec. 234. The largest final sag conditions in Rule 234A are the same conditions provided in Rule 232A. See Rule 232A for a discussion.

Rule 234A2 provides a wind condition for checking horizontal clearance with a 6-lb/ft² wind. The wind condition in Rule 234A2 is the same as in Rule 233A except Rule 234A2 only requires checking the 60°F final sag wind condition. Rule 233A requires checking both the 60°F final and initial wind conditions. See Rule 233A for a discussion. The conductor temperature and loading conditions for measuring vertical and horizontal clearance of conductors to buildings and other installations are shown in Fig. 234-1.

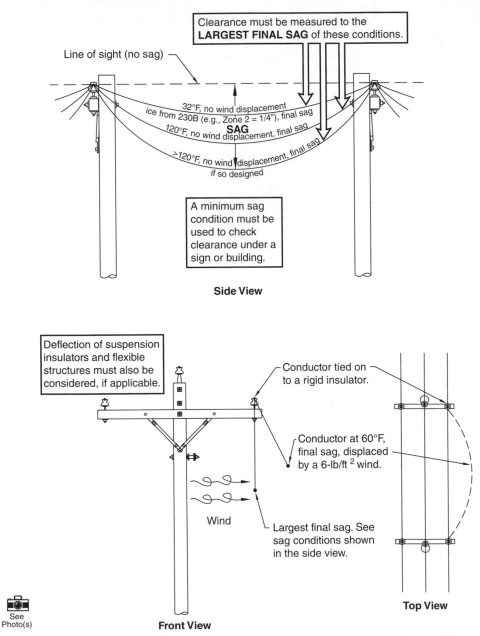

Clearance must be measured to the **LARGEST FINAL SAG** of these conditions.

Line of sight (no sag)

32°F, no wind displacement

ice from 230B (e.g., Zone 2 = 1/4"), final sag

SAG

120°F, no wind displacement, final sag

>120°F, no wind displacement, final sag if so designed

A minimum sag condition must be used to check clearance under a sign or building.

Side View

Deflection of suspension insulators and flexible structures must also be considered, if applicable.

Conductor tied on to a rigid insulator.

Conductor at 60°F, final sag, displaced by a 6-lb/ft^2 wind.

Wind

Largest final sag. See sag conditions shown in the side view.

Top View

See Photo(s)

Front View

Fig. 234-1. Conductor temperature and loading conditions for measuring vertical and horizontal clearance to buildings and other installations (Rule 234A).

The horizontal clearances specified in Rule 234 must be checked at rest and after applying a 6-lb/ft² wind at 60°F, final sag. The wind pressure may be reduced to 4 lb/ft² in sheltered areas with special conditions, but trees are not considered a shelter to a line. See Note 5 in the sample sag and tension chart at the beginning of Sec. 23 for more information on horizontal and vertical sag values. The deflection of suspension insulators must be considered. The deflection of flexible structures must be considered if the highest wire, conductor, or cable attachment is 60 ft or more above grade. Rule 233A1a(2) contains similar requirements for suspension insulators and flexible structures, but unlike Rule 234A2, Rule 233A1a(2) does not specify the 60-ft height requirement. See Fig. 234-2.

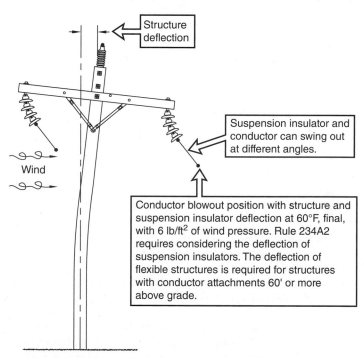

Fig. 234-2. Deflection of suspension insulators and flexible structures (Rule 234A2).

In addition to the largest final sag conditions, Rule 234A1d requires a minimum conductor sag be considered at the minimum conductor temperature with no wind at initial sag. The minimum conductor temperature will be based on the expected minimum ambient temperature. Using the sample sag and tension chart at the beginning of Sec. 23, −20°F, no wind, no ice, initial tension produces a minimum sag of 1.61 ft. This is the smallest sag on the sample sag and tension chart. The clearance under and alongside a building or sign must be checked when the conductor is in this minimum sag position. An example of checking clearance using a minimum sag condition is shown in Fig. 234-3.

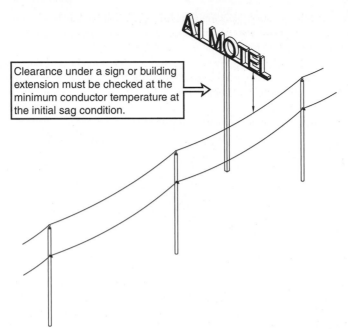

Clearance under a sign or building extension must be checked at the minimum conductor temperature at the initial sag condition.

Fig. 234-3. Example of checking clearance using a minimum sag condition (Rule 234A).

Rule 234A3 describes the method that is used to Transition (T) between Horizontal (H) and Vertical (V) clearances. The T, H, and V values are shown in **NESC** Figs. 234-1(a), (b), and (c). This is one of the few rules in the **NESC** that has graphics to help convey the **Code** requirements. **NESC** Figs. 234-1(a), (b), and (c) are used in conjunction with **NESC** Table 234-1.

234B. Clearances of Wires, Conductors, and Cables from Other Supporting Structures. Rule 234B is used to check clearance of wires, conductors, and cables to street lighting poles, traffic signal poles, antenna support poles, or a pole of another power line or communication line. In these cases, the line in question is not attached to the pole in question. It is important to note that the clearance is from "any part of such structure and associated equipment attached thereto." Examples of parts or equipment that extend from street light and traffic signal pole structures are a photocell mounted on a street light pole and a video camera detector mounted on a traffic signal pole. Examples of parts or equipment that extend from an antenna pole are the antenna panels. OSHA requires nonqualified electrical workers to stay 10 ft or more away from power lines, this is commonly referred to as the "OSHA 10 ft Rule" (OSHA 1910.333). There are instances where the **NESC** clearance requirements of Rule 234B permit less than 10 ft of clearance between a street light, traffic signal, or antenna and a power line. Where this occurs, the worker changing the light bulb on the street light or traffic signal, or the worker installing or maintaining the antenna needs to be a qualified electrical worker trained in the minimum approach distances in **NESC** Sec. 43, **NESC** Sec. 44, OSHA 1910.268, or OSHA 1910.269, or some other applicable standard. Examples of a 12.47/7.2-kV, three-phase, four-wire distribution line adjacent to a street light pole and a traffic signal pole are shown in Figs. 234-4 and 234-5.

Per Rule 234B1b, the horizontal clearance between the phase conductor of a 12.47/7.2 kV, 3Ø, 4W distribution line and a street light pole must be not less than 4.5' with a 6-lb/ft^2 wind applied to the conductor at 60°F final sag. No wind clearance is specified for an effectively grounded neutral (230E1), a secondary (230C3) cable below 750 V, or a communication cable. The clearance is measured after the line is placed in the blowout position.

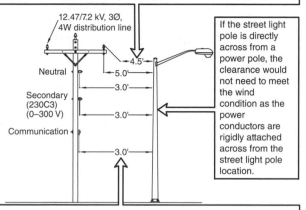

If the street light pole is directly across from a power pole, the clearance would not need to meet the wind condition as the power conductors are rigidly attached across from the street light pole location.

Per Rule 234B1a, the horizontal clearance between the phase conductor of a 12.47/7.2 kV, 3Ø, 4W distribution line and a street pole (without wind) must be not less than 5.0'.

An exception permits the neutral (230E1), secondary (230C3) up to 300 V, and communication cable to have a horizontal clearance (without wind) to be not less than 3.0'.

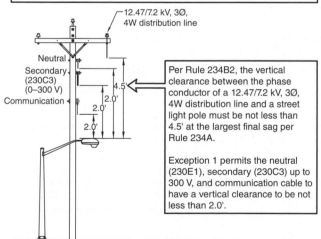

Per Rule 234B2, the vertical clearance between the phase conductor of a 12.47/7.2 kV, 3Ø, 4W distribution line and a street light pole must be not less than 4.5' at the largest final sag per Rule 234A.

Exception 1 permits the neutral (230E1), secondary (230C3) up to 300 V, and communication cable to have a vertical clearance to be not less than 2.0'.

See
Photo(s)

Fig. 234-4. Example of clearance of a conductor to a street lighting pole (Rule 234B).

There are times when a power line and street lighting compete for the same right of way and the only solution to meeting Rule 234B is bending the street light poles. See Fig. 234-6.

A note in Rule 234B clarifies that the clearances in Rule 234B are not intended to apply to personnel working in the vicinity of antennas. Rule 420Q and Rule 410A6

Per Rule 234B1b, the horizontal clearance between the phase conductor of a 12.47/7.2 kV, 3Ø, 4W distribution line and a traffic signal pole must be not less than 4.5' with a 6-lb/ft^2 wind applied to the conductor at 60°F final sag. No wind clearance is specified for an effectively grounded neutral (230E1), a secondary (230C3) cable below 750 V, or a communication cable. The clearance is measured after the line is placed in the blowout position.

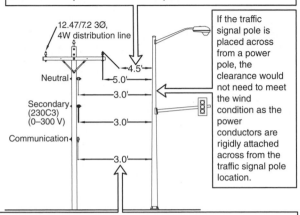

If the traffic signal pole is placed across from a power pole, the clearance would not need to meet the wind condition as the power conductors are rigidly attached across from the traffic signal pole location.

Per Rule 234B1a, the horizontal clearance between the phase conductor of a 12.47/7.2 kV, 3Ø, 4W distribution line and a traffic signal pole (without wind) must be not less than 5.0'.

An exception permits the neutral (230E1), secondary (230C3) up to 300 V, and communication cable to have a horizontal clearance (without wind) to be not less than 3.0'.

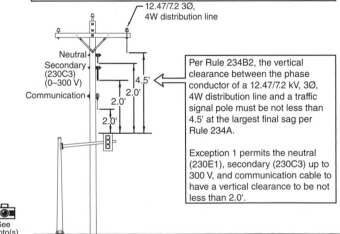

Per Rule 234B2, the vertical clearance between the phase conductor of a 12.47/7.2 kV, 3Ø, 4W distribution line and a traffic signal pole must be not less than 4.5' at the largest final sag per Rule 234A.

Exception 1 permits the neutral (230E1), secondary (230C3) up to 300 V, and communication cable to have a vertical clearance to be not less than 2.0'.

See Photo(s)

Fig. 234-5. Example of clearance of a conductor to a traffic signal pole (Rule 234B).

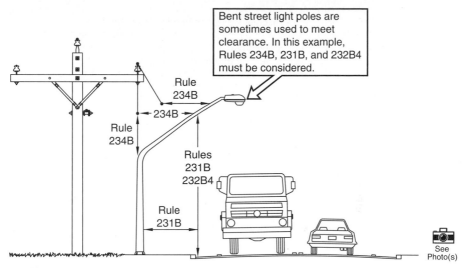

Fig. 234-6. Application of Rules 234B, 231B, and 232B4 to bent street lighting poles (Rule 234B).

address radio frequency issues. An example of a 12.47/7.2-kV, three-phase, four-wire distribution line adjacent to a small cell antenna pole is shown in Fig. 234-7.

Another application of Rule 234B is a transmission line with a distribution under-build that has "skip-span" construction. Skip-span construction is also sometimes used on a distribution line with a joint-use communication line attached to it. An example of a transmission line with a distribution underbuild that has skip-span construction is shown in Fig. 234-8.

The clearance adders for voltages above 22 kV are provided in Rule 234G. Alternate clearances for voltages exceeding 98 kV AC to ground or 139 kV DC to ground are provided in Rule 234H. The horizontal clearance in Rule 234B1 requires clearance adders for voltages greater than 22 kV for conductors at rest and greater than 22 kV for conductors displaced by wind. Rule 234B1 references **NESC** Table 234-7 for horizontal wind clearances. **NESC** Table 234-7 does not address 230E1 (neutral conductors), 230C3 cables below 750 V (e.g., 120/240 V secondary triplex conductors), or communication cables. The vertical clearance in Rule 234B2 also requires clearance adders for voltages greater than 22 kV. An exception applies to vertical clearances in Rule 234B2 when the wires and supporting structure are operated and maintained by the same utility.

234C. Clearances of Wires, Conductors, Cables, and Rigid Live Parts from Buildings, Signs, Billboards, Chimneys, Antennas, Tanks, Flagpoles and Associated Flags, Banners, and Other Installations Except Bridges and Supporting Structures.   Clearance from an energized conductor to a building or other installation is just as important as clearance to the surface below the line. As the title to Rule 234C states, this rule includes not only clearance to buildings, but also signs, billboards, chimneys, antennas, tanks, flagpoles and associated flags, banners, and other installations except bridges and supporting structures. This is one of the few rules in the **NESC** that has graphics to help convey the **Code** requirements.

Per Rule 234B1b, the horizontal clearance between the phase conductor of a 12.47/7.2 kV, 3Ø, 4W distribution line and an antenna pole must be not less than 4.5' with a 6-lb/ft^2 wind applied to the conductor at 60°F final sag. No wind clearance is specified for an effectively grounded neutral (230E1), a secondary (230C3) cable below 750V, or a communication cable. The clearance is measured after the line is placed in the blowout position.

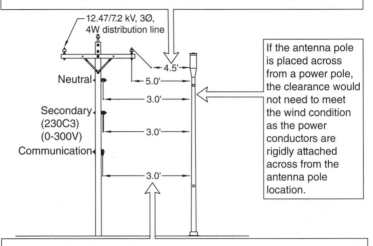

If the antenna pole is placed across from a power pole, the clearance would not need to meet the wind condition as the power conductors are rigidly attached across from the antenna pole location.

Per Rule 234B1a, the horizontal clearance between the phase conductor of a 12.47/7.2 kV, 3Ø, 4W distribution line and an antenna pole (without wind) must be not less than 5.0'. Workers must not be exposed to excessive radiation levels per Rule 420Q.

An exception permits the neutral (230E1), secondary (230C3) up to 300 V, and communication cable to have a horizontal clearance (without wind) to be not less than 3.0'.

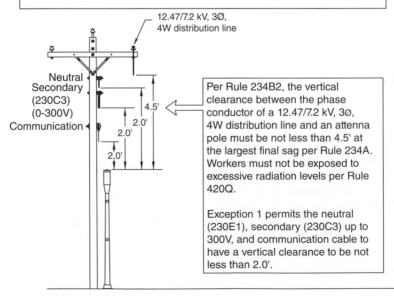

Per Rule 234B2, the vertical clearance between the phase conductor of a 12.47/7.2 kV, 3Ø, 4W distribution line and an antenna pole must be not less than 4.5' at the largest final sag per Rule 234A. Workers must not be exposed to excessive radiation levels per Rule 420Q.

Exception 1 permits the neutral (230E1), secondary (230C3) up to 300V, and communication cable to have a vertical clearance to be not less than 2.0'.

Fig. 234-7. Example of clearance of a conductor to an antenna pole (Rule 234B).

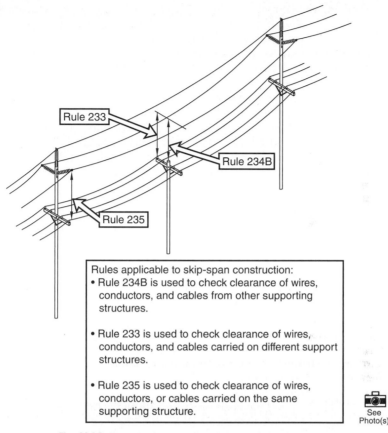

Fig. 234-8. Example of skip-span construction (Rule 234B).

The first complete sentence of Rule 234C has four very important words that are not in the title of Rule 234C. The words are "…and any projections therefrom." This statement means that clearance must be maintained not just to the building structure, but to a gutter, awning, or any other projection from a building, sign, billboard, chimney, antenna, tank, flagpole and associated flag, banner, or other installation except bridges and supporting structures. The "except bridges" statement is due to the fact that bridges have their own rule, Rule 234D. The "except supporting structures" statement is due to the fact that supporting structures have their own rule, Rule 234B. Also, not mentioned are swimming pools and swimming areas covered in Rule 234E, grain bins covered in Rule 234F, rail cars covered in Rule 234I, and equipment covered in Rule 234J.

The vertical and horizontal clearances for Rule 234C are determined by using **NESC** Table 234-1 with the aid of **NESC** Figs. 234-1(a), 234-1(b), and 234-1(c). Normally, the **Code** requires you to meet all applicable rules, not one rule or another. However, in the case of building clearances, it is evident from the H, V, and T clearance envelope around the building in **NESC** Fig 234-1(a) that a power or communication line that meets the required vertical clearance above a building does not need to meet any required horizontal clearance to the side of a building and a power or communication line that meets the required horizontal clearance to the side of a building does not need to meet any required vertical clearance above a building. Other issues such as aerial trespass above a building or property line setback from the side of a building may arise when positioning power and communication lines adjacent to buildings but these issues are not **NESC** issues. **NESC** Table 234-1 that is used in conjunction with **NESC** Fig. 234-1(a) address both wires (e.g., a 12.47 kV phase wire) and unguarded rigid live parts (e.g., an energized transformer bushing) in the same table.

It is important to carefully read the title of **NESC** Table 234-1, "Clearance of Wires, Conductors, Cables, and Unguarded Rigid Live Parts Adjacent but Not Attached to Buildings and Other Installations Except Bridges." The important wording is "...adjacent but not attached to..." This table is for conductors passing by a building, not a service that is attached to a building. Building service drops for supply (power) conductors are covered in Rule 234C3. Building service drops for communication conductors are covered in Rule 234C4.

The clearances in **NESC** Table 234-1 for conductors vertically over a building or other installation must be measured with the conductors at the largest final sag condition per Rule 234A. The vertical clearance for conductors under a sign or building projection must be measured with the conductors at the minimum sag condition as defined by Rule 234A. The clearances in **NESC** Table 234-1 for conductors to the side of (horizontal to) a building or other installation are at rest (no wind displacement) values. The horizontal clearance must also be checked using the wind condition in Rule 234A2. The horizontal clearances to conductors displaced by wind are provided in **NESC** Table 234-7. **NESC** Table 234-7 does not address 230E1 (neutral conductors), 230C3 cables below 750 V (e.g., 120/240 V secondary triplex conductors), or communication cables.

Examples of building clearance, billboard clearance, and flagpole clearance are shown in Figs. 234-9 through 234-13.

Rule 234C2 allows guarding as an option where **NESC** Table 234-1 clearances cannot be obtained. The note to this rule states that 230C1a cables are considered guarded. See the figure in Rule 230C for the details of a 230C1a cable.

Footnote 2 to **NESC** Table 234-1 permits reduced clearance to buildings. Footnote 2 permits clearance reductions for covered conductors. Covered conductor or "tree wire" is discussed in Rule 230D. Covered conductor applications along streets and in alleyways have become popular due to cramped space.

There are times when meeting the **NESC** clearance to a building can be challenging. However, several practical methods exist to meet building clearances. These methods include but are not limited to:

- Burying the line
- Rerouting the line/poles
- Installing a taller pole(s)
- Using alley arm or side arm framing
- Using post insulator framing on one side of the pole
- Using poles on each side of an alleyway and spanning between the poles with a bridge arm
- Using covered conductors
- Using covered bundled conductors

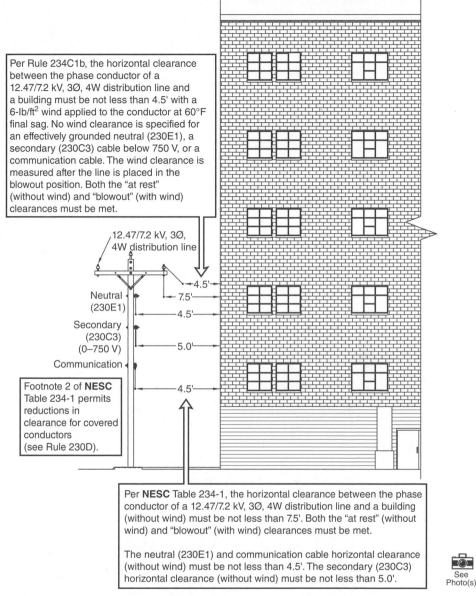

Per Rule 234C1b, the horizontal clearance between the phase conductor of a 12.47/7.2 kV, 3Ø, 4W distribution line and a building must be not less than 4.5' with a 6-lb/ft² wind applied to the conductor at 60°F final sag. No wind clearance is specified for an effectively grounded neutral (230E1), a secondary (230C3) cable below 750 V, or a communication cable. The wind clearance is measured after the line is placed in the blowout position. Both the "at rest" (without wind) and "blowout" (with wind) clearances must be met.

12.47/7.2 kV, 3Ø, 4W distribution line

—4.5'—

Neutral (230E1) — 7.5'—

— 4.5'—

Secondary (230C3) (0–750 V)

Communication

—5.0'—

Footnote 2 of **NESC** Table 234-1 permits reductions in clearance for covered conductors (see Rule 230D).

—4.5'—

Per **NESC** Table 234-1, the horizontal clearance between the phase conductor of a 12.47/7.2 kV, 3Ø, 4W distribution line and a building (without wind) must be not less than 7.5'. Both the "at rest" (without wind) and "blowout" (with wind) clearances must be met.

The neutral (230E1) and communication cable horizontal clearance (without wind) must be not less than 4.5'. The secondary (230C3) horizontal clearance (without wind) must be not less than 5.0'.

See Photo(s)

Fig. 234-9. Example of horizontal clearance of a conductor to a building (Rule 234C1).

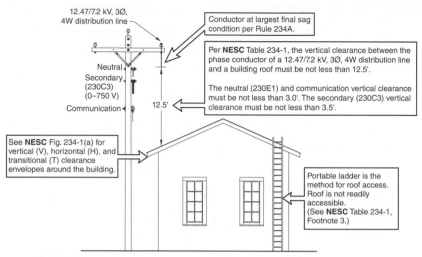

12.47/7.2 kV, 3Ø,
4W distribution line

Conductor at largest final sag condition per Rule 234A.

Per **NESC** Table 234-1, the vertical clearance between the phase conductor of a 12.47/7.2 kV, 3Ø, 4W distribution line and a building roof must be not less than 12.5'.

Neutral

Secondary
(230C3)
(0–750 V)

The neutral (230E1) and communication vertical clearance must be not less than 3.0'. The secondary (230C3) vertical clearance must be not less than 3.5'.

Communication

12.5'

See **NESC** Fig. 234-1(a) for vertical (V), horizontal (H), and transitional (T) clearance envelopes around the building.

Portable ladder is the method for roof access. Roof is not readily accessible. (See **NESC** Table 234-1, Footnote 3.)

Fig. 234-10. Example of vertical clearance of a conductor to a building (Rule 234C1).

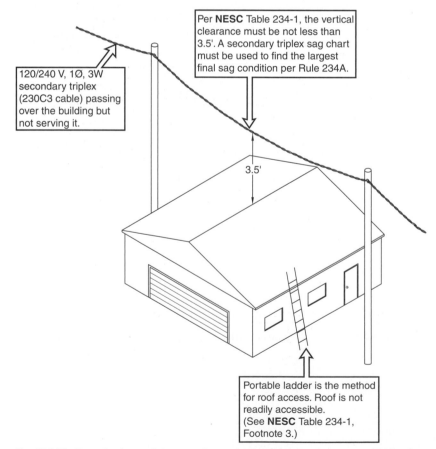

Per **NESC** Table 234-1, the vertical clearance must be not less than 3.5'. A secondary triplex sag chart must be used to find the largest final sag condition per Rule 234A.

120/240 V, 1Ø, 3W secondary triplex (230C3 cable) passing over the building but not serving it.

3.5'

Portable ladder is the method for roof access. Roof is not readily accessible. (See **NESC** Table 234-1, Footnote 3.)

See
Photo(s)

Fig. 234-11. Example of vertical clearance of a supply service drop conductor over a building but not serving it (Rule 234C1).

Per Rule 234C1b, the horizontal clearance between the phase conductor of a 12.47/7.2 kV, 3Ø, 4W distribution line and a billboard must be not less than 4.5' with a 6-lb/ft² wind applied to the conductor at 60°F final sag. No wind clearance is specified for an effectively grounded neutral (230E1), a secondary (230C3) cable below 750 V, or a communication cable. The clearance is determined after the line is placed in the blowout position.

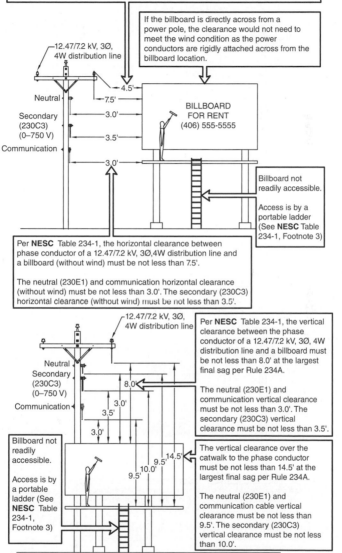

If the billboard is directly across from a power pole, the clearance would not need to meet the wind condition as the power conductors are rigidly attached across from the billboard location.

−12.47/7.2 kV, 3Ø, 4W distribution line

Neutral

Secondary (230C3) (0–750 V)

Communication

BILLBOARD FOR RENT (406) 555-5555

4.5'
7.5'
3.0'
3.5'
3.0'

Billboard not readily accessible.

Access is by a portable ladder (See **NESC** Table 234-1, Footnote 3)

Per **NESC** Table 234-1, the horizontal clearance between phase conductor of a 12.47/7.2 kV, 3Ø,4W distribution line and a billboard (without wind) must be not less than 7.5'.

The neutral (230E1) and communication horizontal clearance (without wind) must be not less than 3.0'. The secondary (230C3) horizontal clearance (without wind) must be not less than 3.5'.

−12.47/7.2 kV, 3Ø, 4W distribution line

Neutral
Secondary (230C3) (0–750 V)

Communication

8.0'
3.0'
3.5'
3.0'

Per **NESC** Table 234-1, the vertical clearance between the phase conductor of a 12.47/7.2 kV, 3Ø, 4W distribution line and a billboard must be not less than 8.0' at the largest final sag per Rule 234A.

The neutral (230E1) and communication vertical clearance must be not less than 3.0'. The secondary (230C3) vertical clearance must be not less than 3.5'.

Billboard not readily accessible.

Access is by a portable ladder (See **NESC** Table 234-1, Footnote 3)

14.5'
9.5'
10.0'
9.5'

The vertical clearance over the catwalk to the phase conductor must be not less than 14.5' at the largest final sag per Rule 234A.

The neutral (230E1) and communication cable vertical clearance must be not less than 9.5'. The secondary (230C3) vertical clearance must be not less than 10.0'.

See Photo(s)

Fig. 234-12. Example of clearance of a conductor to a billboard (Rule 234C1).

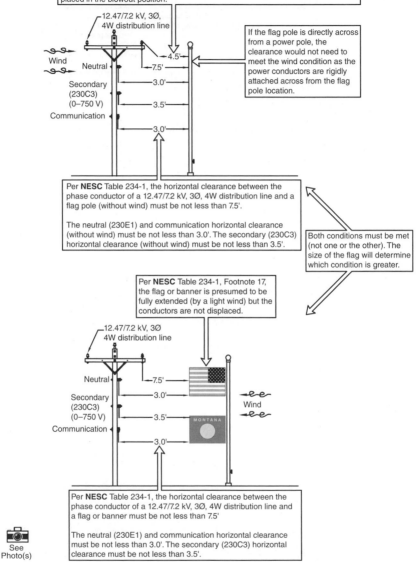

Per Rule 234C1b, the horizontal clearance between the phase conductor of a 12.47/7.2 kV, 3Ø, 4W distribution line and a flag pole must be not less than 4.5' with a 6-lb/ft^2 wind applied to the conductor at 60°F final sag. No wind clearance is specified for an effectively grounded neutral (230E1), a secondary (230C3) cable below 750 V, or a communication cable. The clearance is determined after the line is placed in the blowout position.

12.47/7.2 kV, 3Ø, 4W distribution line

If the flag pole is directly across from a power pole, the clearance would not need to meet the wind condition as the power conductors are rigidly attached across from the flag pole location.

Wind

Neutral — 4.5'
— 7.5'

Secondary (230C3) (0–750 V) — 3.0'
— 3.5'

Communication — 3.0'

Per **NESC** Table 234-1, the horizontal clearance between the phase conductor of a 12.47/7.2 kV, 3Ø, 4W distribution line and a flag pole (without wind) must be not less than 7.5'.

The neutral (230E1) and communication horizontal clearance (without wind) must be not less than 3.0'. The secondary (230C3) horizontal clearance (without wind) must be not less than 3.5'.

Both conditions must be met (not one or the other). The size of the flag will determine which condition is greater.

Per **NESC** Table 234-1, Footnote 17, the flag or banner is presumed to be fully extended (by a light wind) but the conductors are not displaced.

12.47/7.2 kV, 3Ø 4W distribution line

Neutral — 7.5'
Secondary (230C3) (0–750 V) — 3.0'
— 3.5'

Wind

Communication — 3.0'

MONTANA

Per **NESC** Table 234-1, the horizontal clearance between the phase conductor of a 12.47/7.2 kV, 3Ø, 4W distribution line and a flag or banner must be not less than 7.5'

The neutral (230E1) and communication horizontal clearance must be not less than 3.0'. The secondary (230C3) horizontal clearance must be not less than 3.5'.

See Photo(s)

Fig. 234-13. Example of horizontal clearance of a conductor to a flagpole and flag (Rule 234C1).

- Using aerial insulated conductors
- Increasing conductor tension for wind-related clearances
- Shortening span lengths for wind-related clearances
- A combination of the above items

The **NESC** does not address clearance to buildings under construction. The Occupational Safety and Health Administration (OSHA) regulations apply to building construction. See the discussion in Rule 232B for buildings being transported on a roadway (e.g., house moves). Mobile cranes used in the vicinity of power lines are addressed in the OSHA regulations in OSHA 1926.1408, which is part of OSHA 1926 Subpart CC. OSHA requires nonqualified electrical workers and any conductive object that they are handling to stay 10 ft or more away from power lines (at 50 kV—greater distance is required for higher voltage), this is commonly referred to as the "OSHA 10 ft Rule" (OSHA Standard 1910.333). The **NESC** permits a 12.47/7.2 kV distribution line to be not less than 7.5 ft from a building. The 7.5 ft clearance is the "at rest" value from **NESC** Table 234-1, meeting the "with wind" clearance from Rule 234C1b may increase the "at rest" requirement to greater than 7.5 ft. There can be instances where the **NESC** permits less than 10 ft of clearance to a building or sign. If a nonqualified electrical worker, for example a painter, is painting a building and is working more than 10 ft away from the power line, both the **NESC** and OSHA requirements are met. If the painter needs to work closer than 10 ft to the power line, the painter must contact the electric utility to work out a plan. The electric utility may decide to de-energize the power line, reposition the power line on temporary arms, or select some other method for the painter to work in accordance with the OSHA 10-ft requirement. The OSHA 10 ft Rule is a work rule that applies to the painter and the **NESC** 7.5 ft clearance (or greater with wind) is a clearance rule that applies to the electric utility. The OSHA 10 ft Rule does not require an electric utility to increase the **NESC** clearance, and it does not excuse an electric utility from meeting the **NESC** clearance. Some States have high-voltage safety acts with wording similar to the OSHA 10-ft requirement.

In addition to the **NESC** clearances for "tanks" in **NESC** Table 234-1, the Liquefied Petroleum Gas Code (NFPA 58) specifies clearances to LP-Gas (propane) tanks that are located under power lines. See Fig. 234-14.

Chimneys and antennas are listed in **NESC** Table 234-1, Row 2. Footnote 5 to **NESC** Table 234-1 clarifies that chimneys, or other items in Row 2, that are part of or attached to a residential building shall have the clearances required for buildings (and their projections). Footnote 5 to **NESC** Table 234-1 also clarifies that Rule 234B applies to antennas installed on a support structure solely for the purpose of supporting the antenna. See Rule 234B for a discussion and a figure.

The **NESC** does not specifically address clearance between a power line and irrigation equipment except for a general note in Rule 232B1. **NESC** Table 234-1 can be used if the irrigation equipment is considered an "other installation not classified as buildings or bridges." In addition to the clearance between a power line and the irrigation piping, questions commonly arise as to the required clearance between a power line and the irrigation water stream. Two documents are available for use as accepted good practice for determining the clearance between a power line and the irrigation water stream. They are "Guidelines for the Installation and Operation of Irrigation Systems Near High Voltage

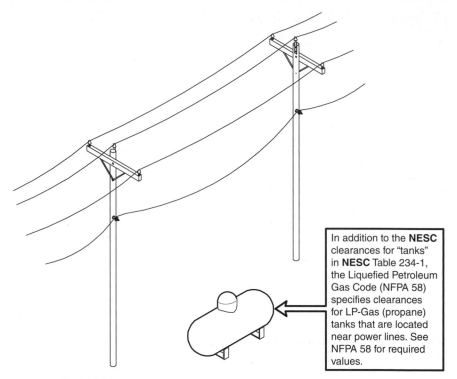

In addition to the **NESC** clearances for "tanks" in **NESC** Table 234-1, the Liquefied Petroleum Gas Code (NFPA 58) specifies clearances for LP-Gas (propane) tanks that are located near power lines. See NFPA 58 for required values.

Fig. 234-14. Liquefied Petroleum Gas Code (NFPA 58) requirements (Rule N/A).

Transmission Lines" published by the Bonneville Power Administration (BPA) and "IEEE Guide for Installation, Maintenance and Operation of Irrigation Equipment Located Near or Under Power Lines" published by the Institute of Electrical and Electronics Engineers (IEEE Std. 1542). See Fig. 234-15.

The **NESC** does not specifically address clearance over fences or walls that are not part of a building structure. Rule 012C, which requires accepted good practice, must be used. **NESC** Table 234-1 can be used if the fence or wall is considered an "other installation not classified as buildings." For clearance over a wall that is wide enough to stand on, the vertical clearance over "surfaces upon which personnel walk" may be appropriate. For clearance over a fence that is not wide enough to stand on, the vertical clearance over "other portions of such installations" may be appropriate. **NESC** Table 234-1, Footnote 14 (which applies to conductors adjacent but not attached to buildings) and **NESC** Rule 234C3d(2) (which applies to supply conductors attached to buildings) address clearance above railings, walls, or parapets around balconies, decks, or roofs. The clearances for wires, conductors, and cables aboveground, roadway, rail, or water surfaces adjacent to the fence must also be met per **NESC** Table 232-1. See Fig. 234-16.

Rule 234C3 covers supply conductors attached to buildings. **NESC** Figs. 234-2(a), 234-2(b), and 234-2(c) are provided in the **Code** to aid the application of supply (power) service drops. The **NESC** and the National Electrical Code (NEC)

The **NESC** does not specifically address clearance between a power line and irrigation equipment except for a general note in Rule 232B1.

NESC Table 234-1 can be used if the irrigation equipment is considered an "other installation not classified as buildings or bridges".

NESC Rule 012C, which requires accepted good practice, must be used for the water stream. Common references are "Guidelines for the Installation and Operation of Irrigation Systems Near High Voltage Transmission Lines" published by the Bonneville Power Administration (BPA) and "IEEE Guide for Installation, Maintenance and Operation of Irrigation Equipment Located Near or Under Power Lines" published by the Institute of Electrical and Electronics Engineers (IEEE Std. 1542).

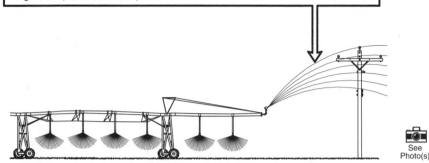

See Photo(s)

Fig. 234-15. Example of clearance between a power line and irrigation equipment (Rule 234C).

overlap at the service entrance point. The NEC has similar rules for service entrance conductors and depending on the authority having jurisdiction (e.g., a city or state electrical inspector), the NEC rules may govern. Examples of supply conductors attached and connecting to a service entrance into a building are shown in Figs. 234-17 through 234-20.

Rule 234C4 contains the requirements for communication service drops attached to buildings. It provides clearances for readily accessible roofs, balconies, attached decks, fire escapes, etc. Rule 234C3 provides a detailed description of readily accessible. For communication service conductors, no clearance is specified for roofs that are not readily accessible. An example of a communications service drop over a deck is shown in Fig. 234-19.

Rule 234C5 requires a clear space for fire-fighting ladders for buildings exceeding three stories or 50 ft in height. A 6-ft (minimum) wide zone should exist adjacent to the building or beginning not over 8 ft from the building. Traditionally, the measurement for the clear space or zone that is at least 6 ft adjacent to the building is from the end of the crossarm to the building and the measurement for the 8-ft distance is from the opposite end of the crossarm to the building. The 6-ft zone beginning not over 8 ft from the building is not practical if 8-ft crossarms are used for a three-phase distribution line and a 7.5-ft clearance is required between the conductor near the edge of the crossarm and the building. Traditionally, this rule assumes the line is paralleling the building. An exception applies if the fire department does not use ladders in alleys near supply conductors. It is a good practice for supply

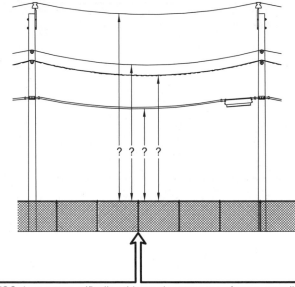

The **NESC** does not specifically address clearance over fences or walls that are not part of a building structure. Rule 012C, which requires accepted good practice, must be used. Examples of accepted good practice are:

• **NESC** Table 234-1 can be used if the fence or wall is considered "other installations not classified as buildings." For clearances over a wall that is wide enough to walk on, the vertical clearance over "surfaces upon which personnel walk" may be appropriate. For clearances over a fence that is not wide enough to walk on, the vertical clearance over "other portions of such installations" may be appropriate.

• **NESC** Table 234-1, Footnote 14 (which applies to conductors adjacent but not attached to buildings) and **NESC** Rule 234C3d(2) (which applies to supply conductors attached to buildings) address clearances above railings, walls, or parapets around balconies, decks, or roofs.

The clearances for wires, conductors, and cables aboveground, roadway, rail, or water surfaces adjacent to the fence must also be met per **NESC** Table 232-1.

Fig. 234-16. Example of clearance of power and communication lines above a fence (Rule 234C).

utilities and the local fire department to communicate with each other to discuss safety issues related to overhead lines.

The clearance adders for voltages above 22 kV are provided in Rule 234G. Alternate clearances for voltages exceeding 98 kV AC to ground or 139 kV DC to ground are provided in Rule 234H.

234D. Clearance of Wires, Conductors, Cables, and Unguarded Rigid Live Parts from Bridges. Depending on the size and design of a bridge structure, power and communication lines and equipment may be located adjacent to or

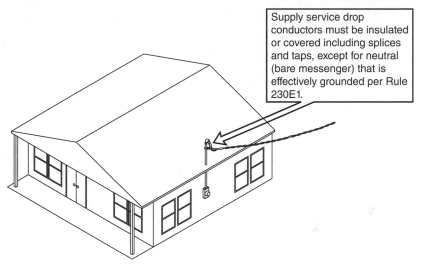

Supply service drop conductors must be insulated or covered including splices and taps, except for neutral (bare messenger) that is effectively grounded per Rule 230E1.

See Photo(s)

Fig. 234-17. Example of clearance of supply conductors (service drops) attached to buildings (Rule 234C3).

within a bridge structure. **NESC** Table 234-2 applies to bridge structures. **NESC** Table 234-2 addresses clearances over bridges when conductors are attached and when not attached. The table also addresses clearance beside, under, and within the bridge structure. The clearances in Table 234-2 are for the at-rest (no wind) conditions outlined in Rule 234A. The main concerns addressed in **NESC** Table 234-2 are outlined in Fig. 234-21.

NESC Figs. 234-1(a) and (b) which focus on horizontal and vertical clearances for signs and buildings, are also applicable to bridges. Bridges, unlike buildings and signs, do not have any required clearance to insulated communication cables and neutrals meeting Rule 230E1.

The horizontal clearance to bridges with wind displacement is outlined in **NESC** Table 234-7. **NESC** Table 234-7 does not address 230E1 (neutral conductors), 230C3 cables below 750 V (e.g., 120/240-V secondary triplex conductors), or communication cables.

The clearance adders for voltages above 22 kV are provided in Rule 234G. Alternate clearances for voltages exceeding 98 kV AC to ground or 139 kV DC to ground are provided in Rule 234H.

Underground lines that have conduits attached to bridges are covered in Rules 320A, 322B, and 351C.

234E. Clearance of Wires, Conductors, Cables, or Unguarded Rigid Live Parts Installed over or near Swimming Areas with No Wind Displacement. Rule 234E applies not only to swimming pools but also to swimming areas and beaches. This rule applies to both in-ground swimming pools and permanently installed above-ground swimming pools. **NESC** Table 234-3 and **NESC** Figs. 234-3(a), 234-3(b), and 234-3(c) are provided in the **Code** to check clearance to both in-ground and above-ground swimming pools. Note 1 in Rule 234E clarifies the application of

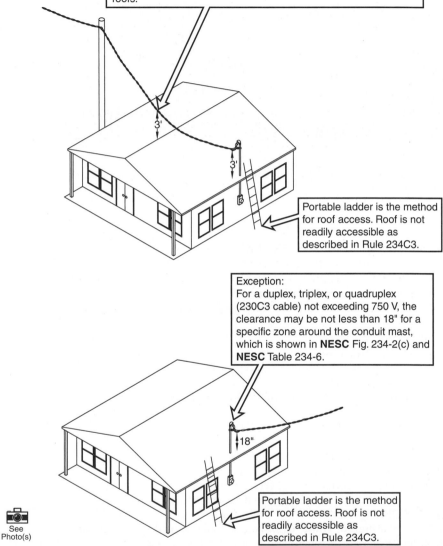

Exception:
For a duplex, triplex, or quadruplex (230C3 cable) not exceeding 750 V, the clearance may be not less than 3' if the roof is not readily accessible. The basic clearance is 10' for readily accessible roofs.

Portable ladder is the method for roof access. Roof is not readily accessible as described in Rule 234C3.

Exception:
For a duplex, triplex, or quadruplex (230C3 cable) not exceeding 750 V, the clearance may be not less than 18" for a specific zone around the conduit mast, which is shown in **NESC** Fig. 234-2(c) and **NESC** Table 234-6.

Portable ladder is the method for roof access. Roof is not readily accessible as described in Rule 234C3.

See Photo(s)

Fig. 234-18. Example of clearance of supply conductors (service drops) attached to buildings (Rule 234C3).

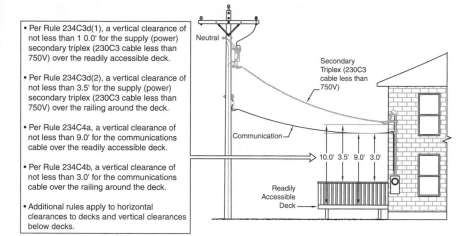

- Per Rule 234C3d(1), a vertical clearance of not less than 1 0.0' for the supply (power) secondary triplex (230C3 cable less than 750V) over the readily accessible deck.

- Per Rule 234C3d(2), a vertical clearance of not less than 3.5' for the supply (power) secondary triplex (230C3 cable less than 750V) over the railing around the deck.

- Per Rule 234C4a, a vertical clearance of not less than 9.0' for the communications cable over the readily accessible deck.

- Per Rule 234C4b, a vertical clearance of not less than 3.0' for the communications cable over the railing around the deck.

- Additional rules apply to horizontal clearances to decks and vertical clearances below decks.

Fig. 234-19. Example of a supply (power) service drop and a communications service drop over a deck (Rules 234C3d(1), 234C3d(2), and 234C4).

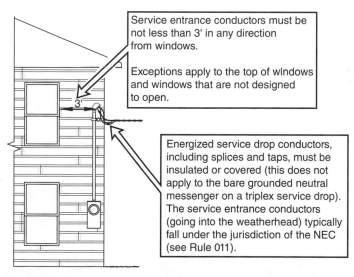

Service entrance conductors must be not less than 3' in any direction from windows.

Exceptions apply to the top of windows and windows that are not designed to open.

Energized service drop conductors, including splices and taps, must be insulated or covered (this does not apply to the bare grounded neutral messenger on a triplex service drop). The service entrance conductors (going into the weatherhead) typically fall under the jurisdiction of the NEC (see Rule 011).

See Photo(s)

Fig. 234-20. Example of clearance of supply conductors (service drops) attached to buildings (Rule 234C3).

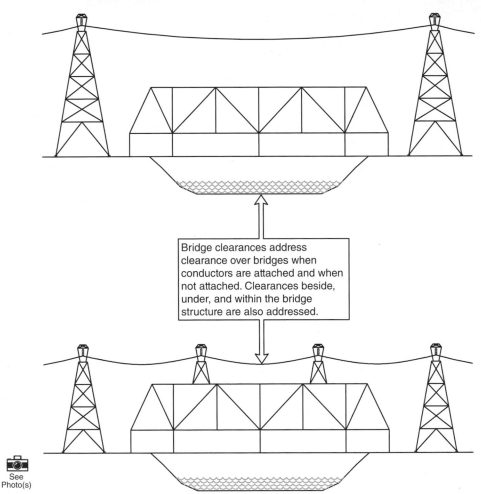

See
Photo(s)

Fig. 234-21. Concerns that are addressed when providing clearance to bridges (Rule 234D).

permanently installed aboveground swimming pools. The main concerns addressed in **NESC** Table 234-3 and **NESC** Figs. 234-3(a), 234-3(b), and 234-3(c) are outlined in Fig. 234-22.

A fully enclosed indoor pool does not apply to Rule 234E1 per Exception 1. A fully enclosed indoor pool is considered to be in a building; therefore, Rule 234C is applicable.

Exception 2 to Rule 234E1 exempts a list of conductors and cables when they are 10 ft or more horizontally from the edge of the pool, diving platform, diving tower, water slide, or other fixed pool-related structures. At that location, Rule 232 clearances apply.

The use of underground power is not required, but it is certainly an acceptable design practice for routing lines near swimming pools. If underground lines are used, Sec. 35, "Direct-buried cable and cable in duct not part of a conduit system," has rules for underground lines near swimming pools in Rule 351C.

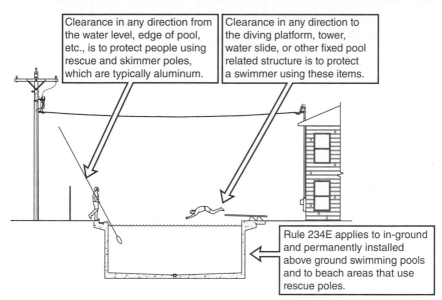

Clearance in any direction from the water level, edge of pool, etc., is to protect people using rescue and skimmer poles, which are typically aluminum.

Clearance in any direction to the diving platform, tower, water slide, or other fixed pool related structure is to protect a swimmer using these items.

Rule 234E applies to in-ground and permanently installed above ground swimming pools and to beach areas that use rescue poles.

Fig. 234-22. Concerns that are addressed when providing clearance to swimming pools (Rule 234E).

Note 2 in Rule 234E1 clarifies the clearance requirements for items that are similar to swimming pools but are not suitable for swimming and therefore do not have skimmer and rescue poles associated with them. These items include whirlpools, hot-tubs, jacuzzis, spas, and wading pools. The note provides guidance as to how to treat each item for clearance purposes.

Swimming areas that are not pools (e.g., beaches, etc.) and have lifeguards using rescue poles must also comply with Table 234-3. The clearances developed for swimming pools were based on the use of the long aluminum rescue and skimmer poles that are used near poolsides. If these poles are also used at a beach area, Table 234-3 applies. If these poles are not used at beach areas, the clearances in Rule 232 (clearance above ground or water) apply. This is also true for water skiing areas. Rule 232 focuses on clearances above water as well as land and considers boating.

The clearance adders for voltages above 22 kV are provided in Rule 234G. Alternate clearances for voltages exceeding 98 kV AC to ground or 139 kV DC to ground are provided in Rule 234H.

234F. Clearance of Wires, Conductors, Cables, and Rigid Live Parts from Grain Bins. Grain bins have unique features that segregate them from Rule 234C (buildings and other installations) and place them in their own dedicated rule. The building clearance rules do apply to grain bins with some very specific modifications. Grain bins are separated into two types, grain bins loaded by permanently installed augers, conveyers, or elevators (Rule 234F1), and grain bins loaded by portable augeurs, conveyers, or elevators (Rule 234F2). **NESC** Figs. 234-4(a) and 234-4(b) are provided in the **Code** to check clearance to grain bins. The main concerns that are addressed in **NESC** Figs. 234-4(a) and 234-4(b) are outlined in Fig. 234-23.

The horizontal clearance to a grain bin loaded by a portable auger can become quite large for tall grain bins as the horizontal clearance is based on the height of the grain bin. See Fig. 234-24.

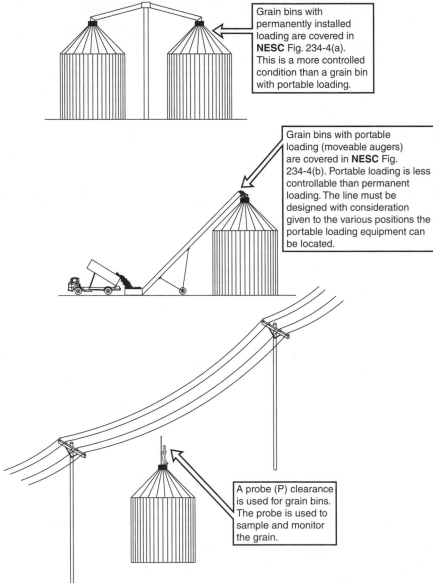

Grain bins with permanently installed loading are covered in **NESC** Fig. 234-4(a). This is a more controlled condition than a grain bin with portable loading.

Grain bins with portable loading (moveable augers) are covered in **NESC** Fig. 234-4(b). Portable loading is less controllable than permanent loading. The line must be designed with consideration given to the various positions the portable loading equipment can be located.

A probe (P) clearance is used for grain bins. The probe is used to sample and monitor the grain.

See Photo(s)

Fig. 234-23. Concerns that are addressed when providing clearance to grain bins (Rule 234F).

The clearance adders for voltages above 22 kV are provided in Rule 234G. Alternate clearances for voltages exceeding 98 kV AC to ground or 139 kV DC to ground are provided in Rule 234H.

234G. Additional Clearances for Voltages Exceeding 22 kV for Wires, Conductors, Cables, and Unguarded Rigid Live Parts of Equipment.   The clearance adders in

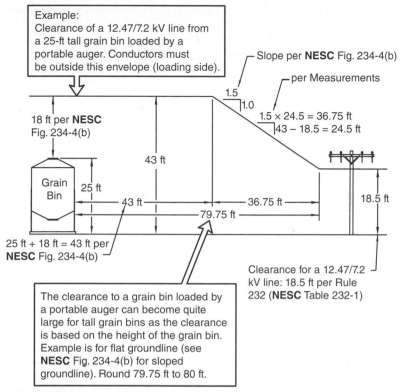

Example:
Clearance of a 12.47/7.2 kV line from a 25-ft tall grain bin loaded by a portable auger. Conductors must be outside this envelope (loading side).

Slope per **NESC** Fig. 234-4(b)

per Measurements

1.5

1.0

18 ft per **NESC** Fig. 234-4(b)

1.5 × 24.5 = 36.75 ft

43 − 18.5 = 24.5 ft

43 ft

Grain Bin   25 ft

43 ft

36.75 ft

18.5 ft

79.75 ft

25 ft + 18 ft = 43 ft per **NESC** Fig. 234-4(b)

Clearance for a 12.47/7.2 kV line: 18.5 ft per Rule 232 (**NESC** Table 232-1)

The clearance to a grain bin loaded by a portable auger can become quite large for tall grain bins as the clearance is based on the height of the grain bin. Example is for flat groundline (see **NESC** Fig. 234-4(b) for sloped groundline). Round 79.75 ft to 80 ft.

Fig. 234-24. Example of clearance to a grain bin loaded by a portable auger (Rule 234F2).

Rule 234G which apply to Rules 234B, 234C, 234D, 234E, 234F, and 234J are similar to the clearance adders used in Rule 232C. See the discussion in Rule 232C.

234H. Alternate Clearances for Voltages Exceeding 98 kV AC to Ground or 139 kV DC to Ground. Alternate clearances may be used for voltages exceeding 98 kV AC to ground or 139 kV DC to ground per Rule 234H. The formulas in Rule 234H must be used for this calculation. The values in **NESC** Table 234-4 can be used to check the proper application of the formulas. See Rule 232D for a discussion of the alternate clearance method.

234I. Clearance of Wires, Conductors, and Cables to Rail Cars. NESC Fig. 234-5 is provided in the Code to aid the application of Rule 234I. Clearance to rail cars is also indirectly covered in Rule 232, NESC Table 232-1, Vertical Clearance over Track Rails of Railroads, and in Rule 231C, Clearance of Supporting Structures from Railroad Tracks.

234J. Clearance of Equipment Mounted on Supporting Structures. Unguarded rigid live parts are not subject to variations in sag. They are required to meet the clearances provided in Rule 234C or 234D, as applicable. Equipment cases, if effectively grounded, may be located on or adjacent to buildings, bridges, or other structures. A grounded meter enclosure is a common example of a grounded equipment case mounted on or adjacent to a building. If an equipment case is not effectively grounded, the clearances in Rules 234C or 234D, as applicable, must be applied. See Rule 215 for overhead equipment grounding requirements and see Sec. 02 for a discussion of the term "effectively grounded." The **NESC** does not address the clearance between a pole and a building. An example of a pole, unguarded rigid live part, and grounded equipment case adjacent to a building is shown in Fig. 234-25.

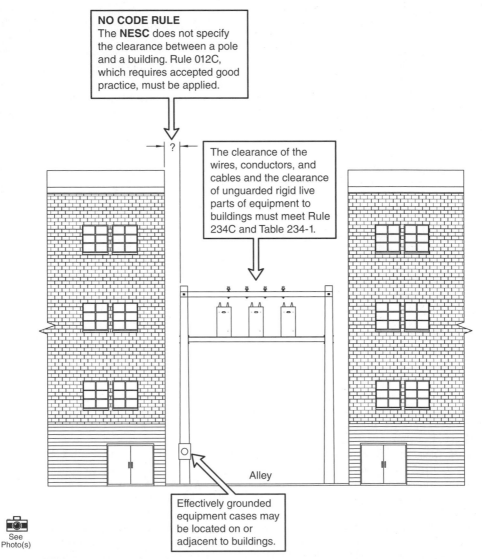

NO CODE RULE
The **NESC** does not specify the clearance between a pole and a building. Rule 012C, which requires accepted good practice, must be applied.

The clearance of the wires, conductors, and cables and the clearance of unguarded rigid live parts of equipment to buildings must meet Rule 234C and Table 234-1.

Alley

Effectively grounded equipment cases may be located on or adjacent to buildings.

See Photo(s)

Fig. 234-25. Example of a pole, conductors, unguarded rigid live parts, and grounded equipment case adjacent to a building (Rule 234J).

235. CLEARANCE FOR WIRES, CONDUCTORS, OR CABLES CARRIED ON THE SAME SUPPORTING STRUCTURE

235A. Application of Rule. Rule 235 can be thought of as the opposite of Rule 233, Clearances between Wires, Conductors, and Cables Carried on Different Supporting Structures. Rule 235 addresses conductors on the same supporting structure (pole). See Rule 233B for a comparison of Rules 233B, 233C, 235B, and 235C. Rule 235 is commonly used for checking vertical clearance on double circuit structures such as a transmission line with a distribution underbuild or a double circuit distribution line. Another common use of this rule is checking vertical clearance between the supply and communication conductors on a joint-use line. Rule 235 should be used in conjunction with Rule 238, Clearances at the Support between Specified Communications and Supply Facilities Located on the Same Structure, when checking joint-use (power and communication) clearances.

Rule 235A outlines the general requirements for checking clearance on the same supporting structure. It does not immediately provide conductor temperature and loading conditions as Rules 232A, 233A, and 234A do. The conductors in Rule 235 are tied onto a structure. Conductor temperature and loading conditions will be needed to check conditions out in the span. Conductor temperature and loading conditions are provided in Rule 235B1b for horizontal clearance and Rule 235C2b(1)(c) for vertical clearance.

Per Rule 235A1, a multiconductor cable meeting Rule 230C or 230D is considered a single conductor for the purposes of this rule. Rule 235A2 provides a list of conductors supported by messengers or span wires that are not subject to Rule 235. See Fig. 235-1.

When comparing line conductors of different circuits, the voltage between the conductors is determined using the greater of the phasor difference or the phase-to-ground voltage of the higher circuit. See Fig. 235-2.

235B. Horizontal Clearance between Line Conductors. Rule 235B applies to the horizontal clearance between line conductors attached to fixed supports (rigid insulators) on the same supporting structure. The horizontal clearance requirements apply between conductors of the same circuit or different circuits on the same supporting structure. For horizontal clearance between circuits of different voltage classifications on the same support arm (crossarm), see the additional requirements in Rule 235F. For horizontal clearance between conductors of different circuits on different supporting structures, Rule 233 applies instead of Rule 235. See Rule 233B for a comparison of Rules 233B, 233C, 235B, and 235C.

The exception to Rule 235B1b states that the NESC does not specify a horizontal clearance between conductors of the same circuit when rated above 50 kV. A similar statement can be found in Rule 235C1 for vertical clearance. Typically, conductor clearance out in the span for transmission lines above 50 kV is checked by performing a galloping analysis. Conductor gallop is a high-amplitude, low-frequency oscillation. See Fig. 235-3.

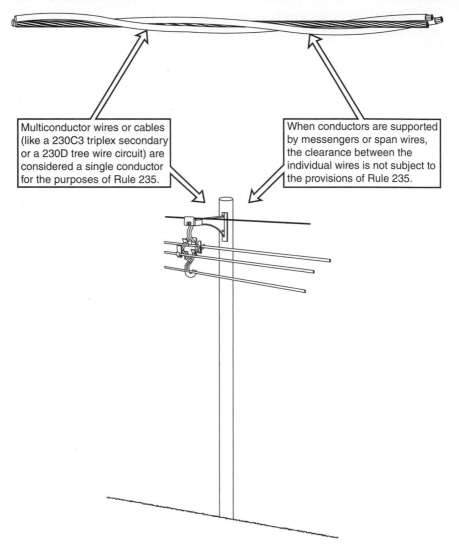

Multiconductor wires or cables (like a 230C3 triplex secondary or a 230D tree wire circuit) are considered a single conductor for the purposes of Rule 235.

When conductors are supported by messengers or span wires, the clearance between the individual wires is not subject to the provisions of Rule 235.

Fig. 235-1. Multiconductor cables and conductors supported by messengers (Rule 235A).

The horizontal clearance must be checked at the support (crossarm) using **NESC** Table 235-1 and out in the span based on sag using either **NESC** Table 235-2 or **NESC** Table 235-3 depending on the conductor size. **NESC** Tables 235-2 and 235-3 have a formula at the bottom of each table. The table or the formula can be used. The formulas are also presented in Rule 235B1b. The horizontal clearance at the structure will typically be controlled by the clearance required out in the span based on sag. See Fig. 235-4.

The values found in **NESC** Tables 235-1, 235-2, and 235-3 are clearances between conductors at supports, not spacing between supports. See Rule 230A3 for a

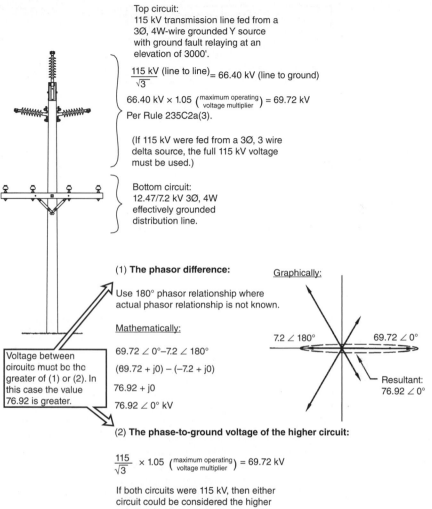

Top circuit:
115 kV transmission line fed from a
3Ø, 4W-wire grounded Y source
with ground fault relaying at an
elevation of 3000'.

$$\frac{115\ kV\ \text{(line to line)}}{\sqrt{3}} = 66.40\ kV\ \text{(line to ground)}$$

$$66.40\ kV \times 1.05\ \left(\substack{\text{maximum operating}\\\text{voltage multiplier}}\right) = 69.72\ kV$$

Per Rule 235C2a(3).

(If 115 kV were fed from a 3Ø, 3 wire
delta source, the full 115 kV voltage
must be used.)

Bottom circuit:
12.47/7.2 kV 3Ø, 4W
effectively grounded
distribution line.

(1) **The phasor difference:**   Graphically:

Use 180° phasor relationship where
actual phasor relationship is not known.

Mathematically:

$7.2 \angle 180°$ $69.72 \angle 0°$

$69.72 \angle 0° - 7.2 \angle 180°$

$(69.72 + j0) - (-7.2 + j0)$

$76.92 + j0$

$76.92 \angle 0°\ kV$

Resultant:
$76.92 \angle 0°$

Voltage between
circuits must be the
greater of (1) or (2). In
this case the value
76.92 is greater.

(2) **The phase-to-ground voltage of the higher circuit:**

$$\frac{115}{\sqrt{3}} \times 1.05\ \left(\substack{\text{maximum operating}\\\text{voltage multiplier}}\right) = 69.72\ kV$$

If both circuits were 115 kV, then either
circuit could be considered the higher
voltage circuit.

Fig. 235-2. Example of the voltage between line conductors of different circuits (Rule 235A).

discussion of spacing and clearance. Live metallic hardware electrically connected to the conductor, like a conductor tie wire on an insulator, must be addressed for the clearance at the structure but not for the clearance at the structure based on the sag out in the span. See Fig. 235-5.

Rule 235B2 has requirements for suspension insulators because **NESC** Table 235-1 is for fixed supports. The rules for suspension insulators require applying a 6-lb/ft^2 wind to the conductor that the suspension insulator is holding. The wind pressure may be reduced to 4 lb/ft^2 in sheltered areas with special conditions, but trees are not considered a shelter to a line. The deflection of flexible structures and fittings

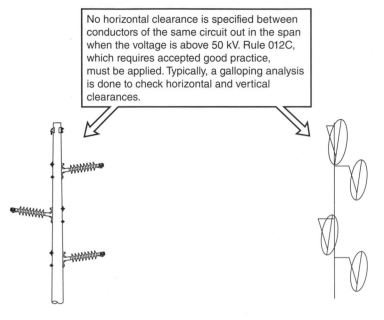

No horizontal clearance is specified between conductors of the same circuit out in the span when the voltage is above 50 kV. Rule 012C, which requires accepted good practice, must be applied. Typically, a galloping analysis is done to check horizontal and vertical clearances.

Transmission Structure   **Galloping Analysis**

Fig. 235-3. Horizontal clearance between conductors of the same circuit rated above 50 kV (Rule 235B1b).

must also be considered if the deflection reduces the clearance in question. See Rules 233A1a(2) and 234A2 for similar requirements.

Rule 235B3 provides alternate clearances for voltages exceeding 98 kV AC to ground or 139 kV DC to ground. The formulas in Rule 235B3 must be used for this calculation. The values in **NESC** Table 235-4 can be used to check the proper application of the formulas. See Rule 232D for a discussion of alternate clearances.

Rule 235B can be applied using two different approaches. One approach is to find the required horizontal clearance of a particular span. Another approach is to determine a maximum span based on the horizontal clearance for a particular structure type, conductor size, conductor sag, and circuit voltage.

235C. Vertical Clearance at the Support for Line Conductors and Service Drops.   Rule 235C applies to the vertical clearance between conductors at the support on the same supporting structure (e.g., pole). Rule 235C applies to line conductors and service drops. The terms line conductors (conductors, line) and service drop are both defined in **NESC** Sec. 02, Definitions of Special Terms. The vertical clearance requirements apply to conductors of the same and different circuits on the same supporting structure. For vertical clearance between conductors of different circuits on different supporting structures, Rule 233 applies instead of Rule 235. See Rule 233B for a comparison of Rules 233B, 233C, 235B, and 235C. Rule 235C1a describes two items that the rule does not cover and therefore **NESC** Table 235-5 does not apply to. First, the **NESC** does not specify a vertical

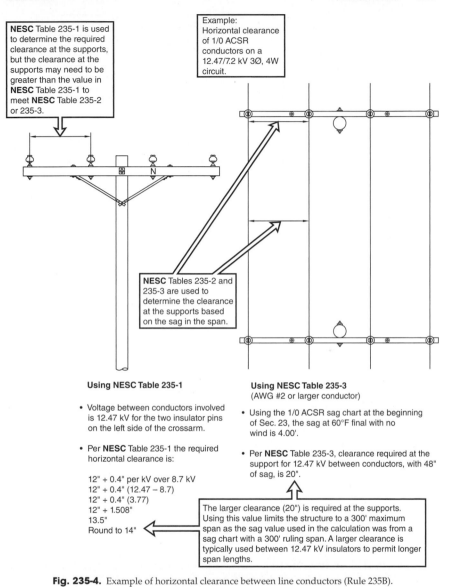

NESC Table 235-1 is used to determine the required clearance at the supports, but the clearance at the supports may need to be greater than the value in NESC Table 235-1 to meet NESC Table 235-2 or 235-3.

Example:
Horizontal clearance of 1/0 ACSR conductors on a 12.47/7.2 kV 3Ø, 4W circuit.

NESC Tables 235-2 and 235-3 are used to determine the clearance at the supports based on the sag in the span.

Using NESC Table 235-1

- Voltage between conductors involved is 12.47 kV for the two insulator pins on the left side of the crossarm.

- Per NESC Table 235-1 the required horizontal clearance is:

  12" + 0.4" per kV over 8.7 kV
  12" + 0.4" (12.47 − 8.7)
  12" + 0.4" (3.77)
  12" + 1.508"
  13.5"
  Round to 14"

Using NESC Table 235-3
(AWG #2 or larger conductor)

- Using the 1/0 ACSR sag chart at the beginning of Sec. 23, the sag at 60°F final with no wind is 4.00'.

- Per NESC Table 235-3, clearance required at the support for 12.47 kV between conductors, with 48" of sag, is 20".

The larger clearance (20") is required at the supports. Using this value limits the structure to a 300' maximum span as the sag value used in the calculation was from a sag chart with a 300' ruling span. A larger clearance is typically used between 12.47 kV insulators to permit longer span lengths.

Fig. 235-4. Example of horizontal clearance between line conductors (Rule 235B).

clearance between conductors of the same circuit when rated above 50 kV. A similar statement can be found in an exception to Rule 235B1b for horizontal clearance out in the span. Second, Rule 235C1a does not specify a vertical clearance between ungrounded open supply conductors (e.g., bare noninsulated phase wires) of the same phase and circuit of the same utility (e.g., a tap off the main line of 0-50 kV). See Fig. 235-6.

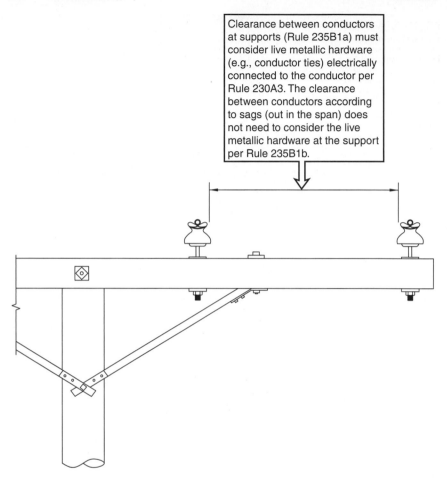

Clearance between conductors at supports (Rule 235B1a) must consider live metallic hardware (e.g., conductor ties) electrically connected to the conductor per Rule 230A3. The clearance between conductors according to sags (out in the span) does not need to consider the live metallic hardware at the support per Rule 235B1b.

Fig. 235-5. Horizontal clearance measurements between conductors at supports (Rule 235B).

The vertical clearance must be checked at the support (e.g., pole) using **NESC** Table 235-5 and out in the span based on sag using Rule 235C2b. Typically, the vertical clearance at the structure will be controlled by the clearance required out in the span based on the upper and lower conductor sags.

Per Rule 235C2b(1)(c), the upper conductor must be checked at 120°F or the maximum operating temperature at final sag while the lower conductor is at the same ambient temperature as the upper conductor without electrical loading at final sag. The upper conductor must also be checked at 32°F with ice from the Zone in Rule 230B at final sag, while the lower conductor is at the same ambient temperature as the upper conductor without electrical loading or ice at final sag.

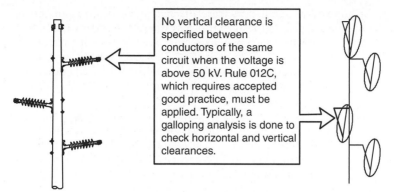

No vertical clearance is specified between conductors of the same circuit when the voltage is above 50 kV. Rule 012C, which requires accepted good practice, must be applied. Typically, a galloping analysis is done to check horizontal and vertical clearances.

Conductors of same circuit above 50 kV

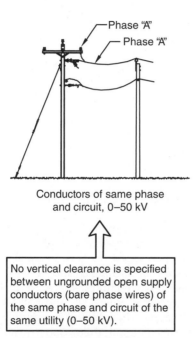

Phase "A"

Phase "A"

Conductors of same phase and circuit, 0–50 kV

No vertical clearance is specified between ungrounded open supply conductors (bare phase wires) of the same phase and same circuit of the same utility (0–50 kV).

Fig. 235-6. Items not covered under vertical clearance between conductors (Rule 235C1a).

Using the 1/0 ACSR sample sag and tension chart at the beginning of Sec. 23, assume the following for an example of applying Rule 235C2b(1)(c)i:

- 1/0 ACSR phase conductor of a 12.47/7.2 kV, 3Ø, 4W, effectively grounded line is loaded to approximately 75% capacity (175 A) at both winter and summer peak periods, medium loading, clearance Zone 2, 300-ft ruling span. See the discussion of this example at the beginning of Sec. 23.
- Upper phase conductor at 167°F, final. Sag is 6.25 ft.
- Lower neutral conductor at 104°F, final. Sag is 5.34 ft.

If the top and bottom conductors were attached at the same height, the top conductor would be sagging 0.91 ft (6.25 ft − 5.34 ft = 0.91 ft) below the bottom conductor. The top conductor attachment must be raised 0.91 ft (11 in) to make the sags level plus 12 in (16 in × 0.75 = 12 in per **NESC** Table 235-5 and Rule 235C2b(1)(a)) to meet the required clearance in the span for a 7.2-kV phase conductor over a neutral conductor. The top conductor attachment must be 23 in (11 in + 12 in = 23 in) higher than the bottom conductor.

Using the 1/0 ACSR sample sag and tension chart at the beginning of Sec. 23, assume the following for an example of applying Rule 235C2b(1)(c)ii:

- 1/0 ACSR phase conductor of a 12.47/7.2 kV, 3Ø, 4W, effectively grounded line is loaded to approximately 75% capacity (175 A) at both winter and summer peak periods, medium loading, clearance Zone 2, 300-ft ruling span. See the discussion of this example at the beginning of Sec. 23.
- Upper phase conductor at 32°F, 0.25 in ice, final. Sag is 4.49 ft.
- Lower neutral conductor at −20°F, no ice, final. Sag is 1.78 ft.

If the top and bottom conductors were attached at the same height, the top conductor would be sagging 2.71 ft (4.49 ft − 1.78 ft = 2.71 ft) below the bottom conductor. The top conductor attachment must be raised 2.71 ft (32.5 in) to make the sags level plus 12 in (16 in × 0.75 = 12 in per **NESC** Table 235-5 and Rule 235C2b(1)(a)) to meet the required clearance in the span for a 7.2-kV phase conductor over a neutral conductor. The top conductor attachment must be 32.5 in + 12 in = 44.5 in higher than the bottom conductor.

The greater vertical clearance at the structure between 23 in and 44.5 in is 44.5 in. A vertical clearance larger than 44.5 in will be needed to permit span lengths longer than the 300-ft ruling span in the sample sag and tension chart.

Exceptions apply to Rule 235C2b(1)(c) that permit ignoring the rule under certain conditions.

Examples of vertical clearance between line conductors are shown in Figs. 235-7 and 235-8.

Rule 235C can be applied using two different approaches. One approach is to find the required vertical clearance of a particular span. Another approach is to determine a maximum span based on the vertical clearance for a particular structure type, conductor size, conductor sag, and circuit voltage.

In addition to powerline-to-powerline clearance, **NESC** Table 235-5 applies to power-to-communication clearance. Rule 235C and **NESC** Table 235-5 specify the conductor-to-conductor clearance. In addition to conductor-to-conductor clearance, a joint-use (power and communication) **Code** review requires equipment-to-equipment and hardware-to-hardware clearance, which are specified in Rule 238. The vertical clearance requirements in Rule 238B and **NESC** Table 235-5 revolve around 40 in. Less than 40 in (e.g., 30 in) is acceptable for certain grounded supply facilities. Greater than 40 in is required for supply voltages above 8.7 kV to ground. The vertical clearance must be checked at the support (e.g., pole) using **NESC** Table 235-5 and out in the span based on sag using Rule 235C2b. Typically, the vertical clearance at the structure will be controlled by the clearance required out in the span based on the upper and lower conductor sags. Depending on what type of conductors are involved, a power conductor sag

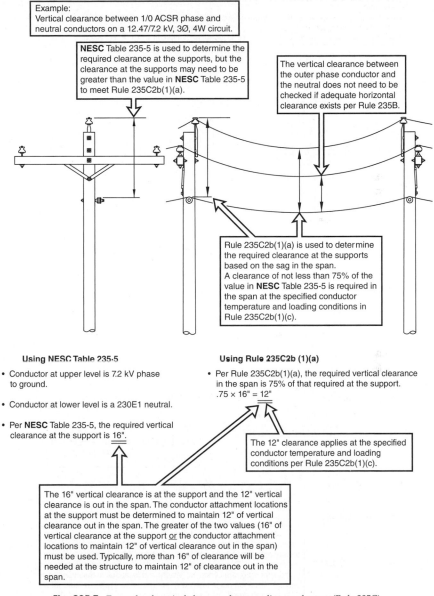

Example:
Vertical clearance between 1/0 ACSR phase and neutral conductors on a 12.47/7.2 kV, 3Ø, 4W circuit.

NESC Table 235-5 is used to determine the required clearance at the supports, but the clearance at the supports may need to be greater than the value in **NESC** Table 235-5 to meet Rule 235C2b(1)(a).

The vertical clearance between the outer phase conductor and the neutral does not need to be checked if adequate horizontal clearance exists per Rule 235B.

Rule 235C2b(1)(a) is used to determine the required clearance at the supports based on the sag in the span. A clearance of not less than 75% of the value in **NESC** Table 235-5 is required in the span at the specified conductor temperature and loading conditions in Rule 235C2b(1)(c).

Using NESC Table 235-5

• Conductor at upper level is 7.2 kV phase to ground.

• Conductor at lower level is a 230E1 neutral.

• Per **NESC** Table 235-5, the required vertical clearance at the support is 16".

Using Rule 235C2b (1)(a)

• Per Rule 235C2b(1)(a), the required vertical clearance in the span is 75% of that required at the support. .75 × 16" = 12"

The 12" clearance applies at the specified conductor temperature and loading conditions per Rule 235C2b(1)(c).

The 16" vertical clearance is at the support and the 12" vertical clearance is out in the span. The conductor attachment locations at the support must be determined to maintain 12" of vertical clearance out in the span. The greater of the two values (16" of vertical clearance at the support or the conductor attachment locations to maintain 12" of vertical clearance out in the span) must be used. Typically, more than 16" of clearance will be needed at the structure to maintain 12" of clearance out in the span.

Fig. 235-7. Example of vertical clearance between line conductors (Rule 235C).

Example:
Vertical clearance between a 115 kV transmission line with ground fault relaying and a 12.47/7.2 kV effectively grounded distribution line. The lines are owned by the same utility and the line is built at an elevation of 3000'. Assume the circuits are 180° out of phase.

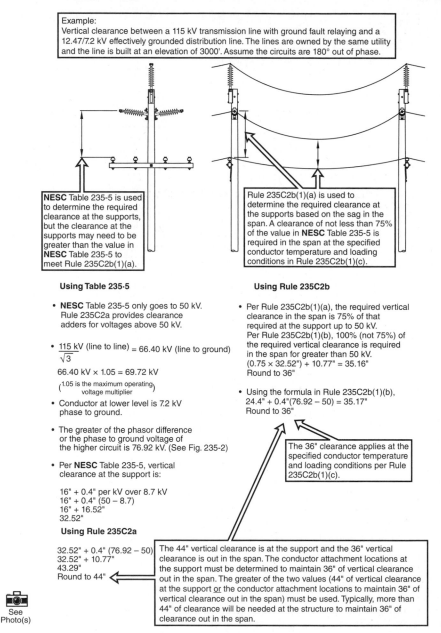

NESC Table 235-5 is used to determine the required clearance at the supports, but the clearance at the supports may need to be greater than the value in **NESC** Table 235-5 to meet Rule 235C2b(1)(a).

Rule 235C2b(1)(a) is used to determine the required clearance at the supports based on the sag in the span. A clearance of not less than 75% of the value in **NESC** Table 235-5 is required in the span at the specified conductor temperature and loading conditions in Rule 235C2b(1)(c).

Using Table 235-5

- **NESC** Table 235-5 only goes to 50 kV. Rule 235C2a provides clearance adders for voltages above 50 kV.

- $\dfrac{115 \text{ kV (line to line)}}{\sqrt{3}} = 66.40$ kV (line to ground)

  66.40 kV × 1.05 = 69.72 kV

  (1.05 is the maximum operating voltage multiplier)

- Conductor at lower level is 7.2 kV phase to ground.

- The greater of the phasor difference or the phase to ground voltage of the higher circuit is 76.92 kV. (See Fig. 235-2)

- Per **NESC** Table 235-5, vertical clearance at the support is:

  16" + 0.4" per kV over 8.7 kV
  16" + 0.4" (50 – 8.7)
  16" + 16.52"
  32.52"

Using Rule 235C2a

32.52" + 0.4" (76.92 – 50)
32.52" + 10.77"
43.29"
Round to 44"

Using Rule 235C2b

- Per Rule 235C2b(1)(a), the required vertical clearance in the span is 75% of that required at the support up to 50 kV. Per Rule 235C2b(1)(b), 100% (not 75%) of the required vertical clearance is required in the span for greater than 50 kV.
  (0.75 × 32.52") + 10.77" = 35.16"
  Round to 36"

- Using the formula in Rule 235C2b(1)(b), 24.4" + 0.4"(76.92 – 50) = 35.17"
  Round to 36"

The 36" clearance applies at the specified conductor temperature and loading conditions per Rule 235C2b(1)(c).

The 44" vertical clearance is at the support and the 36" vertical clearance is out in the span. The conductor attachment locations at the support must be determined to maintain 36" of vertical clearance out in the span. The greater of the two values (44" of vertical clearance at the support or the conductor attachment locations to maintain 36" of vertical clearance out in the span) must be used. Typically, more than 44" of clearance will be needed at the structure to maintain 36" of clearance out in the span.

See Photo(s)

Fig. 235-8. Example of vertical clearance between line conductors (Rule 235C).

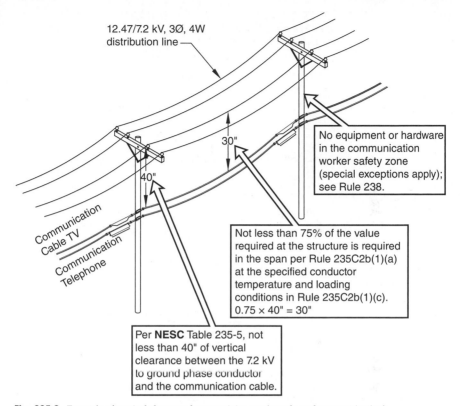

Fig. 235-9. Example of vertical clearance between joint-use (supply and communication) conductors (Rule 235C).

and tension chart and a communications cable sag and tension chart will be needed to check supply-to-communication clearance at midspan. Examples of joint-use (supply and communication) structures are shown in Figs. 235-9 through 235-14.

Exception 1 of Rule 235C1 references Rule 235G which covers conductor spacing on vertical racks or separate brackets. Exceptions 2 and 3 of Rule 235C1 apply to joint-use (supply and communication) construction. See Fig. 235-15.

In addition to the sag-related clearance checks in Rule 235C2b(1), Rule 235C2b(2) requires making sag adjustments when necessary to maintain clearance. When sag is reduced, tension is increased. If sag reductions are made to maintain clearance, the tension limits in Rule 261H1 must not be exceeded. Rule 235C2b(3) requires an additional special clearance check for joint-use (supply and communication) structures. See Fig. 235-16.

Alternate clearances may be used for voltages exceeding 98 kV AC to ground or 139 kV DC to ground. The formulas in Rule 233C3 must be used for this calculation. See Rule 232D for a discussion of the alternate clearance method.

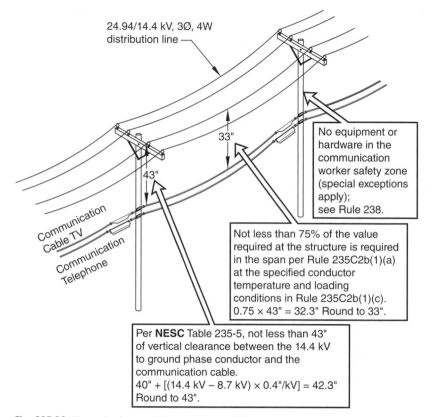

24.94/14.4 kV, 3Ø, 4W
distribution line

33"

43"

Communication
Cable TV

Communication
Telephone

No equipment or
hardware in the
communication
worker safety zone
(special exceptions
apply);
see Rule 238.

Not less than 75% of the value
required at the structure is required
in the span per Rule 235C2b(1)(a)
at the specified conductor
temperature and loading
conditions in Rule 235C2b(1)(c).
0.75 × 43" = 32.3" Round to 33".

Per **NESC** Table 235-5, not less than 43"
of vertical clearance between the 14.4 kV
to ground phase conductor and the
communication cable.
40" + [(14.4 kV − 8.7 kV) × 0.4"/kV] = 42.3"
Round to 43".

Fig. 235-10. Example of vertical clearance between joint-use (supply and communication) conductors (Rule 235C).

Rule 235C4 labels the area between the supply space and the communication space both at the structure and out in the span. The name for this area is the "communication worker safety zone." The work rules in Part 4 of the **NESC** provide the qualifications of a supply employee and a communication employee. Since the communication employee is not trained to work on supply lines, a safety zone exists for the communication employee's protection. If a communication line is positioned below a supply line on an overhead structure, but the required communication worker safety zone clearance does not exist, the communication employee can correct the violation if the communication employee does not violate the minimum approach distances and other requirements in Sec. 43. If the communication employee cannot maintain the minimum approach distances in Sec. 43, the communication employee must contact a supply employee to correct the violation. See Sec. 43 for additional information.

The communication worker safety zone is defined by both Rule 238 and Rule 235C, not one rule or the other but the combined effect of both rules. Rule 235C applies to clearance between wires, conductors, and cables. Rule 238 applies to

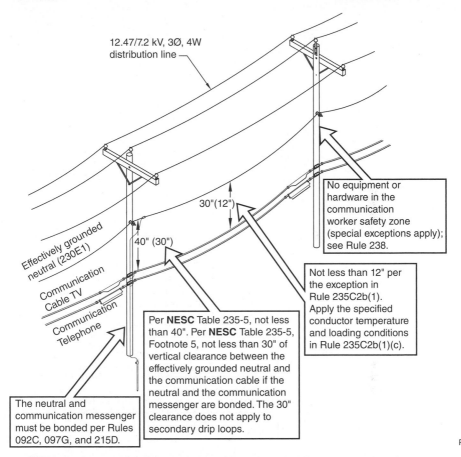

12.47/7.2 kV, 3Ø, 4W distribution line —

Effectively grounded neutral (230E1)

Communication Cable TV

Communication Telephone

30"(12")

40" (30")

No equipment or hardware in the communication worker safety zone (special exceptions apply); see Rule 238.

Not less than 12" per the exception in Rule 235C2b(1). Apply the specified conductor temperature and loading conditions in Rule 235C2b(1)(c).

Per **NESC** Table 235-5, not less than 40". Per **NESC** Table 235-5, Footnote 5, not less than 30" of vertical clearance between the effectively grounded neutral and the communication cable if the neutral and the communication messenger are bonded. The 30" clearance does not apply to secondary drip loops.

The neutral and communication messenger must be bonded per Rules 092C, 097G, and 215D.

See Photo(s)

Fig. 235-11. Example of vertical clearance between joint-use (supply and communication) conductors (Rule 235C).

clearance between supply conductors and communications equipment, between communication conductors and supply equipment, and between supply and communications equipment. Only a few select items are permitted in the communication worker safety zone. The exceptions are the items specified in Rules 238C, 238D, and 239. Rule 238E also addresses the communication worker safety zone. See the communication worker safety zone discussion and figure in Rule 238E.

235D. Clearance between Line Wires, Conductors, and Cables Located at Different Levels in the Same Supporting Structure. Horizontal and vertical clearances are specified in Rule 235. NESC Fig. 235-1 is provided in the Code for determining a clearance envelope for an energized conductor. If both a horizontal

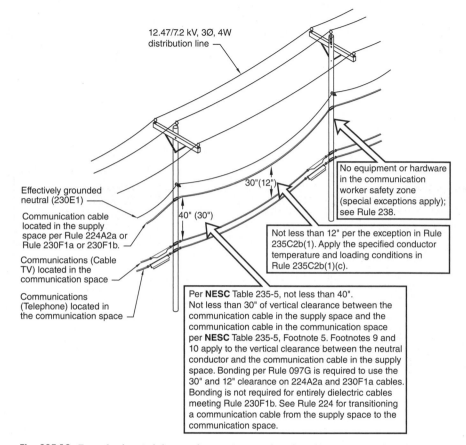

12.47/7.2 kV, 3Ø, 4W
distribution line

Effectively grounded
neutral (230E1)

Communication cable
located in the supply
space per Rule 224A2a or
Rule 230F1a or 230F1b.

Communications (Cable
TV) located in the
communication space

Communications
(Telephone) located in
the communication space

30"(12")

40" (30")

No equipment or hardware
in the communication
worker safety zone
(special exceptions apply);
see Rule 238.

Not less than 12" per the exception in Rule
235C2b(1). Apply the specified conductor
temperature and loading conditions in
Rule 235C2b(1)(c).

Per **NESC** Table 235-5, not less than 40".
Not less than 30" of vertical clearance between the
communication cable in the supply space and the
communication cable in the communication space
per **NESC** Table 235-5, Footnote 5. Footnotes 9 and
10 apply to the vertical clearance between the neutral
conductor and the communication cable in the supply
space. Bonding per Rule 097G is required to use the
30" and 12" clearance on 224A2a and 230F1a cables.
Bonding is not required for entirely dielectric cables
meeting Rule 230F1b. See Rule 224 for transitioning
a communication cable from the supply space to the
communication space.

Fig. 235-12. Example of vertical clearance between joint-use (supply and communication) conductors (Rule 235C).

and a vertical clearance apply to a conductor, a clearance envelope can be determined by drawing a box with the required horizontal clearance from Rule 235B and vertical clearance from Rule 235C.

235E. Clearances in Any Direction from Line Conductors at or Near a Support to Supports, and to Vertical or Lateral Conductors, Service Drops, and Span or Guy Wires Attached to the Same Supporting Structure.   NESC Table 235-6 is used to find clearance in any direction (not just horizontal or vertical) from line conductors (that are at or near a support) to supports and to vertical or lateral conductors, service drops, and span or guy wires attached to the same support. The term line conductors refers to the conductors spanning from pole to pole. Vertical and lateral conductors are discussed in Rule 239. See Rule 239A for a figure showing vertical and lateral conductors. NESC Table 235-6 is commonly used for checking clearance to guys on complicated guying structures. Another common application is clearance of a conductor to the supporting structure (e.g., pole). See Fig. 235-17.

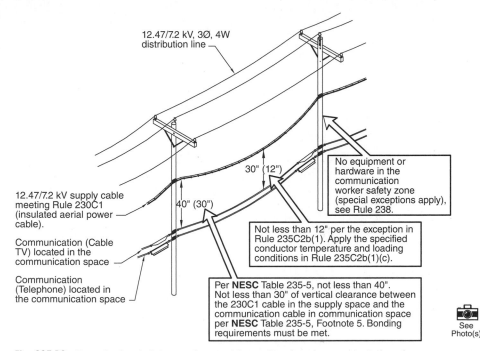

12.47/7.2 kV, 3Ø, 4W distribution line

30" (12")

No equipment or hardware in the communication worker safety zone (special exceptions apply), see Rule 238.

12.47/7.2 kV supply cable meeting Rule 230C1 (insulated aerial power cable).

40" (30")

Communication (Cable TV) located in the communication space

Not less than 12" per the exception in Rule 235C2b(1). Apply the specified conductor temperature and loading conditions in Rule 235C2b(1)(c).

Communication (Telephone) located in the communication space

Per NESC Table 235-5, not less than 40". Not less than 30" of vertical clearance between the 230C1 cable in the supply space and the communication cable in communication space per NESC Table 235-5, Footnote 5. Bonding requirements must be met.

See Photo(s)

Fig. 235-13. Example of vertical clearance between joint-use (supply and communication) conductors (Rule 235C).

NESC Table 235-6 addresses clearance of supply and communication service drops to line conductors. Service drop clearances are also addressed in Rules 232B and 235C. Occasionally, a power service drop angles down past a communications line conductor. An example of a power service drop angling down past a communications line conductor is shown in Fig. 235-18.

The notes in Rule 235E1 direct the reader to Rule 238F for clearance to antennas in the supply space and to Rules 236D1 and 238 for clearance to antennas in the communication space.

Rule 235E2 has requirements for suspension insulators because NESC Table 235-6 is for fixed supports. The rules for suspension insulators require applying a 6-lb/ft^2 wind to the conductor that the suspension insulator is holding. The wind pressure may be reduced to 4 lb/ft^2 in sheltered areas with special conditions, but trees are not considered a shelter to a line. The deflection of flexible structures and fittings must also be considered if the deflection reduces the clearance in question. See Rules 233A1(a)(2) and 234A2 for similar requirements.

Rule 235E3 provides alternate clearances for voltages exceeding 98 kV AC to ground or 139 kV DC to ground. The values in NESC Table 235-7 can be used to check the proper application of the formulas. See Rule 232D for a discussion of alternate clearances.

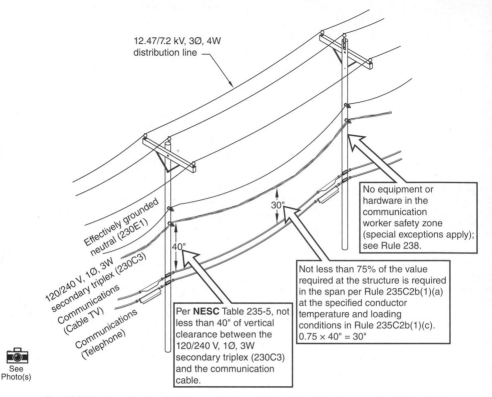

12.47/7.2 kV, 3Ø, 4W distribution line

Effectively grounded neutral (230E1)

120/240 V, 1Ø, 3W secondary triplex (230C3)

Communications (Cable TV)

Communications (Telephone)

See Photo(s)

30"

40"

No equipment or hardware in the communication worker safety zone (special exceptions apply); see Rule 238.

Not less than 75% of the value required at the structure is required in the span per Rule 235C2b(1)(a) at the specified conductor temperature and loading conditions in Rule 235C2b(1)(c). 0.75 × 40" = 30"

Per **NESC** Table 235-5, not less than 40" of vertical clearance between the 120/240 V, 1Ø, 3W secondary triplex (230C3) and the communication cable.

Fig. 235-14. Example of vertical clearance between joint use (supply and communication) conductors (Rule 235C).

235F. Clearances between Circuits of Different Voltage Classifications Located in the Supply Space on the Same Support Arm.   Rule 235F recognizes that circuits of different voltage levels are not always separated vertically on a structure. Rule 220C assumes circuits of different voltage levels will be separated vertically at different heights and does not comment on horizontal separation. Rule 235F must be used in conjunction with Rule 235B, Horizontal Clearance between Line Conductors, and in some cases Rule 236, Climbing Space.

The **NESC** uses the term support arm for a crossarm. The terms bridge arm and sidearm are also used in this rule. A bridge arm is an arm that spans between two poles. A sidearm is an arm to one side of the pole. A sidearm is commonly called an alley arm. Examples of clearance between supply circuits of different voltages on the same support arm are outlined in Fig. 235-19.

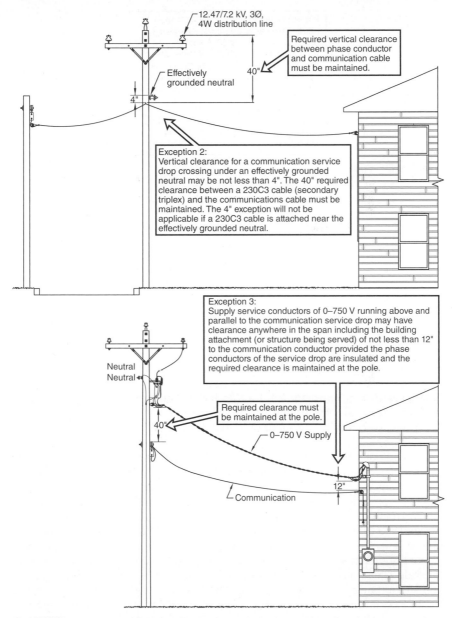

12.47/7.2 kV, 3Ø,
4W distribution line

Required vertical clearance
between phase conductor
and communication cable
must be maintained.

40"

Effectively
grounded neutral

4"

Exception 2:
Vertical clearance for a communication service
drop crossing under an effectively grounded
neutral may be not less than 4". The 40" required
clearance between a 230C3 cable (secondary
triplex) and the communications cable must be
maintained. The 4" exception will not be
applicable if a 230C3 cable is attached near the
effectively grounded neutral.

Exception 3:
Supply service conductors of 0–750 V running above and
parallel to the communication service drop may have
clearance anywhere in the span including the building
attachment (or structure being served) of not less than 12"
to the communication conductor provided the phase
conductors of the service drop are insulated and the
required clearance is maintained at the pole.

Neutral
Neutral

Required clearance must
be maintained at the pole.

40"

0–750 V Supply

12"

Communication

Fig. 235-15. Exceptions to vertical clearance requirements for joint-use (supply and communication) construction (Rule 235C1).

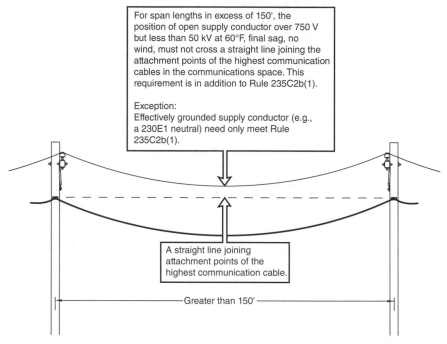

For span lengths in excess of 150', the position of open supply conductor over 750 V but less than 50 kV at 60°F, final sag, no wind, must not cross a straight line joining the attachment points of the highest communication cables in the communications space. This requirement is in addition to Rule 235C2b(1).

Exception:
Effectively grounded supply conductor (e.g., a 230E1 neutral) need only meet Rule 235C2b(1).

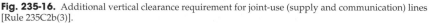

A straight line joining attachment points of the highest communication cable.

Greater than 150'

Fig. 235-16. Additional vertical clearance requirement for joint-use (supply and communication) lines [Rule 235C2b(3)].

235G. Conductor Spacing: Vertical Racks or Separate Brackets.   Reduced vertical clearance between conductors is permitted when vertical racks are used. The rules for conductor spacing on vertical racks are outlined in Figs. 235-20 and 235-21.

235H. Vertical Clearance and Spacing between Communication Conductors, Wires, and Cables, in the Communication Space.   Rule 235H addresses the vertical spacing and vertical clearance between communication cables at the structure and in the span. Rule 235H only addresses vertical clearance and vertical spacing, not horizontal clearance or horizontal spacing. Commonly, communication cables are attached directly to a pole and separated vertically on the pole, but they can also be separated horizontally by placing the cables on opposite sides of the pole (commonly referred to as "boxing" the pole), or on a communications crossarm. Per Rule 235H1, a not less than 12-in vertical spacing is required between messengers supporting communication cables. The 12-in value is a spacing dimension, not a clearance dimension. See Rule 230A3 for a discussion of clearance versus spacing. The 12-in spacing dimension allows through-bolts for supporting communication messengers to be drilled 12 in apart, center-to-center.

Communication cables supported on messengers are installed by first attaching the messenger to the pole and then lashing the communication cable to the

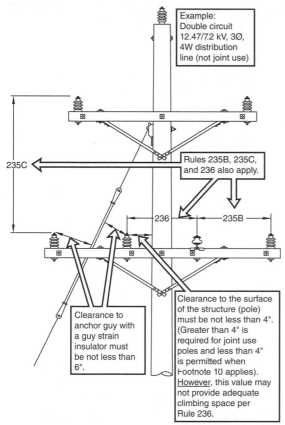

Example:
Double circuit
12.47/7.2 kV, 3Ø,
4W distribution
line (not joint use)

Rules 235B, 235C, and 236 also apply.

235C

236 235B

Clearance to anchor guy with a guy strain insulator must be not less than 6".

Clearance to the surface of the structure (pole) must be not less than 4". (Greater than 4" is required for joint use poles and less than 4" is permitted when Footnote 10 applies). However, this value may not provide adequate climbing space per Rule 236.

Using Table 235-6

• Table voltages are phase to phase.

• Using over 8.7 to 50 kV column and Row 2 (guy wires), the clearance required to the anchor guy is 6" + 0.25" per kV over 8.7 kV.
6" + 0.25" (12.47 − 8.7) = 6.94"
Round to 7"

• If a guy strain insulator is used, Footnote 11 applies and a 25% reduction to 6.94" is applicable.
7"− [7" × 0.25] = 5.25"
Round to 6"

• Using over 8.7 to 50 kV column and Row 4b (all other), the clearance required to the surface of the structure (pole) is 3" + 0.2" per kV over 8.7 kV.
3" + 0.2" (12.47 − 8.7) = 3.75"
Round to 4"

• Greater than 4" is required to the surface of the structure for joint use poles per Row 4a. Less than 4" is permitted when Footnote 10 is applicable.

• Rule 235B must be used to check the horizontal clearance between conductors. Rule 235C must be used to check the vertical clearance between conductors. Rule 236 must be checked for climbing space.

See Photo(s)

Fig. 235-17. Example of clearance of a conductor to an anchor guy and a conductor to a support (Rule 235E).

messenger. The lashing machine travels down the messenger and needs the 12-in spacing for proper operating room. The 12-in spacing between messengers may be reduced by agreement between the parties involved. The parties involved include the communication cable attachees and the pole owner or owners.

Rule 235H2 requires a vertical clearance (not spacing) between conductors and cables of different communication utilities anywhere in the span to be not less than 4 in. The 4-in clearance keeps communication cables owned by one utility separated from the communication cables owned by a different utility. For example, the bottom of a cable TV expansion loop must have 4-in of clearance down to a telephone cable below. The 4-in clearance also keeps a communication circuit owned by one utility from sagging into or below the a communication circuit under it owned by a different utility. Rule 235 does not provide conductor temperature and loading conditions in Rule 235A for use with the entire rule. Rules 232, 233, and 234 use this

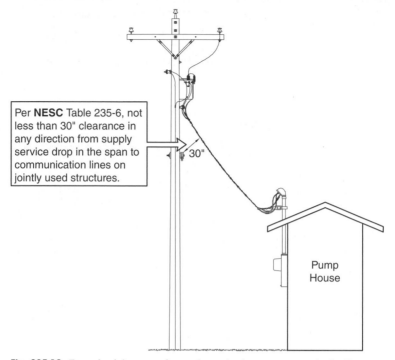

Per **NESC** Table 235-6, not less than 30" clearance in any direction from supply service drop in the span to communication lines on jointly used structures.

30"

Pump House

Fig. 235-18. Example of clearance of a supply service drop to a communication line conductor (Rule 235E).

format, but Rule 235 specifies the conductor temperature and loading conditions in each paragraph. Rule 235H2 only provides one temperature condition (60°F, without wind) to check the 4-in clearance. An initial or final condition is not specified as typically communication cables lashed to steel messengers do not vary much between initial and final sag conditions. Maximum sag conditions (such as 32°F with ice or 120°F) are also not specified. A 60°F condition typically appears near the middle of a sag and tension chart as some days are colder and some days are warmer. The 60°F condition in Rule 235H2 is specified as the ambient air temperature condition as communication cables are not subject to the heating effects of electrical current flow. The 4-in clearance may be reduced (or eliminated) by agreement between the parties involved. The parties involved include the communication cable attachees and the pole owner or owners. The **Code** rules related to clearance and spacing between communication lines are outlined in Fig. 235-22.

If the parties agree to less than a 12-in spacing between communication messengers, the through-bolt spacing on a pole can be less than 12 in. The **NESC** does not specify a minimum distance between through bolts on a wood pole. Rule 012C, which requires accepted good practice, must be used. See Fig. 235-23.

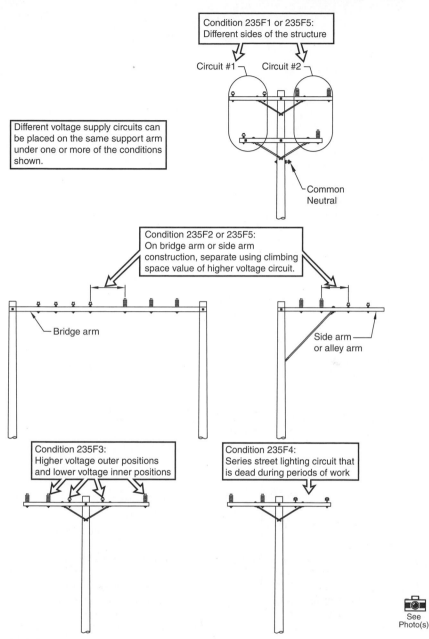

Condition 235F1 or 235F5:
Different sides of the structure

Circuit #1   Circuit #2

Different voltage supply circuits can
be placed on the same support arm
under one or more of the conditions
shown.

Common
Neutral

Condition 235F2 or 235F5:
On bridge arm or side arm
construction, separate using climbing
space value of higher voltage circuit.

Bridge arm

Side arm
or alley arm

Condition 235F3:
Higher voltage outer positions
and lower voltage inner positions

Condition 235F4:
Series street lighting circuit that
is dead during periods of work

See
Photo(s)

Fig. 235-19. Examples of clearance between supply circuits of different voltages on the
same support arm (Rule 235F).

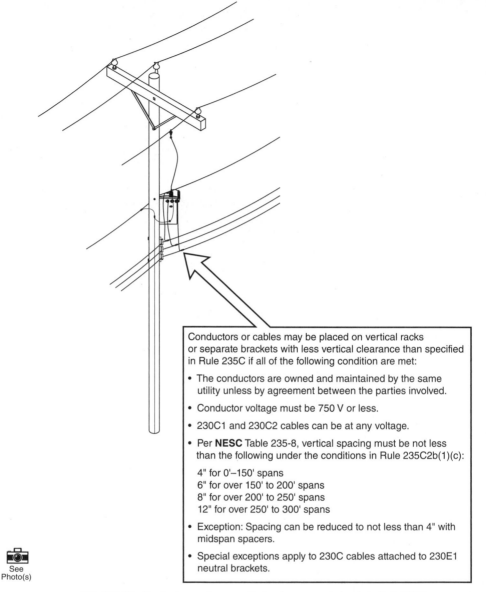

Conductors or cables may be placed on vertical racks
or separate brackets with less vertical clearance than specified
in Rule 235C if all of the following condition are met:

- The conductors are owned and maintained by the same
  utility unless by agreement between the parties involved.

- Conductor voltage must be 750 V or less.

- 230C1 and 230C2 cables can be at any voltage.

- Per **NESC** Table 235-8, vertical spacing must be not less
  than the following under the conditions in Rule 235C2b(1)(c):

  4" for 0'–150' spans
  6" for over 150' to 200' spans
  8" for over 200' to 250' spans
  12" for over 250' to 300' spans

- Exception: Spacing can be reduced to not less than 4" with
  midspan spacers.

- Special exceptions apply to 230C cables attached to 230E1
  neutral brackets.

See
Photo(s)

Fig. 235-20. Conductor spacing on vertical racks or separate brackets (Rule 235G).

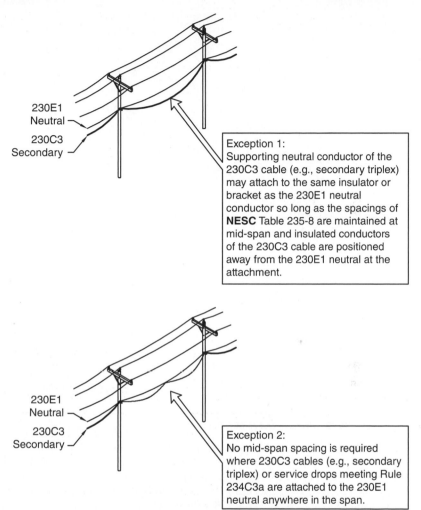

230E1
Neutral

230C3
Secondary

Exception 1:
Supporting neutral conductor of the
230C3 cable (e.g., secondary triplex)
may attach to the same insulator or
bracket as the 230E1 neutral
conductor so long as the spacings of
NESC Table 235-8 are maintained at
mid-span and insulated conductors
of the 230C3 cable are positioned
away from the 230E1 neutral at the
attachment.

230E1
Neutral

230C3
Secondary

Exception 2:
No mid-span spacing is required
where 230C3 cables (e.g., secondary
triplex) or service drops meeting Rule
234C3a are attached to the 230E1
neutral anywhere in the span.

Fig. 235-21. Exceptions to conductor spacing on vertical racks or separate brackets (Rule 235G).

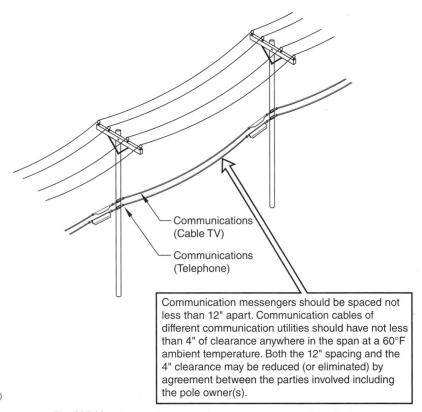

Communications
(Cable TV)

Communications
(Telephone)

Communication messengers should be spaced not
less than 12" apart. Communication cables of
different communication utilities should have not less
than 4" of clearance anywhere in the span at a 60°F
ambient temperature. Both the 12" spacing and the
4" clearance may be reduced (or eliminated) by
agreement between the parties involved including
the pole owner(s).

See
Photo(s)

Fig. 235-22. Clearance and spacing between communication lines (Rule 235H).

Note 1 at the end of Rule 235H allows spacers in the span to help meet the
4 in clearance between communication cables. Note 2 at the end of Rule 235H
points to Rule 235B and **NESC** Table 235-1 for the clearance of open-wire communication conductors. Open-wire communication conductors have typically been
replaced with insulated communication cables or fiber optic cables. The rules for
open-wire communication circuits continue to remain in the **Code**. The horizontal clearance requirements in **NESC** Table 235-1 are only specified for open-wire
communication conductors, not more modern insulated communication cables on
messengers. Rule 012C, which requires accepted good practice, must be used for
the horizontal clearance of communication cables on a communication crossarm.
See Fig. 235-24.

Rule 235H only applies to clearance between communication cables, not
between communication cables and communication equipment. See Rule 238F2b
for clearance between communication cables and antennas in the communication
space.

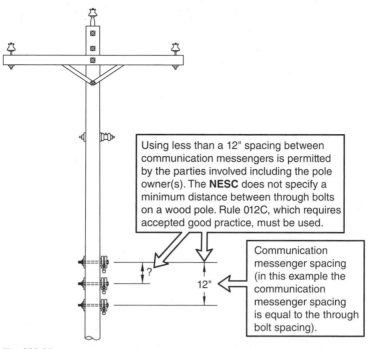

Fig. 235-23. Example of using less than a 12 inch spacing between communication messengers (Rule 235H).

236. CLIMBING SPACE

Rule 236, Climbing Space, and Rule 237, Working Space, may appear complicated but they revolve around one simple idea. The supply and communication worker must have adequate space to climb and work on a pole and its conductors, supports, and equipment.

Rule 236 applies when workers climb the pole. Wood poles are commonly climbed by some utilities and rarely climbed by other utilities that predominantly use bucket trucks. Steel distribution poles without steps are typically not climbed.

Climbing space may be provided by temporarily moving the line conductors using live line tools per Exception 3 of Rule 236E. The NESC work rules in Part 4 must be applied when climbing structures and moving live conductors.

Rule 236 provides requirements for measuring the climbing space in Rule 236B, positioning arms in the climbing space in Rule 236C, and locating equipment outside the climbing space in Rule 236D.

Rule 236E provides the basic horizontal clearances between conductors bounding the climbing space. NESC Table 236-1 provides horizontal clearances between conductors. The horizontal clearances in NESC Table 236-1 are intended to provide 24 in of clear climbing space while the conductors bounding the climbing space are covered with temporary insulation (e.g., insulating line hose, insulating blankets,

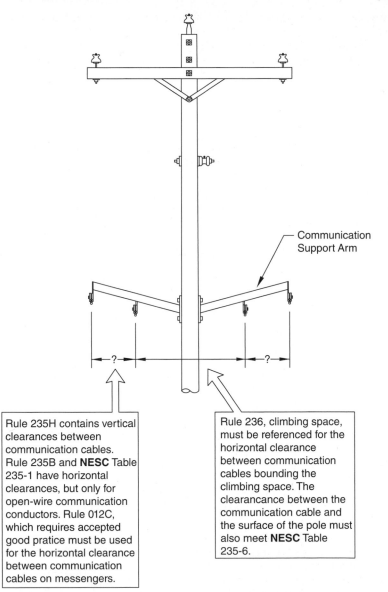

Communication
Support Arm

Rule 235H contains vertical clearances between communication cables. Rule 235B and **NESC** Table 235-1 have horizontal clearances, but only for open-wire communication conductors. Rule 012C, which requires accepted good pratice must be used for the horizontal clearance between communication cables on messengers.

Rule 236, climbing space, must be referenced for the horizontal clearance between communication cables bounding the climbing space. The clearcance between the communication cable and the surface of the pole must also meet **NESC** Table 235-6.

Fig. 235-24. Absence of a Code Rule for the horizontal clearance between communication cables (Rule N/A).

insulator hoods, etc.). For example, a 30-in horizontal clearance between conductors bounding the climbing space in **NESC** Table 236-1 (for 12.47/7.2 kV) assumes the temporary insulation on both sides of the climbing space is 6 in, leaving 24 in of clear climbing space. As the voltages on the table increase, the clearances increase as the temporary insulation is assumed to be thicker and bulkier. Examples of climbing space between conductors are shown in Figs. 236-1 and 236-2.

Methods to provide climbing space on buckarm construction are provided in Rule 236F. A buck arm is a crossarm used to change the direction of all or a part

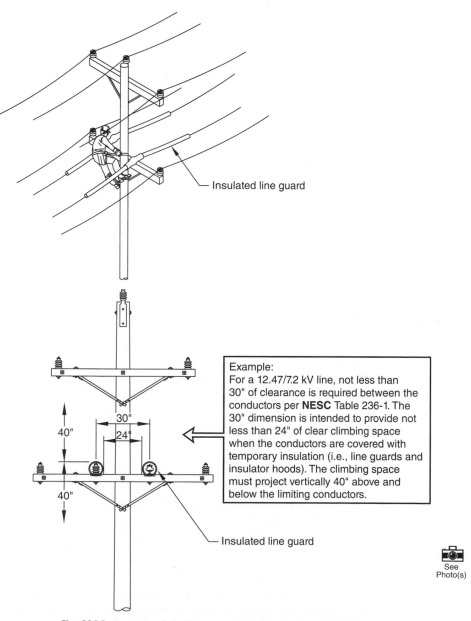

Example:
For a 12.47/7.2 kV line, not less than 30" of clearance is required between the conductors per **NESC** Table 236-1. The 30" dimension is intended to provide not less than 24" of clear climbing space when the conductors are covered with temporary insulation (i.e., line guards and insulator hoods). The climbing space must project vertically 40" above and below the limiting conductors.

Insulated line guard

See Photo(s)

Fig. 236-1. Examples of climbing space between conductors (Rule 236E).

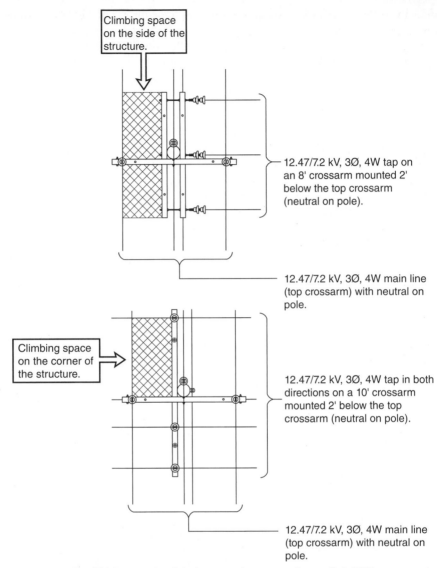

Fig. 236-2. Examples of climbing space between conductors (Rule 236E).

of the conductors on the line. A buck arm is generally placed at right angles to the line arm.

Rule 236G provides rules for climbing space past longitudinal runs not on support arms. The location, size, and quantity of the longitudinal runs not on support arms (e.g., communication cables attached to the pole) must be considered when determining if a qualified worker can climb past them. The **NESC** does not define a specific quantity or size of cables that a qualified climber can climb past. On a typical joint-use (power and communications) pole, it is the power lineworker that has to climb past the communication cables (not the communication lineworker); therefore, it is appropriate for the power lineworker (or power company) to work with the communication lineworker (or communication company) to determine if the location, size, and quantity of the cables permit qualified workers to climb past them. Keeping one side of the pole open and free from communication attachments is one way of assuring that adequate climbing space exists; however, there are times when communication attachments are made to both sides of the pole which is commonly referred to as pole boxing. Pole boxing is not specifically addressed in the **NESC**. The utility responsible for pole replacements typically does not prefer pole boxing (locating communication conductors on both sides of the pole) as it makes the pole replacement more time consuming. The restriction to locate communication cable on one side of a pole may be found in a joint-use contract between the power and communication utilities, but this restriction is not found in the **NESC**. Commonly, communication companies want to use pole boxing to avoid a power-to-communications clearance violation or a communications-to-groundline clearance violation. If it is determined that the location, size, and quantity of the communication cables do not permit qualified workers to climb past them, a communications crossarm can be used to provide adequate climbing space between the communication conductors or adequate climbing space can be provided by having the communication cables attached directly to the pole on one side of the pole and placed on a standoff bracket on the other side of the pole. The crossarm method and standoff bracket method are not needed if it is determined that the location, size, and quantity of the cables (longitudinal runs not on support arms) permit qualified workers to climb past them. Secondary service drops are also addressed in Rule 236G, and **NESC** Fig. 236-1 is provided as a guide.

Per Rule 236H, vertical conductors (risers) in duct, conduit, or other protective covering (e.g., U-guard) that are securely attached to the pole without riser brackets are not considered an obstruction to the climbing space. When using the clearances in **NESC** Table 236-1, a duct, conduit, or U-guard can be located within the clearances provided. Rule 362B, in Sec. 36, Risers, provides an additional requirement stating that the number, size, and location of the risers must be limited to allow adequate access for climbing. No specifics are provided as to the number or size of the conduits, Rule 012C, which requires accepted good practice, must be used. A conduit mounted on offset brackets is considered an obstruction to the climbing space. However, conduit offset brackets are used to organize the conduit risers on a pole and make climbing easier, especially when multiple conduits or large conduits are present. The climbing space is then located on a side or corner of the pole that does not contain the riser conduits on offset brackets. See Rule 362 for additional information and a figure related to standoff brackets.

Rule 236I requires climbing space to the top conductor position where the center phase is on the pole top and the two outer phases are on a crossarm mounted below the pole top.

One item that the **NESC** does not specifically address, but indirectly implies, regards the location of obstructions at the base of the pole where the climbing space starts. Obstructions at the base of the pole are sometimes addressed in joint-use agreements between power and communication utilities. For example, the joint-use agreement may specify that telephone pedestals can be located outside the climbing space at the base of the pole. Other obstructions like fences can sometimes not be avoided, as many times a pole is located along a property line where a fence exists. If an obstruction like a telephone pedestal or fence exists at the base of the climbing space, the climbing space rules can be met if the climber climbs on the opposite side of the pole or rotates position on the way up and down the pole to avoid the obstruction at the base of the pole. See Figs. 236-3 and 236-4.

The fall protection requirements in Rule 420K necessitate the need for some type of wood pole fall restriction device. This device typically clamps or cinches around a wood pole when the pole is being climbed. Rule 420K also applies to transitioning across obstacles on the pole such as power and communication equipment. Power and communication equipment must be mounted outside of the climbing space. Although not required by Rule 236, stand-off brackets can be used to mount the power or communications equipment to provide a method for the pole climber to meet Rule 420K. In other words, the pole climber will only need to transition the wood pole fall restriction device across the stand-off bracket, not the full height of the power or communication equipment. See Fig. 236-5.

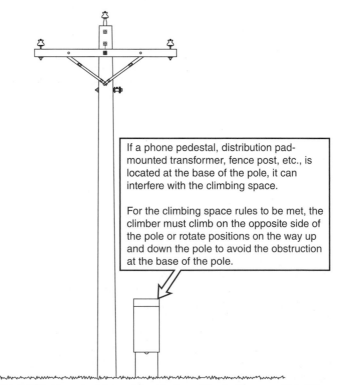

If a phone pedestal, distribution pad-mounted transformer, fence post, etc., is located at the base of the pole, it can interfere with the climbing space.

For the climbing space rules to be met, the climber must climb on the opposite side of the pole or rotate positions on the way up and down the pole to avoid the obstruction at the base of the pole.

See Photo(s)

Fig. 236-3. Example of obstructions at the base of the climbing space (Rule 236).

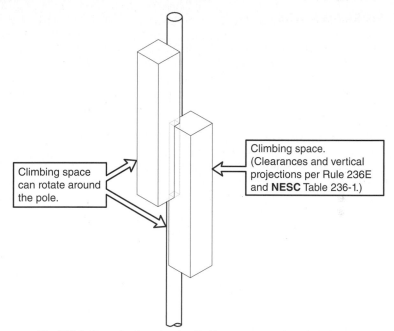

Climbing space can rotate around the pole.

Climbing space. (Clearances and vertical projections per Rule 236E and **NESC** Table 236-1.)

Fig. 236-4. Example of rotating the climbing space around a pole (Rule 236).

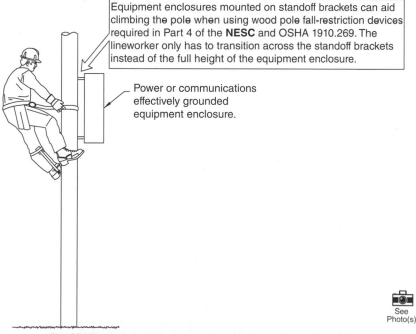

Equipment enclosures mounted on standoff brackets can aid climbing the pole when using wood pole fall-restriction devices required in Part 4 of the **NESC** and OSHA 1910.269. The lineworker only has to transition across the standoff brackets instead of the full height of the equipment enclosure.

Power or communications effectively grounded equipment enclosure.

See Photo(s)

Fig. 236-5. Example of using standoff brackets for equipment (Rule 236).

237. WORKING SPACE

Rule 237, Working Space, and Rule 236, Climbing Space, may appear complicated but they revolve around one simple idea. The supply and communication worker must have adequate space to climb and work on a pole and its conductors, supports, and equipment.

The working space rules are tied into the climbing space rules in Rule 236A. The working space dimensions required in Rule 237B are obtained from the climbing space requirements in Rule 236E and **NESC** Table 236-1. The working space dimension rules are outlined in Fig. 237-1.

Rule 237C provides requirements for working space relative to vertical and lateral conductors. Conductor jumpers and vertical risers that are not in duct or conduit must be located outside the working space.

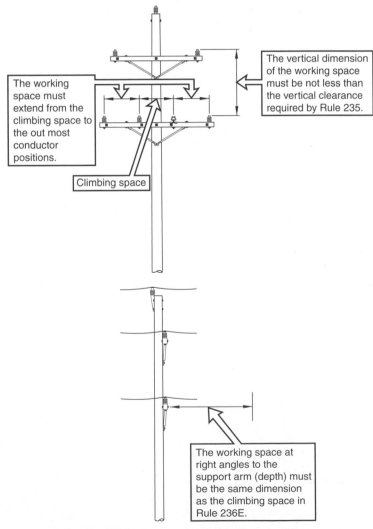

The working space must extend from the climbing space to the out most conductor positions.

The vertical dimension of the working space must be not less than the vertical clearance required by Rule 235.

Climbing space

The working space at right angles to the support arm (depth) must be the same dimension as the climbing space in Rule 236E.

Fig. 237-1. Working space dimensions (Rule 237B).

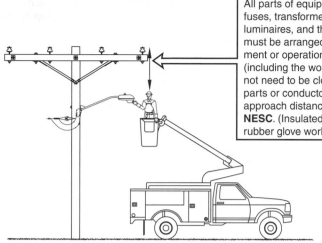

All parts of equipment (e.g., switches, fuses, transformers, surge arresters, luminaires, and their support brackets) must be arranged so that during adjustment or operation, the worker's body (including the worker's hands) does not need to be closer to energized parts or conductors than the minimum approach distances in Part 4 of the **NESC**. (Insulated live line tools and the rubber glove work method may be used.)

See Photo(s)

Fig. 237-2. Example of working clearances from energized equipment (Rule 237F).

Rule 237D provides rules for working space relative to buckarm construction. This rule is to be used with the corresponding climbing space requirements in Rule 236F. **NESC** Fig. 237-1 is provided as a guide.

Energized equipment must be guarded to avoid contact if all the conditions listed in Rule 237E apply.

Rule 237F ties the working space requirements of Rule 237 to the work rules of Part 4. An example of working clearances from energized equipment is shown in Fig. 237-2.

238. CLEARANCES AT THE SUPPORT BETWEEN SPECIFIED COMMUNICATIONS AND SUPPLY FACILITIES LOCATED ON THE SAME STRUCTURE

238A. Equipment. A definition is provided in Rule 238A for equipment as it applies in Rule 238. Since the definition is specific to this rule, it is provided in Rule 238 instead of Sec. 02, "Definitions of Special Terms." The definition of equipment on a typical joint-use (supply and communication) pole is outlined in Fig. 238-1.

A wood crossarm and crossarm brace (for supply conductors) are not considered equipment per the definition of equipment in this rule. A metal crossarm brace (for supply conductors) may be considered equipment per the definition depending on the position of the metal brace and what the metal brace is attached to. Wood and metal crossarm braces supporting communication cables are always considered equipment per the definition of equipment in this rule.

238B. Clearances in General. The vertical clearance between the following joint use facilities must be considered:

- Supply conductors and communication equipment
- Communication conductors and supply equipment
- Supply and communication equipment

NESC Table 238-1 is used to establish the vertical clearance requirements. **NESC** Table 238-1 provides vertical clearances only, not horizontal or any direction

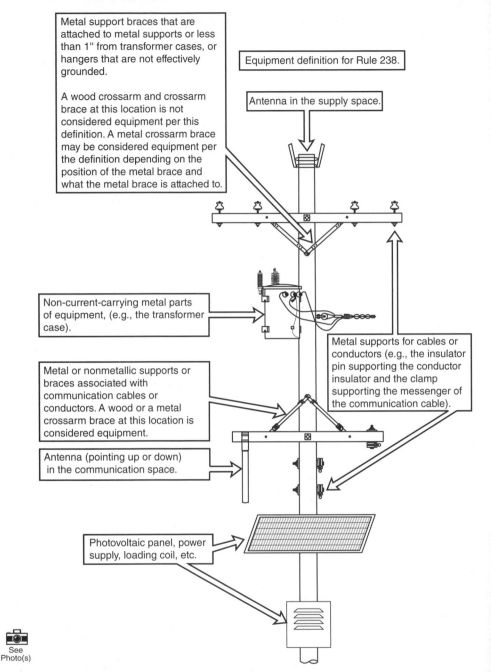

Metal support braces that are attached to metal supports or less than 1" from transformer cases, or hangers that are not effectively grounded.

A wood crossarm and crossarm brace at this location is not considered equipment per this definition. A metal crossarm brace may be considered equipment per the definition depending on the position of the metal brace and what the metal brace is attached to.

Equipment definition for Rule 238.

Antenna in the supply space.

Non-current-carrying metal parts of equipment, (e.g., the transformer case).

Metal supports for cables or conductors (e.g., the insulator pin supporting the conductor insulator and the clamp supporting the messenger of the communication cable).

Metal or nonmetallic supports or braces associated with communication cables or conductors. A wood or a metal crossarm brace at this location is considered equipment.

Antenna (pointing up or down) in the communication space.

Photovoltaic panel, power supply, loading coil, etc.

See Photo(s)

Fig. 238-1. Definition of equipment as it applies to communication and supply facilities on the same structure (Rule 238A).

clearances. The vertical clearance required in Rule 238 is in addition to the vertical clearance required in Rule 235C. The clearance requirements in both Rules 235C and 238 must be met. The clearances required in Rule 235 and 238 may be the same (e.g., 40 in), but the clearance is measured at different locations. In Rule 238, a supply equipment to communication equipment measurement is required. In Rule 235C, a supply conductor to communication cable measurement is required, both at the structure and out in the span. The vertical clearance requirements in Rule 238B and **NESC** Table 238-1 revolve around 40 in. Less than 40 in (e.g., 30 in) is acceptable for certain grounded supply facilities. Greater than 40 in is required for supply voltages above 8.7 kV to ground. Exceptions to the values in **NESC** Rule 238B are provided in Rules 238C, 238D, and 238F. Rule 238C addresses vertical clearances for span wires or brackets supporting luminaires (light fixtures), traffic signals, or trolley conductors. Rule 238D addresses vertical clearances of drip loops associated with luminaires and traffic signals. Rule 238F addresses antennas in the supply space and antennas in the communication space. Examples of vertical clearance between supply and communication equipment on the same structure are shown in Figs. 238-2 through 238-10.

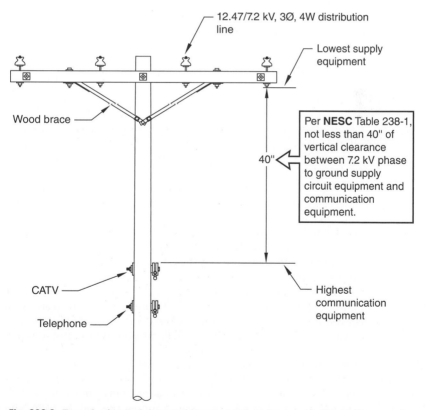

Fig. 238-2. Example of vertical clearance between supply and communication equipment on the same structure (Rule 238B).

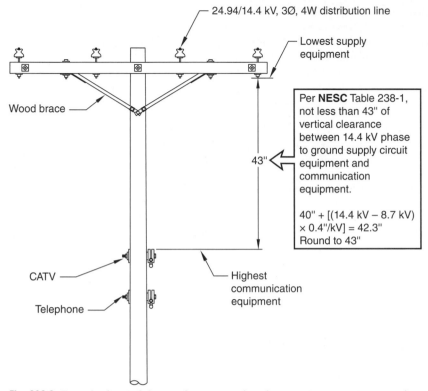

Fig. 238-3. Example of vertical clearance between supply and communications equipment on the same structure (Rule 238B).

The installation of a solar (photovoltaic) panel on a pole may involve both power and communication utilities depending on the location of the solar panel. If the solar panel is located in the supply space, the communication utilities on the pole are not affected. If the solar panel is located just below the communication space, the communication utilities on the pole are affected. A solar panel is considered supply equipment per the definition of electric supply equipment in Sec. 02. Rule 238B, which references **NESC** Table 238-1, applies to the clearance between communication conductors and supply equipment. The solar panel cannot be installed in the communication worker safety zone as it is not one of the exceptions noted in Rule 235C4 or Rule 238E. Several rules apply to mounting a solar panel on a pole. See Fig. 238-11.

238C. Clearances for Span Wires or Brackets.   The space between the supply and communication facilities is called the "communication worker safety zone." Rule 238E which defines the communication worker safety zone allows only a few select items to be located in the communication worker safety zone which are covered in Rules 238C, 238D, and 239. Rule 238C permits span wires or

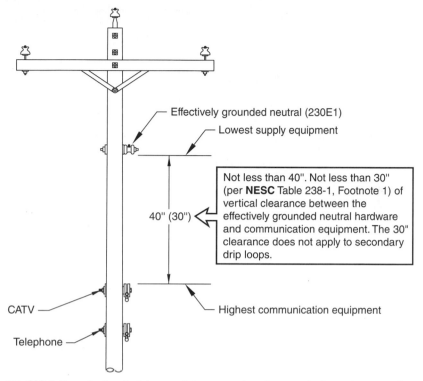

Effectively grounded neutral (230E1)

Lowest supply equipment

> Not less than 40". Not less than 30" (per **NESC** Table 238-1, Footnote 1) of vertical clearance between the effectively grounded neutral hardware and communication equipment. The 30" clearance does not apply to secondary drip loops.

40" (30")

CATV

Highest communication equipment

Telephone

See Photo(s)

Fig. 238-4. Example of vertical clearance between supply and communication equipment on the same structure (Rule 238B).

brackets carrying luminaires (light fixtures), traffic signals, or trolley conductors to be located in the communication worker safety zone. The **Code** recognizes that lighting fixtures and traffic signals provide their own safety function and commonly the appropriate mounting height for span wires or brackets carrying light fixtures, traffic signals, or trolley conductors is in the space between the supply and communication facilities. Rule 238C references **NESC** Table 238-2, which provides the required clearances for these items. If a luminaire or traffic signal is located in the communication worker safety zone, a drip loop is needed for service to the luminaire or traffic signal. Rule 238D applies to the drip loop. See Rule 238D for additional information and related figures.

238D. Clearance of Drip Loops Associated with Luminaires and Traffic Signals. The space between the supply and communication facilities is called the "communication worker safety zone." Rule 238E which defines the communication worker safety zone allows only a few select items to be located in the communication worker safety zone which are covered in Rules 238C, 238D, and 239. Rule 238D recognizes that luminaires (light fixtures), light fixture brackets, and traffic signal brackets are commonly fed from a drip loop positioned below the light or traffic signal bracket.

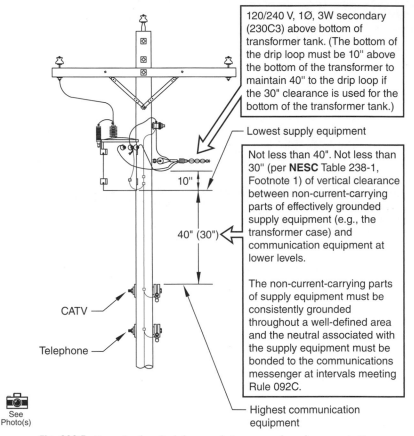

120/240 V, 1Ø, 3W secondary
(230C3) above bottom of
transformer tank. (The bottom of
the drip loop must be 10" above
the bottom of the transformer to
maintain 40" to the drip loop if
the 30" clearance is used for the
bottom of the transformer tank.)

— Lowest supply equipment

Not less than 40". Not less than
30" (per **NESC** Table 238-1,
Footnote 1) of vertical clearance
between non-current-carrying
parts of effectively grounded
supply equipment (e.g., the
transformer case) and
communication equipment at
lower levels.

The non-current-carrying parts
of supply equipment must be
consistently grounded
throughout a well-defined area
and the neutral associated with
the supply equipment must be
bonded to the communications
messenger at intervals meeting
Rule 092C.

10"

40" (30")

CATV

Telephone

See
Photo(s)

— Highest communication
equipment

Fig. 238-5. Example of vertical clearance between supply and communication
equipment on the same structure (Rule 238B).

Special clearances for these drip loops permit them to be located in the communication worker safety zone. An exception applies to Rule 238D that permits the
clearance of a drip loop for a luminaire or traffic signal to be not less than 3 in
above the highest communication cable, through bolt, or other equipment if certain conditions apply. Examples of lighting fixtures mounted above, in, and below
the communication worker safety zone, along with lighting fixture drip loops, are
shown in Figs. 238-12 through 238-15.

It can be easy to lose sight of the fact that the reduced clearance to secondary
drip loops only applies to drip loops feeding luminaires or traffic signals. For
example, even though a 120/240 V circuit can be used to feed a house or a street
lighting luminaire, the 12-in clearance only applies to the luminaire (street light)
drip loop entering the luminaire or luminaire bracket. Luminaires and traffic signals serve their own safety functions so they merit special **Code** consideration.
See Fig. 238-16.

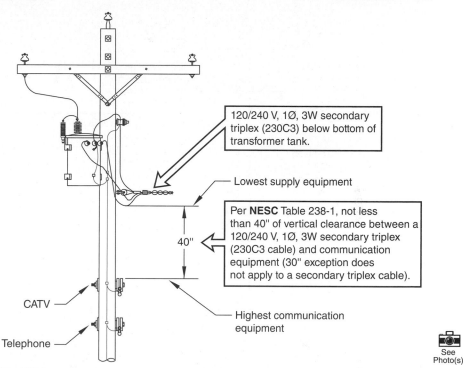

120/240 V, 1Ø, 3W secondary triplex (230C3) below bottom of transformer tank.

Lowest supply equipment

Per **NESC** Table 238-1, not less than 40" of vertical clearance between a 120/240 V, 1Ø, 3W secondary triplex (230C3 cable) and communication equipment (30" exception does not apply to a secondary triplex cable).

40"

CATV

Telephone

Highest communication equipment

See Photo(s)

Fig. 238-6. Example of vertical clearance between supply and communication equipment on the same structure (Rule 238B).

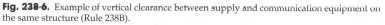

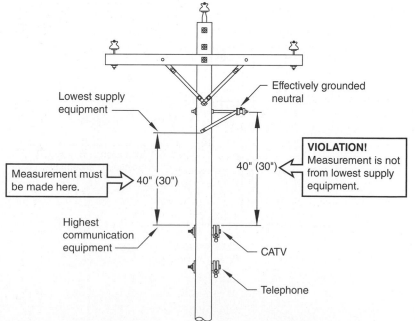

Lowest supply equipment

Effectively grounded neutral

Measurement must be made here.

40" (30")

40" (30")

VIOLATION! Measurement is not from lowest supply equipment.

Highest communication equipment

CATV

Telephone

See Photo(s)

Fig. 238-7. Example of vertical clearance between supply and communication equipment on the same structure (Rule 238B).

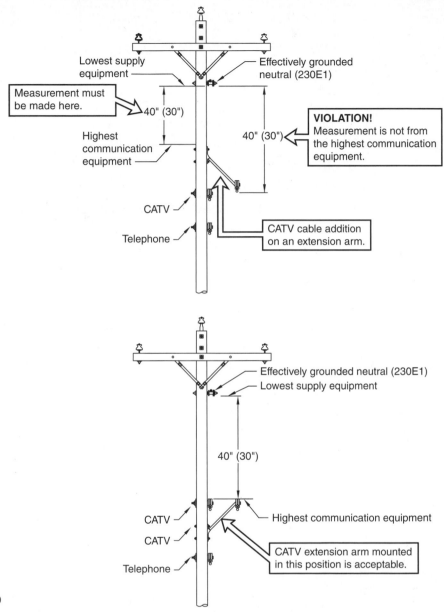

Lowest supply equipment

Effectively grounded neutral (230E1)

Measurement must be made here.

40" (30")

Highest communication equipment

40" (30")

VIOLATION!
Measurement is not from the highest communication equipment.

CATV

CATV cable addition on an extension arm.

Telephone

Effectively grounded neutral (230E1)

Lowest supply equipment

40" (30")

CATV

CATV

Highest communication equipment

CATV extension arm mounted in this position is acceptable.

Telephone

See
Photo(s)

Fig. 238-8. Example of vertical clearance between supply and communication equipment on the same structure (Rule 238B).

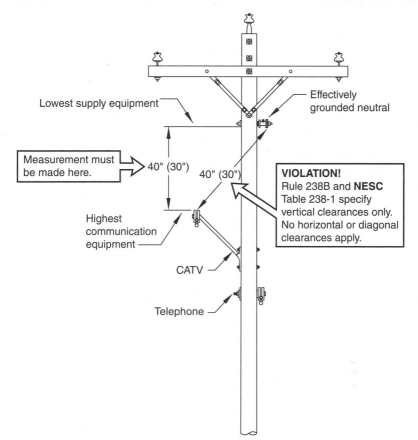

Fig. 238-9. Example of vertical clearance between supply and communication equipment on the same structure (Rule 238B).

238E. Communication Worker Safety Zone. Rule 238E labels the area between the supply space and the communication space both at the structure and out in the span. The name for this area is the "communication worker safety zone." Rule 235C4 also addresses the communication worker safety zone. The work rules in Part 4 of the **NESC** provide the qualifications of a supply employee and a communication employee. Since the communication employee is not trained to work on supply lines, a safety zone exists for the communication employee's protection. If a communication line is positioned below a supply line on an overhead structure, but the required communication worker safety zone clearance does not exist, the communication employee can correct the violation if the communication employee does not violate the minimum approach distances and other requirements in Sec. 43. If the communication employee cannot maintain the minimum approach distances and other requirements in Sec. 43, the communication employee must contact a supply employee to correct the violation. See Sec. 43 for additional information.

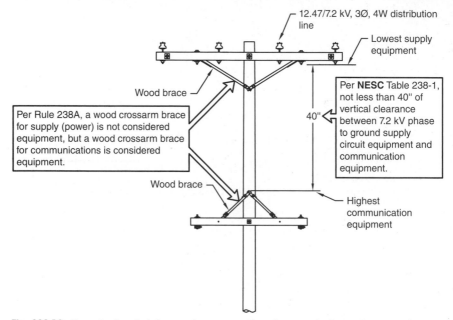

Fig. 238-10. Example of vertical clearance between supply and communication equipment on the same structure (Rule 238B).

The communication worker safety zone is defined by both Rule 238 and Rule 235C, not one rule or the other, but the combined effect of both rules. Rule 235C applies to clearances between wires, conductors, and cables. Rule 238 applies to clearance between supply conductors and communications equipment, between communication conductors and supply equipment, and between supply and communications equipment. The supply space above the communication worker safety zone and the communication space below the communication worker safety zone are defined in Sec. 02, Definitions of Special Terms. Sec. 02 also defines the terms electric supply equipment, communication equipment, conductor (multiple definitions) and lines (multiple definitions). Understanding these terms is helpful to applying the communication worker safety zone specified in both Rules 235C4 and 238E.

The communication worker safety zone clearance revolves around 40 in; however, higher voltages (above 8.7 kV to ground) require greater than 40 in, and exceptions may permit less than 40 in. Knowing where to measure the upper and lower limits of the communication worker safety zone is critical to applying the rule. Only a few select items are permitted in the communication worker safety zone. The exceptions are the items specified in Rule 238C, Rule 238D, and Rule 239 (Vertical Risers). The most common application in Rule 239 is Rule 239G; see Rule 239G for a discussion and a figure. The operating rods covered in Rule 239I are also permitted to pass through the communication worker safety zone. It is possible to locate an antenna in the communication space or the supply space. See Rule 238F for a discussion and a figure of an antenna in both locations. The Communication Worker Safety Zone is outlined in Fig. 238-17.

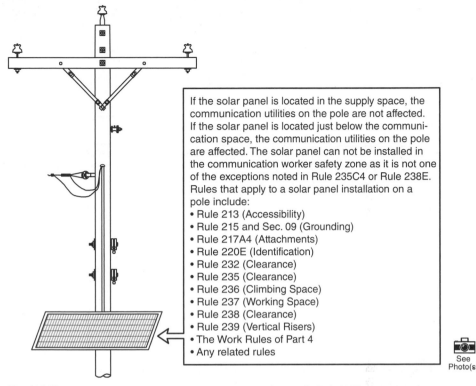

If the solar panel is located in the supply space, the communication utilities on the pole are not affected. If the solar panel is located just below the communication space, the communication utilities on the pole are affected. The solar panel can not be installed in the communication worker safety zone as it is not one of the exceptions noted in Rule 235C4 or Rule 238E. Rules that apply to a solar panel installation on a pole include:
- Rule 213 (Accessibility)
- Rule 215 and Sec. 09 (Grounding)
- Rule 217A4 (Attachments)
- Rule 220E (Identification)
- Rule 232 (Clearance)
- Rule 235 (Clearance)
- Rule 236 (Climbing Space)
- Rule 237 (Working Space)
- Rule 238 (Clearance)
- Rule 239 (Vertical Risers)
- The Work Rules of Part 4
- Any related rules

See Photo(s)

Fig. 238-11. Example of mounting a solar (photovoltaic) panel on a pole (Rule 238B).

238F. Antenna Clearances from Supply and Communication Lines Attached to the Same Supporting Structure. Rule 238F contains requirements for antennas mounted in the supply (power) space in Rule 238F2a, and antennas mounted in the communication space in Rule 238F2b. An antenna cannot be mounted in the communication worker safety zone as it is not one of the exceptions of items that can be located in that space per Rule 238E or Rule 235C4. See Rule 238E for a discussion and a figure outlining the list of exceptions that are permitted in the communication worker safety zone. Rule 238F1 specifies the frequency range of the antennas that are covered in Rule 238F. Antennas found on overhead line poles include, but are not limited to, small cell antennas owned by communication utilities, and antennas used for supervisory control and data acquisition (SCADA) owned by electric utilities. Rule 238F1 contains a requirement that antennas that are located in the supply (power) space must be installed and maintained by a qualified supply worker. Work rules for supply workers can be found in Part 4 of the **NESC**, Sections 42 and 44, and in OSHA 1910.269. In other words, an antenna installed in the supply space must be installed by a power lineworker, not a communications lineworker. See the discussion and a figure in Rule 238A for examples of equipment, including antennas, located in the supply space and the communication space.

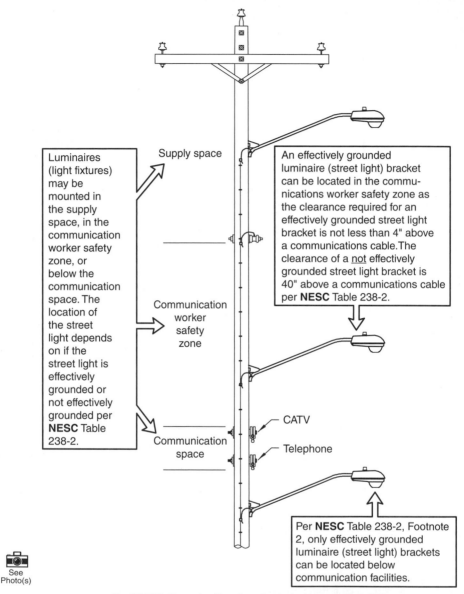

Luminaires (light fixtures) may be mounted in the supply space, in the communication worker safety zone, or below the communication space. The location of the street light depends on if the street light is effectively grounded or not effectively grounded per **NESC** Table 238-2.

Supply space

An effectively grounded luminaire (street light) bracket can be located in the communications worker safety zone as the clearance required for an effectively grounded street light bracket is not less than 4" above a communications cable. The clearance of a <u>not</u> effectively grounded street light bracket is 40" above a communications cable per **NESC** Table 238-2.

Communication worker safety zone

Communication space

CATV

Telephone

Per **NESC** Table 238-2, Footnote 2, only effectively grounded luminaire (street light) brackets can be located below communication facilities.

See Photo(s)

Fig. 238-12. Example of locating a luminaire on a pole (Rule 238).

An antenna mounted in the supply (power) space for cellular communications (sometimes referred to as a small cell antenna) is typically mounted at the top of the pole above the supply conductors. Rule 238F2a references **NESC** Table 238-3 for the clearance between the antenna in the supply space and supply conductors. **NESC** Table 238-3 specifies vertical clearances, but the footnotes to Table 238-3 per-

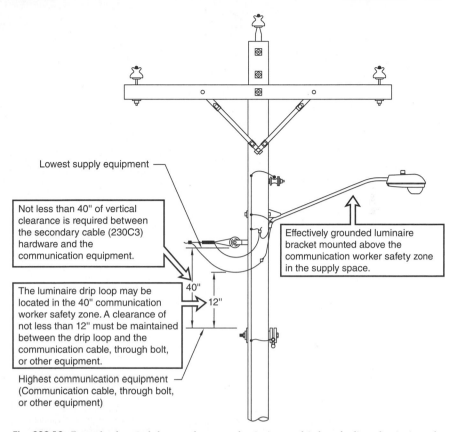

Lowest supply equipment —

Not less than 40" of vertical
clearance is required between
the secondary cable (230C3)
hardware and the
communication equipment.

Effectively grounded luminaire
bracket mounted above the
communication worker safety zone
in the supply space.

The luminaire drip loop may be
located in the 40" communication
worker safety zone. A clearance of
not less than 12" must be maintained
between the drip loop and the
communication cable, through bolt,
or other equipment.

40"

12"

Highest communication equipment —
(Communication cable, through bolt,
or other equipment)

See
Photo(s)

Fig. 238-13. Example of vertical clearance between a luminaire or a drip loop feeding a luminaire and communication equipment (Rules 238C and 238D).

mit reduced clearances in any direction provided the affected parties are in agreement. The footnotes to Table 238-3 also address clearance adders for elevations in excess of 3300 ft above sea level and a requirement to use the maximum operating voltage for voltages above 50 kV. Rule 238F2a specifies a vertical clearance of 40 in from an antenna in the supply space to a communication line in the communication space. The exception to Rule 238F2a allows the 40-in value to be reduced to 30 in for effectively grounded equipment cases. The 40-in clearance must be checked in all cases but will certainly apply when the antenna in the supply space is near the bottom of the supply space. This construction can be found when an antenna is mounted below a recloser or other switching device controlled by a supervisory control and data acquisition (SCADA) system.

A note in Rule 238F2a clarifies that the clearances in this rule are not intended to apply to personnel working in the vicinity of antennas. Rule 420Q addresses radio frequency exposure issues. Additional clearance may be needed to maintain a safe working distance if the antenna is not powered down when working

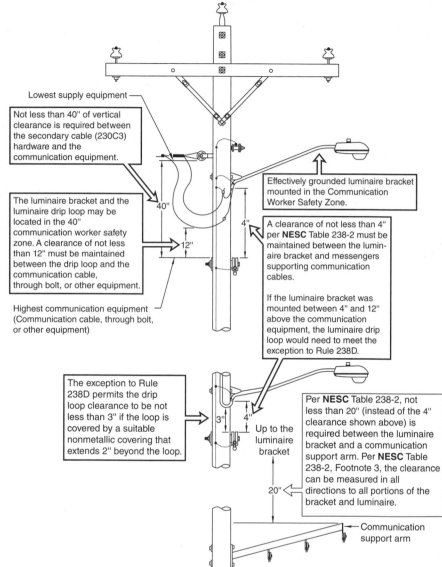

Lowest supply equipment

Not less than 40" of vertical clearance is required between the secondary cable (230C3) hardware and the communication equipment.

The luminaire bracket and the luminaire drip loop may be located in the 40" communication worker safety zone. A clearance of not less than 12" must be maintained between the drip loop and the communication cable, through bolt, or other equipment.

Highest communication equipment (Communication cable, through bolt, or other equipment)

40"

12"

Effectively grounded luminaire bracket mounted in the Communication Worker Safety Zone.

4"

A clearance of not less than 4" per **NESC** Table 238-2 must be maintained between the luminaire bracket and messengers supporting communication cables.

If the luminaire bracket was mounted between 4" and 12" above the communication equipment, the luminaire drip loop would need to meet the exception to Rule 238D.

The exception to Rule 238D permits the drip loop clearance to be not less than 3" if the loop is covered by a suitable nonmetallic covering that extends 2" beyond the loop.

3"

4"

Up to the luminaire bracket

20"

Per **NESC** Table 238-2, not less than 20" (instead of the 4" clearance shown above) is required between the luminaire bracket and a communication support arm. Per **NESC** Table 238-2, Footnote 3, the clearance can be measured in all directions to all portions of the bracket and luminaire.

Communication support arm

See Photo(s)

Fig. 238-14. Example of vertical clearance between a luminaire or a drip loop feeding a luminaire and communication equipment (Rules 238C and 238D).

on the pole. A common reference for determining clearance to a communications antenna based on radio frequency radiation exposure limits is the FCC (Federal Communications Commission) OET (Office of Engineering and Technology) Bulletin 65. FCC OET Bulletin 65 is titled "Evaluating Compliance with FCC

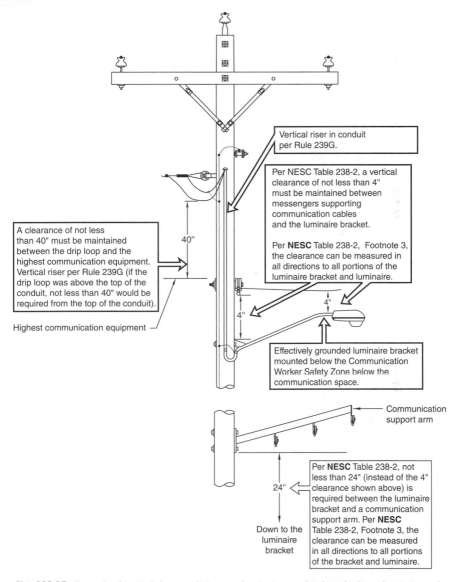

Fig. 238-15. Example of vertical clearance between a luminaire or a drip loop feeding a luminaire and communication equipment (Rules 238C, 238D, and 239G).

Guidelines for Human Exposure to Radiofrequency Electromagnetic Fields." Examples of clearance between supply (power) lines and antennas in the supply space are shown in Fig. 238-18.

Antennas mounted in the communication space can be installed and maintained by communication workers. Rule 238F2b references **NESC** Table 238-1 for

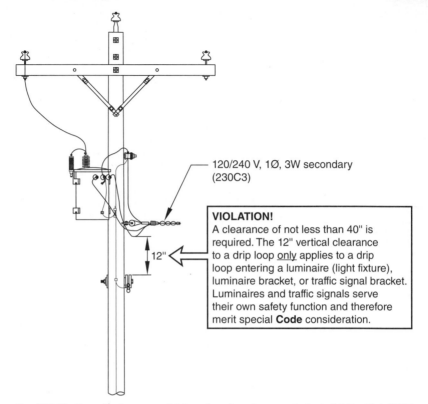

120/240 V, 1Ø, 3W secondary
(230C3)

VIOLATION!
A clearance of not less than 40" is
required. The 12" vertical clearance
to a drip loop <u>only</u> applies to a drip
loop entering a luminaire (light fixture),
luminaire bracket, or traffic signal bracket.
Luminaires and traffic signals serve
their own safety function and therefore
merit special **Code** consideration.

12"

Fig. 238-16. Example of a common joint-use (supply and communication) violation (Rule 238D).

the clearance between the antenna in the communication space and supply conductors. Table 238-1 only provides vertical clearances. Rule 238F2b requires a 3 in
clearance in any direction between the antenna in the communication space (and
its associated equipment cases) and a communication line in the communication
space. The exception in Rule 238F2b apples to an antenna mounted to the strand
of a communication conductor which is commonly referred to as a "strand mount
antenna." Rule 238F2b clarifies that the 3-in clearance does not apply between
span mounted antenna and cable bundle (the strand and its associated lashed
cables) that the antenna is mounted on. The 3-in clearance would apply between
the stand mounted antenna and the communication cable bundle above or below
it. An example of an antenna mounted in the communication space is shown in
Fig. 238-19.

Normally, a communications vertical riser would be located on the pole that
has an antenna. The requirements for vertical communication conductors passing
through the supply space on a pole are found in Rule 239H. In addition, the clearance between the vertical communications riser and a supply line in the supply
space must not be less than the value found in **NESC** Table 235-6, Row 1.

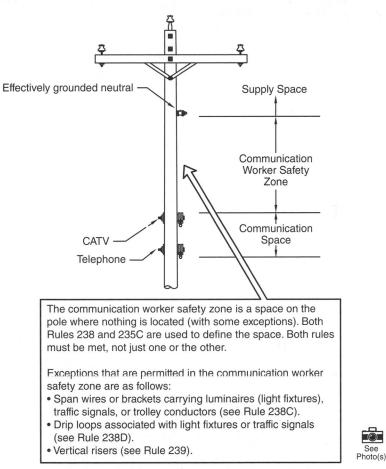

Fig. 238-17. Communication worker safety zone (Rule 238E).

It is important to note the differences between **NESC** Tables 238-1, 238-2, and 238-3. Some of the major differences are outlined below:

- **NESC** Table 238-1 is a phase-to-ground voltage table per the wording in parentheses under the table title.
- **NESC** Table 238-1 specifies vertical clearances per the title of the table.
- **NESC** Table 238-2 does not specify any voltages as the table applies to span wires and brackets supporting luminaires, traffic signals, or trolley conductors.
- **NESC** Table 238-2 specifies vertical clearances per the title of the table, but Footnote 3 addresses clearance in all directions.
- **NESC** Table 238-3 is a phase-to-phase voltage table per the wording under the heading "Supply lines."
- **NESC** Table 238-3 specifies vertical clearances per the wording in the first column of the table, but Footnotes 3, 4, 5, and 6 address reduced clearances in any direction.

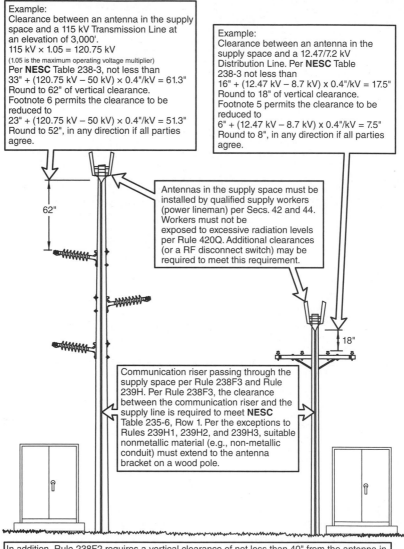

Example:
Clearance between an antenna in the supply space and a 115 kV Transmission Line at an elevation of 3,000'.
115 kV × 1.05 = 120.75 kV
(1.05 is the maximum operating voltage multiplier)
Per **NESC** Table 238-3, not less than
33" + (120.75 kV − 50 kV) × 0.4"/kV = 61.3"
Round to 62" of vertical clearance.
Footnote 6 permits the clearance to be reduced to
23" + (120.75 kV − 50 kV) × 0.4"/kV = 51.3"
Round to 52", in any direction if all parties agree.

Example:
Clearance between an antenna in the supply space and a 12.47/7.2 kV Distribution Line. Per **NESC** Table 238-3 not less than
16" + (12.47 kV − 8.7 kV) x 0.4"/kV = 17.5"
Round to 18" of vertical clearance.
Footnote 5 permits the clearance to be reduced to
6" + (12.47 kV − 8.7 kV) x 0.4"/kV = 7.5"
Round to 8", in any direction if all parties agree.

62"

Antennas in the supply space must be installed by qualified supply workers (power lineman) per Secs. 42 and 44. Workers must not be exposed to excessive radiation levels per Rule 420Q. Additional clearances (or a RF disconnect switch) may be required to meet this requirement.

18"

Communication riser passing through the supply space per Rule 238F3 and Rule 239H. Per Rule 238F3, the clearance between the communication riser and the supply line is required to meet **NESC** Table 235-6, Row 1. Per the exceptions to Rules 239H1, 239H2, and 239H3, suitable nonmetallic material (e.g., non-metallic conduit) must extend to the antenna bracket on a wood pole.

See Photo(s)

In addition, Rule 238F2 requires a vertical clearance of not less than 40" from the antenna in the supply space and a communication line in the communication space. This requirement must be checked in all cases, but will certainly apply when the antenna is at the bottom of the supply space (e.g., below a recloser or other SCADA controlled switch–not shown). Identification tags and warning signs (not shown) must meet the requirements in the **NESC**, OSHA standards, and FCC regulations.

Fig. 238-18. Examples of clearance between supply lines and antennas in the supply space (Rule 238F2a).

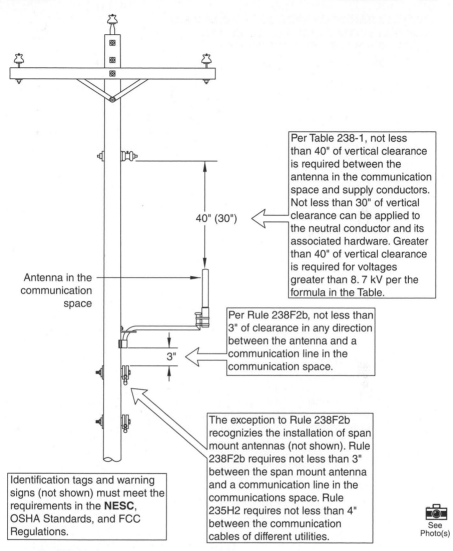

Per Table 238-1, not less than 40" of vertical clearance is required between the antenna in the communication space and supply conductors. Not less than 30" of vertical clearance can be applied to the neutral conductor and its associated hardware. Greater than 40" of vertical clearance is required for voltages greater than 8. 7 kV per the formula in the Table.

40" (30")

Antenna in the communication space

Per Rule 238F2b, not less than 3" of clearance in any direction between the antenna and a communication line in the communication space.

3"

The exception to Rule 238F2b recognizies the installation of span mount antennas (not shown). Rule 238F2b requires not less than 3" between the span mount antenna and a communication line in the communications space. Rule 235H2 requires not less than 4" between the communication cables of different utilities.

Identification tags and warning signs (not shown) must meet the requirements in the **NESC**, OSHA Standards, and FCC Regulations.

See Photo(s)

Fig. 238-19. Example of an antenna mounted in the communication space (Rule 238F2b).

239. CLEARANCE OF VERTICAL AND LATERAL FACILITIES FROM OTHER FACILITIES AND SURFACES ON THE SAME SUPPORTING STRUCTURE

239A. General.   Rule 239A1 provides a list of vertical and lateral conductors that may be placed directly on the supporting structure (pole). A conduit may also be placed directly on the pole. Rule 239D requires guarding of vertical conductors within 8 ft of the ground with certain exceptions. See Rule 239D for a discussion. Examples of vertical and lateral conductors are shown in Fig. 239-1.

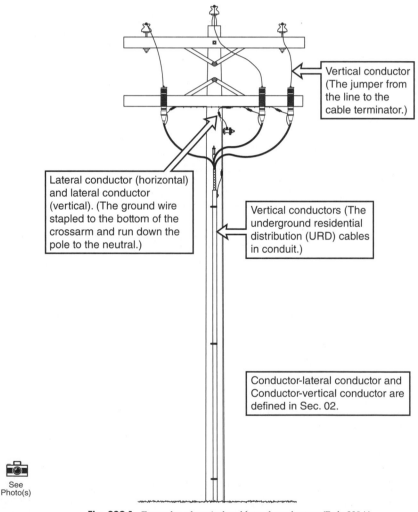

Vertical conductor (The jumper from the line to the cable terminator.)

Lateral conductor (horizontal) and lateral conductor (vertical). (The ground wire stapled to the bottom of the crossarm and run down the pole to the neutral.)

Vertical conductors (The underground residential distribution (URD) cables in conduit.)

Conductor-lateral conductor and Conductor-vertical conductor are defined in Sec. 02.

See Photo(s)

Fig. 239-1. Examples of vertical and lateral conductors (Rule 239A).

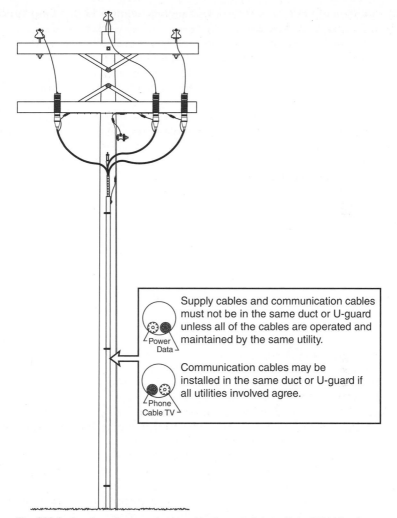

Supply cables and communication cables must not be in the same duct or U-guard unless all of the cables are operated and maintained by the same utility.

Power
Data

Communication cables may be installed in the same duct or U-guard if all utilities involved agree.

Phone
Cable TV

Fig. 239-2. Supply and communication cables in vertical risers (Rules 239A2d and 239A2e).

Rule 239A2 addresses various cables that can be installed in the same vertical riser duct. The rules for supply and communication cables in vertical risers are outlined in Fig. 239-2.

Rules 239A5 and 239A6 address the terms "nonmetallic covering" and "guarding and protection." These terms are used throughout Rule 239 and other locations in the **Code**. Rule 239A5 provides a specific clarification for the term "nonmetallic covering" when the term is used in Rule 239. The **Code** does not specify the type of duct, conduit, or U-guard that is required for risers. See Rule 352 for a discussion of various types of ducts (conduits).

239B. Location of Vertical or Lateral Conductors Relative to Climbing Spaces, Working Spaces, and Pole Steps. Climbing spaces, working spaces, and pole steps must not be obstructed with vertical and lateral conductors. Rule 239B applies to vertical conductors not in conduit. A note to this rule reminds the reader that vertical runs in duct, conduit, or other protective covering (e.g., U-guard) securely attached to the pole surface are not considered to obstruct the climbing space per Rule 236H.

239C. Conductors Not in Duct or Conduit. Vertical and lateral conductors not in duct or conduit must have the same clearances from ducts or conduits (enclosing other conductors) as from other structure surfaces.

239D. Guarding and Protection Near Ground. Vertical conductors (risers) within 8 ft of the ground must be guarded. Duct, conduit, and U-guard are the most common types of guards used to protect the vertical conductors. An exception to Rule 239D1 provides a list of conductors and cables that do not require guarding. See Fig. 239-3.

Rule 239D is very similar to Rule 093D, Guarding and Protection, for grounding conductors. Rule 239D also overlaps with Rule 362, Pole Risers—Additional Requirements. See Rules 093D and 362 for additional information and related figures.

239E. Requirements for Vertical and Lateral Supply Conductors on Supply Line Structures or within Supply Space on Jointly Used Structures. The locations and conductors that apply to Rule 239E are graphically shown in Fig. 239-4.

The general clearances for open (e.g., bare noninsulated) vertical and lateral supply conductors from the surface of the support are provided in NESC Table239-1. NESC Table 239-1 also provides clearance to span, guy, and messenger wires. Clearances in Rule 235E (NESC Table 235-6) must also be met. NESC Table 235-6 provides clearances from line conductors (i.e., the conductors spanning from pole to pole) to vertical and lateral conductors attached to the same support (pole).

NESC Table 239-2 must be applied to special cases. If open line conductors (e.g., bare conductors spanning from pole to pole) are within 4 ft of the pole, vertical conductors must meet two special conditions. The special conditions in Rules 239E2a(1) and 239E2a(2) only apply if open line conductors are within 4 ft of the pole, which is very common, even on an 8-ft crossarm, and if workers climb to work on the line conductors while the vertical conductors in questions are energized. If the vertical conductors in question are de-energized while climbing, or if the pole is worked on from a bucket truck, then the special cases of Rule 239E2 do not apply. NESC Table 239-2 requires clearances larger than NESC Table 239-1 to provide room for workers to climb near energized vertical conductors. NESC Table 239-2 also provides a distance from above and below the line conductor where the clearances apply. If a duct or conduit is used within the distance specified, it must be nonmetallic or protected by a nonmetallic covering to provide additional safety for a worker climbing the pole and working near an energized vertical conductor.

239F. Requirements for Vertical and Lateral Communication Conductors on Communication Line Structures or within the Communication Space on Jointly Used Structures. The locations and conductors that Rule 239F apply to are graphically shown in Fig. 239-5.

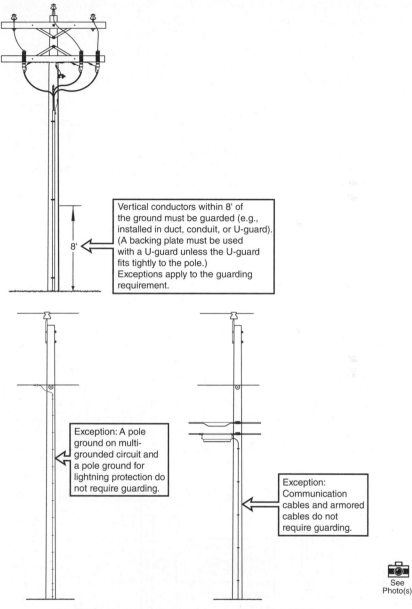

Fig. 239-3. Guarding and protection near ground (Rule 239D).

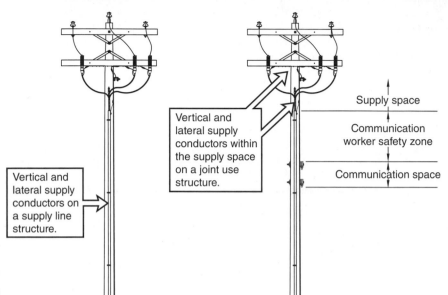

Vertical and
lateral supply
conductors on
a supply line
structure.

Vertical and
lateral supply
conductors within
the supply space
on a joint use
structure.

Supply space

Communication
worker safety zone

Communication space

See
Photo(s)

Fig. 239-4. Location of vertical and lateral supply conductors on supply line structures or within supply space on jointly used structures (Rule 239E).

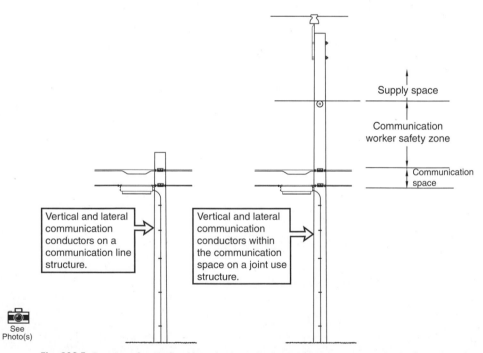

Supply space

Communication
worker safety zone

Communication
space

Vertical and lateral
communication
conductors on a
communication line
structure.

Vertical and lateral
communication
conductors within
the communication
space on a joint use
structure.

See
Photo(s)

Fig. 239-5. Location of vertical and lateral communication conductors on communication line structures or within the communication space on jointly used structures (Rule 239F).

Uninsulated (open wire) vertical and lateral communication conductors are not as common as they once were. Open wire communication has been replaced with insulated communication cables. When an insulated communication cable riser exists on a joint-use (supply and communication) pole, the specified clearance to supply facilities in Rule 239F is essentially the same as required by Rules 235 and 238.

239G. Requirements for Vertical Supply Conductors and Cables Passing through Communication Space on Jointly Used Line Structures.   The locations and conductors to which Rule 239G applies are graphically shown in Fig. 239-6.

Rule 239G applies to a number of common joint-use (supply and communication) conditions. Providing a duct, conduit, or covering (e.g., U-guard) over supply conductors 40 in above the highest communication attachment is a condition that relates to the communication worker safety zone discussed in Rule 238E. Examples

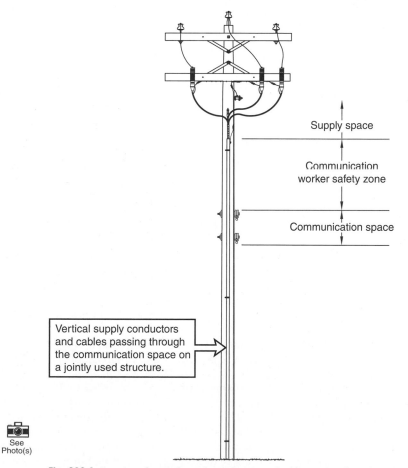

Fig. 239-6. Location of vertical supply conductors and cables passing through the communication space on jointly used line structures (Rule 239G).

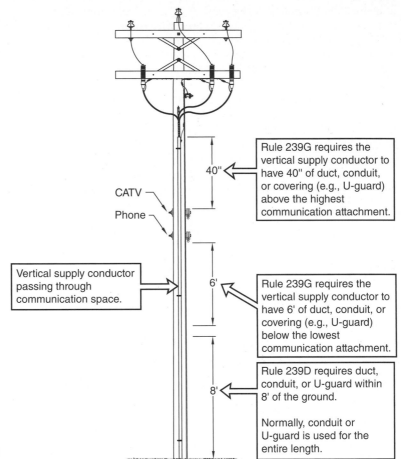

Rule 239G requires the vertical supply conductor to have 40" of duct, conduit, or covering (e.g., U-guard) above the highest communication attachment.

Vertical supply conductor passing through communication space.

Rule 239G requires the vertical supply conductor to have 6' of duct, conduit, or covering (e.g., U-guard) below the lowest communication attachment.

Rule 239D requires duct, conduit, or U-guard within 8' of the ground.

Normally, conduit or U-guard is used for the entire length.

CATV

Phone

See Photo(s)

Fig. 239-7. Example of vertical supply conductors on a joint-use (supply and communication) pole (Rule 239G1).

of vertical supply conductors on joint-use (supply and communication) poles are shown in Figs. 239-7 and 239-8.

Exception 2 to Rule 239G1 permits omitting guarding from a supply grounding conductor (pole ground) on a joint-use (supply and communication) pole if certain conditions are met. See Fig. 239-9.

Rule 239G4 requires a supply cable service drop to be at least 40 in above the highest communications attachment. A similar requirement exists in Rules 235C and 238B. A common example is a 120/240 V, single-phase, three-wire service drop connection at the pole. Rule 239G4 also states that a supply cable service drop can be at least 40 in below the lowest communications attachment (in the communications space). This is a less common location for a supply cable service drop. An example of this installation is a tall pole for a railroad crossing with a railroad signal building located near the base of the tall pole. The service can be run from a transformer in

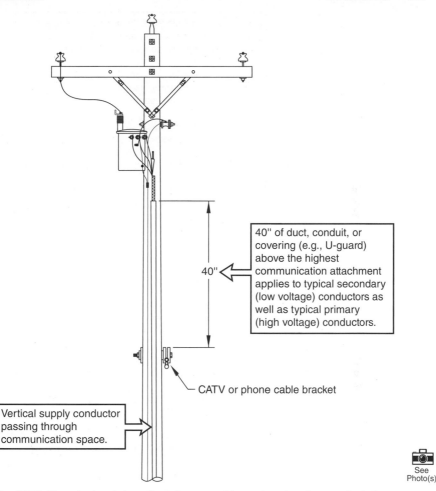

40" of duct, conduit, or
covering (e.g., U-guard)
above the highest
communication attachment
applies to typical secondary
(low voltage) conductors as
well as typical primary
(high voltage) conductors.

40"

CATV or phone cable bracket

Vertical supply conductor
passing through
communication space.

See
Photo(s)

Fig. 239-8. Example of vertical supply conductors on a joint-use (supply and communication) pole (Rule 239G1).

the supply space, down the pole in conduit to at least 40 in below the lowest communications attachment, and then take off as a service drop to the railroad signal building. The supply service drop in the communications space must have insulated splices and connections for the energized phase conductors (not for the grounded neutral which typically serves as the messenger).

239H. Requirements for Vertical Communication Conductors Passing through Supply Space on Jointly Used Structures. The locations and conductors to which Rule 239H applies are graphically shown in Fig. 239-10.

Communication conductors passing through the supply space can be seen on poles with an overhead ground wire (static) with an embedded fiber-optic communication cable and on supply poles with antennas on top. See Rule 238F for additional information on clearance to antennas in the supply space.

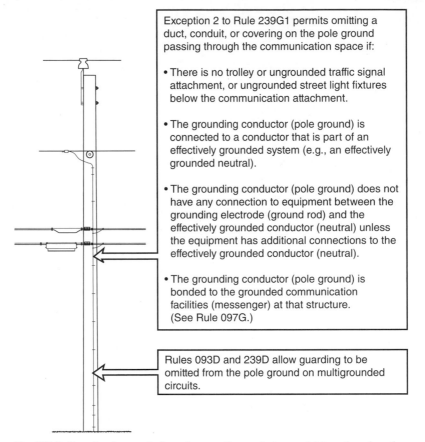

Exception 2 to Rule 239G1 permits omitting a duct, conduit, or covering on the pole ground passing through the communication space if:

- There is no trolley or ungrounded traffic signal attachment, or ungrounded street light fixtures below the communication attachment.

- The grounding conductor (pole ground) is connected to a conductor that is part of an effectively grounded system (e.g., an effectively grounded neutral).

- The grounding conductor (pole ground) does not have any connection to equipment between the grounding electrode (ground rod) and the effectively grounded conductor (neutral) unless the equipment has additional connections to the effectively grounded conductor (neutral).

- The grounding conductor (pole ground) is bonded to the grounded communication facilities (messenger) at that structure. (See Rule 097G.)

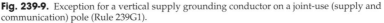

Rules 093D and 239D allow guarding to be omitted from the pole ground on multigrounded circuits.

Fig. 239-9. Exception for a vertical supply grounding conductor on a joint-use (supply and communication) pole (Rule 239G1).

239I. Operating Rods.   Operating rods for switches like a three-phase gang-operated distribution switch can pass through the communication space or the communication worker safety zone, but they must be located outside the climbing space. See Rule 216 for a figure showing an overhead line switch operating rod. The communication worker safety zone is discussed in Rule 235C4 and Rule 238E. See Rule 236 for additional climbing space information.

239J. Additional Rules for Standoff Brackets.   Standoff brackets may be used to support ducts or conduits. Both metallic and nonmetallic ducts or conduits may be supported. The conductors inside the ducts or conduits must have sufficient insulation. Tree wire described in Rule 230D cannot be installed in a conduit riser as it is not fully insulated. Standoff brackets may be used to support cables not in conduit if the cable is a communication cable, 230C1a supply cable, or a supply cable less than 750 V with a single outer jacket or sheath. Traditional duplex, triplex, and quadruplex supply cables do not meet this criteria, nor do

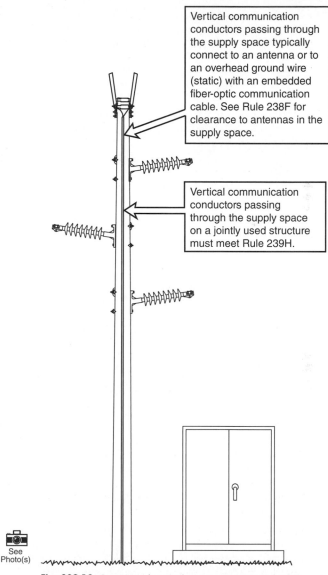

Vertical communication conductors passing through the supply space typically connect to an antenna or to an overhead ground wire (static) with an embedded fiber-optic communication cable. See Rule 238F for clearance to antennas in the supply space.

Vertical communication conductors passing through the supply space on a jointly used structure must meet Rule 239H.

See Photo(s)

Fig. 239-10. Location of vertical communication conductors passing through the supply space on jointly used structures (Rule 239H).

single insulated underground secondary conductors. The rules for positioning riser standoff brackets on a structure to inhibit climbing by unqualified persons are provided in Rule 217A2. Riser standoff brackets are also discussed in Rule 362. Since risers transition from overhead to underground lines, they are covered in both Part 2 (Overhead Lines) and Part 3 (Underground Lines).

Section 24

Grades of Construction

240. GENERAL

Sections 24, 25, and 26 are all interrelated. Section 24, "Grades of Construction," defines the required strength of overhead line construction for safety purposes. Section 25, "Loadings for Grades B and C," defines the physical loads (e.g., ice, wind, and temperature conditions) that overhead line construction must be able to withstand and the load factors that must be applied to the physical loads. Section 26, "Strength Requirements," defines the required strength of the materials used in constructing overhead lines and the strength factors that must be applied to the materials. Line insulators are not addressed in Secs. 24, 25, or 26. Insulators are covered in their own section, Sec. 27.

The details of applying load factors and strength factors are covered in Secs. 25 and 26. The purpose of Sec. 24 is to define the grade of construction that is required for different situations.

This Handbook addresses the **Code** requirements in Secs. 24, 25, 26, and 27. Line design calculations are not presented in this Handbook. Line design calculations can be found in transmission and distribution line design manuals. Large utilities normally develop their own line design manuals. Small rural utilities typically use the transmission and distribution line design manuals published by the Rural Utilities Service (RUS), which in the past was referred to as the Rural Electrification Administration (REA). The RUS distribution line design manuals consist of RUS Bulletins 1724E-150 through 1724E-154. The RUS design manual for high-voltage transmission lines is RUS Bulletin 1724E-200. The RUS also publishes design manuals for communication lines. This

Handbook will aid those individuals writing and using line design manuals and provide a deeper understanding of how the **Code** applies to line design calculations.

Typical strength and loading line design calculations include, but are not limited to, the following:

- Maximum wind span based on wind with ice on conductors
- Maximum weight span based on weight of conductors and ice
- Moment due to wind on pole
- Allowable resisting moment of pole
- Transverse, vertical, and total components of conductor loading
- Total ground line moment on pole
- Ruling span
- Diameter of pole at any point
- Dead-end guying strength
- Bisector guying strength
- Anchor strength
- Weak-link of guy attachment, guy wire, and anchor assembly
- Crossarm strength (vertical)
- Crossarm strength (longitudinal)
- Pole buckling
- Maximum line angle based on insulator strength
- Material deflection
- Equipment loading

Single-pole structures are typically easier to analyze than multiple-pole and lattice structures. Line design calculations also involve clearance calculations in addition to strength and loading calculations (e.g., clearance above ground, clearance to buildings and other structures, horizontal clearance between conductors, vertical clearance between conductors, etc.). The clearance calculations will affect the line design and therefore the selection of pole heights and pole classes. The sample sag and tension chart shown at the beginning of Sec. 23 provides sag for clearance calculations and conductor tension for strength and loading calculations. The "design tension" used for strength and loading calculations is the largest of the conductor tensions calculated after applying Rule 250B (heavy, medium, light, or warm islands loads), Rule 250C (extreme wind loads) if applicable, and Rule 250D (extreme ice with concurrent wind loads) if applicable. The 250B, 250C, and 250D loads are applied to conductors in accordance with Rule 251. The 250B, 250C, and 250D loads are applied to line supports (e.g., structures) in accordance with Rule 252. Rule 250A specifies that where all three rules apply (250B, 250C, and 250D), the required loading shall be the one that has the greatest effect. See Rules 250, 251, and 252 for additional information.

If more than one condition apply to the construction of an overhead line, the condition requiring the higher grade of construction is used. The order of grades of construction is discussed in Rule 241B. A note in Rule 240A references Rule 014 for grades of construction and strength for emergency and temporary installations. Temporary installations are also addressed in **NESC** Table 261-1, Footnote 4.

Section 24 applies to both alternating current (AC) and direct current (DC) circuits. A DC circuit is considered equivalent to the root mean square (rms) values of an AC circuit when applying Sec. 24. This is different from Sec. 23, which requires DC circuits to be equivalent to the AC crest value. See Rule 230G for a discussion of AC rms and crest values.

241. APPLICATION OF GRADES OF CONSTRUCTION TO DIFFERENT SITUATIONS

241A. Supply Cables. For the purposes of determining the grade of construction of a supply cable, supply (not communication) cables are broken into two types. Cable (Type 1) supply cables are cables meeting Rules 230C1, 230C2, or 230C3. See Rule 230C for a discussion of these types of supply cables. The 230C1, 230C2, and 230C3 cables are typically supported on messengers. The strength requirements for supply cable messengers are covered in Rule 261I.

Open (Type 2) supply cables are all other supply cables that do not meet the requirements for 230C1, 230C2, and 230C3 cables. Type 2 supply cables include open wire (e.g., bare) conductors and covered (e.g., tree wire) conductors.

241B. Order of Grades. Rule 241B outlines the order of grades. A general description of each grade of construction is outlined below:

- *Grade B*. The highest grade. The line will be "extra stout." Load and strength factors will be applied to make the strength of the line higher than any other grade. The required pole class will be larger than C and N grades (e.g., a Class 3 pole instead of a Class 5 pole) for the same span length or the required span length for Grade B will be shorter than grades C and N for the same pole class. Crossarms will be stronger than grades C and N, etc.
- *Grade C*. The middle grade. The line will be more "stout" than Grade N but not as "stout" as Grade B. Load and strength factors will be applied, but they will not increase the strength of the line as much as Grade B. Grade C at crossings will be more "stout" than Grade C not at crossings.
- *Grade N*. The lowest grade. The line must be designed to support the anticipated loads but no load or strength factors are specified.

A graphical representation of the grades of construction is shown in Fig. 241-1.

241C. At Crossings. One overhead power line is considered to be at crossings when it crosses another line (in the span or on a common supporting structure), or when it crosses over or overhangs a railroad track, a limited access highway, or a navigable waterway. An exception exists for joint use or collinear construction which in itself is not considered at crossings. See Figs. 241-2 through 241-4.

When two lines cross, the grade of construction for the conductors and supporting structures of the lower line can be determined without considering the upper line. In other words, the grade of construction of the lower line is based on the land use. This is because the lower line cannot fall into the upper line. The upper line, however, can fall into the lower line and cause problems. The conductors and supporting structures of the upper line at the crossing require careful application of Rule 241C4 (Multiple Crossings), Rule 242 (Grades of Construction

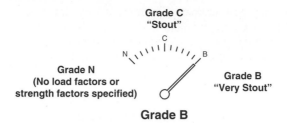

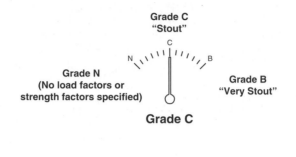

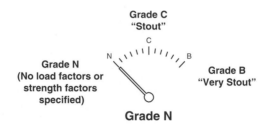

Fig. 241-1. Order of grades of construction (Rule 241B).

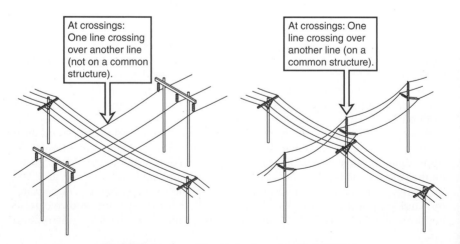

Fig. 241-2. Lines considered to be at crossings (Rule 241C1a).

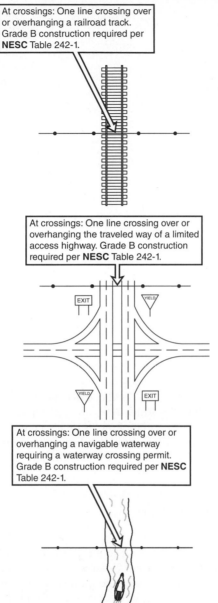

At crossings: One line crossing over or overhanging a railroad track. Grade B construction required per **NESC** Table 242-1.

At crossings: One line crossing over or overhanging the traveled way of a limited access highway. Grade B construction required per **NESC** Table 242-1.

At crossings: One line crossing over or overhanging a navigable waterway requiring a waterway crossing permit. Grade B construction required per **NESC** Table 242-1.

See
Photo(s)

Fig. 241-3. Lines considered to be at crossings (Rule 241C1b).

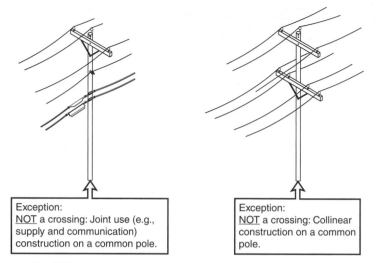

Fig. 241-4. Exceptions to lines considered to be at crossings (Rule 241C1).

for Conductors, **NESC** Table 242-1), and Rule 243 (Grades of Construction for Line Supports).

An example of supply lines at multiple crossings is shown in Fig. 241-5.

An example of communication lines at multiple crossings involving communications, supply, and railroad tracks is shown in Fig. 241-6.

241D. Structure Conflicts.   Structure conflict exists when one line can fall down and conflict with another line. See Rules 220C and 221 for additional information and figures related to structure conflict. Rule 241D also references structure conflict in Rule 243A4 and in Sec. 02, Definitions. For the purposes of determining grades of construction, conflicting lines are treated the same as line crossings since one line can fall on the other.

242. GRADES OF CONSTRUCTION FOR CONDUCTORS

Rule 242 references **NESC** Table 242-1 for determining grades of construction for conductors (and cables). It is important to stress the term conductors. Rule 242 and **NESC** Table 242-1 are used to determine grades of construction for conductors. Rule 243 is used to determine grades of construction for line supports. The grade of construction of a line support is based on the grade of construction of the conductors attached to the line support. Rules 242A through 242F provide specific information for special types of conductors.

There are several important items of discussion related to **NESC** Table 242-1.

- **NESC** Table 242-1 is for conductors and cables (supply and communication) alone, at crossing, or on the same structures with other conductors and cables.

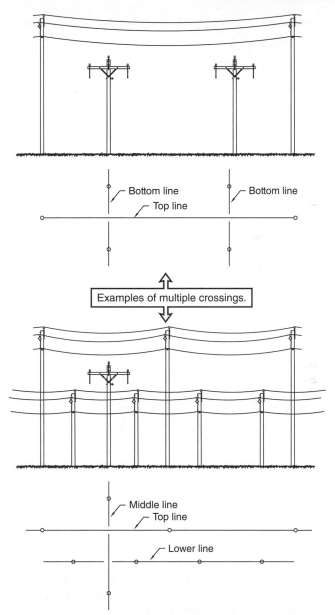

Fig. 241-5. Example of supply lines at multiple crossings (Rule 241C4a).

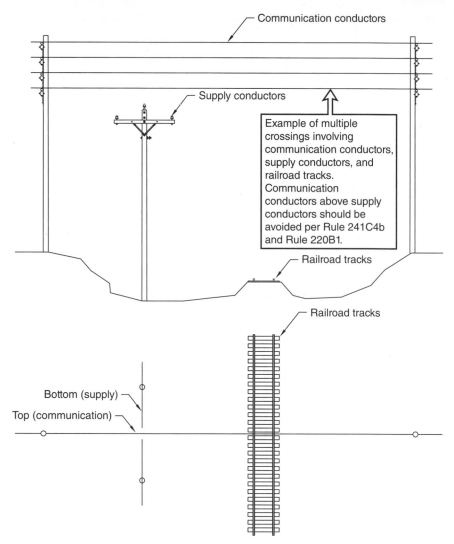

Fig. 241-6. Example of communication lines at multiple crossings (Rule 241C4b).

- Supply or communication lines crossing over or overhanging railroad tracks, limited access highways, or navigable waterways requiring waterway crossing permits must be built to Grade B construction.
- **NESC** Table 242-1 distinguishes between open and cable conductors. Cable conductors are 230C1, 230C2, and 230C3 cables. Open conductors are typically bare (noninsulated) conductors. See Rule 241A for a discussion.
- A 12.47/7.2-kV distribution line (open conductors) on an exclusive private right-of-way can be built to Grade N. This is not a common practice. Typically utilities build to Grade C construction for both common or public rights-of-way and exclusive private rights-of-way.

- An effectively grounded 115-kV transmission line (with ground fault relaying) on a public right-of-way may be built to Grade C (per Footnote 3). This is not a common practice. Typically utilities build transmission lines to Grade B construction.
- A 12.47/7.2-kV distribution line (open conductors) crossing another 12.47/7.2-kV distribution line (open conductors) must be built to Grade C construction. **NESC** Table 253-1 provides separate load factors for Grade C at line crossings and elsewhere.

Per **NESC** Table 242-1, a 12.47/7.2-kV distribution line (open conductors) built joint use with a communication cable must be built to Grade B unless Footnote 6 or 7 can be met. For a 12.47/7.2-kV line, Footnote 6 cannot be applied. Footnote 7 must then be met to construct a Grade C joint use (power and communication) line. Typically utilities build to Grade C construction for this application. If the communication facilities are all-dielectric, Grade C construction may be used. If the communication facilities are not all-dielectric, Footnote 7 (Parts a and b) must be met. The protection of communication equipment required in Footnote 7 (Parts a and b) is also required in Rule 223. The most common method used for the means of protection required in Footnote 7 is insulating the communication conductors (in the form of a communication cable) and bonding the grounded communication messenger to the grounded supply neutral. See Rule 223 for additional information.

A 120/240-V, single-phase, three-wire triplex cable meeting Rule 230C3 falls under "cable" in **NESC** Table 242-1 per Rule 241A. A 120/240-V, single-phase, three-wire triplex cable built joint use with a communication cable can be built to Grade N per **NESC** Table 242-1. Footnote 5 may apply.

It is important to check the Public Service Commission (PSC) rules or Public Utility Commission (PUC) rules in a particular state to determine if the state has developed "pole hardening" rules which sometimes consist of a requirement to design all lines to Grade B.

243. GRADES OF CONSTRUCTION FOR LINE SUPPORTS

Rule 242, Grades of Construction for Conductors, applies to grades of construction for conductors and this rule, Rule 243, applies to grades of construction for line supports. Line supports in this rule include structures, crossarms, insulators, pins, etc., basically the components that support the conductors.

243A. Structures.   The structure (e.g., pole) grade of construction must be the same as the highest grade of conductor supported. In other words, if a pole has Grade B transmission conductors, Grade C distribution conductors, and Grade N communication conductors, all attached to the same pole, then the pole itself must be designed to Grade B construction. There are four modifications to the general rule. See Fig. 243-1.

243B. Crossarms and Support Arms.   The crossarm (or another type of support arm) grade of construction must be the same as the highest grade of conductor that the crossarm supports. If a crossarm carries Grade B conductors, the crossarm

The grade of construction for the structure (pole) must be <u>the highest grade supported on the pole</u>. Four modifications apply to this Rule (Rules 243A1, 2, 3, and 4).

Example: A pole loading calculation for this pole requires Grade B load factors (**NESC** Table 253-1) <u>for every conductor and cable attached to the pole</u> (including the distribution conductors and communication cables). Grade B strength factors (**NESC** Table 261-1) must also be used.

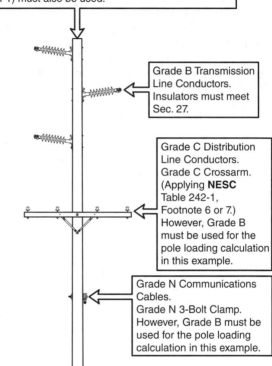

Grade B Transmission Line Conductors. Insulators must meet Sec. 27.

Grade C Distribution Line Conductors. Grade C Crossarm. (Applying **NESC** Table 242-1, Footnote 6 or 7.) However, Grade B must be used for the pole loading calculation in this example.

Grade N Communications Cables. Grade N 3-Bolt Clamp. However, Grade B must be used for the pole loading calculation in this example.

Fig. 243-1. Grades of construction for line supports (Rule 243A).

itself must be designed to Grade B. It is possible to have a Grade C crossarm on a Grade B structure (pole). For example, if a pole has Grade B transmission conductors on a Grade B crossarm and Grade C distribution conductors on a Grade C crossarm, the structure (pole) must be Grade B per Rule 243A but the distribution crossarm can remain Grade C per Rule 243B. There are three modifications to the general rule.

243C. Pins, Armless Construction Brackets, Insulators, and Conductor Fastenings. The pins, armless construction brackets, insulators, and conductor fastenings grade of construction must be the same as the grade of their associated conductor. There are four modifications to the general rule. The most notable is Modification 4 (Rule 243C4). This modification states that insulators used on open conductor supply lines are covered in their own section, Sec. 27.

Section 25

Structural Loadings for Grades B and C

250. GENERAL LOADING REQUIREMENTS AND MAPS

Sections 24, 25, and 26 are all interrelated. Section 24, "Grades of Construction," defines the required strength of overhead line construction for safety purposes. Section 25, "Loadings for Grades B and C," defines the physical loads (e.g., ice, wind, and temperature conditions) that overhead line construction must be able to withstand and the load factors that must be applied to the physical loads. Section 26, "Strength Requirements," defines the required strength of materials used in constructing overhead lines and the strength factors that must be applied to the materials. Line insulators are not addressed in Secs. 24, 25, or 26. Insulators are covered in their own section, Sec. 27.

The purpose of Sec. 24 is to define the grade of construction that is required for different situations. The details of applying load factors and strength factors are covered in Secs. 25 and 26.

This Handbook addresses the **Code** requirements in Secs. 24, 25, 26, and 27. Line design calculations are not presented in this Handbook. Line design calculations can be found in transmission and distribution line design manuals. Large utilities commonly develop their own line design manuals. Some utilities use the transmission and distribution line design manuals published by the Rural Utilities Service (RUS), which in the past was referred to as the Rural Electrification Administration (REA). The RUS distribution line design manuals consist of RUS Bulletins 1724E-150 through 1724E-154. The RUS design manual for high-voltage transmission lines is RUS Bulletin 1724E-200. The RUS also publishes design manuals for communication lines. This Handbook will aid those individuals writing and using line design manuals and provide a deeper understanding of how the **Code** applies to line design calculations.

Typical strength and loading line design calculations include, but are not limited to, the following:

- Maximum wind span based on wind with ice on conductors
- Maximum weight span based on weight of conductors and ice
- Moment due to wind on pole
- Allowable resisting moment of pole
- Transverse, vertical, and total components of conductor loading
- Total ground line moment on pole
- Ruling span
- Diameter of pole at any point
- Dead-end guying strength
- Bisector guying strength
- Anchor strength
- Weak-link of guy attachment, guy wire, and anchor assembly
- Crossarm strength (vertical)
- Crossarm strength (longitudinal)
- Pole buckling
- Maximum line angle based on insulator strength
- Material deflection
- Equipment loading

Single pole structures are typically easier to analyze than multiple pole and lattice structures. Line design calculations also involve clearance calculations in addition to strength and loading calculations (e.g., clearance above ground, clearance to buildings and other structures, horizontal clearance between conductors, vertical clearance between conductors, etc.). The clearance calculations will affect the line design and therefore the selection of pole heights and pole classes. The sample sag and tension chart shown at the beginning of Sec. 23 provides sag for clearance calculations and conductor tension for strength and loading calculations. The "design tension" used for strength and loading calculations is the largest of the conductor tensions calculated after applying Rule 250B (heavy, medium, light, or warm islands loads), Rule 250C (extreme wind loads) if applicable, and Rule 250D (extreme ice with concurrent wind loads) if applicable. The 250B, 250C, and 250D loads are applied to conductors in accordance with Rule 251. The 250B, 250C, and 250D loads are applied to line supports (e.g., structures) in accordance with Rule 252. Rule 250A specifies that where all three rules apply (250B, 250C, and 250D), the required loading shall be the one that has the greatest effect. See Rules 250, 251, and 252 for additional information.

250A. General.   If the grade of construction required in Sec. 24, "Grades of Construction," is B or C, Sec. 25 will define the physical loadings in the form of ice, wind, and load factors that increase the physical loads associated with overhead line construction. The terms "load specified in Rule 250," "Rule 250B loads," "Rule 250C loads," and "Rule 250D loads" will be used throughout Secs. 25 and 26.

When the loads in Rule 250B (heavy, medium, light, or warm islands loading), Rule 250C (extreme wind loading), and Rule 250D (extreme ice with concurrent wind) must all be considered, the required loading must be the one that has the greatest effect. Rule 250 also specifies that the intent is to apply wind in an essentially horizontal plane. In other words, the intent is not to apply upwinds or updrafts or downwinds or downdrafts. If some local terrain condition did warrant applying upwinds or downwinds, this practice is certainly not prohibited. A loading map for overhead lines in the United States is provided in NESC Fig. 250-1. The United States is divided into heavy, medium, light, and warm islands loading districts. Since only four loading districts are defined, the NESC recognizes that heavier loads than specified in the map may exist for a specific region. Using a higher loading than the map shows is certainly acceptable. Using a lower loading requires approval of an administrative authority (e.g., the State Public Service Commission).

The weather loadings specified in Rules 250B, 250C, and 250D may not be sufficient for the forces imposed during construction and maintenance. Additional loads may need to be considered to adjust for this condition.

Rule 250A4 states that if lines are designed to meet Sec. 25 (loadings for Grades B and C) and Sec. 26 (strength requirements), the lines will have sufficient capability to resist earthquake ground motions.

250B. Combined Ice and Wind District Loading. NESC Fig. 250-1 (map of the USA) and NESC Table 250-1 are the two basic references to determine ice and wind loading for the heavy, medium, light, and warm islands loading districts. NESC Fig. 250-1 and NESC Table 250-1 provide a listing of the most predominate warm islands (Hawaii, Puerto Rico, Guam, Virgin Islands, and American Samoa). The statement "the loads of Rule 250B" is used throughout Secs. 25 and 26. When ice is specified, it is specified as a radial value. For example, a conductor with 1/4 in of radial ice (medium loading) adds 1/2 in to the diameter of the conductor (1/4 in of ice plus conductor diameter plus 1/4 in of ice).

The requirements for heavy, medium, light, and warm islands loading districts are outlined in Fig. 250-1.

250C. Extreme Wind Loading. NESC Fig. 250-2 (250-2a for Grade B and 250-2b for Grade C) and NESC Tables 250-1, 250-2, and 250-3 are the basic references for determining extreme wind loading. The statement "the loads of Rule 250C" is used throughout Secs. 25 and 26.

Rule 250C only applies to structures (e.g., poles) or its supported facilities (e.g., conductors, static wires, messengers, cables, etc.) more than 60 ft above ground or water level. See Fig. 250-2.

Rules 261A1c (metal, prestressed, and reinforced concrete poles), 261A2e (wood poles), and 261A3d (fiber-reinforced polymer poles) require the application of extreme wind to the pole, without conductors, for any height pole. Typically, applying the heavy, medium, light, or warm islands loading conditions to a wood pole and its conductors less than 60 ft above ground will result in a worse case design condition than a wood pole less than 60 ft above ground with extreme wind applied to the pole without conductors. The less than 60-ft requirement exists for special cases (but all cases must be checked). If a pole is pre-engineered (e.g., a steel pole or a laminated wood pole), the design strength at the ground line could have a higher resisting moment in the transverse direction than in the longitudinal direction. For a line less than 60 ft above ground, the transverse loading requires application of Rule 250B (heavy, medium, light, or warm islands) loads. The transverse and longitudinal

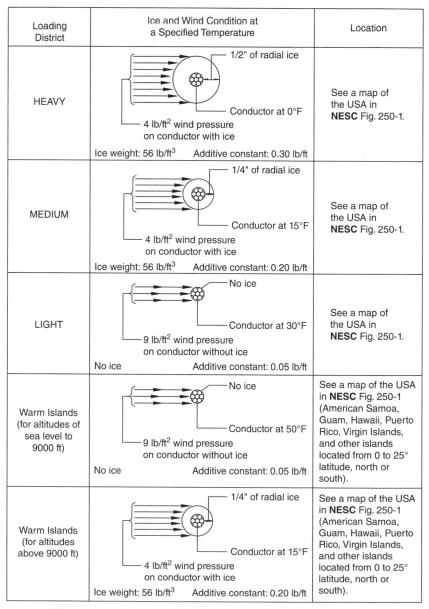

Loading District	Ice and Wind Condition at a Specified Temperature	Location
HEAVY	1/2" of radial ice Conductor at 0°F 4 lb/ft² wind pressure on conductor with ice Ice weight: 56 lb/ft³ Additive constant: 0.30 lb/ft	See a map of the USA in **NESC** Fig. 250-1.
MEDIUM	1/4" of radial ice Conductor at 15°F 4 lb/ft² wind pressure on conductor with ice Ice weight: 56 lb/ft³ Additive constant: 0.20 lb/ft	See a map of the USA in **NESC** Fig. 250-1.
LIGHT	No ice Conductor at 30°F 9 lb/ft² wind pressure on conductor without ice No ice Additive constant: 0.05 lb/ft	See a map of the USA in **NESC** Fig. 250-1.
Warm Islands (for altitudes of sea level to 9000 ft)	No ice Conductor at 50°F 9 lb/ft² wind pressure on conductor without ice No ice Additive constant: 0.05 lb/ft	See a map of the USA in **NESC** Fig. 250-1 (American Samoa, Guam, Hawaii, Puerto Rico, Virgin Islands, and other islands located from 0 to 25° latitude, north or south).
Warm Islands (for altitudes above 9000 ft)	1/4" of radial ice Conductor at 15°F 4 lb/ft² wind pressure on conductor with ice Ice weight: 56 lb/ft³ Additive constant: 0.20 lb/ft	See a map of the USA in **NESC** Fig. 250-1 (American Samoa, Guam, Hawaii, Puerto Rico, Virgin Islands, and other islands located from 0 to 25° latitude, north or south).

Fig. 250-1. Heavy, medium, light, and warm islands ice and wind loading (Rule 250B).

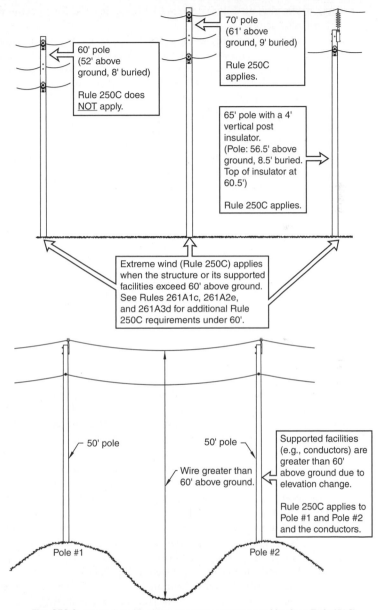

60' pole
(52' above
ground, 8' buried)

Rule 250C does
NOT apply.

70' pole
(61' above
ground, 9' buried)

Rule 250C
applies.

65' pole with a 4'
vertical post
insulator.
(Pole: 56.5' above
ground, 8.5' buried.
Top of insulator at
60.5')

Rule 250C applies.

Extreme wind (Rule 250C) applies
when the structure or its supported
facilities exceed 60' above ground.
See Rules 261A1c, 261A2e,
and 261A3d for additional Rule
250C requirements under 60'.

50' pole

50' pole

Wire greater than
60' above ground.

Supported facilities
(e.g., conductors) are
greater than 60'
above ground due to
elevation change.

Rule 250C applies to
Pole #1 and Pole #2
and the conductors.

Pole #1

Pole #2

Fig. 250-2. Examples of facilities requiring extreme wind loading (Rule 250C).

directions require application of Rule 261A1c, 261A2e, or 261A3d for extreme wind on the pole, without conductors.

Typically the heavy, medium, light, or warm islands loading requirements applied to conductors in the transverse direction will require more pole strength than the extreme wind loading requirements applied to the pole (without conductors) in the longitudinal direction. If a traditional round wood pole is used, the same strength is available in both the transverse and longitudinal directions. Assuming a traditional round wood pole is sized based on the heavy, medium, light, or warm islands loading with conductors in the transverse direction, adequate strength should exist for the extreme wind loading on the pole (without conductors) in any direction. Equal pole strength in all directions may not be accurate for a pre-engineered pole or tower utilizing different strengths in different directions.

If it is determined that Rule 250C applies (i.e., the structures or the supported facilities are over 60 ft), the extreme wind must be applied to the entire structure and supported facilities, not just the portions of the structure above 60 ft. Extreme wind is applied without ice on the conductors and without ice on the structure at a 60°F temperature condition. **NESC** Table 250-1 specifies the temperature at which extreme wind is checked. A formula is provided to calculate the wind load on an object based on the basic wind speed, velocity pressure exposure coefficient, gust response factor, and force coefficient (shape factor).

The basic wind speed is found in **NESC** Fig. 250-2 (250-2a for Grade B and 250-2b for Grade C). The velocity pressure exposure coefficient (k_z) increases with increased height due to wind friction near the earth's surface. Typical k_z values are provided in **NESC** Table 250-2. The formulas to calculate k_z are provided under the table. The gust response factor (G_{RF}) for wires decreases for increased span length as gusts average out over longer span lengths. The gust response factor for structures decreases for increased height as the structure is more flexible and takes longer to respond to wind gusts. Typical G_{RF} values are provided in **NESC** Table 250-3 (250-3a for structures and 250-3b for wires). The formulas to calculate G_{RF} are provided under the tables. Both k_z and G_{RF} have separate values for the structure height and wire height. The values for structure height and wire height for k_z are in the same table (**NESC** Table 250-2). The values for structure height and wire height for G_{RF} are in separate tables (**NESC** Tables 250-3a and 250-3b). Force coefficient (shape factor) is defined in Rule 252B; see Rule 252B for a discussion. A separate factor may be needed for the structure and the wire if the structure is not a round pole. The formulas for k_z and G_{RF} are complex and the heights and span lengths associated with k_z and G_{RF} can vary along the line. Rule 250C provides a note at the end of the rule to simplify the mathematical product of k_z and G_{RF}, if desired. An example of an extreme wind calculation is shown in Fig. 250-3.

The value of 17.53 lb/ft^2 for the wire (calculated in Fig. 250-3) is entered at 60°F on the sample sag and tension chart located at the beginning of Sec. 23. This design condition is only required if the structure or the conductor is greater than 60 ft above ground or water level. A line with 45-ft poles on level ground would not require this condition to be on the conductor sag and tension chart. Additional extreme wind examples can be found in **NESC** Appendix C.

250D. Extreme Ice with Concurrent Wind Loading.   NESC Fig. 250-3 (a through f) and **NESC** Tables 250-1 and 250-4 are the basic references for determining extreme ice with concurrent wind loading. The statement "the loads of 250D" is used

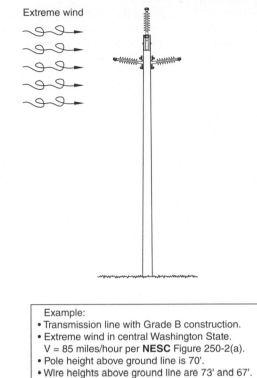

Extreme wind

Example:
- Transmission line with Grade B construction.
- Extreme wind in central Washington State.
  V = 85 miles/hour per **NESC** Figure 250-2(a).
- Pole height above ground line is 70'.
- Wire heights above ground line are 73' and 67'.
- Wind span length is 300'.
- The conductor and the pole have round shapes.

Wind pressure $= 0.00256 \cdot (V)^2 \cdot k_z \cdot G_{RF} \cdot C_f$	
Extreme wind pressure on the pole:	**Extreme wind pressure on the wires:**
• V = 85 miles/hour per **NESC** Figure 250-2a	• V = 85 miles/hour per **NESC** Figure 250-2a
• k_z structure = 1.1 per **NESC** Table 250-2	• k_z wire = 1.2 per **NESC** Table 250-2
• G_{RF} structure = 0.85 per **NESC** Table 250-3a	• G_{RF} wire = 0.79 per **NESC** Table 250-3b
• C_f = 1.0 for cylindrical structures per **NESC** Rule 252B2	• C_f = 1.0 for cylindrical shapes per **NESC** Rule 251A2
Wind pressure = 17.29 lb/ft^2	Wind pressure = 17.53 lb/ft^2

Fig. 250-3. Example of an extreme wind calculation (Rule 250C).

throughout Secs. 25 and 26. Rule 250D only applies to structures (e.g., poles) or their supported facilities (e.g., conductors, static wires, messengers, cables, etc.) more than 60 ft above ground or water level. This same requirement can be found in Rule 250C. See the discussion and figure in Rule 250C for examples of facilities greater than 60' above ground.

If it is determined that Rule 250D applies (i.e., the structure or the supported facilities are over 60 ft), the extreme ice with concurrent wind must be applied to the entire structure and supported facilities, not just the portions of the structure above 60 ft. Extreme ice with concurrent wind is applied to the conductor at a 15°F temperature condition. Only the wind portion of the extreme ice with concurrent wind needs to be applied to the structure (i.e., pole) as only the conductors are required to be covered with ice, not the pole.

NESC Table 250-1 specifies the temperature at which extreme ice with concurrent wind is checked. NESC Table 250-4 provides horizontal wind pressures in lb/ft² based on wind speed in mph. Per the wording under the title of NESC Table 250-4, NESC Table 250-4 is only for use with Rule 250D, not Rule 250C. NESC Table 250-4 does not specify a force coefficient (shape factor). Although it is not stated in the Code, the horizontal wind pressure values in NESC Table 250-4 correspond to cylindrical surfaces. The appropriate force coefficients (shape factors) for flat or latticed structures in Rule 252B2 should be considered for non-cylindrical surfaces. The ice portion for extreme ice with concurrent wind is determined from NESC Fig. 250-3 (a through f) and applied to the conductors in accordance with Rule 251. See Rule 251 for a discussion and figures for applying ice and wind to conductors. Rule 250D2 allows the ice (not wind) to be derated to 0.80 for Grade C construction. An extreme ice with concurrent wind value of 15°F, 0.25 in of ice, and 2.30 lb/ft² of wind is entered on the sample sag and tension chart located at the beginning of Sec. 23. This value was chosen from western Washington State. This design condition on the conductor is only required if the structure or the conductor is greater than 60 ft above ground or water level. A line with 45-ft poles on level ground would not require this condition to be on the conductor sag and tension chart.

251. CONDUCTOR LOADING

251A. General. The ice and wind loads in Rule 250, specifically Rules 250B, 250C, and 250D, must be applied to conductors. Rule 251 specifies how to apply the ice and wind loads to conductors.

If the conductor consists of a cable on a messenger, the ice and wind loads are applied to both the cable and the messenger. Rule 251A2 specifies a method for determining wind loads on conductors and cables without an ice covering. Examples of wind loads on a stranded conductor and on communication cables supported on a messenger are shown in Fig. 251-1.

Rule 251A3 specifies a method for determining ice loads on various conductor and cable configurations. Rule 251A4 requires that testing or a qualified engineering study be performed if the ice loading method in Rule 251A3 is reduced. Examples of ice and wind loads on a stranded conductor and on communication cables supported on a messenger are shown in Fig. 251-2.

251B. Unit Load Components. Rule 251B specifies the individual loads to consider when ice and wind are applied to a conductor. The unit load components are broken into three parts: vertical load, horizontal load, and total load.

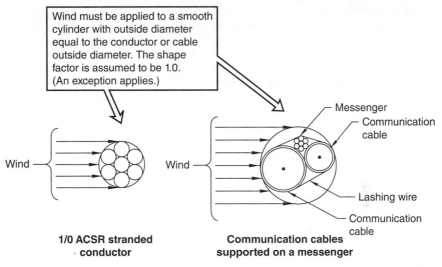

Fig. 251-1. Examples of wind loads on various conductor configurations (Rules 251A1 and 251A2).

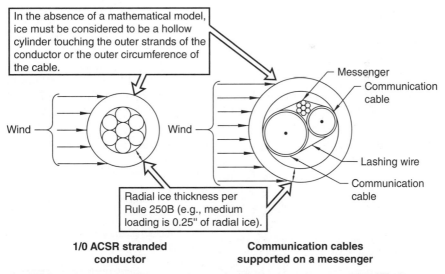

Fig. 251-2. Examples of ice and wind loads on various conductor configurations (Rule 251A3).

The vertical load must consider the weight per unit of the conductor(s) or cable(s) plus the distributed weights of conductor spacers or any equipment that the conductor supports (e.g., Federal Aviation Administration marker balls). These items must be covered with ice to calculate weight for the medium, heavy, and high-elevation portions of warm islands loading districts per Rule 250B or the extreme ice loading per Rule 250D (if applicable).

The horizontal unit load must consider wind on the conductors plus spacers and equipment that the conductor supports (e.g., Federal Aviation Administration marker balls). These items must be covered with ice to calculate the wind force for medium, heavy, and high-elevation portions of warm islands loading districts per Rule 250B or the extreme ice loading per Rule 250D (if applicable). The wind pressures in Rule 250B must be considered and, if applicable, the wind pressures in Rules 250C and 250D must be considered.

The total load is the resultant of the horizontal and vertical loads plus an additive constant (sometimes referred to as a "K-factor") given in **NESC** Table 251-1. The additive constant varies by loading district (i.e., 0.30 lb/ft for heavy, 0.20 lb/ft for medium, 0.05 lb/ft for light, and 0.05 lb/ft or 0.20 lb/ft for warm islands depending on elevation). No constant is added for the extreme wind loading or extreme ice with concurrent wind loading. The conductor or cable messenger tension must be computed from the total load. The conductor vertical load component and conductor horizontal load component are typically calculated on a per foot of conductor basis and then multiplied by the appropriate span lengths described in Rule 252A (e.g., the length for a vertical weight span) and Rule 252B4 (e.g., the length for a horizontal wind span).

An example of the load components for the medium loading district is shown in Fig. 251-3.

The total loads for the medium loading design condition (250B), the extreme wind design condition (250C), and the extreme ice with concurrent wind design condition (250D) are shown in the sample sag and tension chart at the beginning of Sec. 23. The sample sag and tension chart indicates an initial tension of 1366 lb for 15°F, 0.25 in ice, 4 lb/ft^2 wind, and an additive constant of 0.20 lb/ft. This tension is higher than the extreme wind (250C) initial tension of 1146 lb for 60°F and 17.53 lb/ft^2 of wind and the extreme ice with concurrent wind (250D) initial tension of 1070 lb for 15°F, 0.25 in of ice, and 2.30 lb/ft^2 of wind.

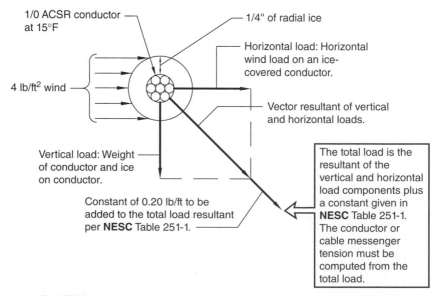

Fig. 251-3. Example of load components for the medium loading district (Rule 251B).

The 1366-lb tension is commonly called the "design tension." It is possible for the 250C (extreme wind) or the 250D (extreme ice with concurrent wind) loads to produce a higher conductor tension than the 250B (heavy, medium, light, or warm islands) loads. If the 250C or 250D loads are applicable (i.e., the structure or conductors are more than 60 ft above ground or water), then the "design tension" must be chosen from the largest conductor tension based on the 250B, 250C, and 250D loads. The 250B, 250C, and 250D loads are applied to conductors in accordance with Rule 251. The 250B, 250C, and 250D loads are applied to line supports (e.g., structures) in accordance with Rule 252. Rule 250A specifies that where all three rules apply (250B, 250C, and 250D), the required loading shall be the one that has the greatest effect.

Rule 261H1 requires the open (e.g., bare noninsulated) supply conductor tension at the Rule 251 condition (using the loads of Rule 250B) to be not more than 60 percent of the conductor rated breaking strength. Rule 261H also specifies tension limits if Rules 250C or 250D are applicable. The sample sag and tension chart at the beginning of Sec. 23 indicates a rated tensile strength of 31.2 percent for the 1366-lb design tension. This percentage meets the not more than 60 percent requirement based on Rule 250B loads. Rule 261H1 also specifies an initial and final tension limit at the applicable temperature in Table 251-1 without external load (i.e., without ice or wind). Rules 261I (supply cable messengers) and 261K2 (communication cable messengers) specify messenger tension limits for the 250B (heavy, medium, light, or warm islands loading), 250C (extreme wind loading), and 250D (extreme ice with concurrent wind loading) conditions applied in accordance with Rule 251.

252. LOADS ON LINE SUPPORTS

252A. Assumed Vertical Loads.   Ice and wind loads are first defined in Rule 250, specifically Rules 250B, 250C, and 250D. How to apply ice and wind loading to conductors is defined in Rule 251. How to apply ice and wind loading on line supports is specified in Rule 252. Rule 250A2 should be reviewed for proper consideration of construction and maintenance loads.

Rule 252A discusses how to apply vertical loads to line supports (i.e., poles, towers, foundations, crossarms, pins, insulators, and conductor fastenings). Rule 252A does not specify any load factors. The load factors are provided in Rule 253. An example of vertical loads on line supports is shown in Fig. 252-1.

Per the definition of span/weight span in Sec. 02, Definitions, a difference in elevation of supports must be considered as it will affect the length of the vertical (weight) span. An example of vertical loads on line supports with different elevations is shown in Fig. 252-2.

It is possible to have a negative vertical load on a line support if the conductor position is above the line support attachment position. This condition can occur on hilly terrain and is commonly referred to as uplift. Vertical loads on wood poles and structures due to guy tensions are discussed in Rule 261A2c.

252B. Assumed Transverse Loads.   Rule 252B discusses how to apply transverse (horizontal) loads to line supports (i.e., poles, towers, foundations, crossarms, pins, insulators, and conductor fastenings). Rule 252B does not specify any load factors. The load factors are provided in Rule 253. An example of transverse loads on line supports is shown in Fig. 252-3.

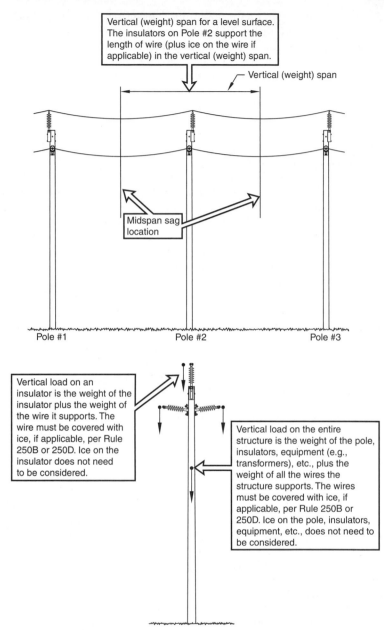

Vertical (weight) span for a level surface. The insulators on Pole #2 support the length of wire (plus ice on the wire if applicable) in the vertical (weight) span.

Vertical (weight) span

Midspan sag location

Pole #1   Pole #2   Pole #3

Vertical load on an insulator is the weight of the insulator plus the weight of the wire it supports. The wire must be covered with ice, if applicable, per Rule 250B or 250D. Ice on the insulator does not need to be considered.

Vertical load on the entire structure is the weight of the pole, insulators, equipment (e.g., transformers), etc., plus the weight of all the wires the structure supports. The wires must be covered with ice, if applicable, per Rule 250B or 250D. Ice on the pole, insulators, equipment, etc., does not need to be considered.

Fig. 252-1. Example of vertical loads on line supports (Rule 252A).

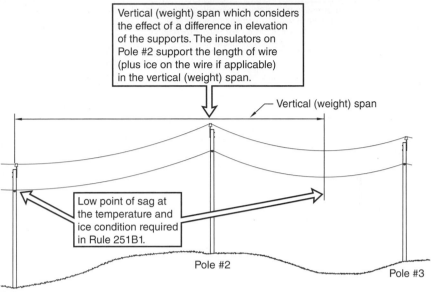

Vertical (weight) span which considers the effect of a difference in elevation of the supports. The insulators on Pole #2 support the length of wire (plus ice on the wire if applicable) in the vertical (weight) span.

Vertical (weight) span

Low point of sag at the temperature and ice condition required in Rule 251B1.

Pole #2

Pole #3

Pole #1

Fig. 252-2. Example of vertical loads on line supports with different elevations (Rule 252A).

Rule 252B2 specifies force coefficients (shape factors) to be used for various surface types. Transverse wind loads on structures are calculated without ice covering. Examples of wind load on various structure types are shown in Fig. 252-4.

Rule 252B3 requires angle structure to consider both the transverse wind load and the wire tension load. The wind direction must be applied to give the maximum resultant load. The angle of the line affects how to apply the wind. Normally, for small and medium angles, an oblique wind in the direction of the bisector of the angle will provide the maximum resultant load. Proper reductions are permissible due to the angularity of the wind on the wire. See Fig. 252-5.

Rule 252B4 states that the calculated transverse load must be based on the wind span. Per the definition, of span/wind span in Sec. 02, Definitions, the wind span is the sum of half of the spans on either side of the supporting structure (e.g., pole). Consideration may be given to line angles and deadends. Rule 252B4 does not require elevation to be considered as Rule 252A does for a vertical span.

252C. Assumed Longitudinal Loading. Longitudinal loads are loads in line with the conductors. The most obvious longitudinal load is at a conductor deadended (single sided deadend) on a pole; however, longitudinal loading can occur on double deadends and on tangent structures under certain conditions. Rule 252C breaks longitudinal loading into six parts (252C1 through 252C6). It does not specify any load factor. The load factors are provided in Rule 253.

Rule 252C1 specifies longitudinal tensions to be used when a Grade B line section is located within a line of a lower grade. A common example is a Grade C

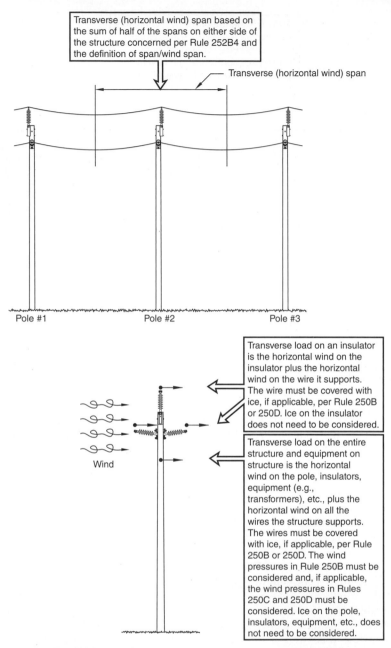

Transverse (horizontal wind) span based on the sum of half of the spans on either side of the structure concerned per Rule 252B4 and the definition of span/wind span.

Transverse (horizontal wind) span

Pole #1   Pole #2   Pole #3

Transverse load on an insulator is the horizontal wind on the insulator plus the horizontal wind on the wire it supports. The wire must be covered with ice, if applicable, per Rule 250B or 250D. Ice on the insulator does not need to be considered.

Transverse load on the entire structure and equipment on structure is the horizontal wind on the pole, insulators, equipment (e.g., transformers), etc., plus the horizontal wind on all the wires the structure supports. The wires must be covered with ice, if applicable, per Rule 250B or 250D. The wind pressures in Rule 250B must be considered and, if applicable, the wind pressures in Rules 250C and 250D must be considered. Ice on the pole, insulators, equipment, etc., does not need to be considered.

Wind

Fig. 252-3. Example of transverse loads on line supports (Rule 252B).

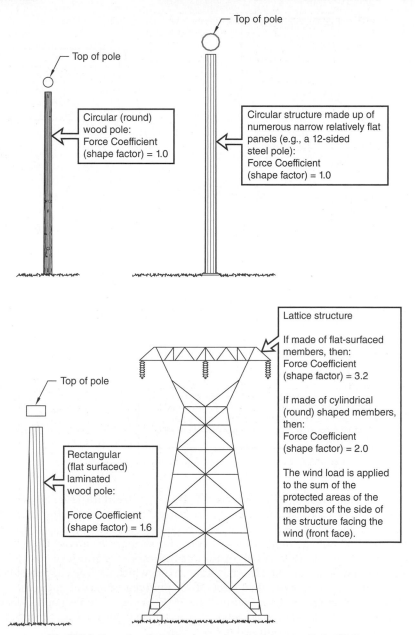

Top of pole

Circular (round)
wood pole:
Force Coefficient
(shape factor) = 1.0

Top of pole

Circular structure made up of
numerous narrow relatively flat
panels (e.g., a 12-sided
steel pole):
Force Coefficient
(shape factor) = 1.0

Top of pole

Rectangular
(flat surfaced)
laminated
wood pole:

Force Coefficient
(shape factor) = 1.6

Lattice structure

If made of flat-surfaced
members, then:
Force Coefficient
(shape factor) = 3.2

If made of cylindrical
(round) shaped members,
then:
Force Coefficient
(shape factor) = 2.0

The wind load is applied
to the sum of the
protected areas of the
members of the side of
the structure facing the
wind (front face).

Fig. 252-4. Examples of wind load on various structure types (Rule 252B2).

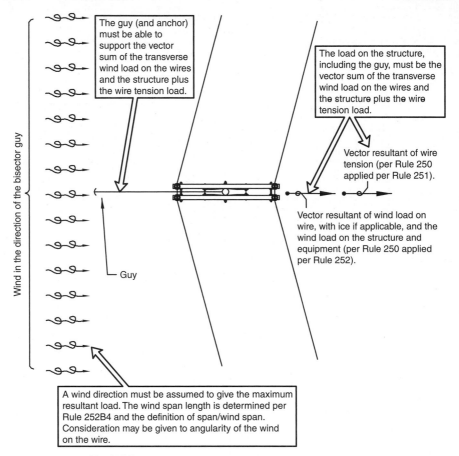

The guy (and anchor) must be able to support the vector sum of the transverse wind load on the wires and the structure plus the wire tension load.

The load on the structure, including the guy, must be the vector sum of the transverse wind load on the wires and the structure plus the wire tension load.

Vector resultant of wire tension (per Rule 250 applied per Rule 251).

Vector resultant of wind load on wire, with ice if applicable, and the wind load on the structure and equipment (per Rule 250 applied per Rule 252).

Wind in the direction of the bisector guy

Guy

A wind direction must be assumed to give the maximum resultant load. The wind span length is determined per Rule 252B4 and the definition of span/wind span. Consideration may be given to angularity of the wind on the wire.

Fig. 252-5. Example of transverse loads at line angles (Rule 252B3).

distribution line crossing a limited access highway, railroad tracks, or a navigable waterway. The Grade C line must be increased to Grade B for the crossing. Rule 252C1 addresses longitudinal loads on poles, towers, and guys. Additional rules for longitudinal strength of structures for changes in grade of construction can be found in Rule 261A5. Additional rules for longitudinal strength of crossarms and braces can be found in Rule 261D5 which includes applied loads. Additional rules for longitudinal strength of pin-type (insulator supports) and conductor fastenings can be found in Rule 261F1 which includes applied loads. The **Code** rules for longitudinal loads where a change in the grade of construction occurs per Rule 252C1 are outlined in Fig. 252-6.

Rule 252C2 addresses the older open-wire communication circuits with multiple conductors on a crossarm. This construction has frequently been replaced with multiple pair insulated cables supported on a messenger and fiber-optic cables.

Rule 252C3 provides rules for longitudinal loading on dead-end structures. See Fig. 252-7.

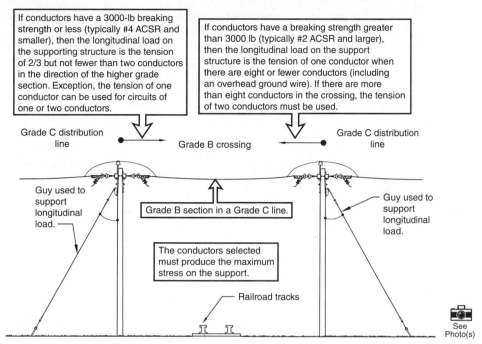

If conductors have a 3000-lb breaking strength or less (typically #4 ACSR and smaller), then the longitudinal load on the supporting structure is the tension of 2/3 but not fewer than two conductors in the direction of the higher grade section. Exception, the tension of one conductor can be used for circuits of one or two conductors.

If conductors have a breaking strength greater than 3000 lb (typically #2 ACSR and larger), then the longitudinal load on the support structure is the tension of one conductor when there are eight or fewer conductors (including an overhead ground wire). If there are more than eight conductors in the crossing, the tension of two conductors must be used.

Grade C distribution line

Grade B crossing

Grade C distribution line

Guy used to support longitudinal load.

Grade B section in a Grade C line.

The conductors selected must produce the maximum stress on the support.

Guy used to support longitudinal load.

Railroad tracks

See Photo(s)

Fig. 252-6. Longitudinal loads at a change in grade of construction (Rule 252C1).

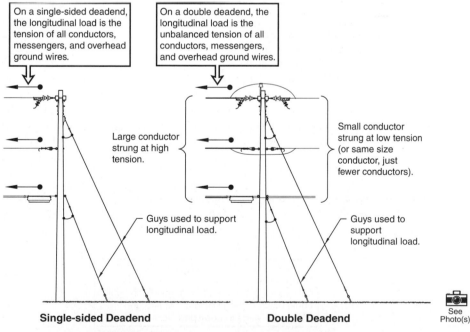

On a single-sided deadend, the longitudinal load is the tension of all conductors, messengers, and overhead ground wires.

On a double deadend, the longitudinal load is the unbalanced tension of all conductors, messengers, and overhead ground wires.

Large conductor strung at high tension.

Small conductor strung at low tension (or same size conductor, just fewer conductors).

Guys used to support longitudinal load.

Guys used to support longitudinal load.

Single-sided Deadend

Double Deadend

See Photo(s)

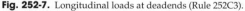

Fig. 252-7. Longitudinal loads at deadends (Rule 252C3).

Rule 252C4 requires that a structure be capable of supporting unbalanced longitudinal loads created by unequal vertical loads or unequal spans. Extreme cases will require a double deadend structure and guying.

Rule 252C5 requires that consideration be given to longitudinal loads during wire stringing operations. Temporary deadend structures and temporary guying are typically used to meet this criterion if permanent facilities do not exist at the wire stringing locations. Rule 252C5 is similar in nature to Rule 250A2, which requires considering construction and maintenance loads.

Rule 252C6 specifies longitudinal tensions for open-wire communication conductors at crossings of railroad tracks, limited access highways and navigable waterways requiring waterway crossing permits. This construction has frequently been replaced with multiple pair insulated cables supported on a messenger and fiber-optic cables.

A recommendation at the end of Rule 252C states that structures with longitudinal load capacity should be provided at reasonable intervals along a line. This rule is very general in nature. The intent of this recommendation is to avoid cascading failures (i.e., a domino effect), which could occur in long stretches of straight lines if the wires in one span were to break. Some degree of longitudinal structure strength exists in a tangent wood pole designed for transverse wind loads. If the conductors break during a nonloaded condition (i.e., no ice or wind), the actual longitudinal loads will be less than if the conductors break during ice and wind conditions. A certain amount of longitudinal strength will also be provided by application of Rules 261A1c, 261A2e, and 261A3d. Typically, the transverse loading and the application of Rules 261A1c, 261A2e, and 261A3d will not be as severe as a longitudinal deadend conductor load. A line with angles, especially 90° turns, has longitudinal strength built in. A line that continues straight for a long distance may require a double deadend structure in the middle of the long straight stretch with enough longitudinal strength to act as a single deadend if the conductors were to break on either side of the pole. When to insert a double deadend pole in a straight line section requires accepted good practice as the Code does not specify a specific distance. See Fig. 252-8.

252D. Simultaneous Application of Loads. A structure may experience a combination of vertical (weight), transverse (horizontal wind and wire tension), and longitudinal (in-line) loads. The structure must be designed to withstand the simultaneous application of these loads when they occur simultaneously. An example of a structure experiencing simultaneous loads is shown in Fig. 252-9.

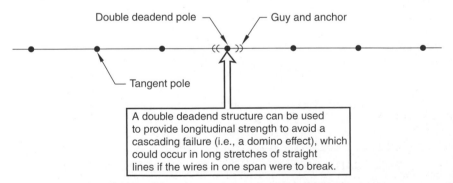

Fig. 252-8. Recommendation to provide longitudinal strength capability (Rule 252C).

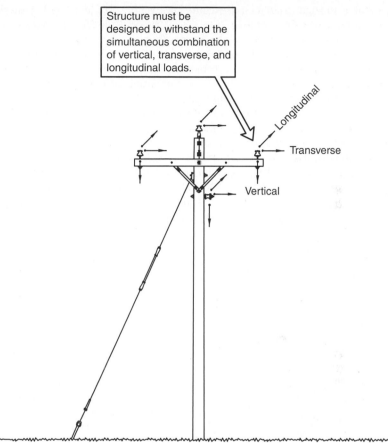

Structure must be designed to withstand the simultaneous combination of vertical, transverse, and longitudinal loads.

Longitudinal

Transverse

Vertical

Fig. 252-9. Simultaneous application of loads (Rule 252D).

253. LOAD FACTORS FOR STRUCTURES, CROSSARMS, SUPPORT HARDWARE, GUYS, FOUNDATIONS, AND ANCHORS

Rule 253 requires load factors to be applied to the loads of Rules 250B, 250C, and 250D. This requirement is also stated in a similar fashion in Rules 261A1a and 261A1b for metal, prestressed-concrete, and reinforced-concrete structures, in Rules 262A2a and 262A2b for wood structures, and in Rules 261A3a and 261A3b for fiber-reinforced polymer structures. Load factors are provided in **NESC** Table 253-1. A load factor is a load multiplier that is greater than or equal to 1.0. Grade C extreme ice with concurrent wind (Rule 250D) has a load factor less than 1.0 for the ice portion only of extreme ice with concurrent wind. The Grade C load factor for Rule 250D is shown in Rule 250D2, not in Table 253-1. A load factor is applied to a base load to increase it. The load factors for Grade B are higher than Grade C with one exception, which is vertical loads. The higher load factors for Grade B produce a line that is stronger than a Grade C line because the Grade B line is required to withstand higher loads. A note in Rule 253 references Rule 250A2 for construction and maintenance loads which may be greater than weather loads.

The load factors in **NESC** Table 253-1 in Sec. 25 must be used in combination with the strength factors in **NESC** Table 261-1 in Sec. 26. A strength factor is a material strength multiplier that is less than or equal to 1.0. It is applied to a material strength to reduce it. The strength factors for Grade B are less than or equal to Grade C. The lower strength factors for Grade B produce a line that is stronger than a Grade C line because less of the materials' full strength is permitted to be used for Grade B construction.

The combined effect of the load factor in **NESC** Table 253-1 and the strength factor in **NESC** Table 261-1 can be determined by dividing the load factor by the strength factor. For example, per **NESC** Table 253-1, the transverse wind load factor of a Grade B line is 2.50. Per **NESC** Table 261-1, the wood pole strength factor of a Grade B line is 0.65. The combined effect of these factors is 2.50 ÷ 0.65 = 3.85. The combined effect of the load factor and the strength factor is sometimes referred to as the overall "safety factor."

An outline of load factors and strength factors is provided in Fig. 253-1.

Examples of applying load and strength factors are provided in Fig. 253-2.

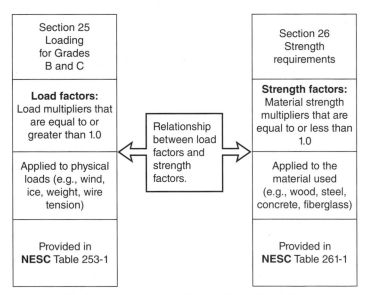

Fig. 253-1. Load and strength factors (Rules 253 and 261).

Example #1	Example #2	Example #3	Example #4
Conditions:	**Conditions:**	**Conditions:**	**Conditions:**
- Transverse wind load on 1/0 ACSR wire	- Transverse wind load on 1/0 ACSR wire	- Transverse wind load on 1/0 ACSR wire	- Transverse wind load on 1/0 ACSR wire
- Medium loading district (Rule 250B) (Rules 250C and 250D are not applicable)	- Medium loading district (Rule 250B) (Rules 250C and 250D are not applicable)	- Medium loading district (Rule 250B) (Rules 250C and 250D are not applicable)	- Medium loading district (Rule 250B) (Rules 250C and 250D are not applicable)
- Wood poles	- Wood poles	- Steel poles	- Steel poles
- Grade B line	- Grade C line	- Grade B line	- Grade C line
Load and Strength Factor:	**Load and Strength Factor:**	**Load and Strength Factor:**	**Load and Strength Factor:**
- Load factor from **NESC** Table 253-1: 2.50	- Load factor from **NESC** Table 253-1: 2.20 (at line crossings) 1.75 (elsewhere)	- Load factor from **NESC** Table 253-1: 2.50	- Load factor from **NESC** Table 253-1: 2.20 (at line crossings) 1.75 (elsewhere)
- Strength factor from **NESC** Table 261-1: 0.65 (Footnote 2 applies to deterioration) (Footnote 4 applies to temporary service)	- Strength factor from **NESC** Table 261-1: 0.85 (Footnote 2 applies to deterioration) (Footnote 4 applies to temporary service)	- Strength factor from **NESC** Table 261-1: 1.0 (Footnote 6 applies to deterioration)	- Strength factor from **NESC** Table 261-1: 1.0 (Footnote 6 applies to deterioration)

Example #5	Example #6	Example #7	Example #8
Conditions:	**Conditions:**	**Conditions:**	**Conditions:**
- Longitudinal wire tension load at a 1/0 ACSR deadend	- Longitudinal wire tension load at a 1/0 ACSR deadend	- Longitudinal wire tension load at a 1/0 ACSR deadend	- Longitudinal wire tension load at a 1/0 ACSR deadend
- Medium loading district (Rule 250B) (Rules 250C and 250D are not applicable)	- Medium loading district (Rule 250B) (Rules 250C and 250D are not applicable)	- Medium loading district (Rule 250B) (Rules 250C and 250D are not applicable)	- Medium loading district (Rule 250B) (Rules 250C and 250D are not applicable)
- Wood poles	- Wood poles	- Steel poles	- Steel poles
- Grade B line	- Grade C line	- Grade B line	- Grade C line
Load and Strength Factor:	**Load and Strength Factor:**	**Load and Strength Factor:**	**Load and Strength Factor:**
- Load factor from **NESC** Table 253-1: 1.65 (Footnote 2 applies to communications only)	- Load factor from **NESC** Table 253-1: 1.30 (Footnote 4 applies to special conditions)	- Load factor from **NESC** Table 253-1: 1.65 (Footnote 2 applies to communications only)	- Load factor from **NESC** Table 253-1: 1.30 (Footnote 4 applies to special conditions)
- Strength factor from **NESC** Table 261-1: 0.65 (Footnote 2 applies to deterioration) (Footnote 4 applies to temporary service)	- Strength factor from **NESC** Table 261-1: 0.85 (Footnote 2 applies to deterioration) (Footnote 4 applies to temporary service)	- Strength factor from **NESC** Table 261-1: 1.0 (Footnote 6 applies to deterioration)	- Strength factor from **NESC** Table 261-1: 1.0 (Footnote 6 applies to deterioration)

Fig. 253-2. Examples of applying load and strength factors (Rules 253 and 261).

Section 26

Strength Requirements

260. GENERAL (SEE ALSO SECTION 20)

Sections 24, 25, and 26 are all interrelated. Section 24, "Grades of Construction," defines the required strength of overhead line construction for safety purposes. Section 25, "Loadings for Grades B and C," defines the physical loads (e.g., ice, wind, and temperature conditions) that overhead line construction must be able to withstand and the load factors that must be applied to the physical loads. Section 26, "Strength Requirements," defines the required strength of materials used in constructing overhead lines and the strength factors that must be applied to the materials. Line insulators are not addressed in Secs. 24, 25, or 26. Insulators are covered in their own section, Sec. 27.

The purpose of Sec. 24 is to define the grade of construction that is required for different situations. The details of applying load factors and strength factors are covered in Secs. 25 and 26.

This Handbook addresses the **Code** requirements in Secs. 24, 25, 26, and 27. Line design calculations are not presented in this Handbook. Line design calculations can be found in transmission and distribution line design manuals. Large utilities commonly develop their own line design manuals. Some utilities use the transmission and distribution line design manuals published by the Rural Utilities Service (RUS), which in the past was referred to as the Rural Electrification Administration (REA). The RUS distribution line design manuals consist of RUS Bulletins 1724E-150 through 1724E-154. The RUS design manual for high-voltage transmission lines is RUS Bulletin 1724E-200. The RUS also publishes design manuals for communication lines. This Handbook will aid those individuals writing and using line design manuals and provide a deeper understanding of how the **Code** applies to line design calculations.

Typical strength and loading line design calculations include, but are not limited to, the following:

- Maximum wind span based on wind with ice on conductors
- Maximum weight span based on weight of conductors and ice
- Moment due to wind on pole
- Allowable resisting moment of pole
- Transverse, vertical, and total components of conductor loading
- Total ground line moment on pole
- Ruling span
- Diameter of pole at any point
- Dead-end guying strength
- Bisector guying strength
- Anchor strength
- Weak-link of guy attachment, guy wire, and anchor assembly
- Crossarm strength (vertical)
- Crossarm strength (longitudinal)
- Pole buckling
- Maximum line angle based on insulator strength
- Material deflection
- Equipment loading

Single pole structures are typically easier to analyze than multiple pole and lattice structures. Line design calculations also involve clearance calculations in addition to strength and loading calculations (e.g., clearance above ground, clearance to buildings and other structures, horizontal clearance between conductors, vertical clearance between conductors, etc.). The clearance calculations will affect the line design and therefore the selection of pole heights and pole classes. The sample sag and tension chart shown at the beginning of Sec. 23 provides sag for clearance calculations and conductor tension for strength and loading calculations. The "design tension" used for strength and loading calculations is the largest of the conductor tensions calculated after applying Rule 250B (heavy, medium, light, or warm islands loads), Rule 250C (extreme wind loads) if applicable, and Rule 250D (extreme ice with concurrent wind loads) if applicable. The 250B, 250C, and 250D loads are applied to conductors in accordance with Rule 251. The 250B, 250C, and 250D loads are applied to line supports (e.g., structures) in accordance with Rule 252. Rule 250A specifies that where all three rules apply (250B, 250C, and 250D), the required loading shall be the one that has the greatest effect. See Rules 250, 251, and 252 for additional information.

260A. Preliminary Assumptions.   Rule 260A1 explains how to account for deformation, deflection, and displacement of parts of a structure when performing strength calculations. Rule 260A1 only addresses deformation, deflection, and displacement when the effects reduce the load. For example, a single-sided deadend structure that is not guyed can deflect into the span, therefore reducing the dead-end tension. Rule 260A1 does not address deformation, deflection, and displacement when the effects increase the load. For example, the weight of a three-phase transformer bank on a leaning pole can increase the effect of the vertical load. Rule 260A1 does not require considering deformation, deflection, and displacement when the effects reduce the load; this rule simply states allowances "may" be made. The decision to consider deformation, deflection, and displacement when performing line design calculations may depend on the method used to perform the calculations. Typically, line design calculations performed by hand or using simple spreadsheets do not account for deformation, deflection, and displacement unless some percentage increase to the pole loading is applied to account for the deformation, deflection, and displacement. Line design calculations performed using software programs make analyzing deformation, deflection, and displacement a simpler task. The term P-Delta is commonly used when referring to structure deformation, deflection, and displacement. Several line design software programs enable the user to perform a nonlinear P-Delta analysis for loading and strength from the ground line to the top of the pole (not just at the ground line). Software programs can also apply fiber strength reductions due to the fiber strength height effect for wood poles (see Rule 261A2 for more information). The **NESC** provides a two-part approach to determine the effects of deformation, deflection, and displacement where the effects reduce the load. First, the ice and wind loads of Rule 250 are applied to the structure without the load factors of Rule 253. These are the loads that are used to determine deformation, deflection, and displacement. Second, from the deformation, deflection, and displacement position, the structure strength is analyzed using the load factors in Rule 253. Examples of allowance for deformation, deflection, or displacement are shown in Fig. 260-1.

If allowance is made for deformation, deflection, or displacement at crossings or conflicts, the calculations used must be subject to mutual agreement of the parties involved.

The note in Rule 260A1 addresses the strength of a structure during the expected life of a structure. The note indicates that guying or bracing may be needed to meet the required strength of the structure over time due to everyday stresses on the structure produced by gravity or tension. For example, a 50-year-old wood dead-end pole that is not guyed may have sufficient strength for some portion of its service life but may not have sufficient strength toward the end of its service life due to the stress of everyday loads.

Per Rule 260A2, the **NESC** recognizes that new materials may become available that are not covered in the strength requirements section. Since the **NESC** is on a 5-year revision cycle, the **Code** permits trial installations of new materials. New materials must be tested and evaluated. See Rule 013 for additional requirements.

260B. Application of Strength Factors. Rule 260B1 explains how to apply **NESC** load and strength factors to line design calculations. An outline of load and strength factors is provided in Fig. 260-2.

NESC Table 261-1 is the main table in Sec. 26 for determining strength factors of various materials. It has strength factors for the loads of 250B (ice and wind in heavy, medium, light, and warm islands loading districts), for the loads of 250C (extreme wind), and for the loads of 250D (extreme ice with concurrent wind). **NESC** Table 261-1 must be used in combination with **NESC** Table 253-1. See Rule 253 for a discussion and examples of applying load and strength factors.

NESC Table 261-1 contains information relating to deterioration of structure strength in the table footnotes. Per **NESC** Table 261-1, Footnote 2, for Rule 250B loads a wood pole must be replaced or rehabilitated when deterioration reduces the

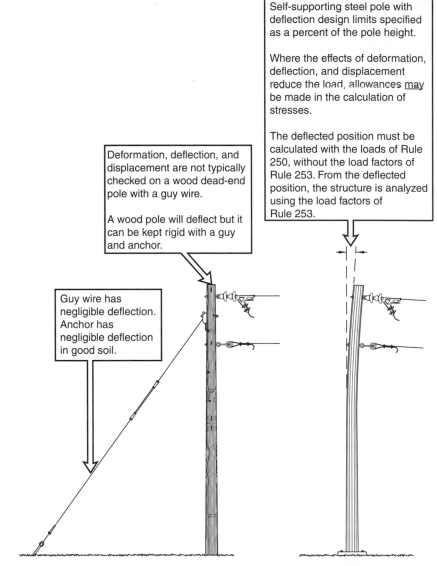

Self-supporting steel pole with deflection design limits specified as a percent of the pole height.

Where the effects of deformation, deflection, and displacement reduce the load, allowances may be made in the calculation of stresses.

The deflected position must be calculated with the loads of Rule 250, without the load factors of Rule 253. From the deflected position, the structure is analyzed using the load factors of Rule 253.

Deformation, deflection, and displacement are not typically checked on a wood dead-end pole with a guy wire.

A wood pole will deflect but it can be kept rigid with a guy and anchor.

Guy wire has negligible deflection. Anchor has negligible deflection in good soil.

Fig. 260-1. Examples of allowance for deformation, deflection, or displacement (Rule 260A).

structure strength to 2/3 of that required when installed. The word "required" is used to distinguish between the required strength of a structure and the actual strength of a structure. If the wood pole class was required to be Class 5, but a Class 4 (larger circumference) pole was installed, the pole must be replaced or rehabilitated when deterioration reduces the structure strength to 2/3 of the Class 5 rating (the required strength rating, not the actual strength rating). The most common

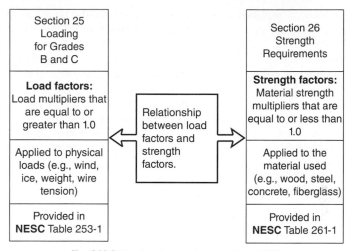

Fig. 260-2. Load and strength factors (Rule 260B).

wood pole deterioration is ground line rot. When wood distribution poles are inspected for ground line deterioration, it is common for the inspector to evaluate the actual strength rating of the pole unless extra steps are taken to determine the required strength rating of the pole. The most common rehabilitation method is pole stubbing using a steel truss or wood stub. If a pole stub is added, the stubbed pole must have strength greater than 2/3 of that required when installed. If the pole is replaced instead of stubbed, the full strength requirements of **NESC** Table 261-1 apply. Footnote 3 of **NESC** Table 261-1 is similar to Footnote 2. Footnote 2 specifies a strength deterioration of 2/3 for Rule 250B (heavy, medium, light, and warm islands) loads. Footnote 3 specifies a deterioration of 3/4 for Rule 250C (extreme wind loads) and Rule 250D (extreme ice with concurrent wind loads). Footnotes 2 and 3 state that the required strength of the structure (e.g., wood pole) must be based on revised loading when new or changed facilities modify the loads on an existing structure. The wording in Footnote 7 of **NESC** Table 261-1 further defines how deterioration applies to existing poles that have new or changed lines or equipment added to them. Footnote 4 of **NESC** Table 261-1 describes the reduced strength of a structure (e.g., wood pole) built for temporary service. This footnote does not use the same wording as Rule 014C which also applies to temporary installations. Footnote 6 addresses deterioration for metal structures (e.g., steel poles) and other items. Per Footnote 6, no deterioration is permitted below the required strength of the structure. If deterioration is anticipated, a strength factor of less than 1.0 must be used. Pole owners typically test wood poles at or near the groundline for rot to determine if the pole strength is adequate or if the wood pole needs to be replaced or rehabilitated (e.g., using a steel truss or wood stub). Visual inspections are typically performed for other structural members such as crossarms and pole hardware and the above ground portion of the pole. If a pole test reveals that the pole requires replacement, typically the crossarm (if one exists) and any associated pole line hardware are also replaced at the same time. If the same pole is reinforced (stubbed) instead of replaced, it is important to evaluate the condition of the pole top and the crossarm (if one exists) as the wood on the pole top and the wood crossarm may be decayed. For example, if stubbing a pole adds 25 years to the life of the pole, the pole top and the wood crossarm may not have 25 years of service life available without deteriorating below the 2/3 strength requirement in **NESC** Table 261-1, Footnote 2. Since testing the wood at the pole top and testing the wood crossarm are not common practices (typically only visual inspections are performed), careful attention should be given to these items on a stubbed pole. It may also be important to evaluate the metal pole line hardware as in some areas, rusting of the pole line hardware can cause failures. See Rule 214 for discussion of inspection and testing. Examples of how loads, materials, and load and strength factors affect the size of a wood pole are shown in Figs. 260-3 and 260-4.

Wood pole deterioration can occur from the outside-in (external decay) or from the inside-out (hollow heart). See Fig. 260-5.

Listings of common wood pole class sizes, examples of pole tags and pole branding information, and an example of a common wood pole burial depth formula are provided in Figs. 260-6 through 260-8.

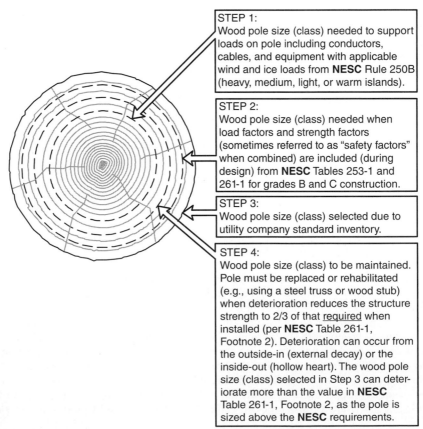

STEP 1:
Wood pole size (class) needed to support loads on pole including conductors, cables, and equipment with applicable wind and ice loads from **NESC** Rule 250B (heavy, medium, light, or warm islands).

STEP 2:
Wood pole size (class) needed when load factors and strength factors (sometimes referred to as "safety factors" when combined) are included (during design) from **NESC** Tables 253-1 and 261-1 for grades B and C construction.

STEP 3:
Wood pole size (class) selected due to utility company standard inventory.

STEP 4:
Wood pole size (class) to be maintained. Pole must be replaced or rehabilitated (e.g., using a steel truss or wood stub) when deterioration reduces the structure strength to 2/3 of that <u>required</u> when installed (per **NESC** Table 261-1, Footnote 2). Deterioration can occur from the outside-in (external decay) or the inside-out (hollow heart). The wood pole size (class) selected in Step 3 can deteriorate more than the value in **NESC** Table 261-1, Footnote 2, as the pole is sized above the **NESC** requirements.

See
Photo(s)

Fig. 260-3. Example of how loads, material, and load and strength factors affect the size of a wood pole (Rule 260B).

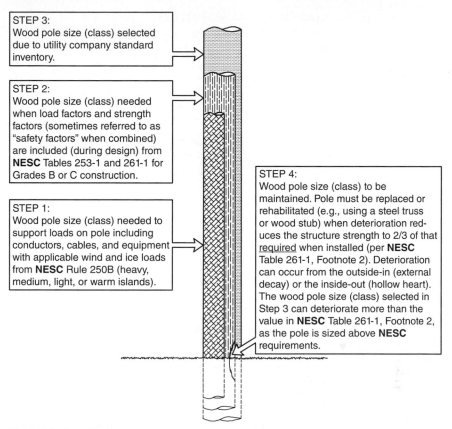

STEP 3:
Wood pole size (class) selected due to utility company standard inventory.

STEP 2:
Wood pole size (class) needed when load factors and strength factors (sometimes referred to as "safety factors" when combined) are included (during design) from **NESC** Tables 253-1 and 261-1 for Grades B or C construction.

STEP 1:
Wood pole size (class) needed to support loads on pole including conductors, cables, and equipment with applicable wind and ice loads from **NESC** Rule 250B (heavy, medium, light, or warm islands).

STEP 4:
Wood pole size (class) to be maintained. Pole must be replaced or rehabilitated (e.g., using a steel truss or wood stub) when deterioration reduces the structure strength to 2/3 of that <u>required</u> when installed (per **NESC** Table 261-1, Footnote 2). Deterioration can occur from the outside-in (external decay) or the inside-out (hollow heart). The wood pole size (class) selected in Step 3 can deteriorate more than the value in **NESC** Table 261-1, Footnote 2, as the pole is sized above **NESC** requirements.

Fig. 260-4. Example of how loads, material, and load and strength factors affect the size of a wood pole (Rule 260B).

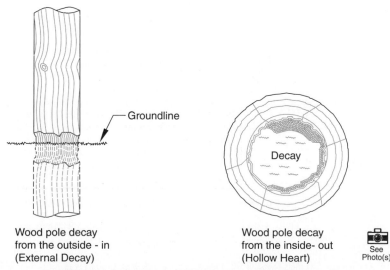

Groundline

Decay

Wood pole decay from the outside - in (External Decay)

Wood pole decay from the inside- out (Hollow Heart)

See Photo(s)

Fig. 260-5. Examples of wood pole deterioration (Rule 260B).

COMMON POLE CLASS RATINGS
(using a 50' pole length)

```
...                          Larger circumference
50'-CLASS H2                  (more stout) (fatter)
50'-CLASS H1
50'-CLASS 1
50'-CLASS 2
50'-CLASS 3
50'-CLASS 4
50'-CLASS 5                   Smaller circumference
...                           (less stout) (thinner)
```

Fig. 260-6. Listing of common wood pole classes
(Rule 260).

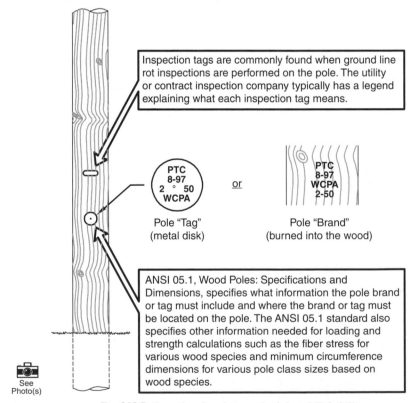

Inspection tags are commonly found when ground line rot inspections are performed on the pole. The utility or contract inspection company typically has a legend explaining what each inspection tag means.

```
   PTC                          PTC
   8-97                         8-97
2  °  50        or             WCPA
  WCPA                         2-50
```

Pole "Tag"   Pole "Brand"
(metal disk)   (burned into the wood)

ANSI 05.1, Wood Poles: Specifications and Dimensions, specifies what information the pole brand or tag must include and where the brand or tag must be located on the pole. The ANSI 05.1 standard also specifies other information needed for loading and strength calculations such as the fiber stress for various wood species and minimum circumference dimensions for various pole class sizes based on wood species.

See
Photo(s)

Fig. 260-7. Examples of a pole tag and pole brand (Rule 260).

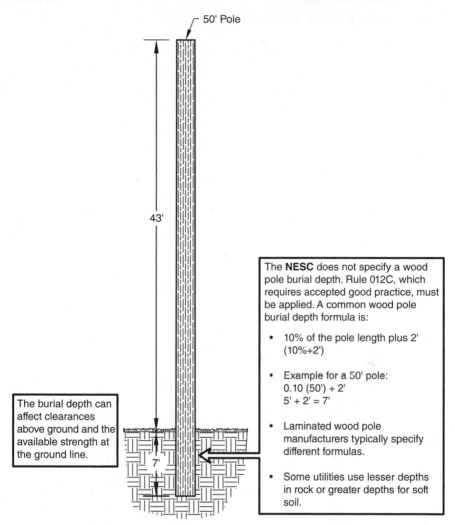

50' Pole

43'

The **NESC** does not specify a wood pole burial depth. Rule 012C, which requires accepted good practice, must be applied. A common wood pole burial depth formula is:

- 10% of the pole length plus 2' (10%+2')

- Example for a 50' pole: 0.10 (50') + 2' 5' + 2' = 7'

- Laminated wood pole manufacturers typically specify different formulas.

- Some utilities use lesser depths in rock or greater depths for soft soil.

The burial depth can affect clearances above ground and the available strength at the ground line.

7'

Fig. 260-8. Example of a common wood pole burial depth formula (Rule 260).

Insulators are covered in their own section, Sec. 27, Line Insulation. **NESC** Table 261-1 does not apply to insulators. Rule 260B1 notes several standards for determining structure design capacity. See Rule 250 or 260 for a discussion of utility line design manuals.

Per Rule 260B2, if Rule 250C (extreme wind) or Rule 250D (extreme ice with concurrent wind) applies and the strength factor is not defined in Rule 261 or Table 261-1, a strength factor of 0.80 must be used for supported facilities.

261. GRADES B AND C CONSTRUCTION

261A. Supporting Structures.   The first sentence in Rule 261A states that the strength requirements of supporting structures (e.g., poles) may be provided by the structure alone or with the aid of guys or braces or both. See Fig. 261-1.

The strength requirements for guys and guyed structures are provided in Rules 261A2a (Exception 1), 261A2c, 261C, and 264. An example of structure strength using a structure alone or a structure with guys is shown in Fig. 261-2.

261A1. Metal, Prestressed-, and Reinforced-Concrete Structures.   The Code rules related to strength requirements of metal, prestressed-, and reinforced-concrete structures are outlined in Fig. 261-3.

261A2. Wood Structures.   The permitted stress level for natural wood poles can vary due to splices, wood grain, knots in the wood, moisture, etc. An ANSI standard, ANSI O5.1, must be used to determine the fiber stress to which strength factors in **NESC** Table 261-1 are applied. Annex A to ANSI Standard O5.1 contains fiber strength reduction factors due to the fiber strength height effect of natural wood poles. Two items can affect the force-resisting moment along the pole: the circumference of the pole which decreases at points

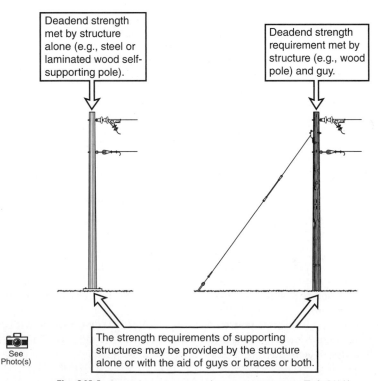

Deadend strength met by structure alone (e.g., steel or laminated wood self-supporting pole).

Deadend strength requirement met by structure (e.g., wood pole) and guy.

The strength requirements of supporting structures may be provided by the structure alone or with the aid of guys or braces or both.

See Photo(s)

Fig. 261-1. Strength requirements of supporting structure (Rule 261A).

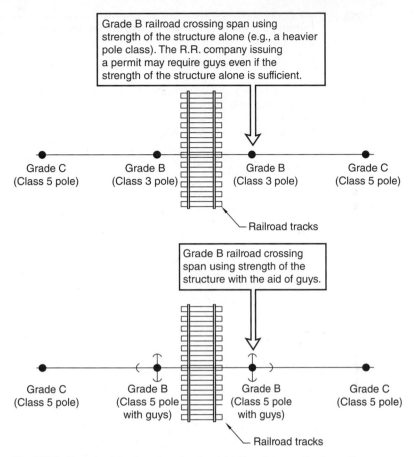

Grade B railroad crossing span using strength of the structure alone (e.g., a heavier pole class). The R.R. company issuing a permit may require guys even if the strength of the structure alone is sufficient.

Grade C
(Class 5 pole)

Grade B
(Class 3 pole)

Grade B
(Class 3 pole)

Grade C
(Class 5 pole)

Railroad tracks

Grade B railroad crossing span using strength of the structure with the aid of guys.

Grade C
(Class 5 pole)

Grade B
(Class 5 pole
with guys)

Grade B
(Class 5 pole
with guys)

Grade C
(Class 5 pole)

Railroad tracks

Fig. 261-2. Example of structure strength using a structure alone or a structure with guys (Rule 261A).

above the ground line due to the taper of the pole and the fiber strength of the pole which also decreases at points above the ground line. ANSI Standard O5.1 is referenced in Rule 261A2b(1) and therefore forms a part of the **NESC** (see the discussion in Sec. 03, References). However, the fiber strength reduction factors in ANSI O5.1 are listed in Annex A of the ANSI O5.1 document and the Annex is for information only and not part of the official Standard. In other words, the **NESC** requires the use of ANSI O5.1 but ANSI O5.1 does not require the use of the fiber strength reduction factors in the Annex. Annex A to the ANSI O5.1 Standard states that for poles 55 ft and shorter, the point of maximum bending stress is usually at or near the ground line. This conclusion corresponds to the 55-ft exception for Rule 261A2a. Using the ANSI O5.1 fiber strength reduction factors in line design calculations is not required, but they are certainly not

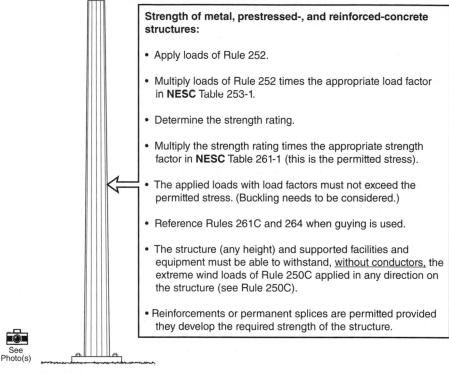

Strength of metal, prestressed-, and reinforced-concrete structures:

- Apply loads of Rule 252.

- Multiply loads of Rule 252 times the appropriate load factor in **NESC** Table 253-1.

- Determine the strength rating.

- Multiply the strength rating times the appropriate strength factor in **NESC** Table 261-1 (this is the permitted stress).

- The applied loads with load factors must not exceed the permitted stress. (Buckling needs to be considered.)

- Reference Rules 261C and 264 when guying is used.

- The structure (any height) and supported facilities and equipment must be able to withstand, <u>without conductors,</u> the extreme wind loads of Rule 250C applied in any direction on the structure (see Rule 250C).

- Reinforcements or permanent splices are permitted provided they develop the required strength of the structure.

See
Photo(s)

Fig. 261-3. Strength requirements of metal, prestressed-, and reinforced-concrete structures (Rule 261A1).

prohibited. See the ANSI Standard O5.1 and Annex A to ANSI Standard O5.1 for more information.

The permitted stress level of sawn or laminated wood structural members, crossarms, and braces is determined by using the ultimate fiber stress of the material in ANSI O5.2 or ANSI O5.3 multiplied by the strength factors in **NESC** Table 261-1.

The **Code** rules related to strength requirements of wood structures are outlined in Figs. 261-4 through 261-7.

261A3. Fiber-Reinforced Polymer Structures. The **Code** rules related to the strength requirements of fiber-reinforced polymer structures are outlined in Fig. 261-8.

261A4. Transverse Strength Requirements for Structures Where Side Guying Is Required, But Can Be Installed Only at a Distance. Rule 261A4 can be applied to various types (e.g., wood, steel, fiber-reinforced polymer, etc.) of supporting structures. The **Code** rules related to transverse structure strength where side guys are required, but can only be installed at a distance, are outlined in Fig. 261-9.

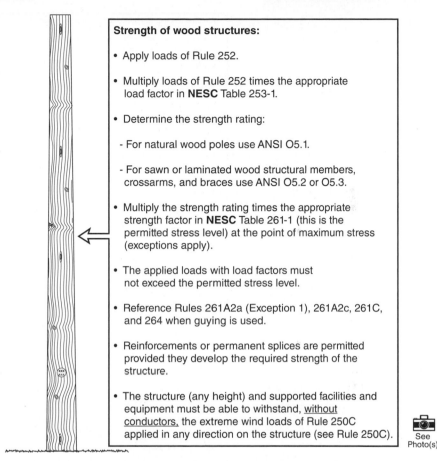

Strength of wood structures:

- Apply loads of Rule 252.

- Multiply loads of Rule 252 times the appropriate load factor in **NESC** Table 253-1.

- Determine the strength rating:

  - For natural wood poles use ANSI O5.1.

  - For sawn or laminated wood structural members, crossarms, and braces use ANSI O5.2 or O5.3.

- Multiply the strength rating times the appropriate strength factor in **NESC** Table 261-1 (this is the permitted stress level) at the point of maximum stress (exceptions apply).

- The applied loads with load factors must not exceed the permitted stress level.

- Reference Rules 261A2a (Exception 1), 261A2c, 261C, and 264 when guying is used.

- Reinforcements or permanent splices are permitted provided they develop the required strength of the structure.

- The structure (any height) and supported facilities and equipment must be able to withstand, <u>without conductors,</u> the extreme wind loads of Rule 250C applied in any direction on the structure (see Rule 250C).

See Photo(s)

Fig. 261-4. Strength requirements of wood structures (Rule 261A2).

261A5. Longitudinal Strength Requirements for Sections of Higher Grade in Lines of a Lower Grade Construction. Rule 261A5 can be applied to various types (e.g., wood, steel, fiber-reinforced polymer, etc.) of supporting structures. The Code rules related to longitudinal strength requirements for sections of higher grade in lines of a lower grade construction are outlined in Fig. 261-10.

Rule 252C1 provides longitudinal loading requirements for sections of Grade B construction when located in lines of lower than Grade B construction.

261B. Strength of Foundations, Settings, and Guy Anchors. The Code rules for strength of foundations, settings, and guy anchors are outlined in Fig. 261-11.

Rule 261B states that design or experience may be used to determine the strength of foundations, settings, and guy anchors. The use of experience is commonly applied to local soil conditions within the service area of a utility. A foundation,

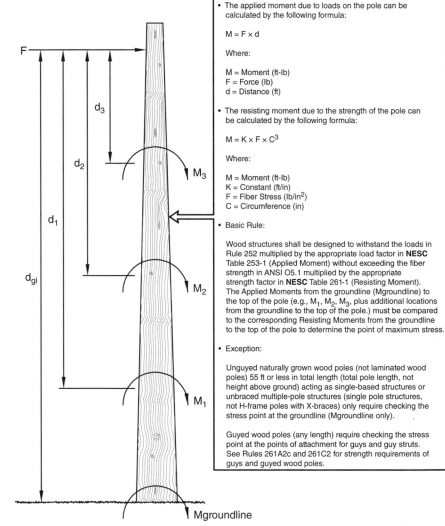

- A moment is measure of rotation force about a point.

- The applied moment due to loads on the pole can be calculated by the following formula:

$$M = F \times d$$

Where:

M = Moment (ft-lb)
F = Force (lb)
d = Distance (ft)

- The resisting moment due to the strength of the pole can be calculated by the following formula:

$$M = K \times F \times C^3$$

Where:

M = Moment (ft-lb)
K = Constant (ft/in)
F = Fiber Stress (lb/in^2)
C = Circumference (in)

- Basic Rule:

Wood structures shall be designed to withstand the loads in Rule 252 multiplied by the appropriate load factor in **NESC** Table 253-1 (Applied Moment) without exceeding the fiber strength in ANSI O5.1 multiplied by the appropriate strength factor in **NESC** Table 261-1 (Resisting Moment). The Applied Moments from the groundline (Mgroundline) to the top of the pole (e.g., M_1, M_2, M_3, plus additional locations from the groundline to the top of the pole.) must be compared to the corresponding Resisting Moments from the groundline to the top of the pole to determine the point of maximum stress.

- Exception:

Unguyed naturally grown wood poles (not laminated wood poles) 55 ft or less in total length (total pole length, not height above ground) acting as single-based structures or unbraced multiple-pole structures (single pole structures, not H-frame poles with X-braces) only require checking the stress point at the groundline (Mgroundline only).

Guyed wood poles (any length) require checking the stress point at the points of attachment for guys and guy struts. See Rules 261A2c and 261C2 for strength requirements of guys and guyed wood poles.

Fig. 261-5. Example of wood pole point of maximum stress (Rules 261A2a and 261A2b).

setting, or guy anchor is only as strong as the soil it is set in. The **NESC** does not specify wood pole burial depths, see Rule 260 for a discussion and a figure. The strength of a guy anchor is determined by both the strength of the anchor rod (see Rule 264F) and the holding capacity of the anchor based on a particular soil classification. If soil types are unknown by experience, soil borings and a geotechnical evaluation can be done to determine soil conditions. Numerous guy anchor

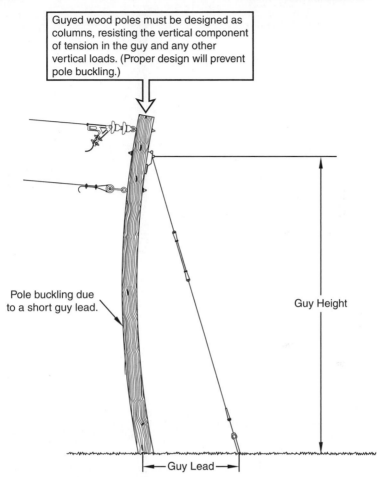

Guyed wood poles must be designed as columns, resisting the vertical component of tension in the guy and any other vertical loads. (Proper design will prevent pole buckling.)

Pole buckling due to a short guy lead.

Guy Height

Guy Lead

Fig. 261-6. Strength of guyed wood poles (Rule 261A2c).

methods exist including expanding, screw, plate, swamp, cone, rock, etc. Each anchor type needs to be evaluated with the soil it is installed in. The **NESC** does not specify a distance between guy anchors. Rule 012C, which requires accepted good practice, must be used. Where two utilities or two separate circuits share a guy anchor, the anchor in soil must be of sufficient strength for the combined loads of the attached guy wires. Examples of multiple guy anchors and multiple guy wires attached to a single guy anchor are shown in Fig. 261-12.

The NOTES in Rule 261B provide a reminder that several factors may reduce clearance and the strength of foundations, settings, and guy anchors. See Fig. 261-13.

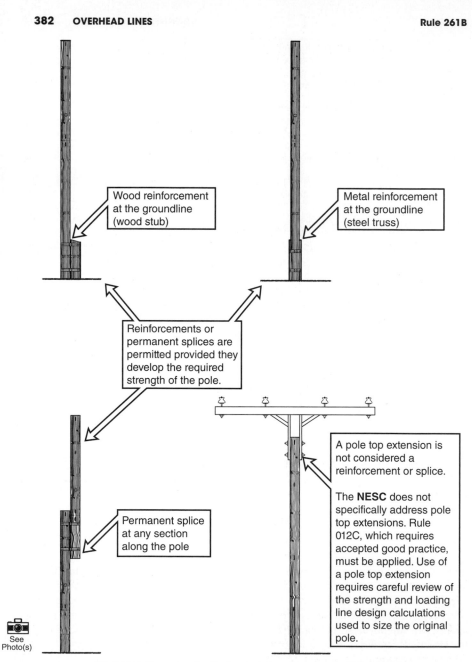

Wood reinforcement at the groundline (wood stub)

Metal reinforcement at the groundline (steel truss)

Reinforcements or permanent splices are permitted provided they develop the required strength of the pole.

A pole top extension is not considered a reinforcement or splice.

The **NESC** does not specifically address pole top extensions. Rule 012C, which requires accepted good practice, must be applied. Use of a pole top extension requires careful review of the strength and loading line design calculations used to size the original pole.

Permanent splice at any section along the pole

See Photo(s)

Fig. 261-7. Examples of spliced and reinforced wood poles (Rule 261A2d).

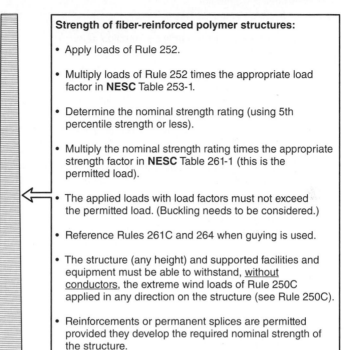

Strength of fiber-reinforced polymer structures:

- Apply loads of Rule 252.

- Multiply loads of Rule 252 times the appropriate load factor in **NESC** Table 253-1.

- Determine the nominal strength rating (using 5th percentile strength or less).

- Multiply the nominal strength rating times the appropriate strength factor in **NESC** Table 261-1 (this is the permitted load).

- The applied loads with load factors must not exceed the permitted load. (Buckling needs to be considered.)

- Reference Rules 261C and 264 when guying is used.

- The structure (any height) and supported facilities and equipment must be able to withstand, <u>without conductors</u>, the extreme wind loads of Rule 250C applied in any direction on the structure (see Rule 250C).

- Reinforcements or permanent splices are permitted provided they develop the required nominal strength of the structure.

See
Photo(s)

Fig. 261-8. Strength requirements of fiber-reinforced polymer structures (Rule 261A3).

261C. Strength of Guys and Guy Insulators.   Guy strength is covered in Rule 264. Guy insulator strength is covered in Rule 279A1c. Rule 261C provides requirements relative to the integration of the guy with the structure to which the guy is attached. Rule 261A2a (Exception 1) and Rule 261A2c provide similar requirements for guys attached to wood poles only. See Fig. 261-14.

The NOTES in Rule 261C are similar to the NOTES in Rule 261B. Excessive movement of a guy (or movement of the anchor attached to the guy) can reduce clearance or structure capacity.

261D. Crossarms and Braces.   Rule 261D covers crossarms and braces made of various materials.

261D1. Concrete and Metal Crossarms and Braces.   The Code rules for concrete and metal crossarms and braces are outlined in Fig. 261-15.

261D2. Wood Crossarms and Braces.   The loads on wood crossarms and braces and the permitted stress levels on wood crossarms and braces are covered in this rule. The permitted stress levels on wood crossarms and braces are also covered in Rule 261A2b(2). The Code rules for wood crossarms and braces in Rule 261D2 are outlined in Fig. 261-16.

Transverse structure strength where side guys are required but can only be installed at a distance:

- The distance between the side guyed structures must not be over 800'.

- The line between the side guyed structures must be substantially straight and the average of the spans must not exceed 150'.

- The side guyed poles must be designed to support the transverse load of the entire section. Intermediate poles are assumed not to carry any transverse load.

- The line between the side guyed structures must be designed to Grade B except for the transverse strength of the poles between the side guyed structures.

- Rule 261A4 applies to Grade B only; it does not apply to Grade C.

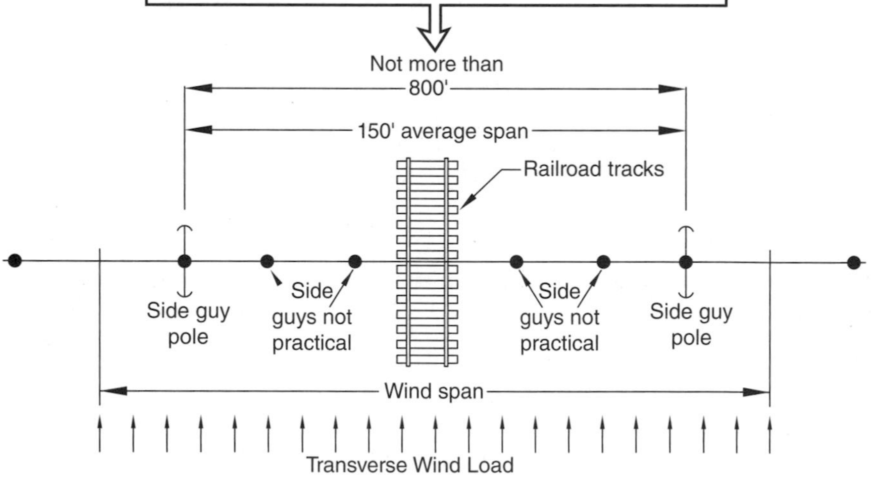

Fig. 261-9. Transverse structure strength where side guys are required but can only be installed at a distance (Rule 261A4).

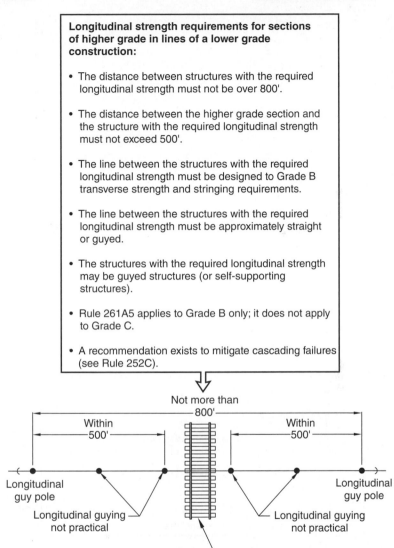

> **Longitudinal strength requirements for sections of higher grade in lines of a lower grade construction:**
>
> - The distance between structures with the required longitudinal strength must not be over 800'.
>
> - The distance between the higher grade section and the structure with the required longitudinal strength must not exceed 500'.
>
> - The line between the structures with the required longitudinal strength must be designed to Grade B transverse strength and stringing requirements.
>
> - The line between the structures with the required longitudinal strength must be approximately straight or guyed.
>
> - The structures with the required longitudinal strength may be guyed structures (or self-supporting structures).
>
> - Rule 261A5 applies to Grade B only; it does not apply to Grade C.
>
> - A recommendation exists to mitigate cascading failures (see Rule 252C).

Not more than
800'

Within
500'

Within
500'

Longitudinal
guy pole

Longitudinal
guy pole

Longitudinal guying
not practical

Longitudinal guying
not practical

Railroad tracks

Fig. 261-10. Longitudinal strength requirements for sections of higher grade in lines of a lower grade construction (Rule 261A5).

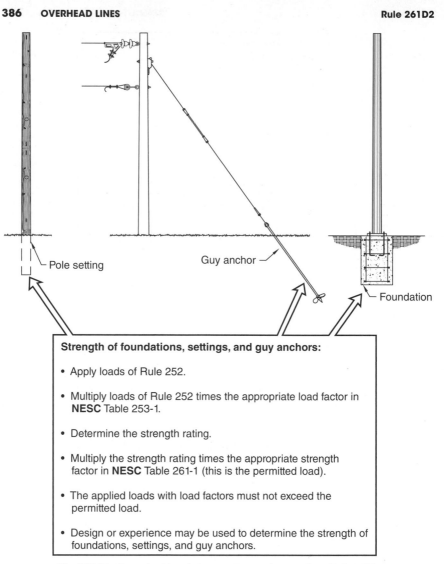

Fig. 261-11. Strength of foundations, settings, and guy anchors (Rule 261B).

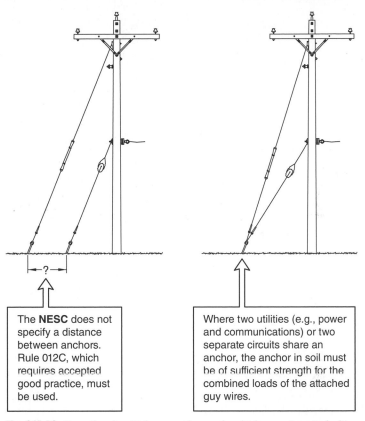

The **NESC** does not specify a distance between anchors. Rule 012C, which requires accepted good practice, must be used.

Where two utilities (e.g., power and communications) or two separate circuits share an anchor, the anchor in soil must be of sufficient strength for the combined loads of the attached guy wires.

Fig. 261-12. Examples of multiple guy anchors and multiple guy wires attached to a single guy anchor (Rule 261B).

In addition to the requirement to use **NESC** Table 253-1 with **NESC** Table 261-1, the **Code** also specifies minimum dimensions for select Southern Pine and Douglas Fir wood crossarms in **NESC** Table 261-2. Crossarms of other wood species may be used if they provide equal strength. See Fig. 261-17.

261D3. Fiber-Reinforced Polymer Crossarms and Braces.   The **Code** rules for fiber-reinforced polymer crossarms and braces are outlined in Fig. 261-18.

261D4. Crossarms and Braces of Other Materials.   Crossarms of other materials not specified in Rule 261D (i.e., other than concrete, metal, wood, and fiber-reinforced polymer) must meet the strength requirements of wood crossarms and braces.

261D5. Additional Requirements.   Rule 261D5 provides additional rules for longitudinal strength of crossarms. Tension values and construction methods are specified. Using double wood crossarms is one method of meeting the longitudinal strength requirements of a Grade B line. A support assembly of equivalent strength (i.e., a pre-engineered crossarm assembly) is also acceptable.

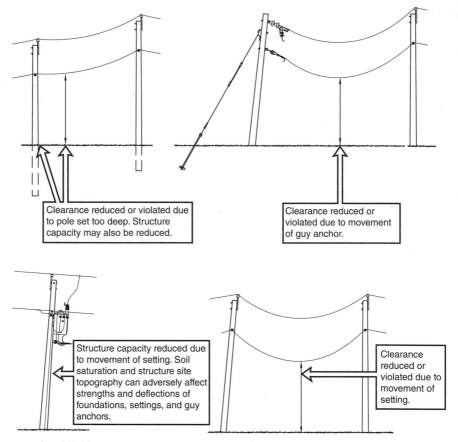

Fig. 261-13. Examples of movement of foundations, settings, and guy anchors (Rule 261B).

261E. Insulators.   Insulators are covered in Sec. 27. See Rule 277 for strengths of line insulators and Rule 279 for strengths of guy and span insulators.

261F. Strength of Pin-Type or Similar Construction and Conductor Fastenings. Per Rule 261F1a, the longitudinal strength of insulator pins and conductor ties can be determined by applying the loads in Rule 252 multiplied by the load factors in Rule 253, or by applying 700 lb to the pin, whichever is greater. A tangent structure can have unbalanced longitudinal loading if the spans on each side of the structure are not equal or if the span on one side is loaded with ice and the span on the other side has dropped its ice. Rule 261F1b specifies a construction method for meeting Rule 261F1a. Using double wood pins and ties is a construction method that is acceptable for meeting the longitudinal strength requirements of Grade B construction. The application of double wood pins in Rule 261F1b corresponds with the application of double wood arms in Rule 261D5a(2). Both rules are for conductors

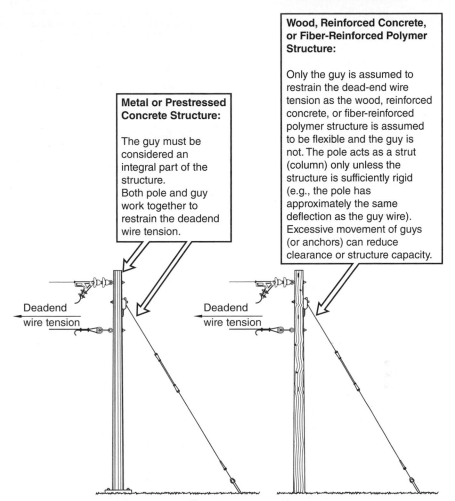

Wood, Reinforced Concrete, or Fiber-Reinforced Polymer Structure:

Only the guy is assumed to restrain the dead-end wire tension as the wood, reinforced concrete, or fiber-reinforced polymer structure is assumed to be flexible and the guy is not. The pole acts as a strut (column) only unless the structure is sufficiently rigid (e.g., the pole has approximately the same deflection as the guy wire). Excessive movement of guys (or anchors) can reduce clearance or structure capacity.

Metal or Prestressed Concrete Structure:

The guy must be considered an integral part of the structure.
Both pole and guy work together to restrain the deadend wire tension.

Deadend
wire tension

Deadend
wire tension

Fig. 261-14. Strength of guys and guy insulators (Rule 261C).

with tensions limited to 2000 lb. Wood insulator pins were once commonly used and occasionally can still be found on older lines. Steel insulator pins are used in new construction. The **NESC** does not specify a construction application using steel pins. It is common to see steel pins used in place of wood pins on double crossarm construction. Rule 261F1c relates to Rule 261A5. Rule 261F1d relates to Rule 261A4. Rule 261F2 relates to Rule 261D5c. Rule 261F3 states that single conductor supports used instead of double wood pins must have the equivalent strength of double wood pins. An example of wood pins is shown in Fig. 261-19.

261G. Armless Construction.   The Code rules for armless construction are outlined in Fig. 261-20.

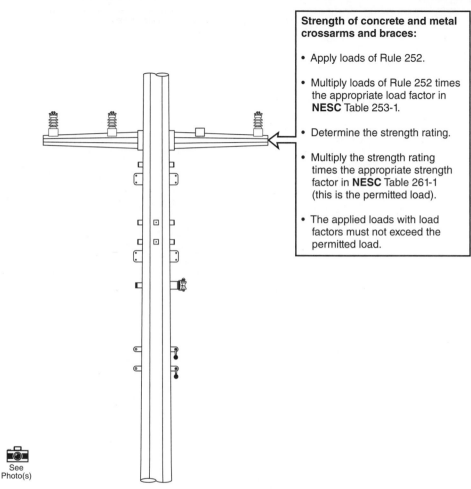

Strength of concrete and metal crossarms and braces:

- Apply loads of Rule 252.

- Multiply loads of Rule 252 times the appropriate load factor in **NESC** Table 253-1.

- Determine the strength rating.

- Multiply the strength rating times the appropriate strength factor in **NESC** Table 261-1 (this is the permitted load).

- The applied loads with load factors must not exceed the permitted load.

See Photo(s)

Fig. 261-15. Strength of concrete and metal crossarms and braces (Rule 261D1).

261H. Open Supply Conductors and Overhead Shield Wires.   Rule 261H1 specifies tension limits for open (e.g., bare) supply conductors and overhead shield wires. The tension limits in Rule 261H1 should be entered into a conductor sag and tension program. Rule 261H1a(1) applies to the loaded condition of the conductor per Rule 250B which uses the heavy, medium, light, and warm island loading districts. Rule 261H1a(2) applies to the loaded condition of the conductor per Rules 250C (extreme wind) and 250D (extreme ice with concurrent wind). Rule 250B applies to all poles. The application of Rules 250C and 250D only applies to structures (e.g., poles) or their supported facilities (e.g., conductors, static wires, messengers, cables, etc.) more than 60 ft above ground or water level. See Rules 250C and 250D for a discussion. Rule 261H1b addresses Aeolian vibration damage

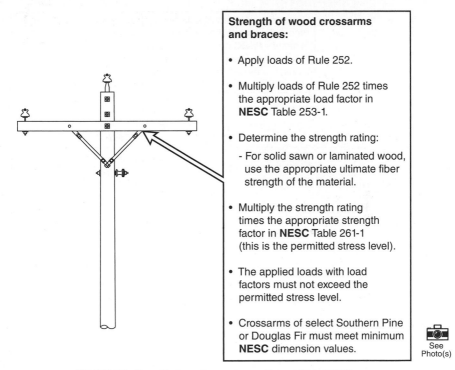

Strength of wood crossarms and braces:

- Apply loads of Rule 252.

- Multiply loads of Rule 252 times the appropriate load factor in **NESC** Table 253-1.

- Determine the strength rating:
  - For solid sawn or laminated wood, use the appropriate ultimate fiber strength of the material.

- Multiply the strength rating times the appropriate strength factor in **NESC** Table 261-1 (this is the permitted stress level).

- The applied loads with load factors must not exceed the permitted stress level.

- Crossarms of select Southern Pine or Douglas Fir must meet minimum **NESC** dimension values.

See Photo(s)

Fig. 261-16. Strength of wood crossarms and braces (Rule 261D2).

to conductors and related hardware. Aeolian vibration is a low-amplitude, high-frequency vibration that can cause failure of the conductor and related hardware. Rule 261H1b lists four Aeolian vibration mitigation methods. Three of the methods (vibration control devices, stress-reduction devices, and self damping conductors or vibration resistant conductors) involve selecting special line hardware or special conductors. The fourth method in Rule 261H1b(4) involves reducing conductor tension and can be applied to various conductor types and does not involve special line hardware. Where Rule 261H1b(4) is determined to sufficiently mitigate Aeolian vibration, Rule 261H1c provides the tension limits and applicable temperatures to be used. The tension limits of Rules 261H1a and 261H1c are outlined in Fig. 261-21.

The conductor manufacturer may also specify tension limits. See the sample sag and tension chart at the beginning of Sec. 23 and the notes associated with the sample sag chart for a discussion of **NESC** tension limits and user-defined tension limits.

Rule 261H2 specifies strength requirements for splices, taps, deadend fittings, and associated hardware. Rules 261F and 261M provide additional requirements. Rule 261H2 provides special requirements for crossing spans and deadends.

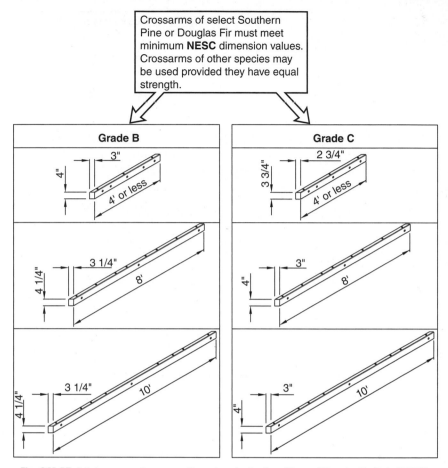

Fig. 261-17. Minimum wood crossarm dimensions for Southern Pine and Douglas Fir (Rule 261D2b).

Crossings include crossing over another line, or crossing over or overhanging railroad tracks, limited access highways and navigable waterways as discussed, with figures, in Rule 241C1. The requirements for location of splices and taps in crossing and adjacent spans are outlined in Fig. 261-22.

261I. Supply Cable Messengers.   The Code rules for strength of supply cable messengers are outlined in Fig. 261-23.

Common examples of supply cables supported on messengers are cables meeting Rules 230C1, 230C2, and 230C3. See Rule 230C for a discussion and a figure.

261J. Open-Wire Communication Conductors.   Open-wire communications circuits have typically been replaced with insulated communication cables or fiberoptic cables. The rules for open-wire communication circuits continue to remain in the Code. Wire tension limits for open-wire communications in Grade B and C

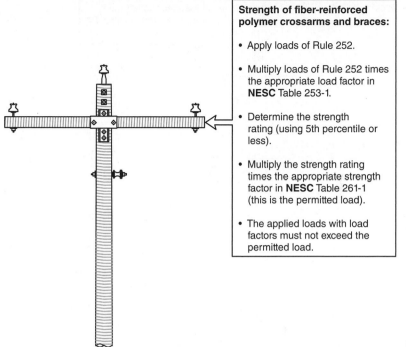

Strength of fiber-reinforced polymer crossarms and braces:

- Apply loads of Rule 252.

- Multiply loads of Rule 252 times the appropriate load factor in **NESC** Table 253-1.

- Determine the strength rating (using 5th percentile or less).

- Multiply the strength rating times the appropriate strength factor in **NESC** Table 261-1 (this is the permitted load).

- The applied loads with load factors must not exceed the permitted load.

See Photo(s)

Fig. 261-18. Strength of fiber-reinforced polymer crossarms and braces (Rule 261D3).

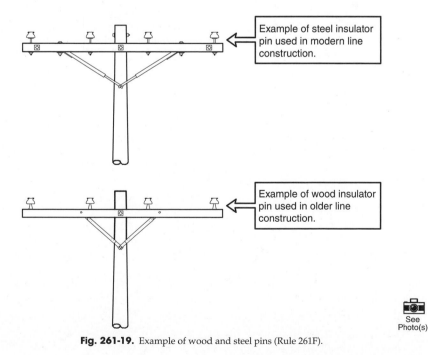

Example of steel insulator pin used in modern line construction.

Example of wood insulator pin used in older line construction.

See Photo(s)

Fig. 261-19. Example of wood and steel pins (Rule 261F).

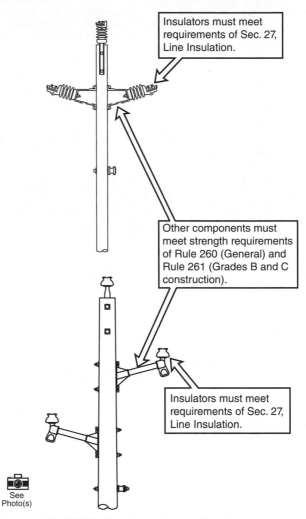

Insulators must meet requirements of Sec. 27, Line Insulation.

Other components must meet strength requirements of Rule 260 (General) and Rule 261 (Grades B and C construction).

Insulators must meet requirements of Sec. 27, Line Insulation.

See Photo(s)

Fig. 261-20. Strength requirements for armless construction (Rule 261G).

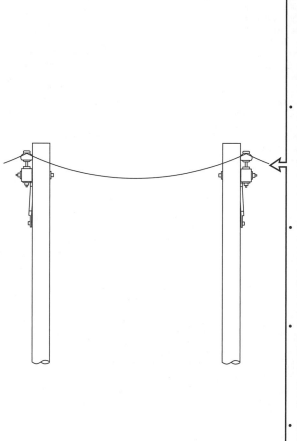

Tension limits for open supply conductors and overhead shield wires:

- The tension in the conductor must not be more than 60% of the rated breaking strength of the conductor when the loads of Rule 250B (heavy, medium, light, or warm islands) in Rule 251 are applied to the supply conductor using a load factor of 1.0.

- The tension in the conductor must not be more than 80% of the rated breaking strength of the conductor when the loads of Rules 250C (extreme wind) and 250D (extreme ice with concurrent wind) in Rule 251 are applied (if applicable) to the supply conductor using a load factor of 1.0.

- Where reducing tension is determined to sufficiently mitigate Aeolian vibration, the tension in the conductor must not be more than 35% initial, at the applicable temperature in **NESC** Table 251-1 without oxtornal loading.

- Where reducing tension is determined to sufficiently miti-gate Aeolian vibration, tension in the conductor must not be more than 25% final, at the applicable temperature in **NESC** Table 251-1 without external loading.

- Notes are provided defining initial tension, final tension, and stating that the above limitations may not protect the conductor from damage due to Aeolian vibration. A separate user-defined tension limit may be needed.

Fig. 261-21. Tension limits for open supply conductors and overhead shield wires (Rules 261H1a and 261H1c).

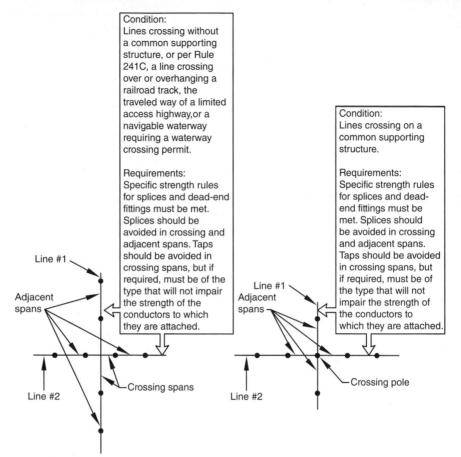

Fig. 261-22. Location of splices and taps in crossing and adjacent spans (Rule 261H2).

construction are the same as the wire tension limits for supply conductors in Rule 261H1. An example of open-wire communication conductors is shown in Fig. 261-24.

261K. Communication Cables and Messengers.   The Code rules for strength of communication cables and messengers are outlined in Fig. 261-25.

The NESC does not use the term "overlash." Rule 261K applies to communication cables lashed to a messenger and communication cables overlashed on a messenger. When a communication cable is overlashed on a messenger, the weight of the resulting cable bundle is increased and the surface area of the cable bundle that is exposed to wind and ice loads is increased. Overlashing a communications cable on a messenger requires a review of the messenger tension (per Rule 261K), a

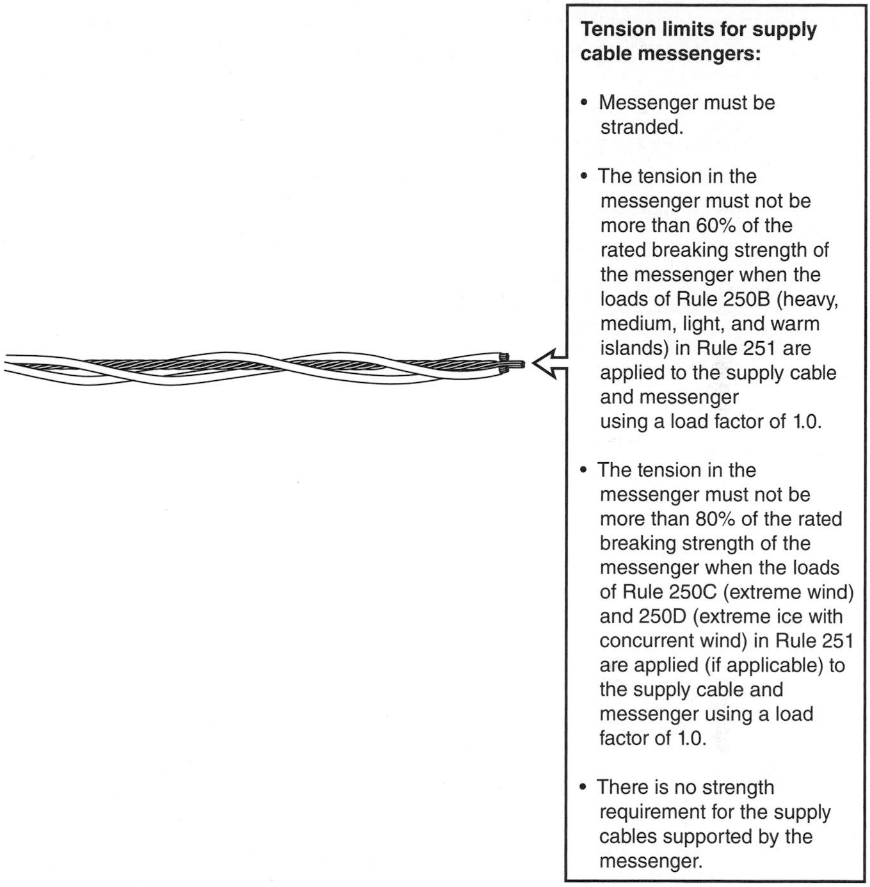

> **Tension limits for supply cable messengers:**
>
> • Messenger must be stranded.
>
> • The tension in the messenger must not be more than 60% of the rated breaking strength of the messenger when the loads of Rule 250B (heavy, medium, light, and warm islands) in Rule 251 are applied to the supply cable and messenger using a load factor of 1.0.
>
> • The tension in the messenger must not be more than 80% of the rated breaking strength of the messenger when the loads of Rule 250C (extreme wind) and 250D (extreme ice with concurrent wind) in Rule 251 are applied (if applicable) to the supply cable and messenger using a load factor of 1.0.
>
> • There is no strength requirement for the supply cables supported by the messenger.

Fig. 261-23. Tension limits for supply cable messengers (Rule 261I).

review of the pole loading (per Secs. 24, 25, and 26) and a review of sag and clearance of the cable bundle (per Sec. 23).

261L. Paired Metallic Communication Conductors. Rule 261L provides requirements for paired metallic communication conductors supported on messengers and paired metallic communication conductors not supported on messengers in addition to the requirements in Rule 261K2. Paired metallic conductors are typically used by telephone utilities.

261M. Support and Attachment Hardware. Rule 261M applies to support and attachment hardware (e.g., nuts, bolts, etc.) not covered in Rule 261F (pin-type or similar construction and conductor fastenings) or Rule 261H2 (splices, taps, dead-end fittings, and associated attachment hardware). Appropriate loads, load factors, and strength factors apply to support hardware.

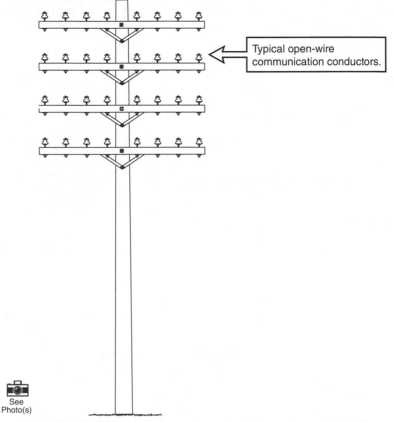

Fig. 261-24. Example of open-wire communication conductors (Rule 261J).

261N. Climbing and Working Steps and Their Attachments to the Structure.   Rule 261N provides strength requirements for climbing devices (e.g., steps, ladders, platforms, and attachments). The load required is 300 lb for the weight of a lineworker including the weight of the items the lineworker carries on a structure unless the owner determines another value. Instead of derating the material strength of the steps, ladders, platforms, etc., a load factor of 2.0 is provided which increases the 300 lb load to 600 lb. The steps, ladders, platforms, etc., must be capable of supporting this load.

Tension limits for communication cables and messengers:

- There is no strength requirement for the communication cables supported by the messenger.

- The tension in the messenger must not be more than 60% of the rated breaking strength of the messenger when the loads of Rule 250B (heavy, medium, light, and warm islands) in Rule 251 are applied to the communications cable and messenger using a load factor of 1.0.

- The tension in the messenger must not be more than 80% of the rated breaking strength of the messenger when the loads of Rule 250C (extreme wind) and 250D (extreme ice with concurrent wind) in Rule 251 are applied (if applicable) to the communications cable and messenger using a load factor of 1.0.

- Self-supporting cables must meet the strength requirements for communication messengers. Note that lesser tensions may be needed to meet self-supporting fiber optic cable operational reliability.

- Metallic paired communication conductors have additional requirements in Rule 261L.

Fig. 261-25. Tension limits for communication cables and messengers (Rule 261K).

262. NUMBER 262 NOT USED IN THIS EDITION

263. GRADE N CONSTRUCTION

Grade N is the lowest grade of construction. Grade N supply and communication lines and structures have to withstand expected loads, including line personnel working on the structure. No load or strength factor is required for Grade N. Said another way, the line or structure is designed to withstand the expected loads and the load factors and strength factors are equal to 1.0. Rules 263A through 263H contain minimum conductor size requirements and requirements for typical construction practices.

264. GUYING AND BRACING

264A. Where Used.   Guys and braces are used when a pole does not have sufficient strength alone. Guys and requirements for guyed poles are also covered in Rules 261A2a (Exception 1), 261A2c, and 261C. Guy insulators are covered in Rule 279. Conductors and power and communication cables on messengers cannot be used as guy wires. Conductors and power and communication cables supported on messengers add vertical (weight) and horizontal (transverse wind) loads to structures and additional horizontal (transverse) loads at angle structures. It is possible for the loads due to line angles of multiple conductors on the same pole to cancel or offset each other but using conductors or power and communication cables on messengers as guy wires is not an acceptable practice. Guys are used to limit the increase in conductor sags (which can cause clearance to be decreased), and guys are used to provide support for unbalanced loads at the following locations:

- Corners
- Angles
- Deadends
- Large differences in span lengths
- Changes in grades of construction

Examples of guying and bracing are shown in Fig. 264-1.

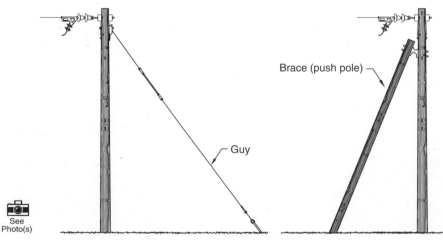

Fig. 264-1. Examples of guying and bracing (Rule 264A).

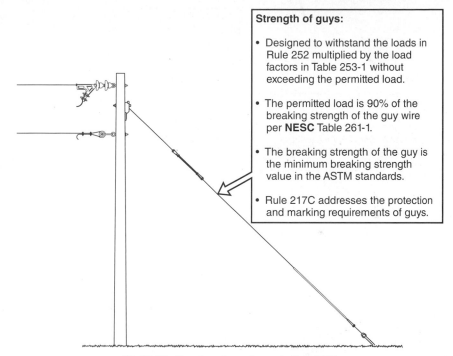

Strength of guys:

- Designed to withstand the loads in Rule 252 multiplied by the load factors in Table 253-1 without exceeding the permitted load.

- The permitted load is 90% of the breaking strength of the guy wire per **NESC** Table 261-1.

- The breaking strength of the guy is the minimum breaking strength value in the ASTM standards.

- Rule 217C addresses the protection and marking requirements of guys.

Fig. 264-2. Guy strength requirements (Rule 264B).

264B. Strength.   Rule 264B applies to the strength of guys. Rule 217C addresses the protection and marking requirements of guys. The **Code** rules for guy strength are outlined in Fig. 264-2.

264C. Point of Attachment.   The guy or brace should be attached to the structure as near as practical to the center of the conductor load. An individual guy for each conductor is typically not practical for small conductors or small angles but may be required for large conductors or large angles. Special consideration is given to insulation reduction for lines exceeding 8.7 kV. Guy insulators used exclusively for BIL insulation are discussed in Rule 279A2. The **Code** rules for the point of attachment of the guy or brace are outlined in Fig. 264-3.

264D. Guy Fastenings.   The strength rating of a guy fastening is just as important as the strength rating of the guy wire itself. Numerous guy attachment methods exist including guy plates, pole bands, wrap guys, etc. The **Code** requires the use of guy thimbles and guy shims for certain guying conditions. The entire guy and anchor system, including guy fastenings, should be reviewed to determine the weak link in the guy and anchor system. A guy is only as strong as the weakest link connected to the guy wire. An example of the weakest link in a guy and anchor system is shown in Fig. 264-4.

264E. Electrolysis.   Electrolytic corrosion of the anchor or anchor rod can jeopardize the strength of the structure. Corrosion normally occurs when two dissimilar metals exist in the soil (e.g., a copper ground rod and a steel anchor rod).

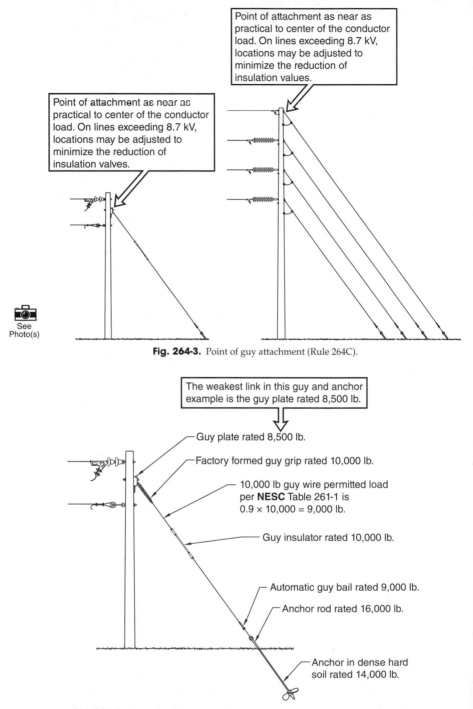

Point of attachment as near as practical to center of the conductor load. On lines exceeding 8.7 kV, locations may be adjusted to minimize the reduction of insulation values.

Point of attachment as near as practical to center of the conductor load. On lines exceeding 8.7 kV, locations may be adjusted to minimize the reduction of insulation valves.

See Photo(s)

Fig. 264-3. Point of guy attachment (Rule 264C).

The weakest link in this guy and anchor example is the guy plate rated 8,500 lb.

Guy plate rated 8,500 lb.

Factory formed guy grip rated 10,000 lb.

10,000 lb guy wire permitted load per **NESC** Table 261-1 is $0.9 \times 10,000 = 9,000$ lb.

Guy insulator rated 10,000 lb.

Automatic guy bail rated 9,000 lb.

Anchor rod rated 16,000 lb.

Anchor in dense hard soil rated 14,000 lb.

Fig. 264-4. Example of the weakest link in a guy and anchor system (Rule 264D).

The soil conditions can have an effect on the rate of corrosion. Typical solutions to the problem are using guy insulators or using steel-coated copper ground rods. Cathodic protection in the form of a sacrificial anode is another solution. Guy insulators used exclusively for limiting corrosion are discussed in Rule 279A2.

264F. Anchor Rods. Rule 264F1 applies to the anchor rod only. Rule 264F2 applies to the anchor and rod assembly. Rules 261B and 264B also relate to the strength of guy anchors. The **Code** rules for strength of anchor rods and anchors provided in Rule 264F are outlined in Fig. 264-5.

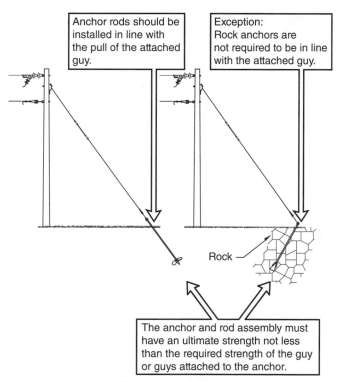

Anchor rods should be installed in line with the pull of the attached guy.

Exception: Rock anchors are not required to be in line with the attached guy.

Rock

The anchor and rod assembly must have an ultimate strength not less than the required strength of the guy or guys attached to the anchor.

Fig. 264-5. Anchor rods (Rule 264F).

Section 27

Line Insulation

270. APPLICATION OF RULE

Line insulation in Sec. 27 applies to insulators for open conductor supply lines only, not for communication lines or secondary duplex, triplex, or quadruplex (230C3) cables. Section 27 does not apply to substation insulators as Sec. 27 is in Part 2, Overhead Lines. Note 1 references Rule 243C4. Rule 243C4 states that the strength requirements in Sec. 27 apply to all grades of construction. There are not separate Grade B and Grade C strength factors for insulators. Note 2 references Rule 242E for insulation requirements of neutral conductors. Rule 242E states that supply neutrals, which are effectively grounded and not located above supply conductors of more than 750 V to ground, do not need to meet any insulation requirement. Even though effectively grounded neutrals do not need to meet any insulation requirement, it is common to see neutral conductors attached to insulators for support and termination purposes. Also, even though the messengers of secondary duplex, triplex, and quadruplex (230C3) cables do not need to meet any insulation requirement, it is common to see the messengers of these cables attached to insulators for support and termination purposes. The Code rules related to the application of line insulation are outlined in Fig. 270-1.

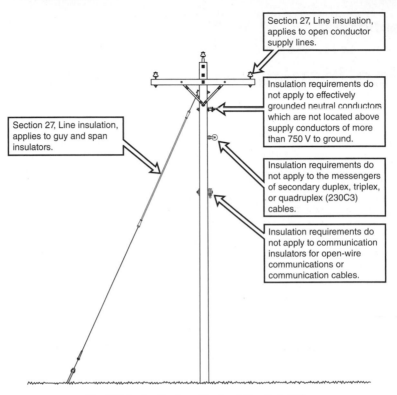

Fig. 270-1. Application of line insulation (Rule 270).

271. MATERIAL AND MARKING

Supply circuit insulators must be manufactured and marked in accordance with the ANSI/NEMA C29 series of standards. An exception applies if performance and marking are appropriate for the application. Examples of supply circuit insulators are shown in Fig. 271-1.

Examples of identification marking on supply circuit insulators are shown in Fig. 271-2.

272. RATIO OF FLASHOVER TO PUNCTURE VOLTAGE

Rule 272 requires insulators to meet conditions and standards for flashover to puncture voltage ratios. A list of applicable standards is provided. When a standard does not exist for the flashover to puncture voltage ratio, Rule 272 specifies a not-to-exceed value of 75 percent with an exception that permits not more than 80 percent in areas of high atmospheric contamination. The flashover to puncture

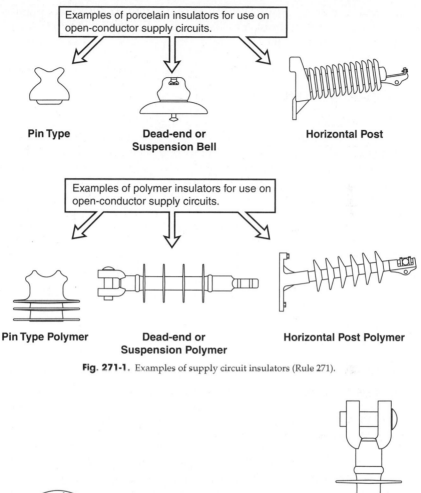

Fig. 271-1. Examples of supply circuit insulators (Rule 271).

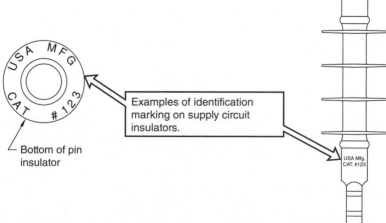

Fig. 271-2. Examples of identification marking on supply circuit insulators (Rule 271).

voltage ratio is used by insulator manufacturers, but it is not always published in insulator catalog data. Insulator catalog data typically does include the low-frequency dry flashover value.

273. INSULATION LEVEL

Rule 273 specifies insulator dry flashover voltages in **NESC** Table 273-1 which must be compared to dry flashover tests done in accordance with ANSI Standard C29.1. Using dry flashover voltage ratings lower than the values in **NESC** Table 273-1 requires a qualified engineering study. Values higher than shown in **NESC** Table 273-1 must be used for areas with severe lightning, high atmospheric contamination (e.g., salt water fog, industrial plant pollution, etc.), or other unfavorable circumstances. The low-frequency dry flashover rating is typically the most common value referenced when comparing insulator specifications. Additional ratings are commonly published in insulator catalog sheets and in insulator specifications. Typical insulation level ratings are listed in Fig. 273-1.

Rating	Values specified in **NESC**
• Low frequency 60-Hz dry flashover	Yes (see **NESC** Table 273-1)
• Low frequency 60-Hz wet flashover	No
• Critical impulse flashover - positive	No
• Critical impulse flashover - negative	No
• Total leakage distance	No

Fig. 273-1. Typical insulation level ratings (Rule 273).

In addition to electrical ratings, the **NESC** specifies mechanical strength ratings for insulators in Rule 277.

274. FACTORY TESTS

Insulator factory tests per ANSI Standards are required for each insulator or insulating part at or above 2.3 kV. Accepted good practice may be used where standards do not exist. An exception exists for guy insulators.

275. SPECIAL INSULATOR APPLICATIONS

Special consideration must be given to insulators for use on constant-current circuits (e.g., series street lighting). The voltage rating of the insulator must be based on the full load rated voltage of the supply transformer, not just the operating voltage of the circuit. See Rule 230H for a discussion and figure related to constant-current circuits.

Single-phase lines fed from three-phase lines are required to use insulators based on the three-phase line phase-to-phase voltage. See Rule 230G for an example and figure related to phase-to-phase and phase-to-ground voltages. Insulation requirements of single-phase circuits connected to three-phase circuits are shown in Fig. 275-1.

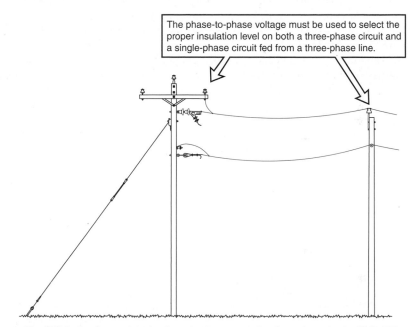

The phase-to-phase voltage must be used to select the proper insulation level on both a three-phase circuit and a single-phase circuit fed from a three-phase line.

Fig. 275-1. Insulators of single-phase circuits connected to three-phase circuits (Rule 275).

276. NUMBER 276 NOT USED IN THIS EDITION

277. MECHANICAL STRENGTH OF INSULATORS

Rule 277 specifies the mechanical strength requirements for insulators. This is done in the form of strength factors that are applied to the insulators. Load factors for wire tension, wire vertical (weight) loads, and wire transverse (wind) loads are not applicable. Rule 277 requires insulators to withstand the loads in Rules 250, 251, and 252. Load factors are in Rule 253, and Rule 253 is not specified in Rule 277. Said another way, the load factor for the loads on insulators is 1.0. The strength factor percentages provided in **NESC** Table 277-1 do not distinguish between Grade B and Grade C construction. This is one reason why insulators have their own section (Sec. 27) rather than being included in Rule 261 and **NESC** Table 261-1. The hardware for insulators is covered in Rules 261M, 261F, 261H2, and **NESC** Table 261-1. The percent of strength ratings provided in **NESC** Table 277-1 are for use with Rule 250B (heavy, medium, light, and warm islands) loads and Rule 250C (extreme wind) and Rule 250D (extreme ice with concurrent wind) loads.

The strength factor percentages listed in **NESC** Table 277-1 are based on manufacturers ratings when determined in accordance with specific ANSI Standards for the type of insulator, the insulator material, and the type of load on the insulator. Per Footnote 4 to **NESC** Table 277-1, some insulators do not currently have a standard related to strength. In these cases, Rule 012C, which requires accepted good practice, must be applied.

The appropriate mechanical strength of an insulator that needs to be derated may not be very clear on an insulator catalog sheet. Catalog sheets may use terms like routine test load, maximum design load, average failing load, average breaking load, or maximum working load. The insulator manufacturer should be consulted or the ANSI Standards should be reviewed to verify that the percentages provided in **NESC** Table 277-1 are being applied to the proper mechanical strength rating.

Examples of cantilever, compression, and tension loads on ceramic line post insulators are shown in Fig. 277-1.

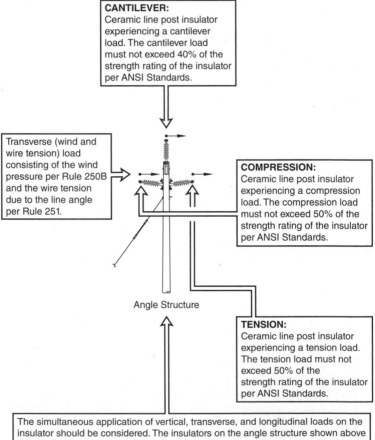

CANTILEVER:
Ceramic line post insulator experiencing a cantilever load. The cantilever load must not exceed 40% of the strength rating of the insulator per ANSI Standards.

Transverse (wind and wire tension) load consisting of the wind pressure per Rule 250B and the wire tension due to the line angle per Rule 251.

COMPRESSION:
Ceramic line post insulator experiencing a compression load. The compression load must not exceed 50% of the strength rating of the insulator per ANSI Standards.

Angle Structure

TENSION:
Ceramic line post insulator experiencing a tension load. The tension load must not exceed 50% of the strength rating of the insulator per ANSI Standards.

The simultaneous application of vertical, transverse, and longitudinal loads on the insulator should be considered. The insulators on the angle structure shown above are experiencing vertical (weight) loads in addition to transverse loads. The vertical (weight) load of the wire and ice per Rule 250B applies a cantilever load on the two bottom insulators and a compression load on the top insulator. Differences in ice loading on the spans of the angle structure can also create longitudinal loads. Insulator manufacturers typically provide a method for checking simultaneous application of loads.

Fig. 277-1. Examples of cantilever, compression, and tension loads on ceramic line post insulators (Rule 277).

278. AERIAL CABLE SYSTEMS

Aerial cable systems primarily consist of covered conductors described in Rule 230D. The insulators supporting aerial cables must meet Rule 273. The covered conductors of aerial cable systems are considered bare conductors for electrical insulation requirements. The insulators and spacers of an aerial cable system must meet Rule 277. Rule 278B2 requires that the insulating spacers used in aerial cable systems withstand the loads specified in Sec. 25. Section 25 includes Rule 253, which contains load factors. The 50% of ultimate rated strength requirement in Rule 278B2 is for use with Rule 250B (heavy, medium, light, and warm islands) loads and the appropriate overload factors in Rule 253. The loads of 250C (extreme winds) and 250D (extreme ice with concurrent wind) are not addressed; therefore, Rule 012C, which requires accepted good practice, must be applied.

279. GUY AND SPAN INSULATORS

279A. Insulators. Per Rules 215C2a and 215C2b, anchor and span guys must be effectively grounded or insulated (per the exceptions). If grounded, Rule 092C2 provides the methods for the grounding of guys. If insulated, the exceptions to Rules 215C2a and 215C2b provide requirements for the location of anchor and span guy insulators. See Fig. 279-1.

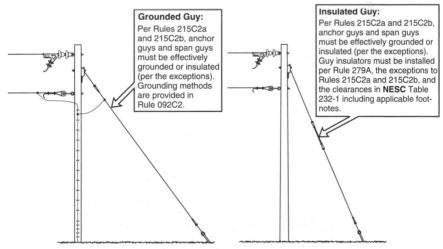

Grounded Guy:
Per Rules 215C2a and 215C2b, anchor guys and span guys must be effectively grounded or insulated (per the exceptions). Grounding methods are provided in Rule 092C2.

Insulated Guy:
Per Rules 215C2a and 215C2b, anchor guys and span guys must be effectively grounded or insulated (per the exceptions). Guy insulators must be installed per Rule 279A, the exceptions to Rules 215C2a and 215C2b, and the clearances in **NESC** Table 232-1 including applicable footnotes.

Fig. 279-1. Guy and span insulators (Rule 279).

The **Code** rules for the material properties of guy insulators and the electrical and mechanical strength of guy insulators per Rule 279A1 are outlined in Fig. 279-2.

The requirements for mechanical strength of a guy insulator hinge on the word "required" strength of the guy. Guy insulators do not require the application of strength factors like line insulators in Rule 277. Guy insulators indirectly have load factors applied to them via the load factors that are applied to wire tension and wind loading. See Fig. 279-3.

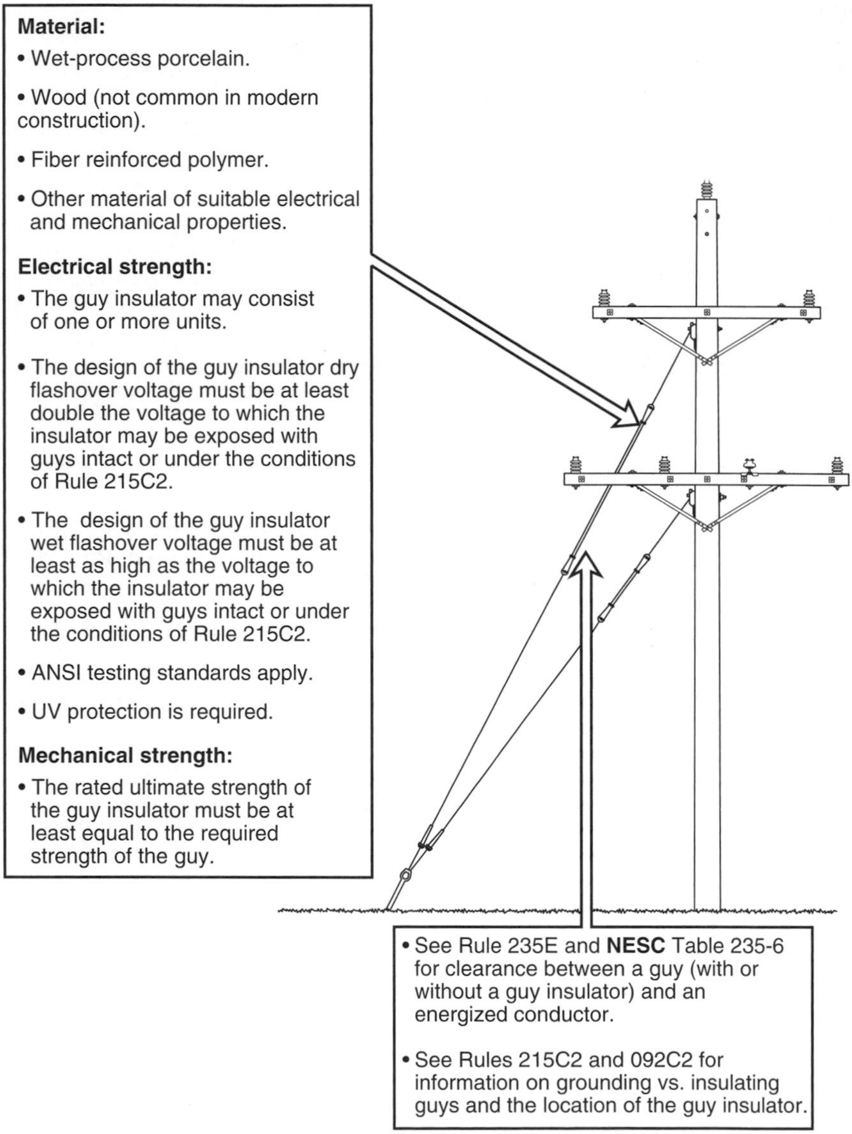

Material:

• Wet-process porcelain.

• Wood (not common in modern construction).

• Fiber reinforced polymer.

• Other material of suitable electrical and mechanical properties.

Electrical strength:

• The guy insulator may consist of one or more units.

• The design of the guy insulator dry flashover voltage must be at least double the voltage to which the insulator may be exposed with guys intact or under the conditions of Rule 215C2.

• The design of the guy insulator wet flashover voltage must be at least as high as the voltage to which the insulator may be exposed with guys intact or under the conditions of Rule 215C2.

• ANSI testing standards apply.

• UV protection is required.

Mechanical strength:

• The rated ultimate strength of the guy insulator must be at least equal to the required strength of the guy.

• See Rule 235E and **NESC** Table 235-6 for clearance between a guy (with or without a guy insulator) and an energized conductor.

• See Rules 215C2 and 092C2 for information on grounding vs. insulating guys and the location of the guy insulator.

See Photo(s)

Fig. 279-2. Properties of guy insulators (Rule 279A1).

Per Rule 279A2a, if a guy insulator is used exclusively for limiting corrosion of an anchor attached to a grounded guy on an effectively grounded system, then it is not classified as a guy insulator. Therefore, the electrical requirements in Rule 279A1b do not apply. The guy in this case is assumed to be grounded per Rule 215C2 using the grounding methods in Rule 092C2. The corrosion protection insulator must

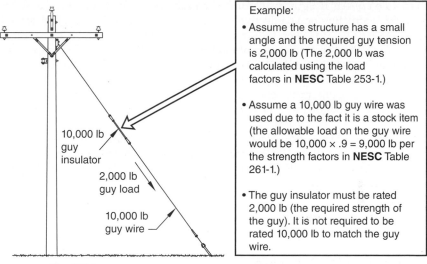

Example:

• Assume the structure has a small angle and the required guy tension is 2,000 lb (The 2,000 lb was calculated using the load factors in **NESC** Table 253-1.)

• Assume a 10,000 lb guy wire was used due to the fact it is a stock item (the allowable load on the guy wire would be 10,000 × .9 = 9,000 lb per the strength factors in **NESC** Table 261-1.)

• The guy insulator must be rated 2,000 lb (the required strength of the guy). It is not required to be rated 10,000 lb to match the guy wire.

10,000 lb guy insulator

2,000 lb guy load

10,000 lb guy wire

Fig. 279-3. Example of the required strength of a guy insulator (Rule 279A1c).

have the mechanical strength at least equal to the required strength of the guy. This wording also appears in Rule 279A1c. Additional requirements for guy insulators used to limit galvanic corrosion can be found in Rule 215C4. An example of a guy insulator used exclusively for limiting galvanic corrosion is shown in Fig. 279-4.

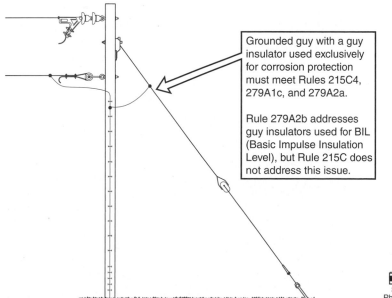

Grounded guy with a guy insulator used exclusively for corrosion protection must meet Rules 215C4, 279A1c, and 279A2a.

Rule 279A2b addresses guy insulators used for BIL (Basic Impulse Insulation Level), but Rule 215C does not address this issue.

See Photo(s)

Fig. 279-4. Example of a guy insulator used to limit galvanic corrosion (Rule 279A2a).

Per Rule 279A2b, if a guy insulator is used exclusively for meeting BIL requirements for a structure on an effectively grounded system, then it is not classified as a guy insulator. However, the strength requirement of the guy insulator must meet Rule 279A1c, and either a combination of Rules 215C2 and 279A1 or a combination of Rules 092C2 and 215C2 must be met. A note in Rule 279A2b refers to **NESC** Fig. 279-1 as one option for applying this rule.

The clearance requirement between a guy (with or without a guy insulator) and an energized conductor on the same structure (pole) is provided in **NESC** Table 235-6. See Rule 235E for a discussion.

279B. Span-Wire Insulators.   The requirements of span-wire insulators are similar to guy insulators. Rule 215C3 is referenced for application and Rule 273 is referenced for insulation levels. The rated ultimate strength of a span-wire insulator is based on the required strength of the span wire in which it is located.

SECTION NUMBER 28 NOT USED IN THIS EDITION

SECTION NUMBER 29 NOT USED IN THIS EDITION

Part 3

Safety Rules
for the Installation
and Maintenance of
Underground Electric
Supply and
Communication Lines

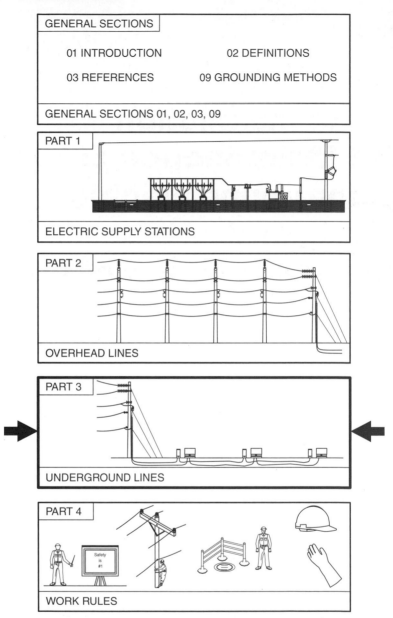

GENERAL SECTIONS

01 INTRODUCTION   02 DEFINITIONS

03 REFERENCES   09 GROUNDING METHODS

GENERAL SECTIONS 01, 02, 03, 09

PART 1

ELECTRIC SUPPLY STATIONS

PART 2

OVERHEAD LINES

PART 3

UNDERGROUND LINES

PART 4

Safety is #1

WORK RULES

Section 30

Purpose, Scope, and Application of Rules

300. PURPOSE

The purpose of Part 3, Underground Lines, is similar to the purpose of the entire NESC outlined in Rule 010, except Rule 300 which is specific to underground supply and communication lines and equipment. Part 3 of the NESC focuses on the practical safeguarding of persons during the installation, operation, and maintenance of underground supply (power) and communication lines and equipment.

301. SCOPE

The scope of Part 3, Underground Lines, includes supply (power) and communications (phone, cable TV, etc.) cables and equipment in underground or buried applications. The term underground equipment covers buried equipment and pad-mounted (aboveground) equipment used with underground conductors or cables. Separate rules apply to Electric Supply Stations (Part 1) and Overhead Lines (Part 2). Part 3 covers underground structural arrangements and extensions into buildings.

There is some overlap between Underground Lines (Part 3) and Overhead Lines (Part 2). The overlap is at the underground riser on the overhead pole. Risers are covered in Sec. 36 of Part 3 and in Sec. 23, Rule 239 of Part 2. See Fig. 301-1.

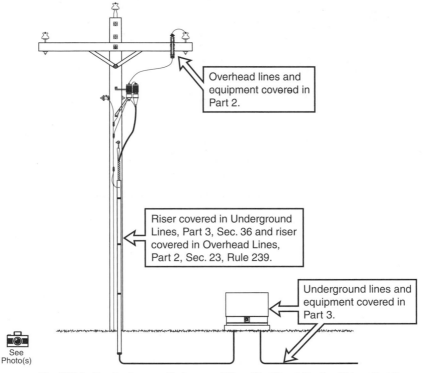

Fig. 301-1. Overlap between Underground Lines (Part 3) and Overhead Lines (Part 2) (Rule 301).

There are distinct differences between Underground Lines (Part 3) and Electric Supply Stations (Part 1). Underground lines outside the electric supply station are covered in Part 3. Conductors and conduit inside the electric supply station are covered in Sec. 16 of Part 1. See Fig. 301-2.

Rule 301 clarifies the application of the **NESC** versus the National Electrical Code (NEC) to underground supply and communication lines. Rule 011 discusses the scope of the **NESC** and the NEC. The **NESC** covers conductors and equipment when they are serving a utility function (not an office building wiring function). The **NESC** underground line rules cover utility functions. There are instances when utilization wiring may be associated with the utility function of underground lines. For example, utilization wiring may be needed for lighting and ventilation in a vault. The **NESC** does not provide specific rules for utilization wiring. Rule 012C, which requires accepted good practice, must be applied when specific conditions are not covered. The NEC is an excellent reference for accepted good practice in this case.

When underground supply equipment is fenced, the rules of Part 1, Electric Supply Stations, may apply. See Rule 110A for a discussion.

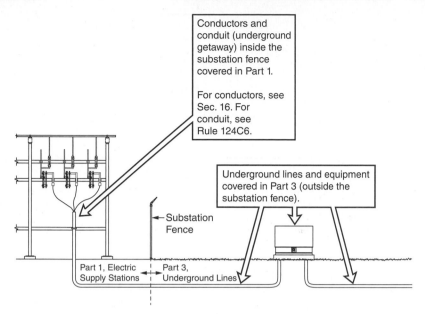

Fig. 301-2. Differences between Underground Lines (Part 3) and Electric Supply Stations (Part 1) (Rule 301).

302. APPLICATION OF RULE

Rule 302 references Rule 013 for the general application of **Code** rules. See Rule 013 for a discussion.

Rule 302 defines how the **NESC** uses the following terms in Part 3, Underground Lines:

- Duct
- Conduit
- Conduit system

The common trade use of these terms can be slightly different from the **NESC** definitions. The use of these terms as they apply to the **NESC** is shown in Fig. 302-1.

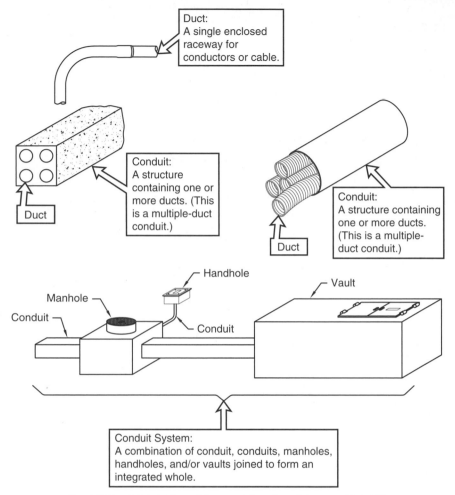

Fig. 302-1. NESC definitions of duct, conduit, and conduit system (Rule 302).

Section 31

General Requirements Applying to Underground Lines

310. REFERENCED SECTIONS

This rule references four sections related to Part 3, Underground Lines, so that rules do not have to be duplicated and the reader of the **Code** realizes that other sections are related to the information provided in Part 3. The related sections are:
- Introduction, Sec. 01
- Definitions, Sec. 02
- References, Sec. 03
- Grounding Methods, Sec. 09

The rules in Part 3, predominantly Rules 314, 342, 374, and 384, will provide the requirements for grounding underground lines and equipment. The grounding methods are provided in Sec. 09.

311. INSTALLATION AND MAINTENANCE

This rule requires that supply and communication utilities be able to locate their underground facilities. Locates are typically done as shown in the example in Fig. 311-1.

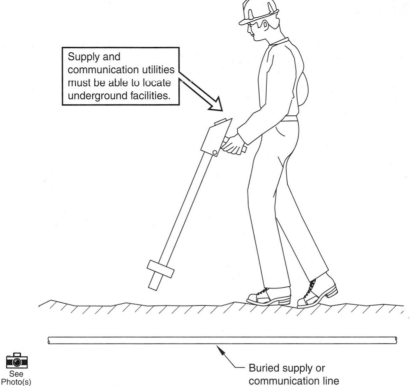

Supply and communication utilities must be able to locate underground facilities.

See Photo(s)

Buried supply or communication line

Fig. 311-1. Example of worker locating underground lines (Rule 311).

Color-coded spray paint markings are typically used to identify power, communication, and other utilities. Most localities have a "call before you dig" system in place to aid location of underground facilities. The **NESC** does not limit locates to supply and communication cables. Any underground facility (e.g., equipment, conduit systems, duct banks, vaults, etc.) must be located. Proper mapping of facilities can aid the location of facilities and is required in Part 4, Work Rules, Rule 411A2.

The **NESC** requires advance notice (a time is not specified) to owners or operators of nearby facilities that may be adversely affected by underground digging or construction by supply or communication utilities.

Rule 311C addresses emergency installations and permits laying certain supply (power) and communication cables directly on the ground. This same permission can be found in Rule 230A2d. See the discussion and figure in Rule 230A. Rule 311C is also referenced in Rule 014B2.

312. ACCESSIBILITY

Rule 312 is the underground equivalent to overhead Rule 213.

Rule 312 generally states the requirement that underground parts needing examination or adjustment during operation must be accessible to authorized supply or communication workers. This rule requires the following, all of which are discussed in more detail throughout Part 3.

- Working spaces
- Working facilities
- Clearances

Working spaces are specifically discussed in Rule 323B. This rule dimensions the working space in manholes. Various rules throughout Part 3 discuss clearances and separations between supply and communication conductors and other facilities.

The **NESC** does not specify a clear working space dimension in front of a pad-mounted transformer or primary junction box. The general requirements of this rule apply, which are that adequate working space must exist. Working space is a challenge for utilities when it comes to service transformers or other equipment on private property. Some utilities apply a notification sign on the front of pad-mounted equipment that asks the landowner to keep the area in front of the equipment clear (this sign is not a **Code** requirement). An example of the need for adequate working space is shown in Fig. 312-1.

An example of a notification sign used on pad-mounted equipment is shown in Fig. 312-2.

Additional requirements for equipment, including pad mounted equipment, can be found in Sec. 38.

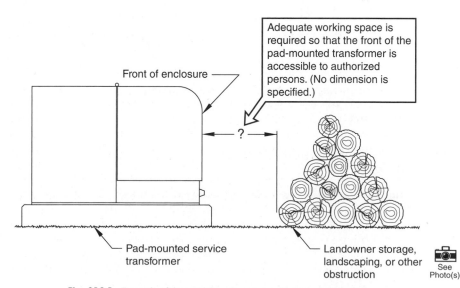

Fig. 312-1. Example of the need for adequate working space (Rule 312).

NOTICE

Adequate working space is required. Please keep shrubs, structures, stored materials, fences, etc., at least 10 ft away from this side and 2 ft away from other sides.

See Photo(s)

For information call USA Power Co. at 1-800-555-1212.

Fig. 312-2. Example of a notification sign used on pad-mounted equipment (Rule 312).

313. INSPECTION AND TESTS OF LINES AND EQUIPMENT

This rule is the underground equivalent of overhead Rule 214. There are slight wording differences between the overhead and underground inspection and testing rules, but the intent is similar.

Cutouts, switches, reclosers, circuit breakers, etc., are used to switch overhead lines and determine if a line is in or out of service. These same overhead devices can be used for underground switching in addition to pad-mounted switches and the termination or standoff of URD elbows.

Rule 313, like its overhead counterpart Rule 214, does not contain a specific inspection or testing checklist, it does not contain any inspection interval (e.g., every year, 2 years, 5 years, 10 years, etc.), and it does not contain specific time periods for making corrections to NESC violations (e.g., 1 day, 1 week, 1 year, etc.). When the Code is not specific, accepted good practice must be applied per Rule 012C. Rule 313A2 has a note similar to Rule 214A2 stating that inspections may be performed in a separate operation or while performing other duties, as desired. See Rule 214 for a discussion and additional information on inspections.

314. GROUNDING OF CIRCUITS AND EQUIPMENT

Rule 314 is the underground equivalent of overhead Rule 215. There is some overlap between Rule 314 and Rule 215 related to grounding riser conduits. See Rule 215.

Rule 314 is broken into three main paragraphs. Rule 314A reminds us that the methods of grounding are specified in Sec. 09, and the requirements of grounding are provided in this rule for circuits and conductive parts to be grounded. Rule 314B focuses on the requirements for grounding conductive parts. Rule 314C focuses on the requirements for grounding circuits. See Sec. 02, "Definitions," for a discussion of the term "effectively grounded." Not all of the grounding requirements for Part 3 are listed in this rule. Various other grounding requirements appear throughout Part 3. One important example is the grounding and bonding requirement for aboveground apparatus (pad-mounted equipment) found in Rule 384. Other Part 3 grounding rules include Rules 342 and 374.

Rule 314B provides a list of conductive parts to be grounded and includes an exception for special cases. The list does not contain conductive manhole or handhole lids that are flush with the surface of a roadway, walkway, lawn, etc. Therefore Rule 012C, which requires accepted good practice, must be used. Methods to effectively ground manhole and handhole lids do exist in the industry. The conductive parts grounding requirements of Rule 314B are outlined in Fig. 314-1.

The circuit grounding requirements of Rule 314C which are broken down into neutrals, other conductors, surge arresters, and use of earth, are outlined in Fig. 314-2.

Typical URD cable.

Cable jacket

Insulation shield (semiconducting material that evens out voltage stress) must be effectively grounded.

Conductor shield (semi-conducting material that evens out voltage stresses) not grounded.

Phase conductor

Insulation

The cable sheath (a concentric neutral conductor on a typical URD cable) must be effectively grounded.

Equipment frames and cases must be effectively grounded.

Conductive (metal) lighting poles must be effectively grounded.

Conductive (metal) handhole covers on nonconductive (fiberglass, composite, etc.) lighting poles must be effectively grounded.

Conductive riser guard (e.g., metal conduit) enclosing supply conductors must be effectively grounded.

Conductive riser guards exposed to open supply conductors (e.g., a metal conduit for a communication riser on a joint-use power and communications pole) must be effectively grounded.

There is some overlap between Rule 314 and Rule 215 related to grounding riser conduits. See Rule 215.

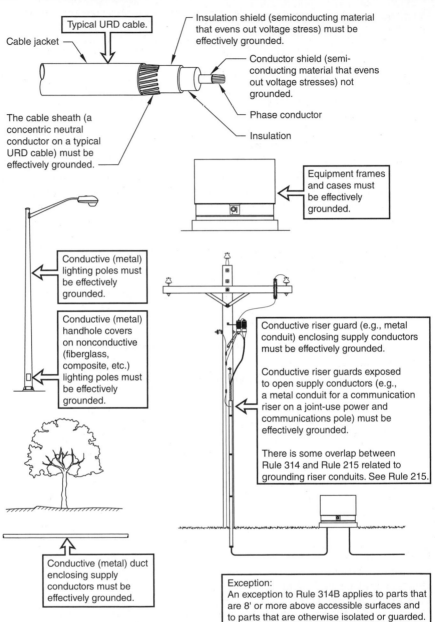

Conductive (metal) duct enclosing supply conductors must be effectively grounded.

Exception:
An exception to Rule 314B applies to parts that are 8' or more above accessible surfaces and to parts that are otherwise isolated or guarded.

See Photo(s)

Fig. 314-1. Conductive parts to be grounded (Rule 314B).

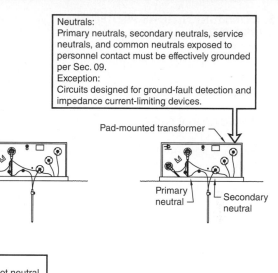

Neutrals:
Primary neutrals, secondary neutrals, service neutrals, and common neutrals exposed to personnel contact must be effectively grounded per Sec. 09.
Exception:
Circuits designed for ground-fault detection and impedance current-limiting devices.

Surge arresters:
Surge arresters must be effectively grounded per Sec. 09.

Elbow surge arrester

Pad-mounted transformer

Primary neutral

Secondary neutral

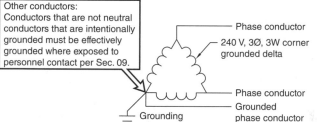

Other conductors:
Conductors that are not neutral conductors that are intentionally grounded must be effectively grounded where exposed to personnel contact per Sec. 09.

Phase conductor

240 V, 3Ø, 3W corner grounded delta

Phase conductor

Grounded phase conductor

Grounding connection

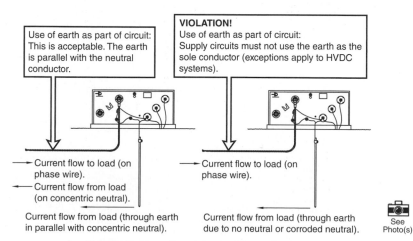

Use of earth as part of circuit:
This is acceptable. The earth is parallel with the neutral conductor.

VIOLATION!
Use of earth as part of circuit:
Supply circuits must not use the earth as the sole conductor (exceptions apply to HVDC systems).

⟶ Current flow to load (on phase wire).

⟵ Current flow from load (on concentric neutral).

Current flow from load (through earth in parallel with concentric neutral).

⟶ Current flow to load (on phase wire).

Current flow from load (through earth due to no neutral or corroded neutral).

See Photo(s)

Fig. 314-2. Grounding of circuits (Rule 314C).

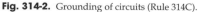

315. COMMUNICATIONS PROTECTIVE REQUIREMENTS

Rule 315 is the underground equivalent to overhead Rule 223. See Rule 223 for a discussion.

316. INDUCED VOLTAGE

Rule 316 is the underground equivalent to overhead Rule 212. See Rule 212 for a discussion.

Section 32

Underground Conduit Systems

As the title indicates, this section is directed to Underground Conduit Systems, which was defined and discussed in Rule 302. A typical conduit system (a combination of conduit, conduits, manholes, handholes, and/or vaults joined to form an integrated whole) can be found under the streets and sidewalks of a large city. The note under the title of Sec. 32 references Rule 350G for supply and communication cables installed in ducts that are not part of a conduit system. Rule 350G states that Sec. 35 (Direct-Buried Cable and Cable in Duct Not Part of a Conduit System) applies to supply and communication cables that are installed in duct that is not part of a conduit system. A typical duct that is not part of a conduit system can be found in a residential subdivision. See Fig. 302-1 in Rule 302 and Fig. 32-1.

320. LOCATION

Section 32 applies to the underground structure only, not the cables in the structure. Section 33 applies to supply cables. Section 34 applies to the installation of cables in an underground structure. Rule 320 deals with the location of underground conduit systems, where they can be routed, and how far they must be separated from other underground installations.

320A. Routing. The general requirements for conduit system routing are outlined below:

- Subject conduit system to the least disturbance practical.
- When conduit system is parallel to another structure, do not locate directly over or under the other structure, if practical.
- Provide sufficient bending radius to limit cable damage.

The **NESC** does not contain specific burial depths for conduit systems. Direct-buried supply cables and cables in duct not part of a conduit system covered in

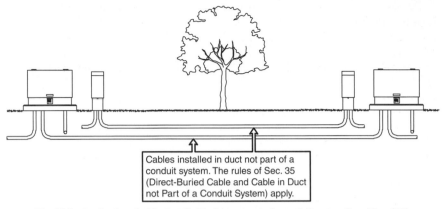

Fig. 32-1. Application of cables installed in duct not part of a conduit system (Secs. 32 and 35).

Sec. 35 do have to meet specified burial depths (covered in Rule 352 and **NESC** Table 352-1); however, no such burial requirement exists for cables installed in conduit systems. The conduit system design requires a structural review using the surface loads specified in Rule 322A3 (for the ducts and joints) and Rule 323A (for the manholes, handholes, and vaults) to determine proper burial depths. The only specific burial depth called out in Sec. 32 is for a conduit system crossing under railroad tracks. The fact that the **NESC** does not contain specific burial depths for conduit systems, and the relationship between several Sec. 32 and Sec. 35 rules is shown in Fig. 320-1.

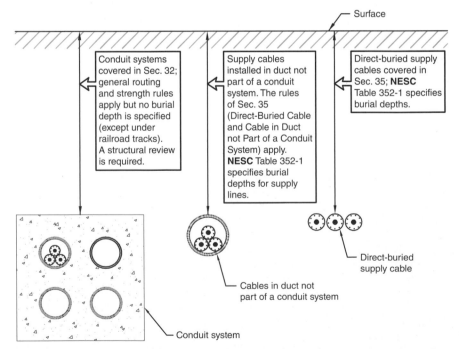

Fig. 320-1. Burial depths for conduit systems, supply cable in duct not part of a conduit system, and direct-buried supply cables (Rules 320, 322A3, 323, 350G, and 352).

The **NESC** is not specific on conduit bending radius requirements. Rule 320A simply states that a sufficient bending radius is needed. Rule 012C, which requires accepted good practice, must be used. The National Electrical Code (NEC) provides much more detail than the **NESC** on conduit bending radius, and the NEC can be used as a reference for accepted good practice. Typically, field bends require a larger bending radius than factory bends. A bend with too tight a radius will deform or damage the conduit material and require excessive forces to pull in the cable. In addition to bending radius requirements, the NEC can be referenced as accepted good practice for conduit support, conduit fill (i.e., the number of conductors in any one conduit), and various other conduit application methods. The details of how conduit bending radius is typically measured are shown in Fig. 320-2.

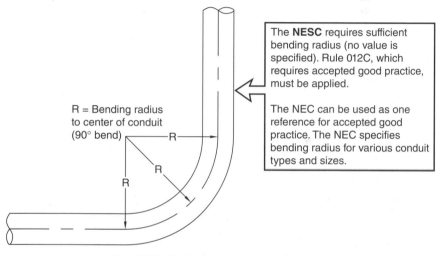

The **NESC** requires sufficient bending radius (no value is specified). Rule 012C, which requires accepted good practice, must be applied.

The NEC can be used as one reference for accepted good practice. The NEC specifies bending radius for various conduit types and sizes.

R = Bending radius to center of conduit (90° bend)

See Photo(s)

Fig. 320-2. Conduit bending radius (Rule 320A1b).

The total bending radius of a conduit run can be determined by adding the angles of each bend. To determine how many bends are acceptable in a conduit run, a cable pulling calculation should be done per Rule 341. See Fig. 320-3.

Natural hazards due to unstable or corrosive soil should be avoided or proper construction methods should be used to minimize the hazard.

If a conduit can be installed outside a roadway, no interference with the road would occur. However, if the conduit is installed longitudinally under the road, using the shoulder or just one lane of traffic will make both installation and maintenance of the conduit conflict less with roadway traffic. The **NESC** does not specify a conduit system burial depth under a road for a longitudinal run or for a road crossing. The road department or highway department having jurisdiction may require specific depths or locations to obtain a permit to install utilities under the road. The rules related to conduit routing under highways and streets are outlined in Fig. 320-4.

Conduit system routing in or on bridges and tunnels should not be damaged by traffic and should be located to provide safe access. Underground conduit system attachments to bridges are covered in Rules 320A4 and 322B5. Rule 351C6

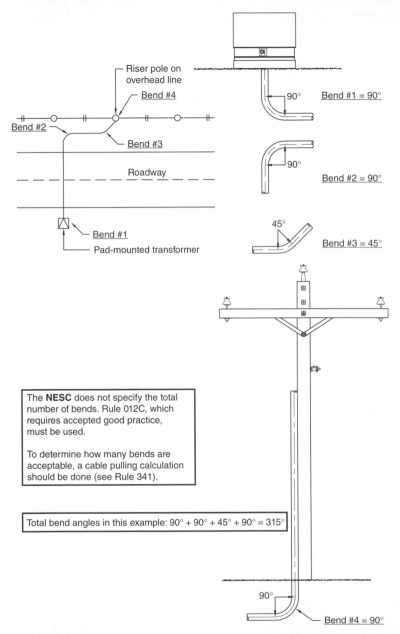

The **NESC** does not specify the total number of bends. Rule 012C, which requires accepted good practice, must be used.

To determine how many bends are acceptable, a cable pulling calculation should be done (see Rule 341).

Total bend angles in this example: 90° + 90° + 45° + 90° = 315°

Fig. 320-3. Example of determining the total number of bend angles of a conduit run (Rules 320A1b and 341).

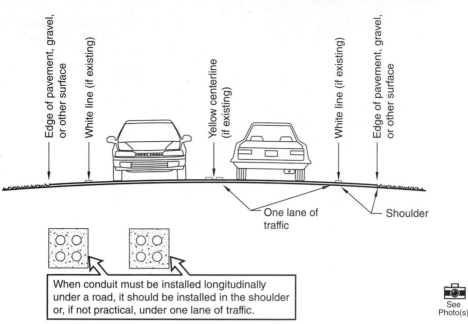

Fig. 320-4. Conduit routing under highways and streets (Rule 320A3).

applies to attaching a duct not part of a conduit system to a bridge. Tunnels are also addressed in Sec. 39.

Conduit systems routing near railroad tracks have a specific burial requirement. This is the only burial depth requirement in Sec. 32 that contains a specific distance. The burial rules for a conduit system under a railroad track are outlined in Fig. 320-5.

Conduit systems crossing water are discussed in Rule 320A6 and in Sec. 35, "Direct Buried Cable and Cable in Duct Not Part of a Conduit System," Rule 351C5. Water crossings are sometimes referred to as submarine crossings. An underwater crossing can typically be found on lake bottoms, river bottoms, or ocean bottoms. A special cable called a "submarine cable" is commonly used for underwater crossings. Conduit can be used for the entire crossing, but it is typically used near the shorelines. An example of a three-phase, 15-kV submarine cable is shown in Fig. 320-6.

Higher-voltage submarine cables are also in use and can consist of oil-filled cable with a lead outer jacket. There are no specific conduit requirements in this rule other than protecting the submarine crossing from erosion by tides or currents and locating the crossing away from where ships normally anchor. Submarine crossings frequently use conduit for some portion of the distance into the water, and then the submarine cable is exposed and lies on or trenched into the bottom of the body of water. An example of a submarine crossing is shown in Fig. 320-7.

Rule 096C, which requires at least four grounds in each mile, has an exception that states the rule does not apply to underwater crossings if other conditions are met.

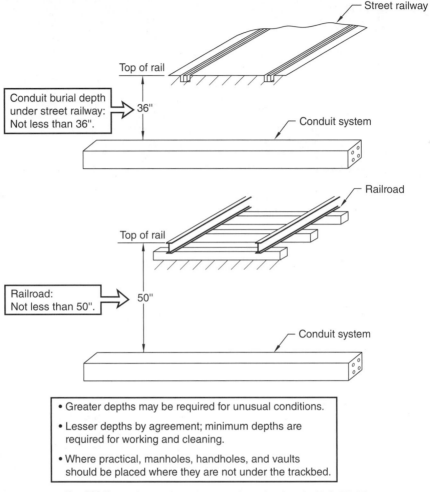

Fig. 320-5. Conduit system crossings under railroad tracks (Rule 320A5).

320B. Separation from Other Underground Installations.   The general requirements for radial separation between conduit systems and other underground
structures are outlined below:
- Provide radial separation to permit maintenance while limiting the likelihood
  of damage to either structure.
- The parties involved should determine separations.
- Exception: Conduits may be supported from roofs of manholes, vaults, and
  subway tunnels with concurrence of the parties involved.

The term structure in this rule is used loosely to apply to many types of buried
structures including building foundations, other conduit systems, and other utility
lines such as sewer lines (sanitary and storm), water lines, gas and other lines that
transport flammable material, steam lines, cryogenic lines, etc.

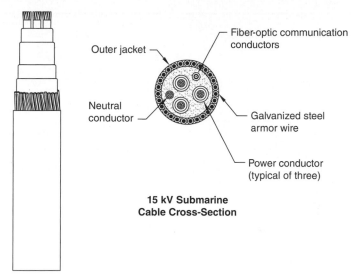

15-kV Submarine Cable

Fig. 320-6. Example of a three-phase, 15-kV "submarine cable" (Rule 320A6).

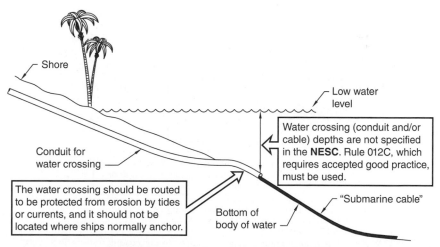

Fig. 320-7. Example of a water crossing (Rule 320A6).

The **Code** provides specific values for separations between supply and communication conduit systems. The separations in Rule 320B2 are between supply and communication conduit systems, not the ducts in the conduit systems. The duct material and separations of the ducts are covered in Rule 322A. The radial separation between supply and communication conduit systems is outlined in Fig. 320-8.

The requirements for conduits crossing and paralleling sanitary and storm sewers, water lines, steam lines (hot), and cryogenic lines (cold) are very general.

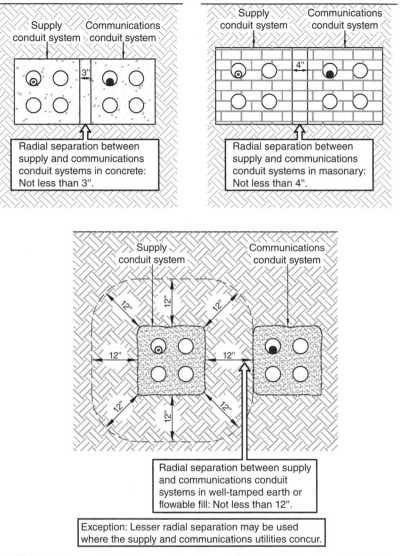

Fig. 320-8. Radial separation between supply and communication conduit systems (Rule 320B2).

No specific dimension is provided. The **Code** is primarily concerned with providing enough separation for maintenance and avoiding damage to the lines involved. See Fig. 320-9.

The **Code** provides a specific value for separation between a conduit and a gas or other line that transports flammable material. In the case of gas lines or other lines that transport flammable material, the measurement is made to the duct in the conduit. See Fig. 320-10.

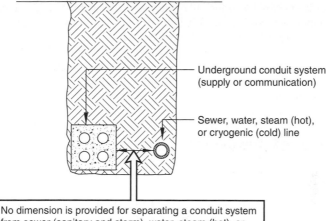

Underground conduit system
(supply or communication)

Sewer, water, steam (hot),
or cryogenic (cold) line

No dimension is provided for separating a conduit system
from sewer (sanitary and storm), water, steam (hot), or
cryogenic (cold) lines. General separation rules apply. The
separations should be determined by the parties involved.

Fig. 320-9. Radial separation from sewer (sanitary and storm), water, and steam
or cryogenic lines (Rules 320B3, 320B4, and 320B6).

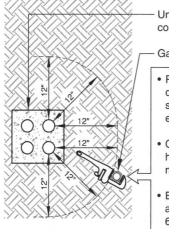

Underground conduit (supply or
communications).

Gas or flammable material line.

• Radial separation measured from the nearest
duct in the conduit not less than 12" <u>AND</u>
sufficient separation for pipe maintenance
equipment.

• Conduit shall not enter the same manhole,
handhole, or vault with gas or flammable
material lines.

• Exceptions apply to communication cables
and supply cables operating at not more than
600 V between conductors. Supplemental
mechanical protection must be evaluated for
supply cables and the utilities involved must
agree.

• See Rule 095B2 for separation between
supply system grounds and high pressure
gas lines.

See
Photo(s)

Fig. 320-10. Radial separation from gas or other lines that transport flammable material
(Rule 320B5).

The conduit system locations discussed in Rule 320A indicate that vertical placement of parallel underground structures is not preferred. The conduit system separations discussed in Rule 320B1 require adequate separation between paralleling structures without any reference to the horizontal or vertical placement.

For separation between direct-buried supply and communication cables (or duct that is not part of a conduit system) and other underground structures, see Rules 353 and 354.

321. EXCAVATION AND BACKFILL

The trench or excavation and backfill rules for conduit systems are outlined in Fig. 321-1.

For trench and backfill requirements related to direct-buried supply and communication cables and cable in duct not part of a conduit system, see Rule 352.

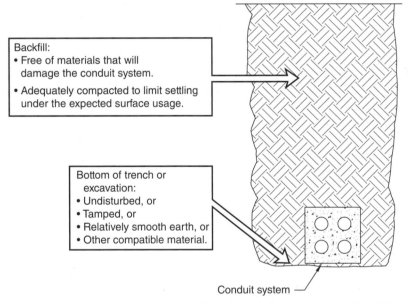

Backfill:
• Free of materials that will damage the conduit system.
• Adequately compacted to limit settling under the expected surface usage.

Bottom of trench or excavation:
• Undisturbed, or
• Tamped, or
• Relatively smooth earth, or
• Other compatible material.

Conduit system

Fig. 321-1.   Trench or excavation and backfill requirements for conduit systems (Rule 321).

322. CONDUIT, DUCTS, AND JOINTS

322A. General.   The general rules for conduit and ducts require the duct material and the construction of the conduit (e.g., a concrete duct bank) be designed so that a cable fault in one duct will not damage the conduit (i.e., duct bank) to an extent that would cause damage to cables in an adjacent duct. Specific separation distances between ducts in the conduit (e.g., concrete duct bank) are not specified. Rule 012C, which requires accepted good practice, must be used. The general rules for conduit include surface loading requirements which are also used to determine the conduit type, as well as the burial depth. The surface loads for manholes, handholes, and vaults must also be analyzed per Rule 323A. The general rules for conduit, ducts, and joints are outlined in Fig. 322-1.

322B. Installation.   The installation rules for conduit, ducts, and joints are outlined in Fig. 322-2.

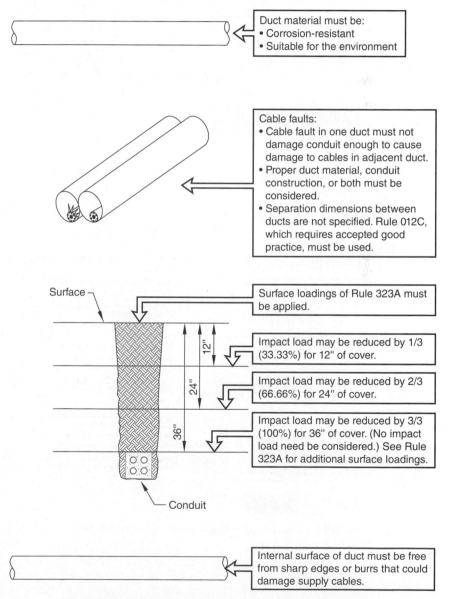

Fig. 322-1. General requirements for conduit, ducts, and joints (Rule 322A).

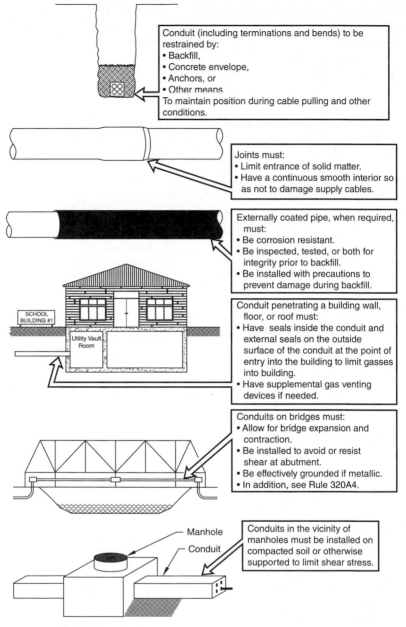

Conduit (including terminations and bends) to be restrained by:
• Backfill,
• Concrete envelope,
• Anchors, or
• Other means
To maintain position during cable pulling and other conditions.

Joints must:
• Limit entrance of solid matter.
• Have a continuous smooth interior so as not to damage supply cables.

Externally coated pipe, when required, must:
• Be corrosion resistant.
• Be inspected, tested, or both for integrity prior to backfill.
• Be installed with precautions to prevent damage during backfill.

SCHOOL BUILDING #1

Utility Vault Room

Conduit penetrating a building wall, floor, or roof must:
• Have seals inside the conduit and external seals on the outside surface of the conduit at the point of entry into the building to limit gasses into building.
• Have supplemental gas venting devices if needed.

Conduits on bridges must:
• Allow for bridge expansion and contraction.
• Be installed to avoid or resist shear at abutment.
• Be effectively grounded if metallic.
• In addition, see Rule 320A4.

Manhole

Conduit

Conduits in the vicinity of manholes must be installed on compacted soil or otherwise supported to limit shear stress.

Fig. 322-2. Installation requirements for ducts and joints (Rule 322B).

323. MANHOLES, HANDHOLES, AND VAULTS

323A. Strength.   Manholes, handholes, and vaults must be designed to withstand a variety of loads as outlined in Fig. 323-1.

In general, dead loads are the weight of the structure, constant weight on the structure, and permanent attachments to the structure. In general, live loads are loads that are not permanent, such as a truck passing over the top of a manhole. Impact loads are generally applied in a similar fashion as live loads and are expressed as a percentage of live loads.

For roadway areas the **Code** provides **NESC** Fig. 323-1 and **NESC** Fig. 323-2 of a typical 10-wheel semitractor trailer truck with associated loads to use for live-load calculations. A 10-wheel semi is used as opposed to an 18-wheeler, as a 10-wheel semi will have a larger force (weight) allocated to each wheel. A note in Rule 323A1 is also provided, reminding the designer that road construction equipment may exceed the typical truck loads. For areas not subject to vehicle traffic, the **Code** requires that the design live load be not less than 300 lb/ft^2.

The **Code** requires that the live loads be increased 30 percent for impact. Impact loads can be reduced for specific burial depths (see Rule 322A).

The weight of the manhole, handhole, or vault must be sufficient to withstand hydraulic, frost, or other uplift forces. If the structure weight is not sufficient, some type of restraint or anchorage must be provided. The weight of equipment inside the structure cannot be considered, as the equipment may be removed or modified.

Examples of live loads in roadways and live loads in areas not subject to vehicular loading are shown in Fig. 323-2.

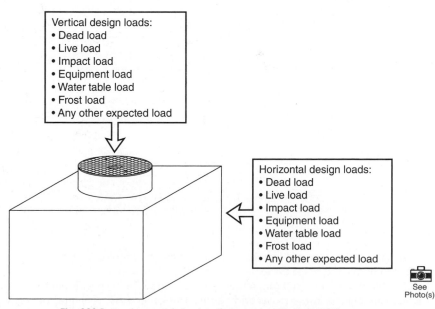

Vertical design loads:
• Dead load
• Live load
• Impact load
• Equipment load
• Water table load
• Frost load
• Any other expected load

Horizontal design loads:
• Dead load
• Live load
• Impact load
• Equipment load
• Water table load
• Frost load
• Any other expected load

See Photo(s)

Fig. 323-1. Loads on manholes, handholes, and vaults (Rule 323A).

Live load in roadway areas must consider moving tractor-semitrailer truck with vehicle wheel loads per **NESC** Figs. 323-1 and 323-2.

Road construction loads may exceed the completed roadway loads.

Live loads must be increased by 30% for impact.

Consider hydraulic, frost, or other uplift.

Live load in areas not subject to vehicular loading must be not less than 300 lb/ft^2.

Live loads must be increased by 30% for impact.

Consider hydraulic, frost, or other uplift.

See
Photo(s)

Fig. 323-2. Examples of live loads in roadways and live loads in areas not subject to vehicular loading (Rules 323A1, 323A2, 323A3, and 323A4).

When a pulling iron is installed in a manhole, handhold, or vault, it must be rated to withstand twice the expected load. See Fig. 323-3.

323B. Manhole and Vault Dimensions.   The Code is very specific about working space dimensions in manholes and vaults. The general requirement for working space in Part 3 is found in Rule 312. Rule 323B is one of the few rules in Part 3 that actually dimensions the working space.

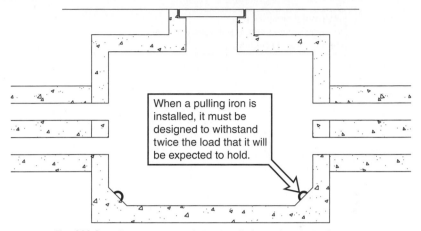

When a pulling iron is installed, it must be designed to withstand twice the load that it will be expected to hold.

Fig. 323-3. Pulling irons in manholes, handholes, and vaults (Rule 323A5).

The rules for working space in manholes and vaults are outlined in Figs. 323-4 and 323-5.

It is important to note that the clear working space dimensions are measured after cables and equipment are placed in the manhole. They are not the dimensions to the bare manhole wall.

323C. Manhole Access.   The rules for manhole access are outlined in Figs. 323-6 through 323-11.

The access dimensions required in Rule 323C are for the manhole opening. They are not the dimensions of the lid or cover. The lid or cover could be larger than the manhole opening.

Additional rules related to manhole access require the openings to be free of protrusions that will injure workers or prevent quick egress. Where practical, they should be located outside of highways, intersections, and crosswalks to reduce traffic hazards to the workers.

The manhole access openings should not be located directly over cables or equipment. If this requirement is not practical, the manhole opening located over cables (not equipment) must have a safety sign or a protective barrier over the cables or a fixed ladder.

A manhole greater than 4-ft deep must be designed so it can be entered by means of a ladder or other suitable climbing device. If proper working space exists in the manhole per Rule 323B and the manhole opening is above the working space, ladder access should not be difficult. The ladder is not required to be permanent. Ladder requirements are also covered in Rule 323F. The **Code** does not consider the cables and equipment in the manhole to be suitable climbing devices.

323D. Covers and Gratings.   The rules for manhole, handhole, and vault covers and gratings are outlined in Fig. 323-12.

Rules for manholes, handholes, and vaults are found in Sec. 32, Underground Conduit Systems. Handholes can be found in underground subdivision design and construction, but Sec. 35, Direct Buried Cable and Cable in Duct Not Part of a

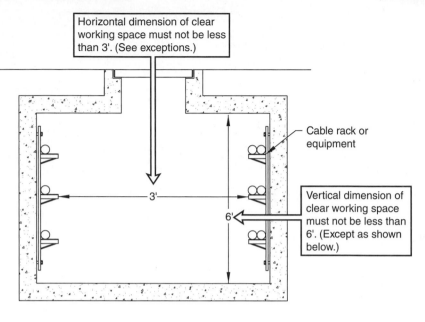

Horizontal dimension of clear working space must not be less than 3'. (See exceptions.)

Cable rack or equipment

Vertical dimension of clear working space must not be less than 6'. (Except as shown below.)

The 6' vertical dimension may be reduced (no value specified) if the manhole or vault opening is within 1' horizontally of the adjacent interior side wall.

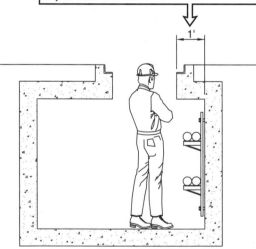

Fig. 323-4. Manhole and vault working space dimensions (Rule 323B).

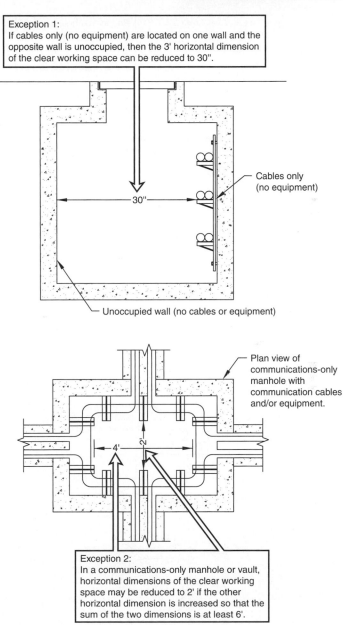

Exception 1:
If cables only (no equipment) are located on one wall and the opposite wall is unoccupied, then the 3' horizontal dimension of the clear working space can be reduced to 30".

Cables only
(no equipment)

30"

Unoccupied wall (no cables or equipment)

Plan view of communications-only manhole with communication cables and/or equipment.

4'

2'

Exception 2:
In a communications-only manhole or vault, horizontal dimensions of the clear working space may be reduced to 2' if the other horizontal dimension is increased so that the sum of the two dimensions is at least 6'.

Fig. 323-5. Exceptions to manhole and vault working space dimensions (Rule 323B).

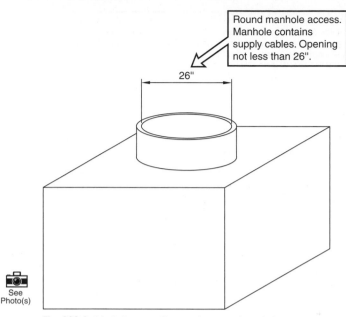

Fig. 323-6. Manhole access diameter for a round manhole opening in a manhole containing supply cables (Rule 323C1).

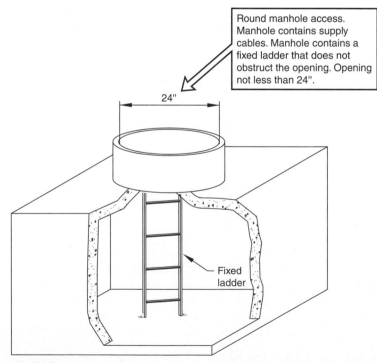

Fig. 323-7. Manhole access diameter for a round manhole opening in a manhole containing supply cables (Rule 323C1).

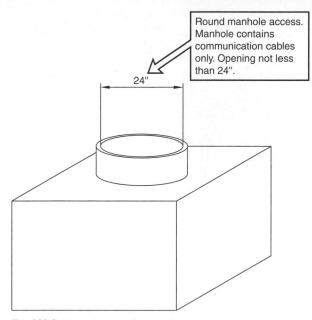

Round manhole access. Manhole contains communication cables only. Opening not less than 24".

24"

Fig. 323-8. Manhole access diameter for round manhole opening in a manhole containing communication cables only (Rule 323C1).

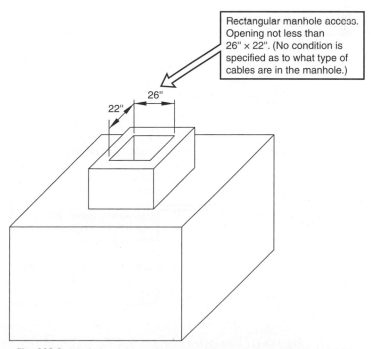

Rectangular manhole access. Opening not less than 26" × 22". (No condition is specified as to what type of cables are in the manhole.)

26"

22"

Fig. 323-9. Manhole access size for a rectangular manhole opening (Rule 323C1).

Openings free from protrusions.

Manhole opening located so that safe access can be provided.

In highway, located outside the paved roadway when practical.

Located outside of street intersections and crosswalks where practical.

Personnel access not located over the cable or equipment. If not practical, openings can be over the cable if a safety sign or protective barrier or fixed ladder is used.

Fig. 323-10. Manhole access opening requirements (Rules 323C2, 323C3, and 323C4).

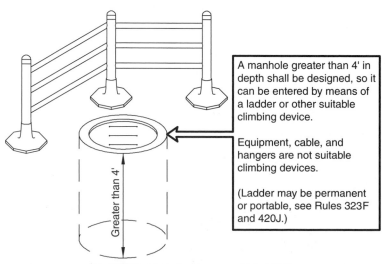

A manhole greater than 4' in depth shall be designed, so it can be entered by means of a ladder or other suitable climbing device.

Equipment, cable, and hangers are not suitable climbing devices.

(Ladder may be permanent or portable, see Rules 323F and 420J.)

Greater than 4'

Fig. 323-11. Manhole entry means (Rule 323C5).

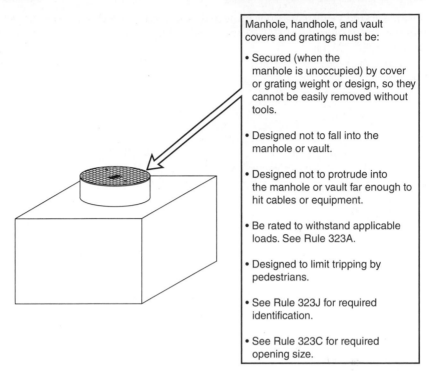

Manhole, handhole, and vault covers and gratings must be:

• Secured (when the manhole is unoccupied) by cover or grating weight or design, so they cannot be easily removed without tools.

• Designed not to fall into the manhole or vault.

• Designed not to protrude into the manhole or vault far enough to hit cables or equipment.

• Be rated to withstand applicable loads. See Rule 323A.

• Designed to limit tripping by pedestrians.

• See Rule 323J for required identification.

• See Rule 323C for required opening size.

Fig. 323-12. Requirements for manhole, handhole, and vault covers and gratings (Rule 323D).

Conduit System does not have any manhole, handhole, or vault requirements. If Sec. 35 applies, using the manhole, handhole, and vault requirements in Sec. 32 is a method of applying accepted good practice (Rule 012C).

Rule 323D4 has the wording, "Covers and gratings should be designed to limit the likelihood of tripping by pedestrians." No specifics are provided as to how much of a change in elevation constitutes a tripping hazard. The Americans with Disabilities Act (ADA) does contain information related to sidewalk tripping dimensions.

323E. Vault Access.   The general access rules for vaults are similar to the general manhole access rules discussed in Rule 323C. No dimension is provided for vault access openings. Accepted good practice must be applied. The manhole access dimensions are certainly one of the best references for accepted good practice in this case. Typically, manholes are thought to have openings or lids, and vaults are thought to have openings, lids, or doors. Utility vaults can be found in basements and on the ground floor of buildings and even on other floors in high-rise buildings. The vault access rules provide requirements for doors accessible to the public including locking of the doors and posting of safety signs. If an above ground vault has a ventilation opening that is not protected, unguarded energized parts and controls must be located using the safety clearance zone requirements of Part 1, Electric Supply Stations, Rule 110A and **NESC** Table 110-1. Essentially, a vault with exposed energized parts is treated like a substation and a ventilation opening in the vault that is not protected is treated like a chain-link fence and requires application of Rule 110A4 and **NESC** Table 110-1. Examples of vaults are shown in Fig. 323-13.

See Sec. 39, Rules 390 and 391, for the requirements related to installations in tunnels.

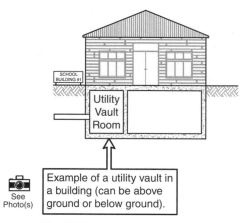

Fig. 323-13. Example of a vault (Rule 323E).

323F. Ladder Requirements. Fixed ladders in manholes and vaults are required to be corrosion-resistant. Portable ladders must meet the work rules in Part 4, Rule 420J. Additional ladder requirements related to manhole access can be found in Rule 323C. The work rules in Part 4, Rule 423B require testing for flammable gases and oxygen deficiency before entering a manhole or an unventilated vault.

323G. Drainage. Drainage for manholes, handholes, and vaults is not required to drain into sewers. Many times, natural drainage is provided by utilizing a dry well or gravel base. When drainage into sewers is desired, a trap is a common method that can be used to keep sewer gas out of the manhole or vault. Sewer gas poses a significant explosion hazard, as well as an asphyxiation and/or fire hazard, when workers enter the manhole or vault. See Fig. 323-14.

323H. Ventilation. Ventilation must be provided if the public can somehow be affected by the gases that collect in manholes, vaults, and tunnels. An exception applies to areas under water or other impractical locations. If ventilation is not needed (typical of most manholes in city streets), the work rules in Part 4, Rule 423B require an adequate air supply for the employee while the employee is working in the manhole or unventilated vault.

323I. Mechanical Protection. If a manhole, handhole, or vault has a grated cover that allows objects to fall on the supply cables and equipment, the supply cables and equipment should be located (positioned) or guarded (physically protected) to prevent damage. This rule does not mention communication cables or

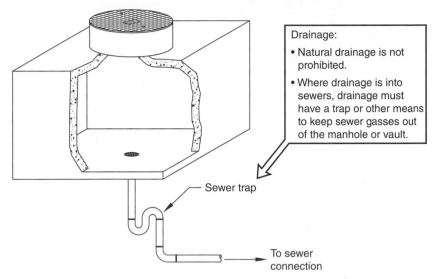

Drainage:
• Natural drainage is not prohibited.
• Where drainage is into sewers, drainage must have a trap or other means to keep sewer gasses out of the manhole or vault.

Sewer trap

To sewer connection

Fig. 323-14. Drainage requirements for manholes, vaults, and tunnels (Rule 323G).

equipment. Grated covers are typically used for venting. Cables and equipment must not be located directly under any type of manhole opening for personnel access per Rule 323C. See Rule 323C for a discussion.

323J. Identification.   The rules for identifying manhole and handhole covers are outlined in Fig. 323-15.

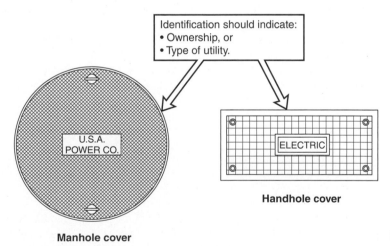

Identification should indicate:
• Ownership, or
• Type of utility.

U.S.A.
POWER CO.

ELECTRIC

Handhole cover

See Photo(s)

Manhole cover

Fig. 323-15. Requirements for identification of manhole and handhole covers (Rule 323J).

Section 33

Supply Cable

330. GENERAL

Section 33 is only applicable to supply cable. It does not address communication cables. Rule 330 starts out with a requirement that supply cables be tested in accordance with applicable standards. The rule also provides additional general requirements that are applicable to both the supply cable specifier and the supply cable manufacturer. The supply cable should be rated for the particular application and environment. The cable must also be able to withstand the fault current to which it will be subjected. Common underground supply cable ratings for distribution lines are 15 kV or 25 kV or 35 kV, 100% insulation level or 133% insulation level, full concentric neutral or 1/3 concentric neutral, ethylene-propylene rubber (EPR) insulation or cross-linked polyethylene (XLP) insulation or tree-retardant cross-linked polyethylene (TRXLP) insulation, jacketed in various aluminum and copper conductor sizes. Underground supply transmission line cables (above 35 kV) are also available and sometimes used in the industry. Rule 350 also addresses supply cable requirements.

331. SHEATHS AND JACKETS

Cable jackets protect cables from adverse environmental conditions. Older underground residential distribution (URD) cables did not have a cable jacket, and concentric-neutral corrosion became a widespread problem in some soils. Newer URD cables can be purchased with an outer jacket to protect the concentric neutral conductors. URD cables with a jacket do not have the concentric neutral in contact with the earth. The jacket may have to be stripped back to ground the concentric neutral. See Rule 096 for a grounding discussion. An example of a sheath, jacket, and shield on a typical URD supply cable is shown in Fig. 331-1.

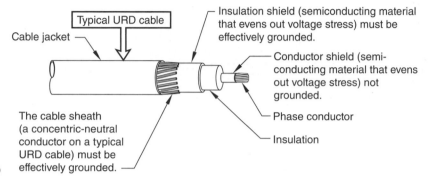

Typical URD cable

Cable jacket

Insulation shield (semiconducting material that evens out voltage stress) must be effectively grounded.

Conductor shield (semi-conducting material that evens out voltage stress) not grounded.

The cable sheath (a concentric-neutral conductor on a typical URD cable) must be effectively grounded.

See Photo(s)

Phase conductor

Insulation

Figure 331-1. Example of a sheath, jacket, and shield on a typical URD supply cable (Rule 331).

332. SHIELDING

General shielding guidelines are provided in this rule. Organizations that specialize in developing cable standards are noted for a reference. See the figure in Rule 331 for a description of the individual components that make up a typical underground residential distribution (URD) supply cable.

333. CABLE ACCESSORIES AND JOINTS

Cable accessories and joints are the elbows, splices, and terminations used on supply cables. These accessories must be able to withstand stresses similar to the cable to which they are connected. The requirements for cable accessories in this rule focus on material specifications. Cable accessories are also discussed in Sec. 37, "Supply Cable Terminations." Section 37 focuses on the installation of the cable terminations. Examples of cable accessories and joints are shown in Fig. 333-1.

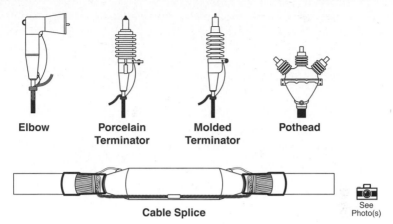

Elbow   **Porcelain Terminator**   **Molded Terminator**   **Pothead**

Cable Splice

See Photo(s)

Figure 333-1. Examples of cable accessories and joints (Rule 333).

Section 34

Cable in Underground Structures

340. GENERAL

Section 34 provides rules for underground cables when they are installed in underground structures (e.g., conduit systems). Section 33 provides rules for the components of a supply cable. Additional rules in Sec. 34 supplement the supply cable rules in Sec. 33. Section 32 provides rules for the conduit system. The rules in Sec. 34 do not apply to direct-buried cables except where the rules in Sec. 35 specifically reference the rules in Sec. 34. See Fig. 340-1.

Per Rule 340B, a supply cable over 2 kV to ground installed in nonmetallic conduit should consider the need for an effectively grounded shield, sheath, or both. When a supply cable is direct-buried or installed in duct not part of a conduit system, Rule 350B requires an effectively grounded shield, sheath, or concentric neutral more emphatically (via the word *shall* instead of *should*), and the requirement is for above 600 V, not 2 kV.

341. INSTALLATION

341A. General. Supply cables are mechanically stressed when they are bent. The NESC does not specify a cable bending radius. Only a general statement requiring bending to be controlled to avoid damage is provided. The cable manufacturer's recommendations should be used. A bending radius is usually expressed as a multiple of the cable diameter. When the supply cable is installed in conduit,

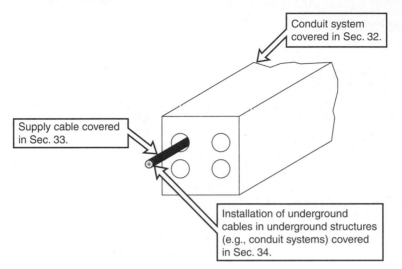

Fig. 340-1. Application of Secs. 32, 33, and 34 (Rule 340).

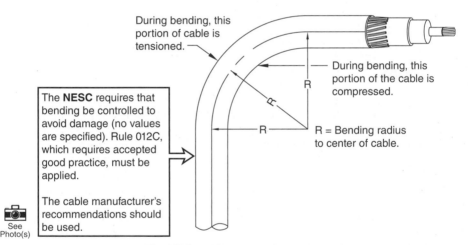

Fig. 341-1. Cable bending radius (Rule 341A1).

the cable bending radius should always be compatible with the conduit bending radius. See Fig. 341-1.

The **NESC** does not specify conduit fill requirements. Conduit fill is the number of conductors or cables that fit in a specific conduit size. Conduit fill is normally limited to some percentage of the conduit cross-sectional area. The National Electrical Code (NEC) is a reference for acceptable good practice in this case, or the cable manufacturer may provide guidance. Many utilities develop conduit

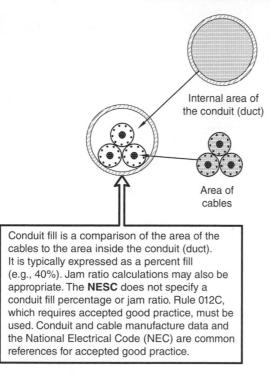

Internal area of
the conduit (duct)

Area of
cables

Conduit fill is a comparison of the area of the
cables to the area inside the conduit (duct).
It is typically expressed as a percent fill
(e.g., 40%). Jam ratio calculations may also be
appropriate. The **NESC** does not specify a
conduit fill percentage or jam ratio. Rule 012C,
which requires accepted good practice, must be
used. Conduit and cable manufacture data and
the National Electrical Code (NEC) are common
references for accepted good practice.

Fig. 341-2. Conduit fill calculations (Rule N/A).

fill tables for the conduits and cables they normally stock. Information regarding conduit fill calculations is shown in Fig. 341-2.

The **Code** requires limiting pulling tensions and sidewall pressures to avoid cable damage. A note suggests using the cable manufacturer's recommendations. Cable manufacturers and cable lubricant suppliers can also supply engineering guidelines and software for making pulling calculations. Cable lubricants are used to make cable pulling easier, therefore reducing the pulling tension. Additional ways to reduce pulling tensions include pulling from an uphill point to a downhill point and starting the pull near a conduit bend instead of away from a conduit bend. Sidewall pressures are the forces that push on the side of the cable around a conduit bend. The same methods used to reduce pulling tension can be used to reduce sidewall pressure. Sidewall pressure is a function of the cable tension at the bend and the radius of the bend. See Fig. 341-3.

The **Code** specifically discusses pulling tensions and sidewall pressures relative to supply cables. The **Code** does not discuss pulling tensions for communication cables, but communication cables require the same considerations.

The **Code** requires ducts to be cleaned of foreign material. This is typically referred to as "swabbing" the duct. The cable lubricants must be safe for the type of conduit and conduit materials. Too little friction can also be a problem. If a cable

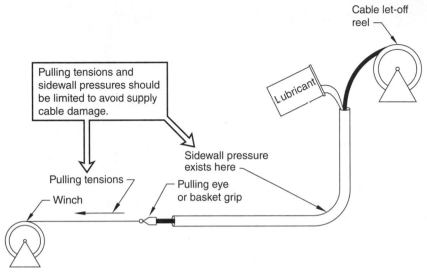

Fig. 341-3. Supply cable pulling tension and sidewall pressure (Rule 341A2).

is installed on a downhill slope or vertical run, the cable may need to be restrained to prevent a cable elbow or termination from being stressed or pulled out.

Rule 320B requires separation between supply and communication conduit systems. Rule 341A6 reinforces that idea by saying that supply and communication cables must not be installed in the same duct unless they are operated and maintained (not necessarily owned) by the same utility. Rule 341A7 permits the installation of communication cables of different communication utilities in the same duct provided the utilities are in agreement. See the beginning of Sec. 32 for definitions of duct, conduit, and conduit system. The requirements of Rules 341A6 and 341A7 related to supply and communication cables in duct are outlined in Fig. 341-4.

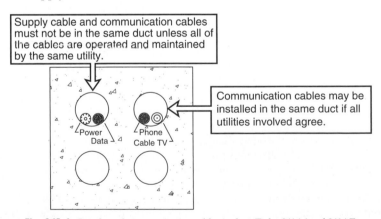

Fig. 341-4. Supply and communication cables in duct (Rules 341A6 and 341A7).

341B. Cable in Manholes and Vaults.   Rule 341B covers the installation of cable in manholes and vaults. The rule is divided into three main parts, supports, clearance, and identification. Rule 341B1 covers supports for cable in manholes and vaults. The rules for supporting cable in manholes and vaults are outlined in Fig. 341-5.

Rule 341B2 covers clearance for cables in manholes and vaults. The rules for clearance of cables in manholes and vaults focus on clearance between joint-use (supply and communication) facilities. Joint-use (supply and communication) manholes and vaults are usually not preferred by either utility. If a joint-use (supply and communication) duct bank is used, the communication conduits can branch off of the duct bank before the duct bank enters the supply manhole. The communication conduits can then enter a separate communication manhole adjacent to the supply manhole. If this system is not used and a joint-use (supply and communication) manhole is desired, it must be done with the concurrence of all the parties involved.

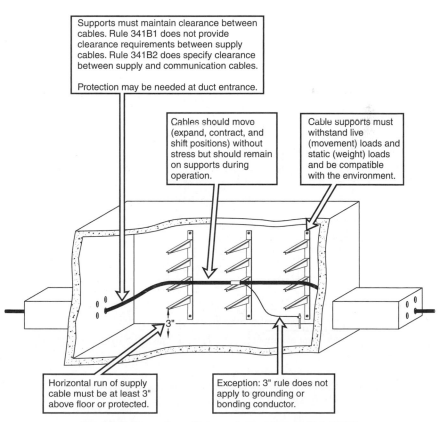

Fig. 341-5. Supporting cable in manholes and vaults (Rule 341B1).

The work rules in Part 4, Rule 433, must be applied when a communications worker is in a joint-use (supply and communication) manhole. The preferred location of supply cables in a joint-use (supply and communication) manhole is below communication cables. This is the opposite position of overhead lines as supply conductors are preferred at the higher level on an overhead line per Rule 220B. The rules for clearance of cables (and equipment per Rule 341B2b) in joint-use (supply and communication) manholes and vaults are outlined in Figs. 341-6 and 341-7.

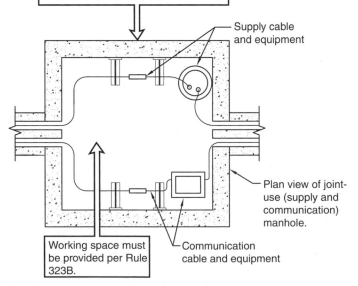

Fig. 341-6. Clearance of cables and equipment in joint-use (supply and communication) manholes and vaults (Rule 341B2).

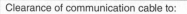

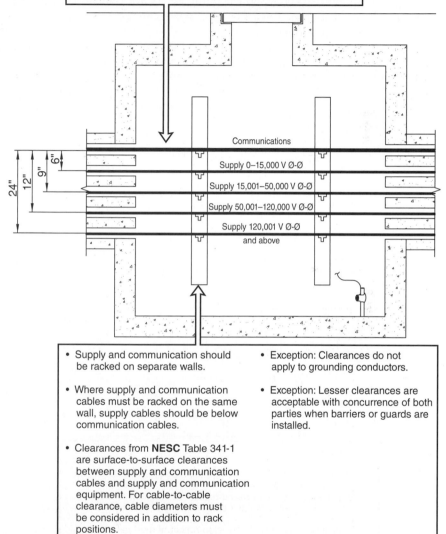

Clearance of communication cable to:
- Supply cable of 0–15,000 V Ø-Ø: Not less than 6".
- Supply cable of 15,001–50,000 V Ø-Ø: Not less than 9".
- Supply cable of 50,001–120,000 V Ø-Ø: Not less than 12".
- Supply cable of 120,001 V Ø-Ø and above: Not less than 24".

Communications

Supply 0–15,000 V Ø-Ø

Supply 15,001–50,000 V Ø-Ø

Supply 50,001–120,000 V Ø-Ø

Supply 120,001 V Ø-Ø
and above

- Supply and communication should be racked on separate walls.

- Where supply and communication cables must be racked on the same wall, supply cables should be below communication cables.

- Clearances from **NESC** Table 341-1 are surface-to-surface clearances between supply and communication cables and supply and communication equipment. For cable-to-cable clearance, cable diameters must be considered in addition to rack positions.

- Exception: Clearances do not apply to grounding conductors.

- Exception: Lesser clearances are acceptable with concurrence of both parties when barriers or guards are installed.

Fig. 341-7. Clearance of cables and equipment in joint-use (supply and communication) manholes and vaults (Rule 341B2).

Rule 341B3 requires the cables in manholes or other access openings of a conduit system (i.e., vaults and handholes) to be identified by tags that are corrosion-resistant and suitable for the environment in which they are installed. The quality of the tag must be good enough to read with portable lighting. Brass tags and plastic compounds are both popular solutions. An exception is provided that eliminates the need for tags if maps or diagrams combined with cable position provide sufficient identification.

For joint-use manholes or vaults, a cable identification tag or marking of some type must be provided denoting the utility name and the type of cable used. This requirement applies to a manhole or vault with multiple utilities of the same kind (i.e., power and power) or joint-use (i.e., power and communication) utilities. The identification requirements of Rule 341B3 are outlined in Fig. 341-8.

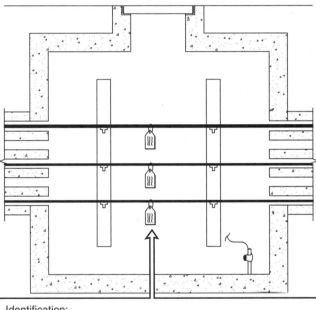

Identification:

- Cables must be permanently identified by tags or other means at manhole or other opening.
  (An exception applies where position, in conjunction with diagrams or maps, provides sufficient identification.)

- Identification must be of corrosion-resistant material suitable for the environment.

- Identification must be readable with portable or fixed lighting.

- Joint-use locations must identify the utility name and type of cable use.

Fig. 341-8. Identification of cables in manholes and vaults (Rule 341B3).

342. GROUNDING AND BONDING

Rule 342 provides grounding requirements in addition to the general grounding requirements in Rule 314. Cable or joints (splices) with bare metallic shields, sheaths, or concentric neutrals that are exposed to personnel contact must be effectively grounded. See the figure in Rule 331 for a description of the individual components that make up a typical underground residential distribution (URD) supply cable. Cable sheaths or shields must have a common ground to keep different cables at equal ground potential, and the grounding materials must be corrosion-resistant or suitably protected.

343. NUMBER 343 NOT USED IN THIS EDITION

344. COMMUNICATION CABLES CONTAINING SPECIAL SUPPLY CIRCUITS

Rule 344 is the underground equivalent to the overhead Rule 224B. Both this rule and Rule 224B provide special conditions for embedding a supply circuit in a communication cable. Rule 224B lists five special conditions (Rule 224B2a through 224B2e) for an overhead cable construction. Rule 344 lists seven special conditions (Rules 344A1 through 344A7) for an underground cable construction. Both rules address the wattage of the power circuit; however, the formatting of the rules is slightly different.

Section 35

Direct-Buried Cable and Cable in Duct Not Part of a Conduit System

As the title indicates, this section addresses Direct-Buried Cable and Cable in Duct Not Part of a Conduit System. A definition of "duct" was discussed in Rule 302. However, the note under the title of Sec. 35 defines how the terms "duct" and "ducts" are used in Sec. 35, saying that anytime either of those terms is used, it specifically does not include duct(s) that are part of a conduit system. This note, plus the note under the title of Sec. 32, clarifies that Sec. 32 applies to conduit systems and Sec. 35 applies to direct-buried cable and cable in duct not part of a conduit system. A typical conduit system as discussed in Rule 302 (a combination of conduit, conduits, manholes, handholes, and/or vaults joined to form an integrated whole) can be found under the streets and sidewalks of a large city. A typical duct that is not part of a conduit system can be found in a residential subdivision. One of the biggest differences between conduit systems (covered in Sec. 32) and direct-buried cable and cable in duct that is not part of a conduit system (covered in Sec. 35) is that Sec. 35 specifies burial depths and Sec. 32 does not. Section 32 requires a structural review to determine burial depths. See the discussion and Fig. 302-1, in Rule 302 and see the discussion and Figs. 32-1, and 320-1 at the beginning of Sec. 32.

350. GENERAL

Rule 350A references Sec. 33 (supply cable) and states that Sec. 33 applies to direct-buried supply cable. Additional rules in Sec. 35 supplement the supply cable rules in Sec. 33. It should be noted that Rule 350A expressly refers to both design and construction. The various installation methods, materials, and locations should be considered in the context of the overall design. For example, direct burial of a supply cable in the vicinity of a heat source like another supply cable or a steam line may significantly impact the dissipation of heat, so the ampacity should be evaluated according to accepted good practice.

Per Rule 350B, a supply cable over 600 V to ground must have an effectively grounded shield, sheath, or concentric neutral that is continuous. See the figure in Rule 331 for a description of the individual components that make up an underground residential distribution (URD) supply cable.

The exception to Rule 350B recognizes that the standard splicing methods used on concentric-neutral cable do not allow the concentric neutral to remain in a concentric pattern across the splice. See Fig. 350-1.

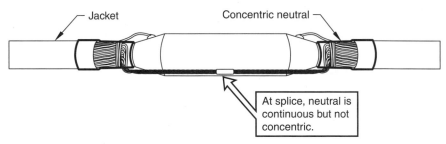

Fig. 350-1. Exception related to splices on concentric-neutral cables (Rule 350B).

Rule 350C states that supply cables meeting Rule 350B of the same supply circuit (e.g., the same three-phase feeder) may be direct-buried or installed in duct with no deliberate separation. See Fig. 350-2.

Rule 350D states that underground direct-buried cables or cables installed in duct below 600 V to ground without an effectively grounded shield or sheath must be placed in close proximity to each other. Underground direct-buried secondary

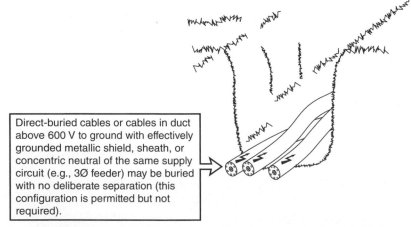

Fig. 350-2. Cables above 600 V to ground with no deliberate separation (Rule 350C).

service conductors typically have an insulated (not bare) neutral. The requirement to place the conductors in close proximity with no intentional separation helps reduce the path that a fault current would need to follow if a cable fault occurred. The cables that meet Rule 350B are permitted to be buried with deliberate separation, but they are not required to be installed in that manner per Rule 350C. Rule 350C differs from Rule 350D as the cables in 350D are required to be installed in close proximity with no intentional separation. See Fig. 350-3.

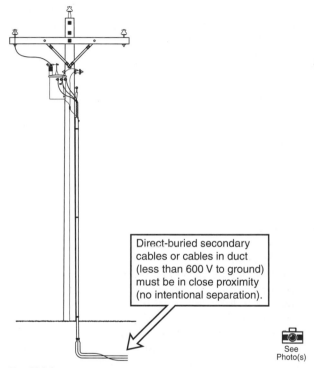

Direct-buried secondary cables or cables in duct (less than 600 V to ground) must be in close proximity (no intentional separation).

See Photo(s)

Fig. 350-3. Cables below 600 V to ground buried in close proximity (Rule 350D).

Rule 350E references Rule 344A for requirements related to communication cables containing special supply circuits. These rules also apply to direct-buried communication cables and cables in duct not part of a conduit system.

Rule 350F requires marking of direct-buried supply cable and direct-buried communication cable and cable in duct not part of a conduit system with the "lightning bolt" for supply (power) and the "telephone handset" for communications. **NESC** Fig. 350-1 provides a diagram of the requirements. The telephone handset is used for all communications, even data and cable TV. This requirement is for direct-buried cables and per Rule 350G and the title of Sec. 35, this requirement also applies to cable in duct not part of a conduit system. Cables installed in conduit systems per Sec. 32 do not need to meet this requirement, and Sec. 32, "Underground Conduit Systems," does not have any conduit labeling or color-coding requirements. A recommendation in Rule 350F states that if color coding is used as an additional method of identifying cable, the American Public Works Association Uniform Color Code for marking underground utility lines is recommended. The recommendation is not recommending color coding as an additional cable identifier; it is simply recommending that if color coding is used that the uniform color code for marking underground lines be applied. The American Public Works Association Uniform Color Code is shown below:

- White—Proposed excavation
- Pink—Temporary survey markings
- Red—Electric power lines, cables, conduit, and lighting cables
- Yellow—Gas, oil, steam, petroleum, or gaseous materials
- Orange—Communication, alarm or signal lines, cables or conduit
- Blue—Potable water
- Purple—Reclaimed water, irrigation and slurry lines
- Green—Sewers and drain lines

Sometimes supply cables are specified with three red stripes in addition to the **Code** required lightning bolt. The three red stripes typically run the length of the cable and are placed an equal distance around the cable. The recommendation does not comment on using stripes, the number of stripes, or the placement of the stripes, only the color of the additional method but not the method itself. Exception 1 to this rule addresses the fact that some cables are too small in diameter to mark or have some other marking constraints. Exception 2 states that unmarked cable from stock can be used to repair existing buried unmarked cables.

Rule 350G is related to the Note at the beginning of Sec. 32 and the Note at the beginning of this section (Sec. 35). See the discussion and Fig. 302-1, in Rule 302 and see the discussion and Figs. 32-1 and 320-1 at the beginning of Sec. 32. Some of the rules in Sec. 35 only use the term cable, but based on Rule 350G, the rules also apply to cable installed in duct (duct that is not part of a conduit system in Sec. 32). The recommendation in Rule 350G is similar to the recommendation in Rule 350F. The recommendation in Rule 350F applies to color coding as a method to identify cables. The recommendation in Rule 350G applies to color coding as a method to identify ducts. It is not recommending color coding as an additional duct identifier; it is simply recommending that if color coding is used that the uniform color code for marking underground lines be applied. If concrete encasement is used around the duct, the **Code** does not contain a rule or recommendation for putting a colored dye in the concrete.

351. LOCATION AND ROUTING

The rules for locating and routing direct-buried cables and cables in duct not part of a conduit system are similar to, but sometimes stricter than, the rules for locating conduit systems in Rule 320.

351A. General.   The general requirements for locating and routing direct-buried cable and cable in duct are outlined below:

- Subject cables to the least disturbance practical.
- When paralleling and directly over or under other buried facilities, the separation requirements in Rule 353 or 354 must be met.
- Install cables as straight as practical.
- Where required, cable bends must be large enough to avoid cable damage. (See Rule 341A for a discussion and figures related to cable bending, conduit

fill, and pulling tensions. See Rule 320A for a discussion and figures related to conduit bending.)

- Route for safe access during construction, inspection, and maintenance.
- Plan route and structure conflicts before trenching, plowing, or boring.

351B. Natural Hazards.   Natural hazards are discussed here for direct-buried cables and in Rule 320 for conduit systems. The cable must be constructed and installed to be protected from damage. If the installation method requires switching from a direct-buried cable or cable in duct installation to a full conduit system installation, Rule 320 would apply.

351C. Other Conditions.   Swimming pools receive a lot of attention in the NESC. Part 2, Overhead Lines, and Part 3, Underground Lines, both have specific rules for swimming pool areas. Rule 351C1 addresses direct-buried supply cable or supply cable in duct (not communications) in the vicinity of in-ground swimming pools. This installation is very common in residential subdivisions in warmer parts of the country. No distance is provided for communications cable or communications cable in duct adjacent to in-ground swimming pools; therefore, the general locating and routing requirements in Rule 351A must be applied. The rules for locating and routing direct-buried supply cable or supply cable in duct in the vicinity of in-ground swimming pools are outlined in Fig. 351-1.

Aboveground swimming pools are addressed in Rule 351C2. Buildings and other structures (including aboveground swimming pools, tanks, tool sheds, etc.) should not have direct-buried cables or cables in duct installed under their foundations. If unavoidable, the building or structure foundation must be properly supported, so it will not damage the cable. If a conduit system is chosen to protect the cables installed under or near a building, Rules 320A1 and 320B1 would apply. See Fig. 351-2.

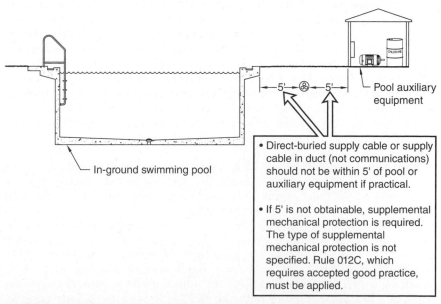

- In-ground swimming pool
- Pool auxiliary equipment

- Direct-buried supply cable or supply cable in duct (not communications) should not be within 5' of pool or auxiliary equipment if practical.

- If 5' is not obtainable, supplemental mechanical protection is required. The type of supplemental mechanical protection is not specified. Rule 012C, which requires accepted good practice, must be applied.

Fig. 351-1. Locating and routing direct-buried supply cable or supply cable in duct near in-ground swimming pools (Rule 351C1).

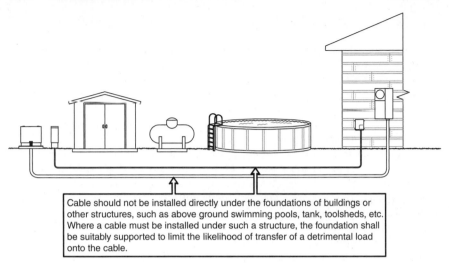

Cable should not be installed directly under the foundations of buildings or other structures, such as above ground swimming pools, tank, toolsheds, etc. Where a cable must be installed under such a structure, the foundation shall be suitably supported to limit the likelihood of transfer of a detrimental load onto the cable.

Fig. 351-2. Cable installed directly under the foundations of buildings or other structures (Rule 351C2).

Rule 351C3 addresses railroad tracks. Direct-buried cable or cable in duct installed longitudinally under the ballast section of railroad tracks is to be avoided. Where it must be done, the direct-buried cable or cable in duct is to be 50 in below the top rail. An exception to Rule 351C3 states that this clearance may be reduced by agreement of the parties concerned. A permit is normally required to cross or parallel a railroad right-of-way, and to obtain the permit, the railroad agency usually requires the utilities to be located as far away from the tracks or as deep as possible. If a direct-buried cable or cable in duct is installed under railroad tracks, the same burial depths required for conduit systems apply. See the figure in Rule 320A.

The **Code** states in Rule 351C4 that the installation of direct-buried cable or cable in duct longitudinally under highways and streets should be avoided. The equivalent rule in the underground conduit systems section (Rule 320A3) does not have the "should be avoided" language as conduit systems are commonly found under roadways in large cities. If a conduit system is not used, cables under roadways are commonly installed in duct, not direct buried. Both Sec. 35, which applies to direct-buried cable and cable in duct, and Sec. 32, which applies to underground conduit systems, define where to locate underground facilities. See the figure in Rule 320A.

Water crossings are addressed in Rule 351C5 in Sec. 35, "Direct-Buried Cable and Cable in Duct Not Part of a Conduit System," and in Rule 320A6 in Sec. 32, "Underground Conduit Systems." The wording is nearly identical. See the water crossing discussion and the figures in Rule 320A.

Rule 351C6 addresses cables in duct (not part of a conduit system) attached to a bridge where permitted by the bridge owner. It contains requirements for location and safe access for inspection or maintenance. Rules 320A4 and 322B5 contain requirements for conduit systems attached to bridges. Obtaining a permit to attach to a bridge from the bridge authority typically requires a detailed review of both the **NESC** rules and the bridge designer requirements.

352. INSTALLATION

Direct-buried cable or cable in duct installations can be done using three primary methods—trenching, plowing, and boring. Installation of underground conduit systems in Sec. 32 focuses on trenching (see Rule 321). Rule 352 for the installation of direct-buried cables or cable in duct focuses on trenching, plowing, and boring. The rules for installation of direct-buried cable or cable in duct by trenching, plowing, and boring are outlined in Figs. 352-1, 352-2, 352-3, and 352-4.

The phrases "sufficient to protect the cable or duct from damage" and "supplemental mechanical protection" are used in this rule and other rules in Sec. 35. The **Code** is not specific about how to accomplish these tasks. Rule 012C, which requires accepted good practice, must be applied. If a conduit system is used for protection, then Sec. 32, "Underground Conduit Systems," applies to the installation. See the discussion and Fig. 302-1, in Rule 302 and see the discussion and Figs. 32-1 and 320-1 at the beginning of Sec. 32.

Rule 352D, Depth of Burial, is one of the most referenced rules in Part 3. The rule starts out by defining how to measure the depth of burial of a direct-buried cable or cable in duct. The measurement is from the surface to the top of the cable or duct, not to the bottom of the trench. The general requirement is that burial depths must be sufficient to protect the cable or duct from damage by the expected surface usage. The expected surface usage may be truck traffic or pedestrian traffic, it may involve lawn mowers or rototillers. The expected surface usage is not typically a backhoe digging a building foundation or construction trench. These types of activities typically require locating buried utilities prior to excavation using a local "call before you dig" or "one-call system." Specific burial depth requirements are presented in **NESC** Table 352-1 for supply cable, conductor, or duct.

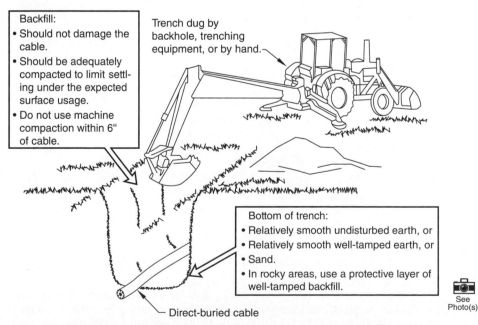

Backfill:
- Should not damage the cable.
- Should be adequately compacted to limit settling under the expected surface usage.
- Do not use machine compaction within 6" of cable.

Trench dug by backhole, trenching equipment, or by hand.

Bottom of trench:
- Relatively smooth undisturbed earth, or
- Relatively smooth well-tamped earth, or
- Sand.
- In rocky areas, use a protective layer of well-tamped backfill.

See Photo(s)

Direct-buried cable

Fig. 352-1. Trenching requirements for direct-buried cable (Rule 352A1).

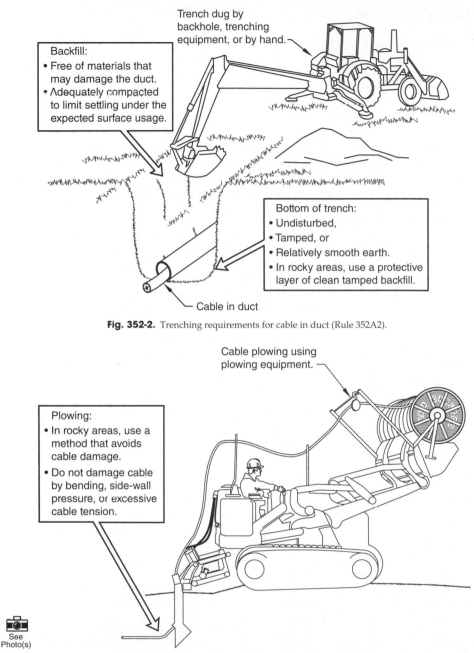

Trench dug by backhole, trenching equipment, or by hand.

Backfill:
• Free of materials that may damage the duct.
• Adequately compacted to limit settling under the expected surface usage.

Bottom of trench:
• Undisturbed,
• Tamped, or
• Relatively smooth earth.
• In rocky areas, use a protective layer of clean tamped backfill.

Cable in duct

Fig. 352-2. Trenching requirements for cable in duct (Rule 352A2).

Cable plowing using plowing equipment.

Plowing:
• In rocky areas, use a method that avoids cable damage.
• Do not damage cable by bending, side-wall pressure, or excessive cable tension.

See Photo(s)

Fig. 352-3. Plowing requirements for direct-buried cables (Rule 352B).

There is not a corresponding table in the **NESC** for communication cable, conductor, or duct burial depth. Communication conductors do fall into the 0- to 600-V range, but **NESC** Table 352-1 is titled Supply Cable, Conductor, or Duct Burial Depth and the voltage column specifies phase-to-phase voltage which implies supply

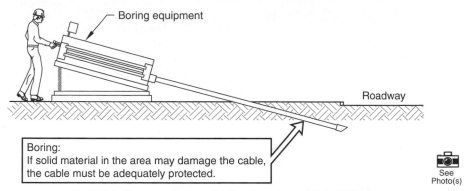

Boring equipment

Roadway

Boring:
If solid material in the area may damage the cable,
the cable must be adequately protected.

See
Photo(s)

Fig. 352-4. Boring requirements for direct-buried cables (Rule 352C).

cables. Depths for communication cable, conductor, or duct must meet the general requirements for burial depth outlined in Rule 352D1. In addition to Rule 352D1, a communication utility must also use Rule 012C, which requires accepted good practice, for communication cable burial depths. Communication cables laid on the ground do not meet the **Code** requirements for burial depth or overhead clearances.

Supply cables laid on the ground do not meet the **Code** requirements for burial depth or overhead clearances. Supply and communication cables may be laid on the ground for emergency (not temporary) installations per Rules 230A2d and 311C. In addition, Rule 014B2 contains information for waiving burial depths (laying a cable on the ground) for emergency installations and an exception to Rule 014C permits laying a cable on the ground for temporary installations under special conditions. The requirements in **NESC** Table 352-1 for supply cable, conductor, or duct are graphically outlined in Fig. 352-5.

Some utilities establish supply cable burial depth values by using the **Code** clearance plus an adder. The adder (e.g., 6 in) can be thought of as a design or construction tolerance adder to maintain the required **Code** burial depth over time. There are several factors that could jeopardize a supply cable burial depth. One factor could be soil erosion. Installing a 15-kV phase-to-phase direct-buried supply cable or duct at 36-in depth can help maintain the **Code**-required 30-in depth over the life of the installation. See Rule 010 for a discussion of clearance adders.

The exception to Rule 352D2 allows lesser burial depths if the burial depths in **NESC** Table 352-1 cannot be met and if supplemental mechanical protection is used. One of the most common reasons for using lesser depths than those specified in **NESC** Table 352-1 is when heavy or solid rock is encountered during trenching. The decision of what to use for supplemental mechanical protection is up to the designer. Rule 012C, which requires accepted good practice, must be applied. Conduit (duct) is a form of supplemental mechanical protection. Concrete encasement of a duct, concrete cover, concrete slabs, and wood planks are also forms of supplemental mechanical protection. The type of duct material (e.g., reinforced thermosetting resin conduit, RTRC, polyvinyl chloride conduit, PVC; high-density polyethylene conduit, HDPE; rigid metal conduit, RMC; intermediate metal conduit, IMC, etc.) and the grade of the conduit (e.g., PVC Type DB, EB, B, C, D, Schedule 40, Schedule 80, 5DR 13.5, SDR 11, SDR 9, etc.) will affect the amount of supplemental mechanical protection. The **Code** does not specify what type of duct **NESC** Table 352-1 applies to or what type of duct can be considered supplemental mechanical protection and be buried less than the depths in **NESC**

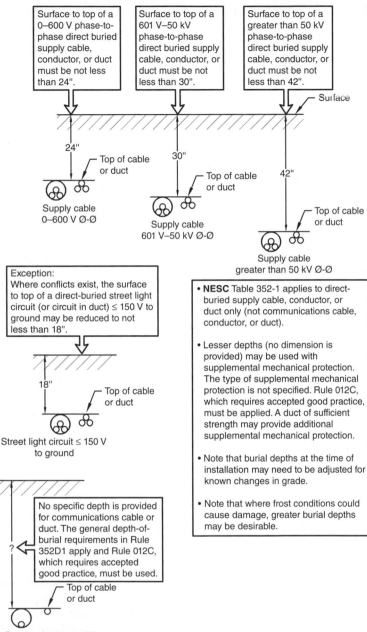

Surface to top of a 0–600 V phase-to-phase direct buried supply cable, conductor, or duct must be not less than 24".

Surface to top of a 601 V–50 kV phase-to-phase direct buried supply cable, conductor, or duct must be not less than 30".

Surface to top of a greater than 50 kV phase-to-phase direct buried supply cable, conductor, or duct must be not less than 42".

Surface

24"

30"

42"

Top of cable or duct

Top of cable or duct

Top of cable or duct

Supply cable 0–600 V Ø-Ø

Supply cable 601 V–50 kV Ø-Ø

Supply cable greater than 50 kV Ø-Ø

Exception:
Where conflicts exist, the surface to top of a direct-buried street light circuit (or circuit in duct) ≤ 150 V to ground may be reduced to not less than 18".

18"

Top of cable or duct

Street light circuit ≤ 150 V to ground

• **NESC** Table 352-1 applies to direct-buried supply cable, conductor, or duct only (not communications cable, conductor, or duct).

• Lesser depths (no dimension is provided) may be used with supplemental mechanical protection. The type of supplemental mechanical protection is not specified. Rule 012C, which requires accepted good practice, must be applied. A duct of sufficient strength may provide additional supplemental mechanical protection.

• Note that burial depths at the time of installation may need to be adjusted for known changes in grade.

• Note that where frost conditions could cause damage, greater burial depths may be desirable.

No specific depth is provided for communications cable or duct. The general depth-of-burial requirements in Rule 352D1 apply and Rule 012C, which requires accepted good practice, must be used.

?

Top of cable or duct

Communications cable

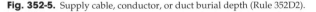

Fig. 352-5. Supply cable, conductor, or duct burial depth (Rule 352D2).

Table 352-1. One reference for accepted good practice for a single duct that is not part of a conduit system is the National Electrical Code (NEC). The NEC specifies burial depths for various types of conduits (ducts) containing circuits at various voltage levels. It also provides methods for reducing conduit (duct) burial depth when rock is encountered. If a conduit system is used, then the rules of Sec. 32 apply. See the discussion and Fig. 302-1, in Rule 302 and see the discussion and Figs. 32-1 and 320-1 at the beginning of Sec. 32.

Rule 352E1 states that supply and communication cables must not be installed in the same duct unless they are operated and maintained (not necessarily owned) by the same utility. Rule 352E2 permits the installation of communication cables of different communication utilities in the same duct provided the utilities are in agreement. Rules 352E1 and 352E2 are similar in nature to Rules 341A6 and 341A7 (Cable in Underground Structures) and to Rules 239A2d and 239A2e (Vertical Risers). The requirements of Rules 352E1 and 352E2 are outlined in Fig. 352-6.

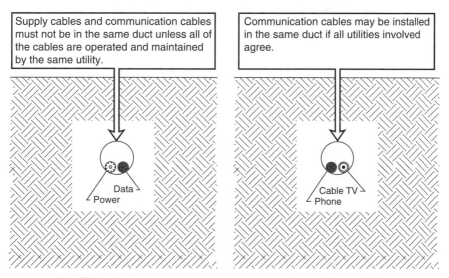

Fig. 352-6. Supply and communication cables in duct (Rules 352E1 and 352E2).

353. DELIBERATE SEPARATIONS—EQUAL TO OR GREATER THAN 12 IN (300 MM) FROM UNDER-GROUND STRUCTURES OR OTHER CABLES

Rules 353 and 354 are closely related. As their titles indicate, Rule 353 applies to 12 in or greater separation and Rule 354 applies to less than 12 in of separation. Rule 353 for direct-buried cable or cable in duct is similar to Rule 320B for underground conduit systems. In Rule 353, the NESC uses the term underground structures to mean sewers, water lines, gas, and other lines that transport flammable material, building foundations, steam lines, etc.

The requirement for radial (any direction) separation between a direct-buried cable or duct and another underground structure or cable in Rule 353 is 12 in or more. This value is assumed to permit maintenance on both facilities without damaging the other. If 12 in of separation is not obtainable, Rule 354 applies. When a direct-buried cable or cable in duct is parallel and directly over or under another underground structure or cable, the parties involved must be in agreement to the method used. No such agreement is specified if a side-by-side configuration is used. Crossings require support or sufficient vertical separation to limit the transfer of detrimental loads from one system to the other. The rules related to the radial separation of direct-buried cables or cable in duct from another underground structure or cable are outlined in Fig. 353-1.

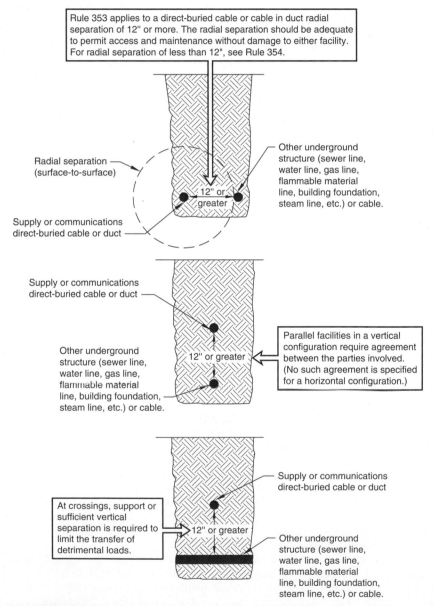

Rule 353 applies to a direct-buried cable or cable in duct radial separation of 12" or more. The radial separation should be adequate to permit access and maintenance without damage to either facility. For radial separation of less than 12", see Rule 354.

Radial separation (surface-to-surface)

Other underground structure (sewer line, water line, gas line, flammable material line, building foundation, steam line, etc.) or cable.

12" or greater

Supply or communications direct-buried cable or duct

Supply or communications direct-buried cable or duct

12" or greater

Other underground structure (sewer line, water line, gas line, flammable material line, building foundation, steam line, etc.) or cable.

Parallel facilities in a vertical configuration require agreement between the parties involved. (No such agreement is specified for a horizontal configuration.)

Supply or communications direct-buried cable or duct

At crossings, support or sufficient vertical separation is required to limit the transfer of detrimental loads.

12" or greater

Other underground structure (sewer line, water line, flammable material line, building foundation, steam line, etc.) or cable.

Fig. 353-1. Deliberate separation of 12 in or more between a direct-buried cable or cable in duct and another underground structure or cable (Rule 353).

Vertical separation between supply and communication lines is typically used by utilities when plowing in direct-buried cables or cable-in-duct. The separations in Rule 353 are from surface to surface, not center to center. Therefore, a 12-in center-to-center spacing of the plow chutes may not provide a 12-in separation between the cables or ducts depending on the cable or duct diameters.

When paralleling or crossing a line that involves hot or cold temperatures like a steam (hot) line or a cryogenic (cold) line, consideration must be given to the heat or cold, so it does not damage the crossing or paralleling line. Adequate separation is to be used. If separation is not obtainable, a thermal barrier is required.

354. RANDOM SEPARATION—SEPARATION LESS THAN 12 IN (300 MM) FROM UNDERGROUND STRUCTURES OR OTHER CABLES

Rules 354 and 353 are closely related. As their titles indicate, Rule 354 applies to less than 12 in of separation and Rule 353 applies to 12 in or greater separation. Random separation is sometimes referred to as random lay. The focus of the random separation rules is to prevent damage to direct-buried communication lines when a fault occurs on direct-buried supply lines. The intent is not to reduce or minimize conductive interference or "noise."

The general rules for random separation are outlined below:

- The rules apply to a direct-buried cable or cable in duct and another underground structure or cable with a radial (any direction) separation less than 12 in.
- The radial separation between a direct-buried supply or communication cable or cable in duct and a steam line, gas or flammable material line, or other line that transports flammable material must not be less than 12 in. (Rule 353 must be used in this case but an exception does apply between steam lines, gas or flammable material lines and supply cables operating at 300 V or below and between gas lines and communication cables.)
- Supply circuits (above 300 V to ground or 600 V phase to phase), when faulted, need to be de-energized by a protective device (i.e., fuse, recloser, circuit breaker, etc.).
- Supply and communication cables and conductors in random separation are treated as one system when considering separation from other conductors. or structures.

The exception to Rule 354A2 permits a 120/240 V, single-phase, three-wire secondary service to be less than 12 in from a gas or flammable material line provided certain conditions are met and provided the utilities agree. A similar exception permits a communication cable to be less than 12" from a steam line or gas or flammable material line provided the utilities agree. The NESC does not specify clearances between gas and electric meters or equipment on the building, only in the trench. See Figs. 354-1 and 354-2.

Supply cables above 600 V to ground with an effectively grounded continuous metallic shield, sheath, or concentric neutral of the same supply circuit may be buried in random separation (no deliberate separation), see Rule 350C. Supply cables of the same circuit operating below 600 V to ground without an

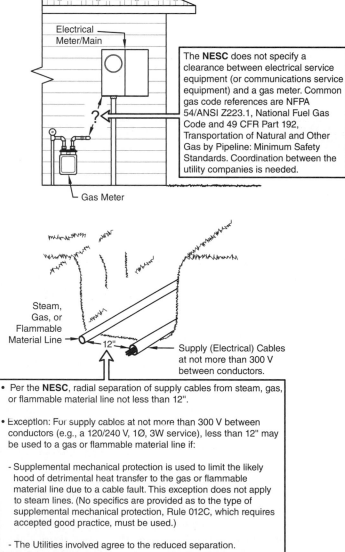

Electrical Meter/Main

The **NESC** does not specify a clearance between electrical service equipment (or communications service equipment) and a gas meter. Common gas code references are NFPA 54/ANSI Z223.1, National Fuel Gas Code and 49 CFR Part 192, Transportation of Natural and Other Gas by Pipeline: Minimum Safety Standards. Coordination between the utility companies is needed.

Gas Meter

Steam, Gas, or Flammable Material Line

12"

Supply (Electrical) Cables at not more than 300 V between conductors.

• Per the **NESC**, radial separation of supply cables from steam, gas, or flammable material line not less than 12".

• Exception: For supply cables at not more than 300 V between conductors (e.g., a 120/240 V, 1Ø, 3W service), less than 12" may be used to a gas or flammable material line if:

- Supplemental mechanical protection is used to limit the likely hood of detrimental heat transfer to the gas or flammable material line due to a cable fault. This exception does not apply to steam lines. (No specifics are provided as to the type of supplemental mechanical protection, Rule 012C, which requires accepted good practice, must be used.)

See Photo(s)

- The Utilities involved agree to the reduced separation.

Fig. 354-1. Supply line and steam, gas, or flammable material line in random separation (Rule 354A2).

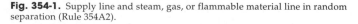

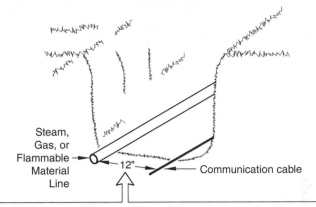

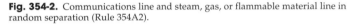

Steam, Gas, or Flammable Material Line

12"

Communication cable

- Radial separation of communications cables from steam, gas, or flammable material line not less than 12".

- Exception: For communication cables, less than 12" may be used if the utilities involved agree to the reduced separation. This exception applies to steam, gas, or flammable material lines.

Fig. 354-2. Communications line and steam, gas, or flammable material line in random separation (Rule 354A2).

effectively grounded shield or sheath must be buried in random separation (in close proximity with no deliberate separation), see Rule 350D.

Per Rule 354B, supply cables of multiple supply circuits may be buried in random separation (no deliberate separation) if all parties involved are in agreement, see Fig. 354-3.

Per Rule 354C, communication cables of multiple communication circuits may be buried in random separation (no deliberate separation) if all parties involved are in agreement. See Fig. 354-4.

Rule 354D focuses on supply and communication cables or conductors that are buried less than 12 in apart. If supply and communication cables are direct-buried less than 12 in apart, a variety of rules must be met. The rules include special consideration of voltage limitations, grounding and bonding requirements, protection requirements, type of cable jacket, and concentric-neutral material and size. The steps needed to install direct-buried supply and communication conductors in random separation (less than 12 in apart) are outlined in Fig. 354-5.

When following the steps in Rule 354D, one common construction method is to meet Rule 354D1 and Rule 354D4. When applying these Rules, a 12.47/7.2 kV effectively grounded supply cable in nonmetallic conduit (duct) can be random-laid (less than 12 in separation) with communication cables. Rule 354D4 requires that the supply cables meet Rule 354D3. Rule 353D3 does not address a typical 120/240 V underground secondary service to a residence. Rule 012C, which requires accepted good practice, must be used for 120/240 V underground secondary service installations.

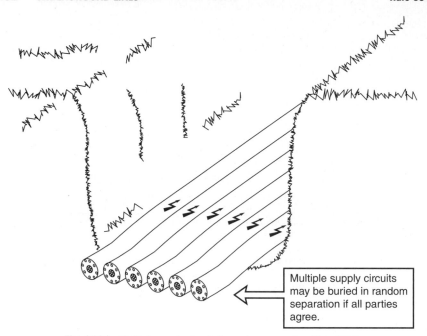

Fig. 354-3. Multiple supply circuits in random separation (Rule 354B).

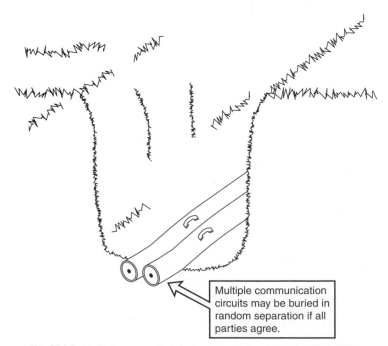

Fig. 354-4. Multiple communication circuits in random separation (Rule 354C).

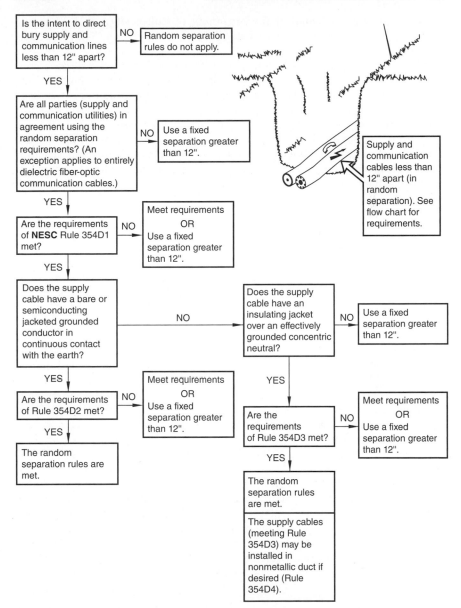

Fig. 354-5. Direct-buried supply and communication random separation requirements (Rule 354D).

An exception is provided at the beginning of Rule 354D for all-dielectric fiber-optic cables. The exception to Rule 354D must meet Rules 354D1a through 354D1d.

Rule 354E provides rules for less than 12 in of separation between a direct-buried supply or communication cable and a nonmetallic water or sewer line. The rule requires that all parties involved are in agreement. Metallic water or sewer lines must have 12 in or more separation (per Rule 353) as they are not addressed in Rule 354. Building foundations are also not specifically addressed in Rule 354 and therefore must meet the equal to or greater than 12 in requirement in Rule 353. If a conduit (duct) penetrates a building foundation to serve the building or if a conduit (duct) rises up a building foundation to serve a meter on the building, it seems reasonable that the 12-in requirement does not apply.

355. ADDITIONAL RULES FOR DUCT NOT PART OF A CONDUIT SYSTEM

The requirements in this rule for duct not part of a conduit system are similar to some (not all) of the requirements for ducts and joints in Rule 322. See Rules 322A and 322B for a discussion and related figures.

Section 35 does not address conduit (duct) bending radius, conduit (duct) fill, or pulling calculations. Rule 012C, which requires accepted good practice, must be applied. See Rule 341A for a discussion and figures related to cable bending, conduit fill, and pulling tensions. See Rule 320A for a discussion and figures related to conduit bending.

Ducts not part of a conduit system may be connected to handholes. The strength and cover requirements for handholes are not addressed in Sec. 35. Rules 323A and 323D can be used as accepted good practice for handholes connected to duct that is not part of a conduit system.

Rule 355D is similar to Rule 322B4. It is more common to see a conduit system enter a building vault than it is to see a duct not part of a conduit system enter a building. For example, for a typical underground residential or commercial electrical service, the utility company duct (covered by the **NESC**) stops at the meter or meter/main enclosure on the exterior of the building and a building wiring conduit (covered by the NEC) enters the building. In this example, the National Electrical Code (NEC) applies to the conduit entering the building, not the **NESC**, and the NEC rules need to be referenced for sealing the conduit entering the building.

Section 36

Risers

360. GENERAL

The rules for risers overlap in Part 3, Underground Lines, and Part 2, Overhead Lines. Rule 239D is referenced in Rule 360A as Rule 239D focuses on mechanical protection of a riser on an overhead pole. See Rule 239 for additional information. Rule 360 extends the protection required in Rule 239D at least 1 ft below grade. Cable bending must be considered when transitioning from a buried horizontal position to a vertical riser position. See Rule 341 for a discussion of cable bending. Risers consisting of metallic duct or under metallic guards (i.e., U-Guard) need to have the metallic duct or metallic guard effectively grounded in accordance with Rule 314B. Rule 360C does not require risers containing supply conductors to be metal, but if they are metal, they must be effectively grounded. The general requirements for risers are outlined in Fig. 360-1.

361. INSTALLATION

Rule 361 provides general installation rules for risers. Rule 362 provides additional rules for pole risers, and Rule 363 provides additional rules for risers entering pad-mounted equipment. The general rules for the installation of risers are outlined in Fig. 361-1.

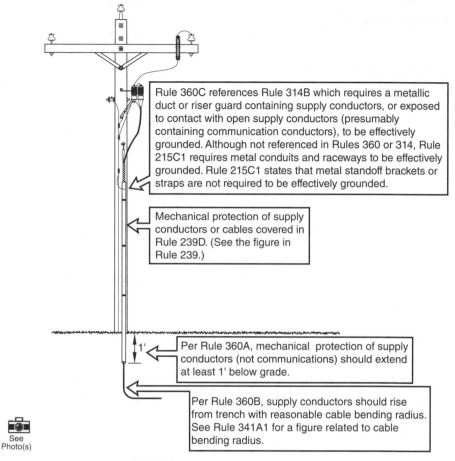

Rule 360C references Rule 314B which requires a metallic duct or riser guard containing supply conductors, or exposed to contact with open supply conductors (presumably containing communication conductors), to be effectively grounded. Although not referenced in Rules 360 or 314, Rule 215C1 requires metal conduits and raceways to be effectively grounded. Rule 215C1 states that metal standoff brackets or straps are not required to be effectively grounded.

Mechanical protection of supply conductors or cables covered in Rule 239D. (See the figure in Rule 239.)

Per Rule 360A, mechanical protection of supply conductors (not communications) should extend at least 1' below grade.

Per Rule 360B, supply conductors should rise from trench with reasonable cable bending radius. See Rule 341A1 for a figure related to cable bending radius.

See
Photo(s)

Fig. 360-1. General requirements for risers (Rule 360).

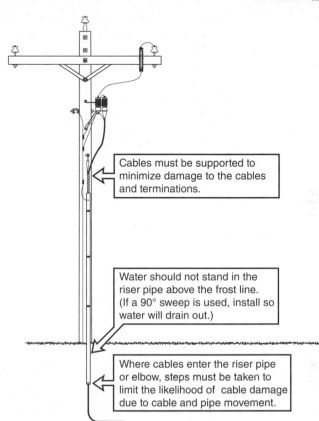

Cables must be supported to minimize damage to the cables and terminations.

Water should not stand in the riser pipe above the frost line. (If a 90° sweep is used, install so water will drain out.)

Where cables enter the riser pipe or elbow, steps must be taken to limit the likelihood of cable damage due to cable and pipe movement.

Fig. 361-1. Riser general installation requirements (Rule 361).

362. POLE RISERS—ADDITIONAL REQUIREMENTS

Rules 360 and 361 apply to risers in general. Rule 362 applies specifically to pole risers. The rules for pole risers are outlined in Fig. 362-1.

Standoff brackets are used by some utilities to organize pole risers and to make climbing easier. If standoff brackets are used on risers, Rule 217A2c must be met. See the figure showing the arrangement of standoff brackets in Rule 217 and a discussion of riser conduits and climbing space in Rule 236. Examples of using standoff brackets to organize conduit risers are shown in Fig. 362-2.

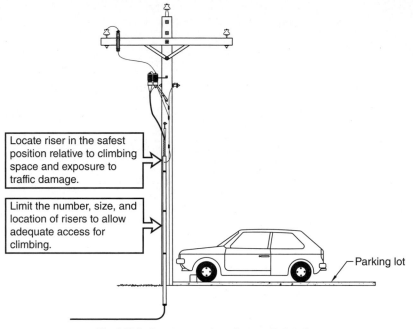

Locate riser in the safest position relative to climbing space and exposure to traffic damage.

Limit the number, size, and location of risers to allow adequate access for climbing.

Parking lot

Fig. 362-1. Requirements for pole risers (Rule 362).

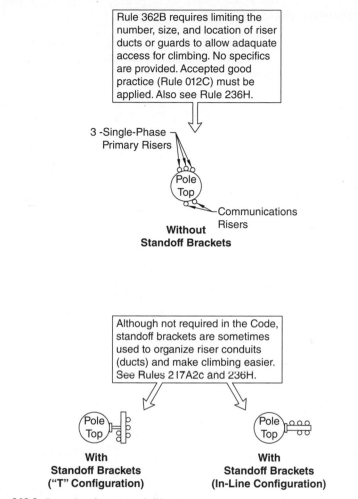

Rule 362B requires limiting the number, size, and location of riser ducts or guards to allow adaquate access for climbing. No specifics are provided. Accepted good practice (Rule 012C) must be applied. Also see Rule 236H.

3 -Single-Phase Primary Risers

Pole Top

Communications Risers

Without Standoff Brackets

Although not required in the Code, standoff brackets are sometimes used to organize riser conduits (ducts) and make climbing easier. See Rules 2I7A2c and 236H.

Pole Top

Pole Top

With Standoff Brackets ("T" Configuration)

With Standoff Brackets (In-Line Configuration)

Fig. 362-2. Examples of using standoff brackets to organize conduit (duct) risers (Rule 362B).

363. PAD-MOUNTED INSTALLATIONS

Rules 360 and 361 apply to risers in general. Rule 363 applies specifically to underground cables rising from a horizontal position up to pad-mounted equipment. The rules for pad-mounted installations are outlined in Fig. 363-1.

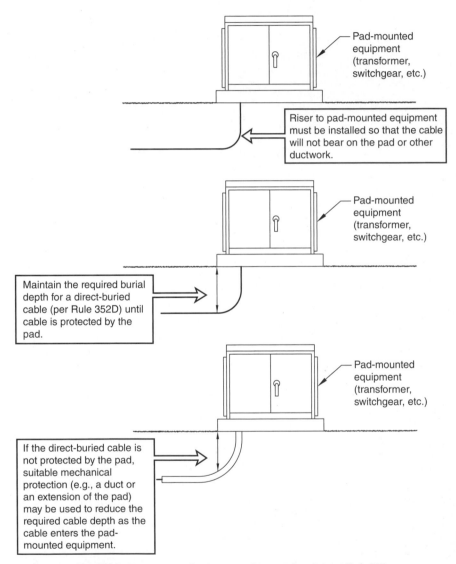

Pad-mounted equipment (transformer, switchgear, etc.)

Riser to pad-mounted equipment must be installed so that the cable will not bear on the pad or other ductwork.

Pad-mounted equipment (transformer, switchgear, etc.)

Maintain the required burial depth for a direct-buried cable (per Rule 352D) until cable is protected by the pad.

Pad-mounted equipment (transformer, switchgear, etc.)

If the direct-buried cable is not protected by the pad, suitable mechanical protection (e.g., a duct or an extension of the pad) may be used to reduce the required cable depth as the cable enters the pad-mounted equipment.

Fig. 363-1. Requirements for risers to pad-mounted equipment (Rule 363).

Section 37

Supply Cable Terminations

370. GENERAL

Supply cable terminations are the various items used to terminate underground cables such as elbows, cable terminators, potheads, etc. A reference to Rule 333 is made to tie Sec. 37 to Rule 333. Rule 333 focuses on the material specifications of the termination. Section 37 focuses on the installation of the termination. Cable terminators protect conductors and conductor insulation from mechanical damage, moisture, and electrical stress. Examples of supply cable terminations are shown in Fig. 370-1.

If a cable terminator is located inside a vault or pad-mounted equipment, the rules of Part 3, Underground Lines, apply. If the cable terminator is on a riser that terminates overhead inside a substation, the clearance rules found in Part 1, Electric Supply Stations, apply. If the cable terminator is on a pole outside of a substation, the clearance rules in Part 2, Overhead Lines, apply.

Equipment manufacturers determine clearance between cable terminators inside pad-mounted equipment by referencing ANSI and NEMA standards. The NESC provides a general statement (no specific distances) that suitable clearance must exist for the voltage and basic impulse level (BIL). Fully insulated terminations (i.e., elbows) or insulated barriers may be used to meet the necessary requirements.

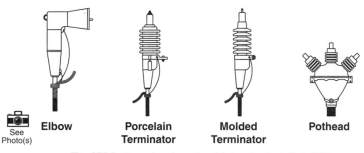

See
Photo(s)

Elbow   Porcelain   Molded   Pothead
               Terminator   Terminator

Fig. 370-1. Examples of supply cable terminations (Rule 370).

371. SUPPORT AT TERMINATIONS

Cable terminators must be installed to maintain their position. Underground residential distribution (URD) elbows must not pull out of the bushing wells they are installed in and riser pole cable terminators must not be moving around on the pole. The cable attached to the URD elbow or riser pole terminator may need to be supported to minimize stress on the cable terminator. Requirements for supporting cables are also stated in Rules 341A5 and 361B.

Supply utilities commonly use both porcelain cable terminators that are supported on a bracket or crossarm and lightweight polymer terminators that are not supported independent of the cable. The **Code** does not specify exactly how to support the cable termination, it only states that the termination must maintain its installed position.

372. IDENTIFICATION

Identification of a cable at a termination point is a simple method to maintain an organized electrical system, and it is a **Code** requirement for supply cables. Rule 372 does not apply to communication cables, as it is in Sec. 37, which is titled "Supply Cable Terminations." An exception is provided that eliminates the need for circuit identification if the position of the termination combined with maps or diagrams provides sufficient identification.

Additional identification rules for underground lines can be found in Rules 311 (underground locates), 323J (manhole and handhole covers), 341B3 (cables in manholes), 350F (marking of cables), and 385 (equipment operating in multiple). An example of supply cable identification at termination points is shown in Fig. 372-1.

The **NESC** does not contain a rule requiring the use of underground warning ribbons or tapes. The **NESC** also does not have a requirement for cable route markers. The **NESC** certainly does not prohibit the use of these items and these items are installed by some utilities. The lack of a **NESC** rule related to underground warning ribbons or tapes and cable route markers is shown in Fig. 372-2.

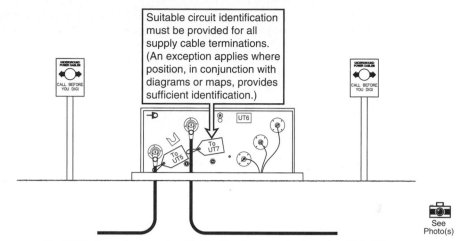

Fig. 372-1. Example of supply cable identification at termination points (Rule 372).

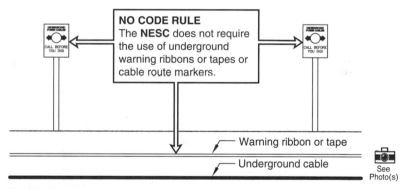

Fig. 372-2. Absence of Code rule related to underground warning ribbons or tapes and cable route markers (Rule N/A).

373. CLEARANCES IN ENCLOSURES OR VAULTS

Clearance between terminators depends on numerous factors including the type of terminator, the insulation method, and the voltage level.

This rule provides only a general statement that adequate clearance is required between supply terminations and between terminations and ground. No specific distances are provided. If terminators are in an enclosure (like a pad-mounted switch, junction box, transformer enclosure, etc.), the manufacturer will reference industry standards for the type of equipment. Adequate clearance or insulated barriers are needed for live parts in a manufactured enclosure. The burden is on the manufacturer to determine the specific clearances. The same is true for live parts in a vault, only this time the burden is on the utility designer. Guarding or isolating live parts in a vault is required. Using dead-front fittings and terminations or metal-clad switchgear are methods of guarding or isolating live parts. Adequate physical clearance above the floor is also a method of guarding or isolating live parts. Since no specific dimension is provided, Rule 012C, which requires accepted good practice, applies. The electric supply station rules in Part 1 of the NESC, specifically Rule 124, would be a good starting point for the accepted good practice of guarding or isolating live parts in a vault.

374. GROUNDING

Rule 374 has grounding requirements specifically for supply cable terminations in addition to the general grounding requirements in Rule 314. The rules for grounding supply cable terminations are outlined in Fig. 374-1.

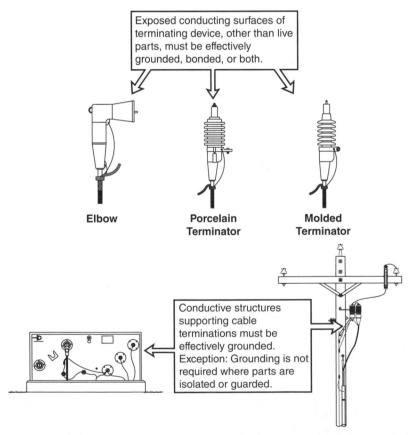

Fig. 374-1. Grounding supply cable terminations (Rule 374).

Section 38

Equipment

380. GENERAL

Rule 380 starts out by providing examples of supply and communication equipment relevant to this section. See Fig. 380-1.

Equipment located in a joint-use manhole can only be installed with the concurrence of the parties concerned. Similar wording appears in Rule 341B2b(1) for cables in joint-use (supply and communication) manholes. The term joint-use in this rule refers to simultaneous use by two or more utilities. Rule 341B2b(1) is for joint-use (supply and communication) installations. If the parties do not concur, separate conduit systems and manholes can be used or a joint conduit system can be used where the conduits leave the joint conduit system and enter separate manholes or pedestals at every termination point or pulling location.

The pads, supports, and foundations used to support equipment must be rated for the load and stress of the equipment and the equipment's operation. For example, the forces associated with installing and removing underground residential distribution (URD) elbows must be considered.

Rule 380D is similar to Rule 231A. Rule 380D specifies a distance from pad-mounted equipment to a fire hydrant. Rule 231A specifies a distance from a supporting structure (e.g., pole) to a fire hydrant. The main purpose of this rule is to allow adequate working space for the fire department to connect wrenches, hoses, and equipment to the fire hydrant. See Fig. 380-2.

The **Code** does not address clearance from pad-mounted equipment to a roadway or clearance of an oil-filled pad-mounted transformer to a building. Since the **Code** does not provide dimensions for these installation conditions, accepted good practice (Rule 012C) must be used. The absence of a **Code** rule related to pad-mounted equipment location adjacent to roads and buildings is outlined in Fig. 380-3.

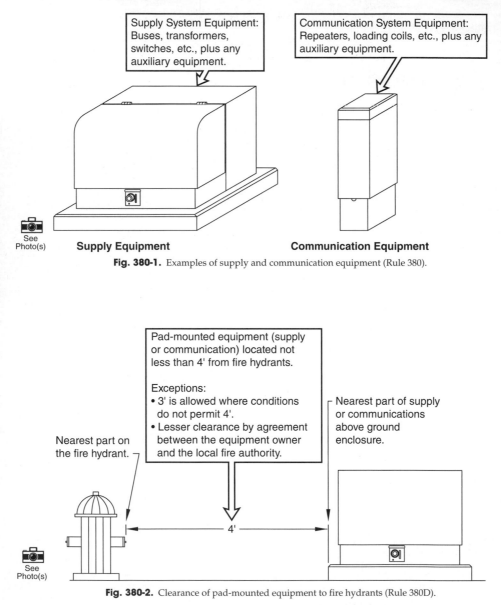

Supply System Equipment:
Buses, transformers,
switches, etc., plus any
auxiliary equipment.

Communication System Equipment:
Repeaters, loading coils, etc., plus any
auxiliary equipment.

See
Photo(s)

Supply Equipment   **Communication Equipment**

Fig. 380-1. Examples of supply and communication equipment (Rule 380).

Pad-mounted equipment (supply
or communication) located not
less than 4' from fire hydrants.

Exceptions:
• 3' is allowed where conditions
  do not permit 4'.
• Lesser clearance by agreement
  between the equipment owner
  and the local fire authority.

Nearest part of supply
or communications
above ground
enclosure.

Nearest part on
the fire hydrant.

See
Photo(s)

4'

Fig. 380-2. Clearance of pad-mounted equipment to fire hydrants (Rule 380D).

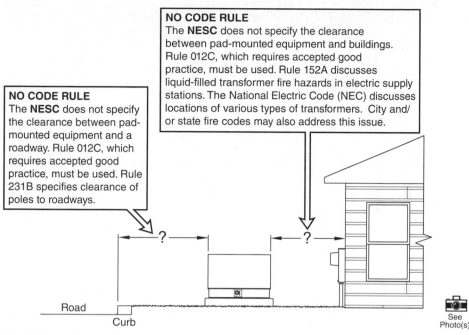

NO CODE RULE
The **NESC** does not specify the clearance
between pad-mounted equipment and buildings.
Rule 012C, which requires accepted good
practice, must be used. Rule 152A discusses
liquid-filled transformer fire hazards in electric supply
stations. The National Electric Code (NEC) discusses
locations of various types of transformers. City and/
or state fire codes may also address this issue.

NO CODE RULE
The **NESC** does not specify
the clearance between pad-
mounted equipment and a
roadway. Rule 012C, which
requires accepted good
practice, must be used. Rule
231B specifies clearance of
poles to roadways.

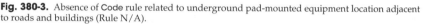

Fig. 380-3. Absence of Code rule related to underground pad-mounted equipment location adjacent to roads and buildings (Rule N/A).

381. DESIGN

Equipment design and mounting must consider the following:
- Thermal conditions
- Chemical conditions
- Mechanical conditions
- Environmental conditions

Equipment and auxiliary devices must also be rated for the expected:
- Normal conditions
- Emergency conditions
- Fault conditions

The requirement for switching underground equipment is similar in nature to switching requirements for overhead switches discussed in Rule 216. Switches for underground lines must provide a clear indication of the switch contact position, and the switch handle must be marked with operating directions. The rules for underground switches are outlined in Fig. 381-1.

Remotely controlled or automatic switching devices such as pad-mounted vacuum fault interrupt switches must have provisions on the equipment to render remote or automatic controls inoperable. This feature protects workers from accidental operation or energization during maintenance.

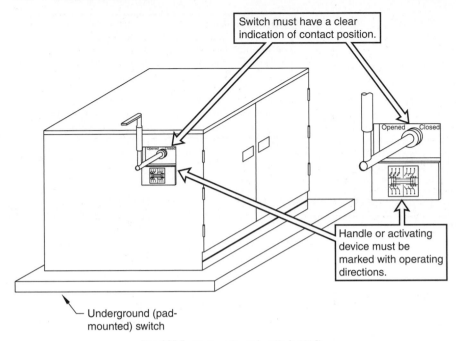

Switch must have a clear indication of contact position.

Handle or activating device must be marked with operating directions.

Underground (pad-mounted) switch

Fig. 381-1. Design of switches (Rule 381C).

When applying equipment containing fuses and interrupting contacts, the following must be considered during operation:

- Normal conditions
- Emergency conditions
- Fault conditions

When tools such as insulated shotgun sticks are used to handle energized devices in underground equipment, physical space or barriers must provide adequate clearance from ground or between phases.

Rule 381G provides the locking and access requirements for pad-mounted and other aboveground equipment. Pad-mounted and other aboveground equipment is typically not fenced-in like substation equipment, and it is not elevated like overhead equipment. Pad-mounted equipment has more exposure to the public than most other supply and communication facilities. Rule 381G1 applies to both supply and communication pad-mounted equipment. Pad-mounted and other aboveground equipment must have an enclosure that is locked or otherwise secured. The purpose of this is to keep out unauthorized persons (i.e., the public). Typical locking or securing methods used by both supply and communication utilities include keyed padlocks, disposable locks that can be cut with bolt cutters, penta head bolts, tamperproof bolts, etc. Rule 381G2 applies to supply pad-mounted equipment over 600 V. This requirement applies to almost all supply utility transformers, switches, junction boxes, etc., except for low-voltage secondary enclosures. Rule 381G2 requires two separate conscious acts to access exposed live parts. The first act is opening the pad-mounted enclosure that was locked or secured in Rule 381G1. It does not matter how many locks or penta head bolts are handled to open the enclosure; opening the enclosure is the first conscious act. The second conscious act must be the opening of a door (not the enclosure door in the first conscious act) or the removal of a barrier.

A pad-mounted switch with fuses is a good example of a device that requires two conscious acts to access exposed live parts. After opening the enclosure door (the first conscious act), a separate steel door or a separate fiberglass barrier exists that must be opened or removed (the second conscious act) before the live parts are exposed. A slightly different example is a typical pad-mounted service transformer that uses load break elbows for terminating the high-voltage cables. This type of transformer does not have any exposed live parts over 600 V and is usually referred to as "dead front." Opening the enclosure door is the first conscious act. No exposed live parts in excess of 600 V exist when the transformer enclosure is opened. The transformer secondary lugs (i.e., 120/240 V) may be exposed, but they are under 600 V. The pulling of the insulated elbow is the second conscious act. The barrier for the second conscious act is the insulated elbow itself. See Fig. 381-2.

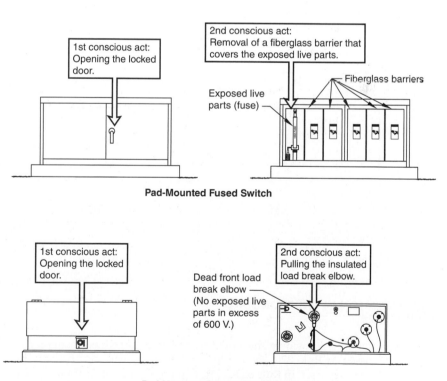

Pad-Mounted Service Transformer

Fig. 381-2. Examples of access to exposed live parts in excess of 600 V (Rule 381G).

Rule 381G2 states that a safety sign should be visible when the first door or barrier is opened or removed. The ANSI Z535 series of signing standards are referenced in a note. Signage on pad-mounted equipment is just as critical as signage for an electric supply station. Some utilities use a warning sign on the outside of the pad-mounted enclosure and a danger sign on the inside of the pad-mounted enclosure. This approach uses the ANSI Z535 philosophy that warning is appropriate on the outer barrier (enclosure) and if that barrier is breached, a danger sign is then appropriate. Utilities should consult the ANSI documents, federal or state regulatory agencies, and the utility's insurance company for signing applications. It is important to note that **NESC** Rule 381G2 does not require a safety sign on the outside of pad-mounted equipment. Safety signs are only required for the inside of the equipment. See the discussion and the figure in Rule 110A for additional information on ANSI signing requirements. The safety sign requirements for pad-mounted equipment are outlined in Fig. 381-3.

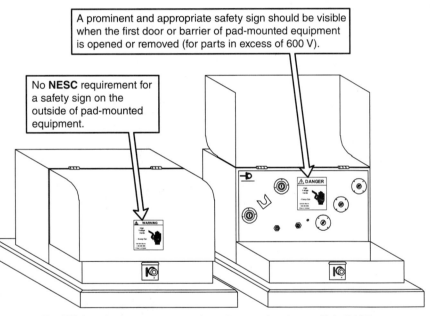

See
Photo(s)

Fig. 381-3. Safety sign requirements for pad-mounted equipment (Rule 381G2).

382. LOCATION IN UNDERGROUND STRUCTURES

When equipment is located in the underground structures (i.e., manholes and vaults) discussed in Rule 320, the equipment must not obstruct the personnel access openings discussed in Rule 323C. The equipment must also not impede the egress of a person working in the manhole or vault per Rule 382A.

The only dimension provided in Rule 382 is the 8-in clearance from equipment to the back of a fixed ladder. Fixed ladders are not required except for the special conditions specified in Rules 323C1 and 323C4. Ladder requirements are also discussed in Rules 323C5 and 323F. No dimension is given from equipment to the front of the ladder. The clearance to the front of the ladder must meet the general requirement that the equipment must not interfere with the proper use of the ladder. The remaining clearance requirements in this rule are also general wording requirement without any stated dimension. Equipment arrangement must consider installation, operation, and maintenance requirements. Switching equipment must be operable from a safe position. Equipment must not interfere with the drainage or ventilation of the underground structure.

Although Rule 382 does not reference Rule 341B2 and **NESC** Table 341-1, the clearances in this table apply between joint-use (supply and communication) equipment.

383. INSTALLATION

The installation requirements for underground equipment (both pad-mounted equipment and equipment installed in manholes and vaults) are very general. No dimension is provided for any of the requirements. Equipment installation must consider the following:

- Equipment weight (lifting, rolling, and mounting considerations)
- Guarding or isolating live parts from persons
- Easy access to operate, inspect, and test facilities
- Isolating or protecting live parts from conductive liquids or other materials
- Locking or securing operating controls of supply equipment

384. GROUNDING AND BONDING

Rules 384A and 384B have grounding requirements similar to the general grounding requirements in Rule 314.

Rule 384C requires bonding between aboveground metallic supply and communications enclosures that are 6 ft or less apart. The intent of this rule is to prevent different potentials between adjacent metallic enclosures. Bonding between fiberglass enclosures or a metallic enclosure and a fiberglass enclosure is not required. A size is not specified for the bonding jumper. Rule 012C, which requires accepted good practice, must be used. Rule 099C, which requires an AWG No. 6 copper bond, is one option for the accepted good practice in this case. Rules 342 and 093C7 provide additional information on bonding conductors. The 6-ft spacing is an average reach for an adult person. Rule 384C does not address (or require) bonding of dozens of other metallic items (e.g., metal conduit risers, steel poles, fire hydrants, sign posts, fences, etc.) that can be adjacent to aboveground metallic supply and communications enclosures. Rule 384C clarifies that a supply pole ground is not required to be bonded to an adjacent metallic communications enclosure. A note to the rule clarifies that bonding a supply pole ground to an adjacent metallic communications pedestal is not required, but it is not prohibited if the parties agree. The rules for bonding aboveground metallic supply and communications enclosures are outlined in Fig. 384-1.

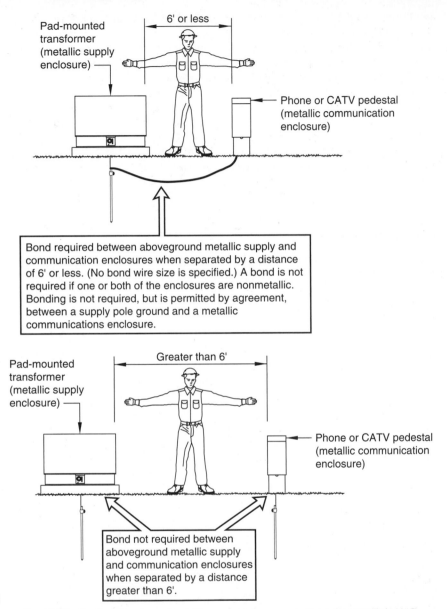

Fig. 384-1. Bonding aboveground metallic supply and communication enclosures (Rule 384C).

385. IDENTIFICATION

Identification of equipment that operates in multiple eliminates confusion and adds safety.

Section 39

Installation in Tunnels

390. GENERAL

Installation in tunnels must meet applicable rules found in Part 3 and the additional rules of this section. Rules 320A4 and 323H provide additional requirements related to tunnels. If unqualified persons (i.e., the general public) have access to the tunnel, the applicable requirements of Part 2 must be met. The **Code** is referring to a tunnel that could be a roadway traffic tunnel with a utility line passing through it, or the tunnel could be a dedicated utility tunnel under the surface of the ground. Utility tunnels can be found connecting buildings in urban areas or on university campuses, for example. In some cases, the utility tunnel doubles as a public walkway. In other cases, the tunnels are used just for utilities accessible to qualified personnel. If the access door to the utility tunnel is accessible to the public, the door must be locked or attended. A safety sign is required when utility tunnels contain exposed live parts. In addition to the **NESC** rules, the National Electrical Code (NEC) can be referenced for accepted good practice. A tunnel used as a public walkway between buildings will require adherence to the NEC. No matter what type of tunnel is involved, all parties concerned must agree on the design of the tunnel structure and the design of the utilities within the structure. An example of a utility tunnel between buildings is shown in Fig. 390-1.

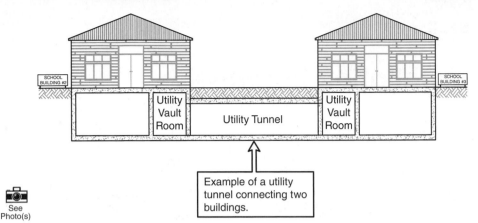

See Photo(s)

Fig. 390-1. Example of a utility tunnel between buildings (Rule 390).

391. ENVIRONMENT

If the tunnel is accessible to the public or workers, the environment within the tunnel must be suitable for people. If the tunnel is not accessible to the public or workers, the construction would be similar to duct bank construction where cables are pulled in and out of ducts but access is only obtainable at pulling or splice locations. Rule 391A provides requirements for general environmental safety, egress, and working space. Rule 391B provides additional rules for joint-use (supply and communication) tunnels.

Part 4

Work Rules for the Operation of Electric Supply and Communications Lines and Equipment

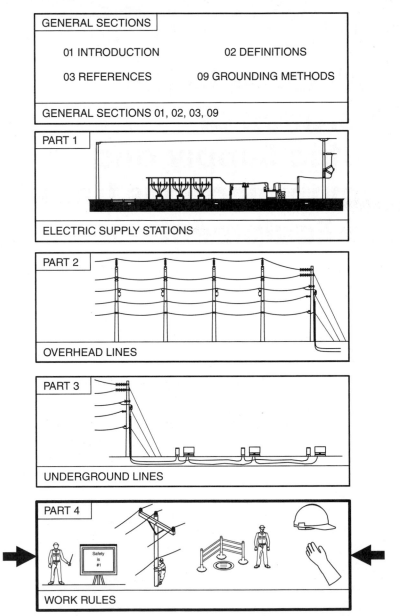

GENERAL SECTIONS

01 INTRODUCTION   02 DEFINITIONS

03 REFERENCES   09 GROUNDING METHODS

GENERAL SECTIONS 01, 02, 03, 09

PART 1

ELECTRIC SUPPLY STATIONS

PART 2

OVERHEAD LINES

PART 3

UNDERGROUND LINES

PART 4

WORK RULES

Section 40

Purpose and Scope

400. PURPOSE

The purpose of Part 4, Work Rules, is similar to the purpose of the entire NESC outlined in Rule 010, except that Rule 400 is specific to the work rules for the operation of electric supply and communication lines and equipment. In other words, Part 4 focuses on the actions/behaviors of the workers, while Parts 1-3 are focused on the physical elements of the utility systems. Part 4 of the NESC focuses on practical work rules as a means of safeguarding employees and the public. The Code states that the intent of Part 4 is not to require unreasonable steps to comply with the rules; however, reasonable steps must be taken. A more specific statement is given in Rule 410A4. Rule 410A4 requires that the work rules be used, but if strict enforcement of the work rules seriously impedes safety, the employee in charge may temporarily modify the work rules without increasing hazards. This flexibility is needed because every conceivable situation cannot be covered in the work rules. Flexibility in applying the work rules must be abused in situations where the work rules apply.

401. SCOPE

The scope of Part 4, Work Rules, includes work rules to be used in the installation, operation, and maintenance of both electric supply and communications systems. Part 4 is broken down into four sections that are all interrelated. The four sections are outlined below:

- Section 41, "Supply and Communications Systems—Rules for Employ*ers.*" These rules apply to the supply and communications company.
- Section 42, "General Rules for Employ*ees.*" These rules apply to the supply and communications employee working for the supply and communications company.
- Section 43, "Additional Rules for *Communications* Employ*ees.*" These rules are additional rules for communications employees only. Note that Sec. 42 was for both supply and communications employees.
- Section 44, "Additional Rules for *Supply* Employ*ees.*" These rules are additional rules for the supply employees only. Note that Sec. 42 was for both supply and communications employees.

The titles of **NESC** Secs. 41, 42, 43, and 44 are graphically represented in Fig. 401-1.

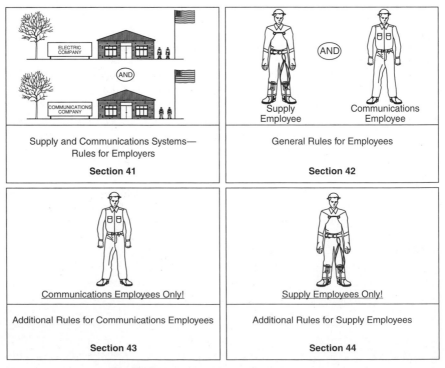

Fig. 401-1. Titles of Secs. 41, 42, 43, and 44 (Rule 401).

The format of this Handbook for Part 4, Work Rules, is different from the format of the other parts of this book. The format for Part 4 summarizes the **Code** text in an outline bulleted list. A graphic is used to visually represent which section, rule, and paragraph the outline applies to. A brief general discussion of each section is provided at the beginning of each section.

Occupational Safety and Health Administration (OSHA) regulations also apply to the operation, maintenance, and construction of electric supply and communications systems. See Rule 402 for a discussion of applicable OSHA standards.

402. REFERENCED SECTIONS

This rule references four **NESC** sections related to Part 4, Work Rules, so that rules do not have to be duplicated and the reader of the **Code** realizes that other sections are related to the information provided in Part 4. The related sections are:
- Introduction, Sec. 01
- Definitions, Sec. 02
- References, Sec. 03
- Grounding Methods, Sec. 09

The Work Rules in Part 4 of the **NESC** are similar to (but they are not exactly the same as) the Occupational Safety and Health Administration (OSHA) regulations related to electric supply and communications systems.

The scope of this **NESC** Handbook does not include specific comments on the OSHA regulations. Many utilities tend to place more emphasis on the OSHA regulations than the work rules in Part 4 of the **NESC**, as OSHA performs both routine inspections and accident investigations. In general, if a state adopts the **NESC** in its entirety, utilities are required to meet the **NESC** rules and the OSHA standards; therefore, both apply, not one or the other, and where there may be inconsistency, the most stringent or safest approach should be taken. The Work Rules in **NESC** Part 4 are tied to the rules in other parts of the **NESC**.

The OSHA standards related most specifically to the **NESC** Work Rules are listed below:
- 1910.268 Telecommunications
- 1910.269 (including Appendix A–G) Electric Power Generation, Transmission, and Distribution (Operation and Maintenance)
- 1926.950 through 1926.968 (including Appendix A–G) 1926 Subpart V, Power Transmission and Distribution (Construction)

The OSHA 1910 series is for general industry applications, such as operation and maintenance. OSHA Standards 1910.268 and 1910.269 are listed under 1910 Subpart R—Special Industries. The OSHA 1926 series is for construction. OSHA Standards 1926.950 through 1926.968 are listed under 1926 Subpart V—Power Transmission and Distribution. OSHA uses the terms electric power and telecommunications instead of the **NESC** terms electric supply and communications. OSHA separates its standards into general industry standards and construction standards, the **NESC** does not. OSHA has both general industry standards and construction standards specific to the electric power industry. OSHA only has a general industry standard that is specific to the telecommunications industry. When OSHA does not have a construction standard for a specific industry, as in the case of telecommunications, the general OSHA construction standards apply. Where specific rules do not exist in the general OSHA construction standards, applying the 1910.268 operation and maintenance standards to construction work, for example, is a prudent way of meeting OSHA's General Duty Clause (OSHA Section 5—Duties of the OSHA Act of 1970).

The OSHA 1910.268 Telecommunications standard became effective in 1975. The OSHA 1910.269 Electric Power Generation, Transmission, and Distribution standard became effective in 1994. The OSHA 1926.950 through 1926.968 (1926 Subpart V) Power Transmission and Distribution standards became effective in 1972. Amendments have been made to the standards over the years. The OSHA 1926.950 through 1926.968 (1926 Subpart V) Power Transmission and Distribution (Construction) standards and the OSHA 1910.269 Electric Power Generation, Transmission, and Distribution (General Industry) standard both received a major revision in 2014.

There are several other OSHA standards referenced throughout 1910.268, 1910.269, and 1926.950 through 1926.968. These related standards are referenced either without amendment or referenced and supplemented with additional information. Some of the most commonly referenced related standards are listed below:

- 1910 Subpart S—Electrical (1910.301 to 1910.399)
- 1910.151—Medical Services and First Aid (Part of 1910 Subpart K)
- 1910.147—The Control of Hazardous Energy (Lockout/Tagout) (Part of 1910 Subpart J)
- 1910 Subpart I—Personal Protective Equipment (1910.132 to 1910.138)
- 1910.137—Electrical Protective Devices (Part of 1910 Subpart I)
- 1926 Subpart M—Fall Protection (1926.500 to 1926.503)
- 1926 Subpart P—Excavations (1926.650 to 1926.652)
- 1910 Subpart D—Walking-Working Surfaces (1910.21 to 1910.30)
- 1910.67 Vehicle-Mounted Elevating and Rotating Work Platforms (Part of 1910 Subpart F)
- 1910 Subpart N—Materials Handling and Storage (1910.176 to 1910.184)
- 1926 Subpart W—Rollover Protective Structures; Overhead Protection (1926.1000 to 1926.1003 and Appendix A)

OSHA standards are very easy to access from the Internet. Each standard also has a list of requested interpretations. Each interpretation lists the question asked and the OSHA response. The phone numbers and addresses of OSHA offices are also easily accessed from the Internet. Individual states have the option to adopt the federal OSHA standards or develop their own OSHA-approved state plan. State-run OSHA plans must be at least as effective as the federal OSHA program. All of this information is available at www.osha.gov. An outline of the OSHA power and communication standards on www.osha.gov is shown in Fig. 402-1.

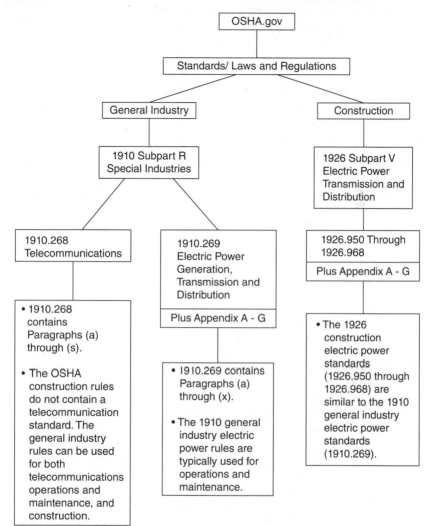

Fig. 402-1. Outline of OSHA 1910.268, OSHA 1910.269, and OSHA 1926 Subpart V (1926.950 through 1926.968) (Rule 402).

Section 41

Supply and Communications Systems—Rules for Employers

The format of this Handbook for Part 4, Work Rules, is different from the format of the other parts of this book. The format for Part 4 summarizes the **Code** text in an outline bulleted list. A graphic is used to visually represent which section, rule, and paragraph the outline applies to. A brief general discussion of each section is provided at the beginning of each section. The following brief general discussion covers selected topics in Secs. 41 and 42. This discussion is repeated at the beginning of Sec. 42.

Both employers and employees have an obligation to safety. Section 41 provides rules for supply and communications employers. Section 42 provides general rules for supply and communications employees. Section 43 provides additional rules for communications employees. Section 44 provides additional rules for supply employees.

Sections 41 and 42 are written to put the responsibility for safety on both the employer (company) and the employee (worker). The employer must designate an employee in charge to represent the company. The employee in charge is responsible for making sure employees adhere to the work rules. The employees must also assume responsibility for following safety rules. This system builds redundancy by putting the safety requirement on both the employer and employee.

Two specific examples of the relationship between Secs. 41 and 42 (employer and employee) are listed below:

- The *employer* must inform each employee of the safety rules (Rule 410A). The *employee* must read and study the safety rules (Rule 420A).
- The *employer* must have an adequate supply of protective devices and equipment (e.g., hard hats, rubber gloves, insulated tools, body belts, etc.) sufficient to enable employees to meet the requirements of the work to be undertaken (Rule 411B).

513

- The *employee* must use personal protective equipment, the protective devices, and the special tools provided for their work and inspect these devices and tools before starting work to verify that they are in good condition (Rule 420H).

When the employee is required to perform a task, for example, inspecting personal protective equipment, the employer must have a designated employee in charge responsible for making sure the employees perform the inspection. The duties of the designated employee in charge are outlined in Rule 421A.

Section 41 (rules for employers) provides a list of typical protective devices and equipment for employees (who are covered in Sec. 42) to use. The **Code** does not dictate what protective devices and equipment must be used for a particular task. The choice of what protective devices and equipment need to be used is site- and task-specific. The **NESC** cannot cover every conceivable specific situation, which is why an assessment is required (see Rule 410A3).

Section 41 (rules for employers), Rule 411A2, requires that diagrams (i.e., maps) be maintained and on file for employees (who are covered in Sec. 42) to use. The diagrams aid the identification of structures required in Part 2, Overhead Lines, Rule 217A3, and the location of underground facilities required in Part 3, Underground Lines, Rule 311A. Accurate diagrams are necessary for minimizing errors and accidents. Inaccurate maps can increase errors and accidents. Diagrams cannot be used as a substitute for applying proper safety procedures.

Section 41 (Rule 410A) requires the employer (the company) to perform an assessment to determine the potential exposure to an electric arc for employees who work on or near energized lines or equipment. The assessment considers the employees assigned tasks and work activities. For example, an engineer entering a substation to read information off of a transformer nameplate has a different potential exposure to an electric arc than a lineworker replacing a pole top transformer on an energized line. In addition to the assessment of the work activity, the employer must perform an arc hazard analysis to determine the effective arc rating of clothing to be worn by the employee. The arc thermal protection value (ATPV) can be calculated in an engineering study or determined using **NESC** Tables 410-1, 410-2, and 410-3. Each table is for a different voltage range, and the assumptions used to create the tables are documented in the footnotes at the bottom of the tables. Section 42 (Rule 420I) requires the employee to wear the clothing that the employer (the company) determined necessary.

Sometimes confusion exists between the application of the **NESC** Work Rules and NFPA 70E (Standard for Electrical Safety in the Workplace). A careful review of the scope of NFPA 70E indicates that NFPA 70E does not apply to supply (electric power) or communications utilities. In general, NFPA 70E applies to electrical work performed by electricians. However, even though the scope of NFPA 70E does not apply to utility companies, it can be a useful reference for accepted good practice. One practice that NFPA 70E addresses that is not commonly found in the electric utility industry is labeling equipment such as circuit breakers, fused switches, etc., with a sign or label indicating the arc hazard incident energy or required level of arc-rated clothing. Utility companies typically use training of their employees to address this issue, not signs or labels. For more information, see the NFPA 70E document in its entirety.

The Work Rules in Part 4 of the **NESC** are similar to (but they are not exactly the same as) the Occupational Safety and Health Administration (OSHA) regulations related to electric supply and communications systems. See Rule 402 for a discussion and a figure describing the OSHA standards that are related to the NESC Work Rules.

410. GENERAL REQUIREMENTS

410A. General

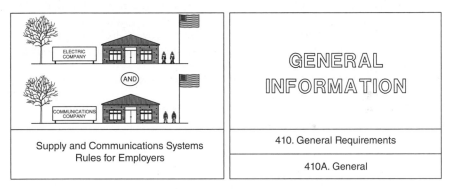

Supply and Communications Systems Rules for Employers

GENERAL INFORMATION

410. General Requirements

410A. General

- Employer must inform each employee of the safety rules.
- Employer must provide a copy of the safety rules (when necessary).
- Employer must train employees.
- Employer must ensure that each employee has demonstrated proficiency in required tasks.
- Employer must retrain employees who are not following work rules.
- Employer must ensure an assessment is performed to determine the potential exposure to electric arcs for employees who work on or near energized lines or equipment.
- If the assessment determines potential employee exposure, the following materials must not be worn (unless arc rated):
  ✓ Acetate
  ✓ Nylon
  ✓ Polyester
  ✓ Polypropylene
- If the assessment determines potential employee exposure, an outer layer of clothing that could ignite and burn must not be worn.
- Assessments that determine a potential employee exposure greater than 2 cal/cm² require a detailed arc hazard analysis or use of **NESC** Tables 410-1 (50-600 V), 410-2 (1.1 kV-46 kV open air), 410-3 (46.1 kV-800 kV open air) or 410-4 (1 kV-36 kV enclosed equipment) to determine the effective arc rating of clothing to be worn by employees. While no table is given for the range from 601 V to 1000 V, a note on Table 410-1 references Rule 012C which requires the use of accepted good practice.
- Assessments that determine a potential employee exposure greater than 2 cal/cm² require the arc hazard analysis calculation of the estimated arc energy to be based on the available fault current, the duration of the arc in cycles, and the distance of the arc to the employee.
- Assessments that determine a potential employee exposure greater than 2 cal/cm² require the employee to cover the entire body with arc rated clothing and equipment that has an effective arc rating at least equal to the anticipated level of arc energy.

- An exception applies if the clothing required creates a hazard greater than the possible exposure to the heat energy of the electric arc, in this case clothing with a lower arc rating than required may be worn.
- Special exceptions apply to the employee's hands.
- Special exceptions apply to the employee's feet.
- Special exceptions apply to the employee's head and face.
- Special exceptions apply to DC systems with voltages from 50 V to 250 V and 8000 A maximum fault current.
- Note that consideration can be given to the employee's assigned tasks and/or work activities when performing the assessment to determine potential exposure to electric arcs.
- Note that multiple layers of arc rated clothing have been shown by tests to block more heat than a single layer.
- Note that clothing includes shirts, pants, jackets, and coveralls in single or multiple layers.
- Note that engineering controls can be used to reduce arc energy levels and work practices can be used to reduce exposure levels.
- Employers must use procedures to assure compliance with safety rules.
- If strict enforcement of rules seriously impedes safety, the employee in charge may temporarily modify rules as long as hazards are not increased.
- If a disagreement on how to apply operating rules occurs, the employer's decision will be final; however, the decision must not be hazardous to the employee.
- Employer must provide training to employees who work around antennas to recognize and mitigate exposure to radio frequencies.
- Regulatory standards in OSHA, FCC, and IEEE can be referenced for radiation level limits.

410B. Emergency and First Aid Procedures

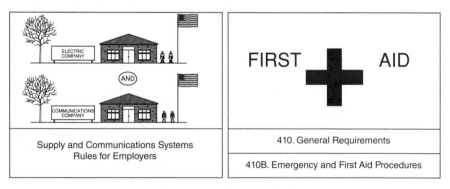

Supply and Communications Systems Rules for Employers	410. General Requirements
	410B. Emergency and First Aid Procedures

- Employer must inform employee of emergency procedures and first aid methods including resuscitation (CPR).
- Copies of emergency procedures and first aid methods must be kept where number of employees and type of work warrants their placement (e.g., in vehicles and other locations).

- Employers must regularly instruct employees on the methods of first aid and emergency procedures (if duties warrant such training).

410C. Responsibility

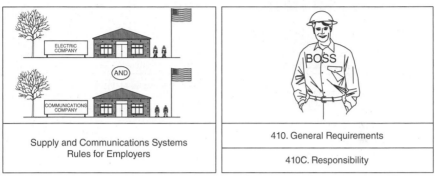

- Employer must select a designated person to be in charge of operations and responsible for safety.
- A crew must have only one person in charge.
- For multiple locations, one person may be in charge at each location.

411. PROTECTIVE METHODS AND DEVICES

411A. Methods

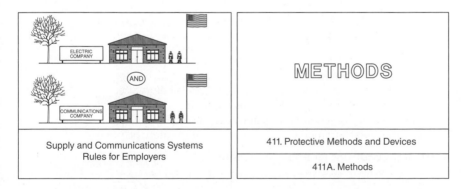

- Employer must restrict access to energized or rotating equipment except by authorized employees.
- Diagrams (i.e., maps) of the electric system must be available to authorized employees.
- Employees are to be instructed before work starts.
- Employees are to be instructed to take additional precautions for unusual hazards.

411B. Devices and Equipment

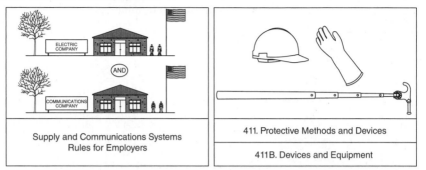

Supply and Communications Systems Rules for Employers

411. Protective Methods and Devices

411B. Devices and Equipment

- Employer must have an adequate supply of protective devices and equipment (e.g., hard hats, rubber gloves, insulated tools, body belts, etc. based on the requirements of the job).
- Employer must have an adequate supply of first aid equipment.
- Protective devices must conform to applicable standards.

411C. Inspection and Testing of Protective Devices and Equipment

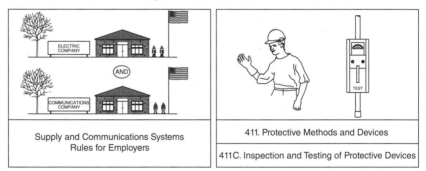

Supply and Communications Systems Rules for Employers

411. Protective Methods and Devices

411C. Inspection and Testing of Protective Devices

- Inspect or test protective devices and equipment.
- Inspect insulating gloves, sleeves, and blankets before use.
- Test insulated gloves and sleeves, as required.
- Inspect climbing and fall protection equipment (e.g., line worker's body belts, lanyards, fall restriction devices, etc.) to ensure safety before use.

411D. Signs and Tags for Employee Safety

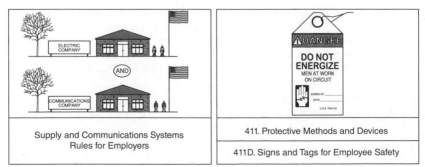

Supply and Communications Systems Rules for Employers

411. Protective Methods and Devices

411D. Signs and Tags for Employee Safety

- Safety signs and tags must meet ANSI Z535 Standards.

411E. Identification and Location

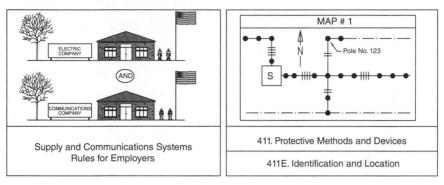

- Provide a means to identify lines before they are worked on.
- Be able to locate underground facilities.

411F. Fall Protection

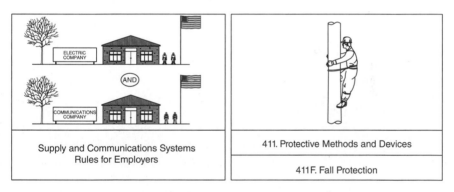

- Employer must develop, implement, and maintain an effective fall protection program.
- The fall protection program must include all of the following:
  - ✓ Training, retraining, and documentation
  - ✓ Guidance on equipment selection, inspection, care, and maintenance
  - ✓ Considerations concerning structural design and integrity, with particular reference to anchorages and their availability
  - ✓ Rescue plans and related training
  - ✓ Hazard recognition
- The employer must not permit the use of 100 percent leather positioning straps or nonlocking snaphooks.

Section 42

General Rules for Employees

The format of this Handbook for Part 4, Work Rules, is different from the format of the other parts of this book. The format for Part 4 summarizes the **Code** text in an outline bulleted list. A graphic is used to visually represent which section, rule, and paragraph the outline applies to. A brief general discussion of each section is provided at the beginning of each section. The following brief general discussion covers selected topics in Secs. 41 and 42. This discussion is repeated at the beginning of Sec. 41.

Both employers and employees have an obligation to safety. Section 41 provides rules for supply and communications employers. Section 42 provides general rules for supply and communications employees. Section 43 provides additional rules for communications employees. Section 44 provides additional rules for supply employees.

Sections 41 and 42 are written to put the responsibility for safety on both the employer (company) and the employee (worker). The employer must designate an employee in charge to represent the company. The employee in charge is responsible for making sure employees adhere to the work rules. The employees must also assume responsibility for following safety rules. This system builds redundancy by putting the safety requirement on both the employer and employee.

Two specific examples of the relationship between Secs. 41 and 42 (employer and employee) are listed below:

- The *employer* must inform each employee of the safety rules (Rule 410A). The *employee* must read and study the safety rules (Rule 420A).
- The *employer* must have an adequate supply of protective devices and equipment (e.g., hard hats, rubber gloves, insulated tools, body belts, etc.) sufficient

to enable employees to meet the requirements of the work to be undertaken (Rule 411B).

- The *employee* must use personal protective equipment, the protective devices, and the special tools provided for their work and inspect these devices and tools before starting work to verify that they are in good condition (Rule 420H).

When the employee is required to perform a task, for example, inspecting personal protective equipment, the employer must have a designated employee in charge responsible for making sure the employees perform the inspection. The duties of the designated employee in charge are outlined in Rule 421A.

Section 41 (rules for employers) provides a list of typical protective devices and equipment for employees (who are covered in Sec. 42) to use. The **Code** does not dictate what protective devices and equipment must be used for a particular task. The choice of what protective devices and equipment need to be used is site-and task-specific. The **NESC** cannot cover every conceivable specific situation, which is why an assessment is required (see Rule 410A3).

Section 41 (rules for employers), Rule 411A2, requires that diagrams (i.e., maps) be maintained and on file for employees (who are covered in Sec. 42) to use. The diagrams aid the identification of structures required in Part 2, Overhead Lines, Rule 217A3, and the location of underground facilities required in Part 3, Underground Lines, Rule 311A. Accurate diagrams are necessary for minimizing errors and accidents. Inaccurate maps can increase errors and accidents. Diagrams cannot be used as a substitute for applying proper safety procedures.

Section 41 (Rule 410A) requires the employer (the company) to perform an assessment to determine the potential exposure to an electric arc for employees who work on or near energized lines or equipment. The assessment considers the employees assigned tasks and work activities. For example, an engineer entering a substation to read information off of a transformer nameplate has a different potential exposure to an electric arc than a lineworker replacing a pole top transformer on an energized line. In addition to the assessment of the work activity, the employer must perform an arc hazard analysis to determine the effective arc rating of clothing to be worn by the employee. The arc thermal protection value (ATPV) can be calculated in an engineering study or determined using **NESC** Tables 410-1, 410-2, and 410-3. Each table is for a different voltage range, and the assumptions used to create the tables are documented in the footnotes at the bottom of the tables. Section 42 (Rule 420I) requires the employee to wear the clothing that the employer (the company) determined necessary.

Sometimes confusion exists between the application of the **NESC** Work Rules and NFPA 70E (Standard for Electrical Safety in the Workplace). A careful review of the scope of NFPA 70E indicates that NFPA 70E does not apply to supply (electric power) or communications utilities. In general, NFPA 70E applies to electrical work performed by electricians. However, even though the scope of NFPA 70E does not apply to utility companies, it can be a useful reference for accepted good practice. One practice that NFPA 70E addresses that is not commonly found in the electric utility industry is labeling equipment such as circuit breakers, fused switches, etc., with a sign or label indicating the arc hazard incident energy or required level of arc-rated clothing. Utility companies typically use training of their employees to address this issue, not signs or labels. For more information, see the NFPA 70E document in its entirety.

The Work Rules in Part 4 of the **NESC** are similar to (but they are not exactly the same as) the Occupational Safety and Health Administration (OSHA) regulations related to electric supply and communications systems. See Rule 402 for a discussion and a figure describing the OSHA standards that are related to the **NESC** Work Rules.

420. GENERAL

420A. Rules and Emergency Methods

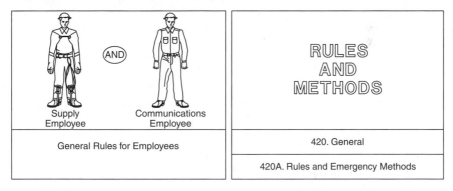

- Read and study safety rules.
- Show knowledge of safety rules.
- Be familiar with first aid, rescue techniques, and fire extinguishing.

420B. Qualification of Employees

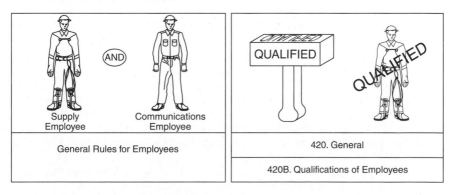

- Employees must only perform tasks for which they are trained, equipped, authorized, and directed.
- Inexperienced employees must work under experienced employees and perform only directed tasks.
- Employees operating mechanical equipment must be qualified to perform those tasks.

- If safety is in doubt, request instructions from supervisor.
- Employees who only occasionally work on electric supply lines can only do work when authorized.

420C. Safeguarding Oneself and Others

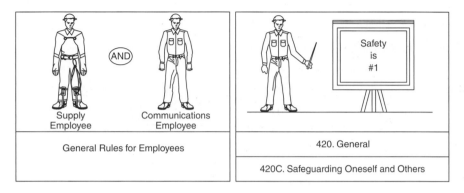

Supply Employee | Communications Employee

General Rules for Employees

420. General

420C. Safeguarding Oneself and Others

- Heed safety signs and signals.
- Warn others who are in danger or near energized lines.
- Report line or equipment defects (e.g., low clearance, broken insulators, etc.).
- Report accidentally energized items.
- Report any defect that may cause danger.
- Employees who do not work on lines and equipment must keep away from them and keep away from worksites with falling objects.
- Employees who work on energized lines must:
  ✓ Consider the effects of their actions.
  ✓ Account for their own safety.
  ✓ Account for the safety of other employees on the job site.
  ✓ Account for the safety of other employees remote from the job site but affected by the work.
  ✓ Account for the property of others.
  ✓ Account for the public.
- Communications employees must not approach energized parts closer than the minimum approach distances shown in Rule 431 (**NESC** Table 431-1).
- Communications employees must not take conductive objects near energized parts closer than the minimum approach distances shown in Rule 431 (**NESC** Table 431-1).
- Supply employees must not approach energized parts closer than the minimum approach distances shown in Rule 441 (**NESC** Table 441-1 or 441-5).
- Supply employees must not take conductive objects near energized parts (without an insulating handle) closer than the minimum approach distances shown in Rule 441 (**NESC** Table 441-1 or 441-5).
- Employees must use care when working with metal ropes, tapes, or wires in the vicinity of energized high-voltage lines due to energization and induced voltages.
- Clearance measurements from energized lines must be done with approved devices (e.g., insulated measuring sticks).

420D. Energized or Unknown Conditions

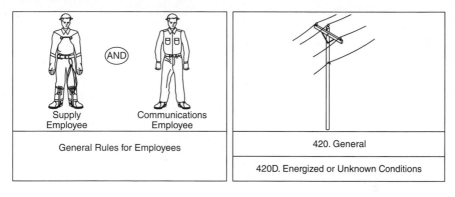

Supply Employee AND Communications Employee

General Rules for Employees

420. General

420D. Energized or Unknown Conditions

- Consider equipment and lines to be energized unless they are positively known to be de-energized.
- Determine existing conditions before starting work by inspection or tests.
- Determine the operating voltage of equipment and lines before starting work.

420E. Ungrounded Metal Parts

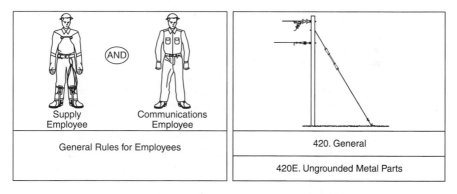

Supply Employee AND Communications Employee

General Rules for Employees

420. General

420E. Ungrounded Metal Parts

- Consider ungrounded metal parts energized at the highest voltage to which they are exposed.

420F. Arcing Conditions

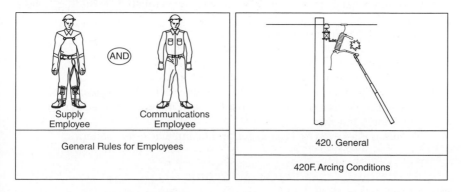

Supply Employee AND Communications Employee

General Rules for Employees

420. General

420F. Arcing Conditions

- Keep body parts far away from devices that produce arcs during operation, such as switches.

420G. Batteries

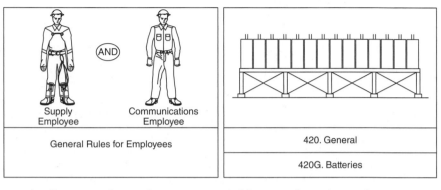

- Applies to: stationary battery arrays, grid storage batteries, and energy storage facilities.
- Does not apply to: small or single batteries or battery packs like those used in portable tools and test equipment.
- Do not handle (i.e., touch with hands or tools) energized parts of batteries (e.g., terminals or bus bars) unless appropriate precautions are taken to prevent short circuits and shocks.
- When working near batteries, avoid contact and make sure that protective barriers are in place (reference Rules 431 and 441).
    - ✓ Protection against inadvertent short circuits or accidental contact may be provided with effective use of guards on terminals and bus bars.
    - ✓ An even greater risk of electric shock is present on solidly grounded systems than for ungrounded or high-resistance ground systems.
- Battery systems may also pose a significant arc flash hazard, and therefore arc flash should be considered with regard to clothing and personal protective equipment as explained in Rule 410A3, with attention to that rule's Exception 5 and Notes 1, 2, and 5.
- When working on or around batteries, workers must remove all exposed conductive articles like chains and jewelry.
- Avoid smoking, open flames, or tools that produce sparks.
- Determine that battery areas are adequately ventilated, including, where applicable:
    - ✓ safe entry indication by sensor.
    - ✓ operational fans.
    - ✓ all vents clear of obstruction.
- Use only insulated tools when working on batteries.
- When working on batteries or working with battery electrolytes, appropriate personal protective equipment must be used, including protection for eyes, face, hands, and skin.
    - ✓ This applies even for sealed, gel, and maintenance-free batteries.

- When work involves contact with battery lids, terminals, or flame arrestors, workers must touch a grounded surface to discharge static on body and clothing prior to contact with batteries.

420H. Tools and Protective Equipment

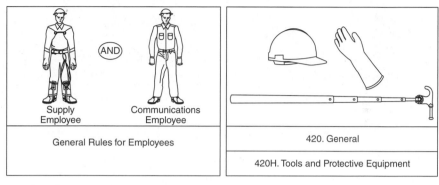

- Use the personal protective equipment, devices, and tools provided for the work.
- Before starting work, carefully inspect the personal protective equipment, devices, and tools to verify they are in good condition.

420I. Clothing

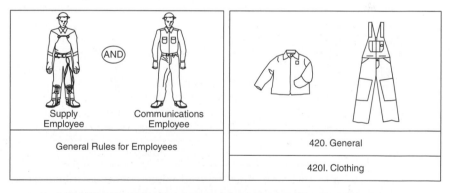

- Wear clothing suitable for the assigned task and work environment.
- Employees exposed to an electric arc must wear clothing or a clothing system in accordance with Rule 410A (**NESC** Tables 410-1, 410-2, and 410-3).
- Avoid wearing exposed metal articles near energized lines.

420J. Ladders and Supports

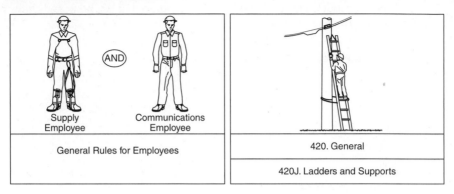

- Verify ladders, aerial lifts, etc., are strong, in good condition, and secure before using them.
- Do not paint portable wood ladders except with a clear nonconductive coating.
- Do not reinforce portable wood ladders with metal.
- Do not use portable metal ladders near energized parts.
- Conductive portable ladders for specialized work must only be used for the work intended.

420K. Fall Protection

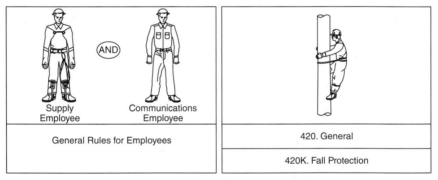

- Appropriate fall protection must be used when climbing, transferring, and transitioning unless doing so is not feasible or creates a greater hazard.
- Work positioning shall be rigged so the climber cannot fall more than 2 ft.
- Anchorages for work positioning must support at least twice the impact load of a fall or 3,000-lb force, whichever is greater.
- Note that tested wood-pole fall-restriction devices are considered to meet the anchorage strength requirements when properly used.
- Note that engineering practices, design specifications, and maintenance procedures may be used to determine anchorage strength requirements, but visual inspections must be performed.
- Note that bolted attachments, step bolts, or other equipment may serve as an anchorage, but visual inspections must be performed.

- Fall protection systems must be used when working above 4 ft on poles, towers, or while working from aerial lifts, helicopters, and cable carts.
- Fall protection equipment must be inspected before use.
- Fall arrest equipment must be suitably anchored.
- Determine that the fall protection system is engaged and secure.
- Be aware of accidental disengagement of the snap hook from the D-ring by foreign objects.
- Be aware of accidental disengagement of the snap hook from the D-ring by rollout.
- Use locking snap hooks and compatible hardware.
- Do not connect snap hooks to each other.
- Do not use 100 percent leather positioning straps or nonlocking snaphooks.
- Use wire rope lanyards where the lanyard could be cut.
- Do not use wire rope lanyards near energized lines.

420L. Fire Extinguishers

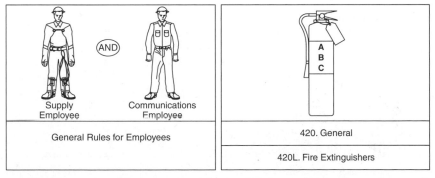

- Use fire extinguishers or materials rated for energized parts or de-energize the parts first.

420M. Machines or Moving Parts

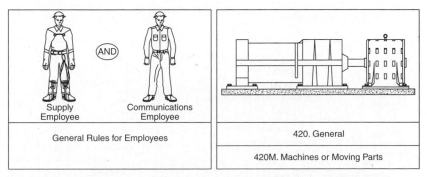

- When working on moving parts, verify accidental startup will not occur by using lockout/tagout procedures.
- When working on automatic switches, stay clear of moving parts.

420N. Fuses

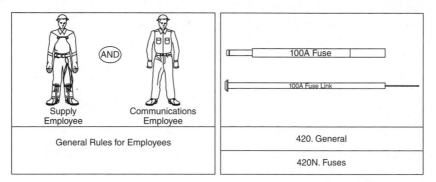

- Use insulated gloves or tools when installing fuses on energized lines.
- Use eye protection and stand clear when installing expulsion-type fuses on energized lines.

4200. Cable Reels

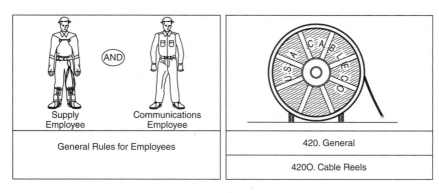

- Block cable reels so they do not accidentally roll.

420P. Street and Area Lighting

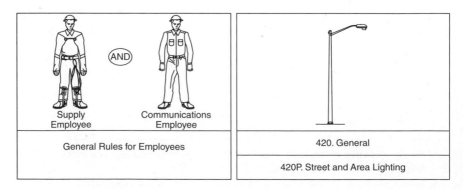

- Periodically examine lowering ropes or chains, supports, and fastenings.
- A device must be provided to safely disconnect each lamp on a series lighting circuit of more than 300 V before the lamp is handled.
- An exception applies when insulated devices or tools are used and the circuit is treated as a full-voltage circuit.

420Q. Antennas

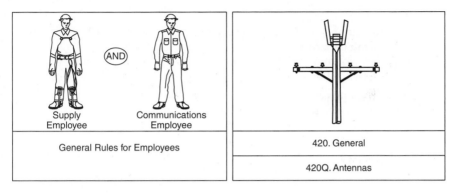

- Use controls to mitigate radio-frequency exposure levels when working around antennas.
- Regulatory standards in OSHA, FCC, and IEEE can be referenced for radiation level limits.

421. GENERAL OPERATING ROUTINES

421A. Duties of a First-Level Supervisor or Person in Charge

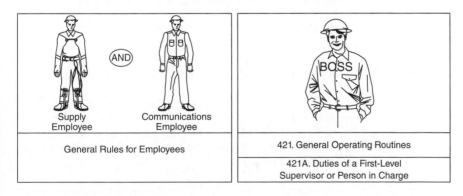

- Duties of the individual in charge:
  - ✓ Adopt precautions to prevent accidents.
  - ✓ See that employees are observing safety rules and operating procedures.
  - ✓ Keep records and make reports.

✓ Prevent unauthorized persons from approaching the workplace.

✓ Do not allow tools or devices unsuitable for the work.

✓ Do not allow tools or devices to be used without testing or inspecting first.

✓ Conduct a job briefing (e.g., tailgate) before each job and discuss:

- Work procedures
- Personal protective equipment
- Energy source controls
- Job hazards
- Special precautions
- Information to respond to emergencies
- Other items as needed
- When working alone, a job briefing is not needed, however, the tasks performed should be planned as if a briefing was required.

421B. Area Protection

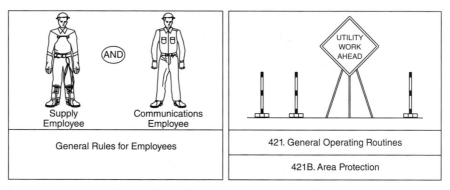

Supply Employee **AND** Communications Employee

General Rules for Employees

421. General Operating Routines

421B. Area Protection

- Areas accessible to vehicular and pedestrian traffic:
  - ✓ Prevent vehicles and pedestrians from approaching the work site.
  - ✓ Warn the public of openings or obstructions.
  - ✓ Openings or obstructions exposed at night must have warning lights and must be enclosed with protective barricades.
- Areas accessible to employees only:
  - ✓ If the work exposes energized or moving parts that are normally protected, safety signs must be displayed.
  - ✓ If the work exposes energized or moving parts that are normally protected, barricades must be erected.
  - ✓ Work on one section of a switchboard with multiple sections or work on one portion of a substation with several portions requires barriers to prevent contact with energized parts.
- Locations with crossed or fallen wires:
  - ✓ If an employee encounters crossed or fallen wires, the employee must remain on guard or use other means to prevent accidents.
  - ✓ The proper authority must be notified.
  - ✓ If qualified, and if safety rules can be met, the employee may correct the condition.

421C. Escort

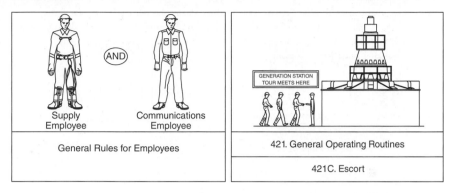

Supply Employee **AND** Communications Employee	GENERATION STATION TOUR MEETS HERE
General Rules for Employees	421. General Operating Routines
	421C. Escort

- An employee responsible for safety must escort nonqualified employees or visitors near electrical lines or equipment.

422. OVERHEAD LINE OPERATING PROCEDURES

422A. Setting, Moving, or Removing Poles in or near Energized Electric Supply Lines

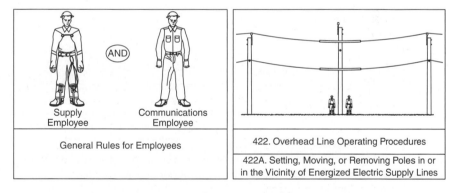

Supply Employee **AND** Communications Employee	
General Rules for Employees	422. Overhead Line Operating Procedures
	422A. Setting, Moving, or Removing Poles in or in the Vicinity of Energized Electric Supply Lines

- Employees working on overhead lines must observe the following rules in addition to other applicable rules in Secs. 43 and 44.
- When setting, moving, or removing poles near energized lines:
    ✓ Avoid direct contact of the pole with the energized conductors.
    ✓ Wear insulating gloves.
    ✓ Do not contact the pole with uninsulated body parts.
    ✓ Avoid touching trucks or other equipment unless wearing suitable protective equipment.

422B. Checking Structures Before Climbing

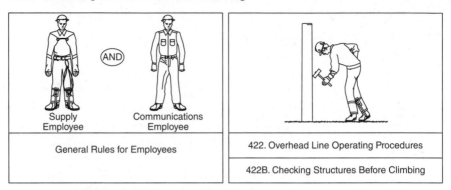

Supply Employee (AND) Communications Employee

General Rules for Employees

422. Overhead Line Operating Procedures

422B. Checking Structures Before Climbing

- Before climbing poles, ladders, etc., verify the structure is capable of handling the additional weight and unbalanced forces.
- Poles must not be climbed if unsafe unless guying, bracing, or other means are used to create a safe condition.

422C. Installing and Removing Wires or Cables

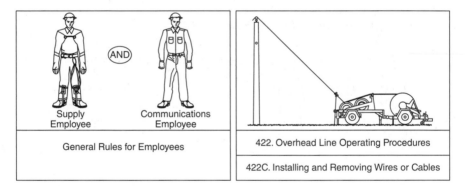

Supply Employee (AND) Communications Employee

General Rules for Employees

422. Overhead Line Operating Procedures

422C. Installing and Removing Wires or Cables

- Wires being installed or removed must be kept clear from energized wires.
- Wires being installed or removed that are not bonded to an effective ground must be considered energized.
- Control sag of wires being installed or removed to prevent pedestrian and vehicle traffic damage.
- Verify that structures can handle the forces associated with installing or removing wires.
- Avoid contact with moving winch lines.
- Consider the effect of a higher voltage line on a lower voltage line. Verify that the line being worked on is free from dangerous leakage and induction voltages or verify that it is effectively grounded.

423. UNDERGROUND LINE OPERATING PROCEDURES

423A. Guarding Manhole and Street Openings

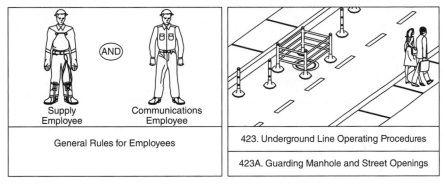

- Employees working on underground lines must observe the following rules in addition to other applicable rules in Secs. 43 and 44.
- Open manholes, handholes, and vaults must be protected with a barrier, temporary cover, or guard.

423B. Testing for Gas in Manholes and Unventilated Vaults

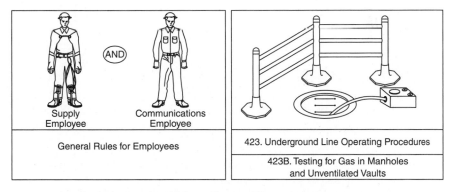

- Test manhole for combustible or flammable gases before entry.
- If combustible or flammable gases exist, ventilate before entry.
- Test for oxygen deficiency.
- Make provisions for a good air supply during work.

423C. Flames

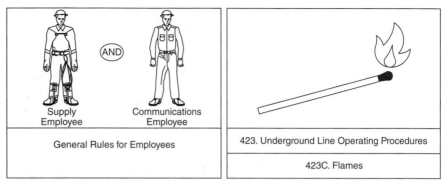

Supply Employee AND Communications Employee

General Rules for Employees

423. Underground Line Operating Procedures

423C. Flames

- Do not smoke in manholes.
- Use extra precaution to ensure ventilation when flames are required for work.
- Test excavation areas for combustible gases or liquids (e.g., near a gasoline service station) before using open flames.
- Provide adequate air space or a barrier to protect lines that transport flammable material when flames are required for work and the lines are exposed.

423D. Excavation

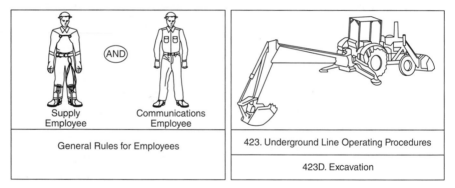

Supply Employee AND Communications Employee

General Rules for Employees

423. Underground Line Operating Procedures

423D. Excavation

- Locate buried utilities prior to excavation.
- Existing utilities should be exposed where the bore path of guided boring or direction drilling machines crosses existing utilities.
- Hand tools used for manual excavation near supply cables must have nonconductive handles.
- Hand digging must be used when close to cables or other utilities.
- If lines that transport flammable material are broken or damaged, the employee must:
  ✓ Leave the excavation open
  ✓ Where safe, eliminate ignition sources
  ✓ Notify the proper authority
  ✓ Keep the public away

- When an employee is working in a trench or excavation in excess of 5 ft deep, or when a trench or excavation presents a cave-in hazard, shoring, sloping, or shielding methods must be used for employee protection.

423E. Identification

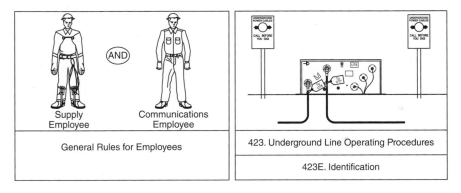

- Identify and protect exposed buried utilities.
- When working on one cable, protect other cables from damage.
- Before cutting a cable or opening a splice, verify its identity.
- Where multiple cables exist, the cable to be worked on must be positively identified.

423F. Operation of Power-Driven Equipment

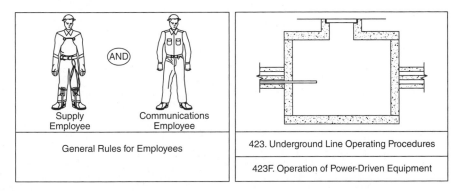

- Keep out of manholes when power rodding.

Section 43

Additional Rules for Communications Employees

The format of this Handbook for Part 4, Work Rules, is different from the format of the other parts of this book. The format for Part 4 summarizes the **Code** text in an outline bulleted list. A graphic is used to visually represent which section, rule, and paragraph the outline applies to. A brief general discussion of each section is provided at the beginning of each section. The following brief general discussion covers selected topics in Secs. 43 and 44. This discussion is repeated at the beginning of Sec. 44.

Sections 43 and 44 provide additional rules for employees. Section 43 provides additional rules for communications employees only. Section 44 provides additional rules for supply employees only. Section 41 provides rules for supply and communications employers, and Sec. 42 provides rules for supply and communications employees.

The additional rules in Sec. 43 for communications employees primarily focus on keeping the communications employee safe when the communication lines are on joint-use (supply and communication) structures or in joint-use (supply and communication) manholes with electric supply conductors. Section 43 requires that communications employees maintain minimum approach distances between the communications employee and electric supply lines and equipment. In addition to the minimum approach distances to electric supply conductors, communications employees must not position themselves above the lowest electric supply conductor exclusive of vertical runs (risers) and street lights.

The additional rules in Sec. 44 for supply employees primarily focus on minimum approach distances to energized parts, switching control procedures, work on energized lines, de-energizing lines, protective grounding, and live-line work. It is interesting to note that for a 12.47/7.2 kV line, the communication worker minimum approach distance found in **NESC** Table 431-1 and the supply worker minimum approach distance found in **NESC** Table 441-1 are both 2 ft-3 in. The difference is that the supply worker gets additional training in Sec. 44 to go closer than the minimum approach distance. The communication worker does not get additional training to go closer than the minimum approach distance.

If a communications line is located in the supply space in accordance with the overhead line rules in Part 2 of the **NESC**, the worker who enters the supply space to work on the communications line must be trained as a supply employee (power lineworker) see **NESC** Rule 224A1.

If a communications line is located below a supply line on an overhead structure and the proper communications worker safety zone clearances in Sec. 23 are met, the worker who is maintaining the communications line must be trained as a communications employee (communications lineworker).

If a communications line is positioned below a supply line on an overhead structure, but the proper clearances in Sec. 23 are not met, the communications employee can correct the violation if the communications employee does not violate the minimum approach distances and other requirements in Sec. 43. If the communications employee cannot maintain the minimum approach distances in Sec. 43, the communications employee must contact a supply employee to correct the violation.

For example, the minimum approach distance rules in Sec. 43 (**NESC** Table 431-1) require the communications worker to have a 2 ft-3 in minimum approach distance to a 12.47/7.2 kV line. A common supply to communications clearance value in Rules 235 and 238 is 40 in (see Rules 235 and 238 for specific applications). This value is intended to provide enough space on the pole for the communications worker to get into position and work on the communications cable while still maintaining the 2 ft-3 in minimum approach distance to the 12.47/7.2 kV line. If a violation exists on the pole and the 40-in clearance requirement between supply and communications is only 38 in, the communications worker may be able to get into position and lower the communications cable down 2 in (from 38 in to 40 in) without violating the 2 ft-3 in minimum approach distance to the 12.47/7.2 kV line. However, if the communications cable was in violation and mounted only 1 ft below the 12.47/7.2 kV line (say right under the crossarm brace), the communications worker could not correct this violation. The communications worker would have to call on a trained supply worker (power lineworker) to fix the problem.

If a communications employee is not a qualified employee, then the worker is considered an unqualified employee and the worker must maintain at least a 10-ft distance from energized lines (more than 10 ft for greater than 50 kV) per OSHA Standard 1910.333. This is commonly referred to as the OSHA 10-ft Rule. The OSHA 10-ft Rule also applies to other workers in the vicinity of power lines such as painters, roofers, etc. The **NESC** does not provide distances for unqualified workers but the OSHA standards do. The bottom line is that communications workers need the proper training or they cannot work on a communications line that is constructed jointly with a power line.

The crane and derrick equipment operation standard (OSHA 1926.1408 which is part of OSHA 1926 Subpart CC) contains minimum clearance distance requirements ranging from 10 ft to 45 ft depending on the voltage of the line with initial evaluations starting at 20 ft. Appropriate measures, such as spotters, should be used to ensure that clearances are maintained.

The OSHA 10-ft Rule (up to 50 kV) in OSHA Standard 1910.333 also applies to vehicles or mechanical equipment capable of having parts of its structure elevated near energized overhead lines. OSHA Standard 1910.333 allows a reduction from 10 ft to 4 ft if the vehicle is in transit with its structure lowered.

The terms qualified, qualified employee, qualified person, qualified worker, and qualified electrical worker are commonly used in the power and communication utility industry. The **NESC** and OSHA both define some of these terms. Supply (power) workers must be trained in OSHA Standard 1910.269. OSHA Standard 1910.269 states in Paragraph 1910.269(a)(2)(ii) that qualified employees shall be trained and competent in:

- The skills and techniques necessary to distinguish exposed live parts from other parts of electric equipment,
- The skills and techniques necessary to determine the nominal voltage of exposed live parts,
- The minimum approach distances specified in this section corresponding to the voltages to which the qualified employee will be exposed and the skills and techniques necessary to maintain those distances,
- The proper use of the special precautionary techniques, personal protective equipment, insulating and shielding materials, and insulated tools for working on or near exposed energized parts of electric equipment, and
- The recognition of electrical hazards to which the employee may be exposed and the skills and techniques necessary to control or avoid these hazards.

The first three bullets above outline the required training necessary for an engineer, technician, lineworker, or related worker at an electric utility to go from a 10 ft (unqualified) employee down to a 2 ft-3 in employee (the minimum approach distance for a 12.47/7.2 kV line or part). The fourth bullet above outlines additional training. This additional training is much more intense for a power lineworker than for an engineer or technician due to the differences in their job duties or assigned tasks. The OSHA Standards require employees to have training for the tasks that their job involves. For example, it is not necessary for an engineer for the power company to be trained in live line work procedures if the engineer does not perform live line work but the engineer does need to know minimum approach distances to energized parts to enter a substation. **NESC** Rule 421C permits nonqualified employees or visitors to be escorted in the vicinity of electric equipment or lines. For example, a qualified employee can escort a nonqualified employee or visitor into a substation. The qualified employee is responsible for safeguarding the people in their care. The fifth bullet above is a general requirement for the safety of electrical workers.

OSHA Standards 1910.332 and 1910.333 have similar training and minimum approach distance requirements for general industry workers (including communication workers) to go from a 10 ft (unqualified) employee down to 2 ft-3 in employee (the minimum approach distance for a 12.47/7.2 kV line or part).

Communication workers must also be trained in OSHA Standard 1910.268. Many people do not think that communication workers need to use insulated gloves, which is true in many cases but OSHA 1910.268 does have at least three work tasks that require a communication worker to wear insulated gloves. These tasks include attaching and removing temporary bonds (a specific sequence must be followed), handling suspension strand that is being installed on poles carrying exposed energized power conductors, and handling poles near energized power conductors when a possibility exists that the pole may contact a power conductor. Many people do not think that communication companies own joint-use (power and communication) poles but some do and some perform pole replacements of these poles. This work must be done by communication workers trained in specific rules for handling poles near energized power conductors, not by communication workers who normally perform cable installation and lashing work or service drop work.

NESC Table 431-1 (for communication lineworkers) has one column of minimum approach distances for the voltage range of 0 to 72.5 kV. NESC Table 441-1 (for power lineworkers) has two columns of minimum approach distances for the voltage range between 0 and 72.5 kV. The first column in NESC Table 441-1 requires the use of the phase-to-phase voltage of the circuit or energized part. The second column in NESC Table 441-1 contains minimum approach distances when an employee has a phase-to-ground exposure to the energized part (e.g., when climbing a pole or ascending a pole in a bucket truck). The third column in NESC Table 441-1 contains minimum approach distances when an employee has a phase-to-phase exposure to the energized part. To understand how to apply the phase-to-phase minimum approach distances, OSHA 1910.269, Appendix B discusses the two types of exposures (phase-to-ground and phase-to-phase). Per OSHA 1910.269, Appendix B, the exposure is phase-to-phase, with respect to an energized part, when an employee is at the potential of another energized part (at a different potential) during live-line barehand work. Live-line barehand work is a special technique in which the worker is energized at the line voltage like a bird on a wire.

The term minimum approach distance (also referred to as MAD) is an industry term for saying "how far to stay away." The minimum approach distance values also apply to how far back a workers hands have to be on a live-line tool, such as an insulated hot stick. The concept of minimum approach distance is outlined in Fig. 43-1.

The Work Rules in Part 4 of the NESC are similar to (but not exactly the same as) the Occupational Safety and Health Administration (OSHA) regulations related to electric supply and communications systems. See Rule 402 for a discussion and a figure describing the OSHA standards that are related to the NESC Work Rules.

Minimum Approach Distance:
- Employees must not approach or bring any conductive object within the minimum approach distance.
- The minimum approach distance contains an electrical component and an inadvertent (i.e., unintentional or unexpected) movement component.
- Communications worker minimum approach distances are found in **NESC** Table 431-1. (A similar table is found in OSHA 1910.268.)
- Supply worker minimum approach distances are found in **NESC** Tables 441-1 through 441-5. (Similar tables are found in OSHA 1910.269.)
- Supply workers can work inside the minimum approach distance if additional work rules are followed.
- Communication workers may not approach closer than the minimum approach distance.

Reach or Extended Reach:
- Reach is defined as the range of anticipated motion of an employee while performing a task.
- Extended reach is defined as the range of anticipated motion of a conductive object being held by an employee while performing a task.
- Anticipated motions include:
  ✓ Adjusting a hardhat
  ✓ Maneuvering tools
  ✓ Reaching for items being passed to the employee
  ✓ Adjusting parts
  ✓ Adjusting work positions, etc.

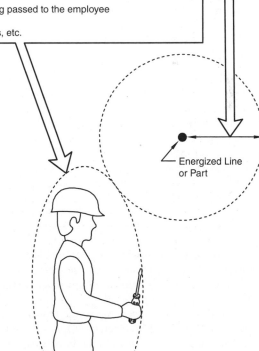

Energized Line
or Part

Fig. 43-1. Minimum approach distance (Sec. 43).

430. GENERAL

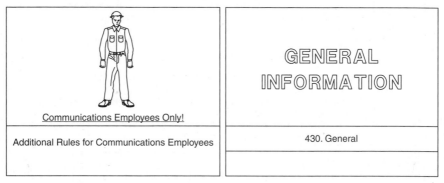

• Section 42, "General Rules for Employees" (both supply and communications) also apply.

431. APPROACH TO ENERGIZED CONDUCTORS OR PARTS

431A. No Employee Shall Approach...

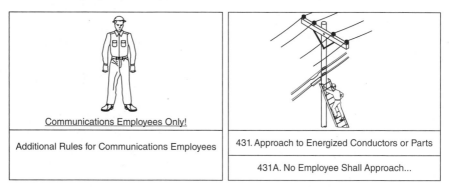

• Communications employees must not approach energized parts closer than the minimum approach distances shown in **NESC** Table 431-1.
• Communications employees must not take conductive objects closer to energized parts than the minimum approach distances shown in **NESC** Table 431-1.
• Communications employees repairing storm damage to communications lines that are joint-use with electric supply lines must:
    ✓ Treat the supply and communications lines as energized to the highest voltage to which they are exposed, or
    ✓ Assure that the electric supply lines are de-energized and grounded per the work rules of **NESC** Sec. 44

431B. Altitude Correction

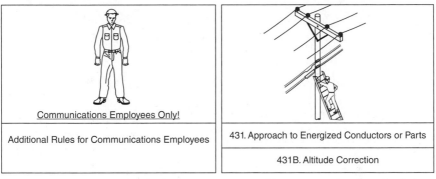

Communications Employees Only!

Additional Rules for Communications Employees

431. Approach to Energized Conductors or Parts

431B. Altitude Correction

- The minimum approach distances in **NESC** Table 431-1 are for altitudes below 12,000 ft.
- **NESC** Table 441-6 provides altitude correction factors, which must be applied to the electrical component of the minimum approach distance for altitudes above 12,000 ft.

431C. When Repairing Underground...

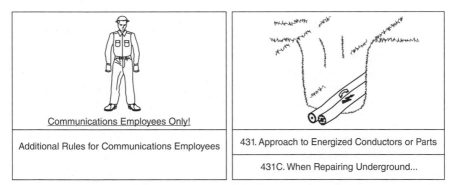

Communications Employees Only!

Additional Rules for Communications Employees

431. Approach to Energized Conductors or Parts

431C. When Repairing Underground...

- Communications employees repairing underground communications lines that are joint-use with damaged electric supply cables must:
  - ✓ Treat the supply and communications lines as energized to the highest voltage to which they are exposed, or
  - ✓ Assure that the electric supply lines are de-energized and grounded per the work rules of **NESC** Sec. 44

432. JOINT-USE STRUCTURES

Communications Employees Only!

Additional Rules for Communications Employees

432. Joint-Use Structures

- On joint-use structures (power and communications), communications employees must not approach energized parts closer than the distances shown in **NESC** Table 431-1.
- On joint-use structures (power and communications), communications employees must not take conductive objects near energized parts closer than the distances shown in **NESC** Table 431-1.
- Communications employees must not position themselves above the lowest electric supply conductor or equipment (not including vertical risers, street lighting, and supply equipment permitted to be below the communications space).
- Note that examples of supply equipment mounted below the communications space include electronic controls, solar panels, etc.
- An exception applies to this rule when fixed rigid barriers are installed between the supply and communications facilities if the supply voltage is 140 kV or below.

433. ATTENDANT ON SURFACE
AT JOINT-USE MANHOLE

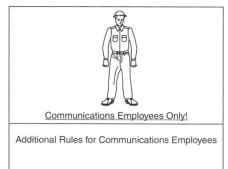

Communications Employees Only!

Additional Rules for Communications Employees

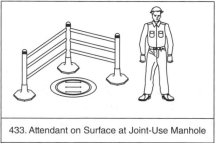

433. Attendant on Surface at Joint-Use Manhole

- Work in joint-use (power and communications) manholes requires an employee to be available on the surface to assist the worker in the manhole.

434. SHEATH CONTINUITY

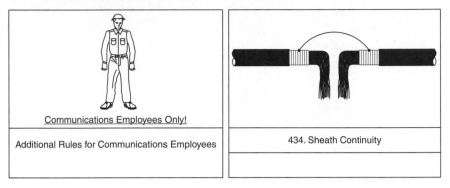

Communications Employees Only!

Additional Rules for Communications Employees

434. Sheath Continuity

- Metallic or semiconductive sheath continuity must be maintained when working on underground cables.

Section 44

Additional Rules for Supply Employees

The format of this Handbook for Part 4, Work Rules, is different from the format of the other parts of this book. The format for Part 4 summarizes the **Code** text in an outline bulleted list. A graphic is used to visually represent which section, rule, and paragraph the outline applies to. A brief general discussion of each section is provided at the beginning of each section. The following brief general discussion covers selected topics in Secs. 44 and 43. This discussion is repeated at the beginning of Sec. 43.

Sections 44 and 43 provide additional rules for employees. Section 44 provides additional rules for supply employees only. Section 43 provides additional rules for communications employees only. Section 41 provides rules for supply and communications employers, and Sec. 42 provides rules for supply and communications employees.

The additional rules in Sec. 44 for supply employees primarily focus on minimum approach distances to energized parts, switching control procedures, work on energized lines, de-energizing lines, protective grounding, and live-line work. It is interesting to note that for a 12.47/7.2 kV line, the communication worker minimum approach distance found in **NESC** Table 431-1 and the supply worker minimum approach distance found in **NESC** Table 441-1 are both 2 ft-3 in. The difference is that the supply worker gets additional training in Sec. 44 to go closer than the minimum approach distance. The communication worker does not get additional training to go closer than the minimum approach distance.

The additional rules in Sec. 43 for communications employees primarily focus on keeping the communications employee safe when the communications lines are on joint-use (supply and communication) structures or in joint-use (supply and communication) manholes with electric supply conductors. Section 43 requires that communications employees maintain minimum approach distances between the communications employee and electric supply lines and equipment. In addition to the minimum approach distances to electric supply conductors, communications employees must not position themselves above the lowest electric supply conductor exclusive of vertical runs (risers) and street lights.

If a communications line is located in the supply space in accordance with the overhead line rules in Part 2 of the **NESC**, the worker who enters the supply space to work on the communications line must be trained as a supply employee (power lineworker) see **NESC** Rule 214A1.

If a communications line is located below a supply line on an overhead structure and the proper communications worker safety zone clearances in Sec. 23 are met, the worker who is maintaining the communications line must be trained as a communications employee (communications lineworker).

If a communications line is positioned below a supply line on an overhead structure, but the proper clearances in Sec. 23 are not met, the communications employee can correct the violation if the communications employee does not violate the minimum approach distances and other requirements in Sec. 43. If the communications employee cannot maintain the minimum approach distances in Sec. 43, the communications employee must contact a supply employee to correct the violation.

For example, the minimum approach distance rules in Sec. 43 (**NESC** Table 431-1) require the communications worker to have a 2 ft-3 in minimum approach distance to a 12.47/7.2 kV line. A common supply to communications clearance value in Rules 235 and 238 is 40 in (see Rules 235 and 238 for specific applications). This value is intended to provide enough space on the pole for the communications worker to get into position and work on the communications cable while still maintaining the 2 ft-3 in minimum approach distance to the 12.47/7.2 kV line. If a violation exists on the pole and the 40-in clearance requirement between supply and communications is only 38 in, the communications worker may be able to get into position and lower the communications cable down 2 in (from 38 in to 40 in) without violating the 2 ft-3 in minimum approach distance to the 12.47/7.2 kV line. However, if the communications cable was in violation and mounted only 1 ft below the 12.47/7.2 kV line (say right under the crossarm brace), the communications worker could not correct this violation. The communications worker would have to call on a trained supply worker (power lineworker) to fix the problem.

If a communications employee is not a qualified employee, then the worker is considered an unqualified employee and the worker must maintain at least a 10-ft distance from energized lines (more than 10 ft for greater than 50 kV) per OSHA Standard 1910.333. This is commonly referred to as the OSHA 10-ft Rule. The OSHA 10-ft Rule also applies to other workers in the vicinity of power lines such as painters, roofers, etc. The **NESC** does not provide distances for unqualified workers but the OSHA standards do. The bottom line is that communications workers need the proper training or they cannot work on a communications line that is constructed jointly with a power line.

The crane and derrick equipment operation standard (OSHA 1926.1408 which is part of OSHA 1926 Subpart CC) contains minimum clearance distance requirements ranging from 10 ft to 45 ft depending on the voltage of the line with initial evaluations starting at 20 ft. Appropriate measures, such as spotters, should be used to ensure that clearances are maintained.

The OSHA 10-ft Rule (up to 50 kV) in OSHA Standard 1910.333 also applies to vehicles or mechanical equipment capable of having parts of its structure elevated near energized overhead lines. OSHA Standard 1910.333 allows a reduction from 10 ft to 4 ft if the vehicle is in transit with its structure lowered.

The terms qualified, qualified employee, qualified person, qualified worker, and qualified electrical worker are commonly used in the power and communication utility industry. The **NESC** and OSHA both define some of these terms. Supply (power) workers must be trained in OSHA Standard 1910.269. OSHA Standard 1910.269 states in Paragraph 1910.269(a)(2)(ii) that qualified employees shall be trained and competent in:

- The skills and techniques necessary to distinguish exposed live parts from other parts of electric equipment,
- The skills and techniques necessary to determine the nominal voltage of exposed live parts,
- The minimum approach distances specified in this section corresponding to the voltages to which the qualified employee will be exposed and the skills and techniques necessary to maintain those distances,
- The proper use of the special precautionary techniques, personal protective equipment, insulating and shielding materials, and insulated tools for working on or near exposed energized parts of electric equipment, and
- The recognition of electrical hazards to which the employee may be exposed and the skills and techniques necessary to control or avoid these hazards.

The first three bullets above outline the required training necessary for an engineer, technician, lineworker, or related worker at an electric utility to go from a 10 ft (unqualified) employee down to a 2 ft-3 in employee (the minimum approach distance for a 12.47/7.2 kV line or part). The fourth bullet above outlines additional training. This additional training is much more intense for a power lineworker than for an engineer or technician due to the differences in their job duties or assigned tasks. The OSHA Standards require employees to have training for the tasks that their job involves. For example, it is not necessary for an engineer for the power company to be trained in live line work procedures if the engineer does not perform live line work but the engineer does need to know minimum approach distances to energized parts to enter a substation. **NESC** Rule 421C permits nonqualified employees or visitors to be escorted in the vicinity of electric equipment or lines. For example, a qualified employee can escort a nonqualified employee or visitor into a substation. The qualified employee is responsible for safeguarding the people in their care. The fifth bullet above is a general requirement for the safety of electrical workers.

OSHA Standards 1910.332 and 1910.333 have similar training and minimum approach distance requirements for general industry workers (including communication workers) to go from a 10 ft (unqualified) employee down to 2 ft-3 in employee (the minimum approach distance for a 12.47/7.2 kV line or part).

Communication workers must also be trained in OSHA Standard 1910.268. Many people do not think that communication workers need to use insulated gloves, which is true in many cases but OSHA 1910.268 does have at least three work tasks that require insulated gloves. These tasks include attaching and removing temporary bonds (a specific sequence must be followed), handling suspension strand that is being installed on poles carrying exposed energized power conductors, and handling poles near energized power conductors when a possibility exists that the pole may contact a power conductor. Many people do not think that communication companies own joint-use (power and communication) poles but some do and some perform pole replacements of these poles. This work must be done by communication workers trained in specific rules for handling poles near energized power conductors, not by communication workers who normally perform cable installation and lashing work or service drop work.

NESC Table 431-1 (for communication lineworkers) has one column of minimum approach distances for the voltage range of 0 to 72.5 kV. **NESC** Table 441-1 (for power lineworkers) has two columns of minimum approach distances for the voltage range between 0 and 72.5 kV. The first column in **NESC** Table 441-1 requires the use of the phase-to-phase voltage of the circuit or energized part. The second column in **NESC** Table 441-1 contains minimum approach distances when an employee has a phase-to-ground exposure to the energized part (e.g., when climbing a pole or ascending a pole in a bucket truck). The third column in **NESC** Table 441-1 contains minimum approach distances when an employee has a phase-to-phase exposure to the energized part. To understand how to apply the phase-to-phase minimum approach distances, OSHA 1910.269, Appendix B discusses the two types of exposures (phase-to-ground and phase-to-phase). Per OSHA 1910.269, Appendix B, the exposure is phase-to-phase, with respect to an energized part, when an employee is at the potential of another energized part (at a different potential) during live-line barehand work. Live-line barehand work is a special technique in which the worker is energized at the line voltage like a bird on a wire.

The term minimum approach distance (also referred to as MAD) is an industry term for saying "how far to stay away." The minimum approach distance values also apply to how far back a workers hands have to be on a live-line tool, such as a hot stick. The concept of minimum approach distance is outlined in Fig. 44-1.

The Work Rules in Part 4 of the **NESC** are similar to (but not exactly the same as) the Occupational Safety and Health Administration (OSHA) regulations related to electric supply and communications systems. See Rule 402 for a discussion and a figure describing the OSHA standards that are related to the **NESC** Work Rules.

Minimum Approach Distance:
• Employees must not approach or bring any conductive object within the minimum approach distance.
• The minimum approach distance contains an electrical component and an inadvertent (i.e., unintentional or unexpected) movement component.
• Communications worker minimum approach distances are found in **NESC** Table 431-1. (A similar table is found in OSHA 1910.268.)
• Supply worker minimum approach distances are found in **NESC** Tables 441-1 through 441-5. (Similar tables are found in OSHA 1910.269.)
• Supply workers can work inside the minimum approach distance if additional work rules are followed.
• Communication workers may not approach closer than the minimum approach distance.

Reach or Extended Reach:
• Reach is defined as the range of anticipated motion of an employee while performing a task.
• Extended reach is defined as the range of anticipated motion of a conductive object being held by an employee while performing a task.
• Anticipated motions include:
✓ Adjusting a hardhat
✓ Maneuvering tools
✓ Reaching for items being passed to the employee
✓ Adjusting parts
✓ Adjusting work positions, etc.

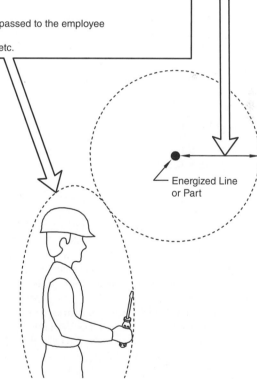

Energized Line or Part

Fig. 44-1. Minimum approach distance (Sec. 44).

440. GENERAL

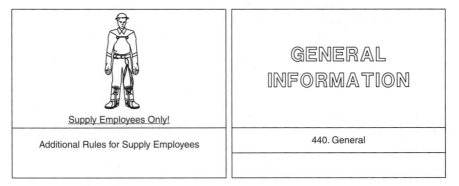

Supply Employees Only!	**GENERAL INFORMATION**
Additional Rules for Supply Employees	440. General

- Section 42, "General Rules for Employees" (both supply and communications) also apply.

441. ENERGIZED CONDUCTORS OR PARTS

441A. Minimum Approach Distance to Live Parts

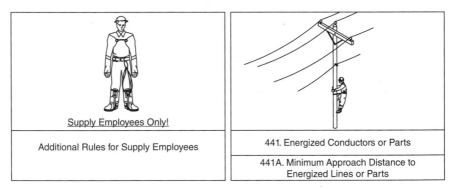

Supply Employees Only!	441. Energized Conductors or Parts
Additional Rules for Supply Employees	441A. Minimum Approach Distance to Energized Lines or Parts

- Supply employees must not approach energized parts or take conductive objects near exposed energized lines or parts closer than the minimum approach distances listed in **NESC** Table 441-1 or 441-5 unless one of the following conditions is met:
  - ✓ The line or part is de-energized and grounded per Rule 444D (special exceptions apply for voltages less than 600 V).
  - ✓ The employee is insulated from the energized line or part using insulated tools, rubber gloves, or rubber gloves with sleeves rated for the applicable voltage.
  - ✓ The energized line or part is insulated from the employee and any other line or part at a different voltage.
  - ✓ The employee is performing bare-hand live-line work in accordance with Rule 446.

- Note that minimum approach distances contain an electrical component and an inadvertent movement component.
  - ✓ Minimum approach distances calculated for 0.301 kV to 0.750 kV contain an electrical component plus a 1-ft inadvertent movement component.
  - ✓ Minimum approach distances calculated for 0.751 kV to 72.5 kV contain an electrical component plus a 2-ft inadvertent movement component.
  - ✓ Minimum approach distances calculated for voltages above 72.5 kV contain an electrical component plus a 1-ft inadvertent movement component.
- Note that the method used for calculating minimum approach distances was taken from OSHA 1910.269, Appendix B.
- Note that voltage ranges are contained in ANSI C84.1, Table 1.
- Note that for the purpose of Sec. 44, "reach" is defined as "the range of anticipated motion of an employee while performing a task" and "extended reach" is defined as "the range of anticipated motion of a conductive object being held by an employee while performing a task."
- Precaution for approaching voltages from 51 to 300 V:
  - ✓ Do not contact exposed energized parts unless the above conditions have been met.
- Precautions for approaching voltages from 301 V to 72.5 kV:
  - ✓ Employees must be protected from phase-to-phase and phase-to-ground differences in voltage (**NESC** Table 441-1 or 441-5 applies).
  - ✓ Exposed grounded lines, conductors, neutrals, parts, messengers, or guys in the work area must be guarded or insulated (special exceptions apply for voltages between 300 V and 750 V).
  - ✓ Rubber insulating gloves must be insulated for the maximum use voltage in **NESC** Table 441-7 and must be worn whenever the employee is within the reach or extended reach of the minimum approach distances in **NESC** Table 441-1 or 441-5 (special exceptions apply).
  - ✓ When the rubber glove method is used, it must be used with one of the two following methods:
    - Rubber insulating sleeves which are insulated for the maximum use voltage in **NESC** Table 441-7 (special exceptions apply for voltages at 750 V or less).
    - Insulating exposed energized lines or parts within the employee's reach or extended reach (this does not apply to the part being worked on under positive control).
  - ✓ When the rubber glove method is used on voltages above 15 kV phase-to-phase, an insulated aerial device, insulated structure-mounted platform, or other supplementary insulation must be used to support the worker.
  - ✓ Insulating cover-up shall be rated for the phase-to-phase voltage of the circuit or the phase-to-ground voltage of the circuit depending on the exposure of the circuit being worked.
  - ✓ Determination of the phase-to-phase or phase-to-ground exposure must consider factors such as work rules, conductor spacing, worker position, tasks being performed, and any other relative factors.
  - ✓ When insulated cover-up is used, it must be applied as the employee first approaches the energized line and it must be removed in the reverse order.
  - ✓ Insulated cover-up must extend beyond the reach of the employee's anticipated work position or extended reach position.

- Precautions for approaching voltages above 72.5 kV:
  - ✓ Employees must position themselves so that they are not within the reach or extended reach of the applicable minimum approach distance.
  - ✓ The minimum approach distances in **NESC** Table 441-1 apply.
  - ✓ In lieu of using the minimum approach distances in **NESC** Table 441-1, the minimum approach distances in **NESC** Tables 441-2 through 441-4 may be used if the per unit transient overvoltage value (T) has been determined using an engineering analysis considering system design, expected operating conditions, and control measures.
  - ✓ Note that control measures include blocking reclosing, prohibiting switching during work, using protective air gaps, use of closing resistors and surge arresters, etc.
  - ✓ Note that IEEE Std. 516 and OSHA 1910.269 Appendix B contain information that may be used to perform an engineering analysis to determine "T". "T" may be determined on a system basis or per-line basis.
- A temporary (transient) over voltage control device (TTOCD) which is designed and tested for installation adjacent to a work site to limit the TOV at the work site may be used to obtain a lower value of "T". (Example formulas are provided in the rule, and **NESC** Appendix D provides additional information on per-unit overvoltage factors.)
- The minimum approach distances in **NESC** Tables 441-1 through 441-5 must be increased for altitudes above 3000 ft.
- **NESC** Table 441-6 provides altitude correction factors, which must be applied to the electrical component of the minimum approach distance.

441B. Additional Approach Requirements

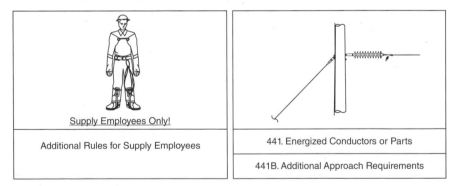

Supply Employees Only!

Additional Rules for Supply Employees

441. Energized Conductors or Parts

441B. Additional Approach Requirements

- The clear insulation distance of insulators is defined as the shortest straight-line air-gap distance from the nearest energized part to the nearest grounded part.
- When working on insulators using rubber gloves or insulated tools, the clear insulation distance must not be less than the straight-line distance with live-line tools required by Rule 441A4.

- To work on the grounded end of an open switch, all of the following conditions must be met:
  - ✓ The full air-gap distance of the switch must be maintained.
  - ✓ The air-gap distance of the switch must not be less than the electrical component of the minimum approach distance (inadvertent movement components are not required).
  - ✓ The minimum approach distance to the energized portion of the switch must be maintained.
- Special rules apply to working on insulator assemblies operating above 72.5 kV.

441C. Live-Line Tool Clear Insulation Length

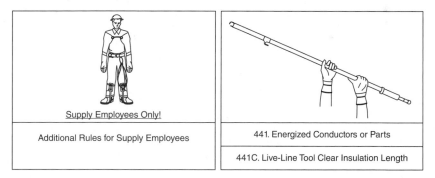

Supply Employees Only!

Additional Rules for Supply Employees

441. Energized Conductors or Parts

441C. Live-Line Tool Clear Insulation Length

- For live-line tools (i.e., hot sticks), the minimum approach distances with tools required by Rule 441A4 must be maintained or exceeded between the conductive end of the tool and the employee's hands or other body parts.
- Insulated conductor support tools may be used if the clear insulation distance is at least as long as the insulator string or the distances specified in Rule 441A.
- When installing insulated conductor support tools, employee minimum approach distances must be maintained.
- Note that the conductive portion of an insulated tool can decrease the insulation value of the tool more than just the length of the conductive portion.

442. SWITCHING CONTROL PROCEDURES

442A. Designated Person

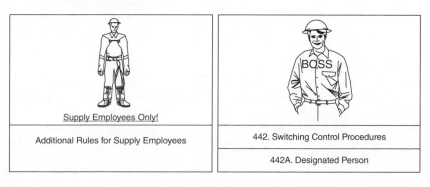

Supply Employees Only!

Additional Rules for Supply Employees

442. Switching Control Procedures

442A. Designated Person

- A designated person must authorize switching.
- The designated person must:
  - ✓ Keep informed of operating conditions to maintain safety
  - ✓ Maintain suitable records
  - ✓ Issue or deny authorization for switching to maintain safety

442B. Specific Work

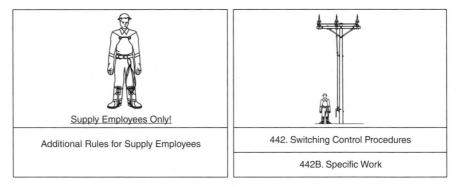

- The designated person shall give authorization before beginning work.
- The designated person shall be notified when work ends.
- Exceptions to the switching control procedures exist for emergencies and catastrophic service disruptions.

442C. Operations at Stations

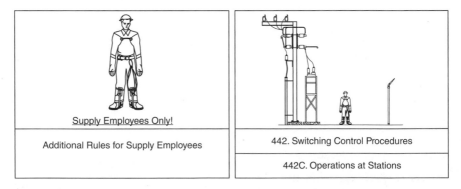

- Qualified employees must obtain authorization from the designated person before switching.
- Specific operating schedules may be used.
- If specific operating schedules do not exist, authorization from the designated person must be obtained for switching or starting and stopping of equipment.
- Exceptions exist for switching sections of distribution circuits and for emergency situations.

442D. Re-energizing after Work

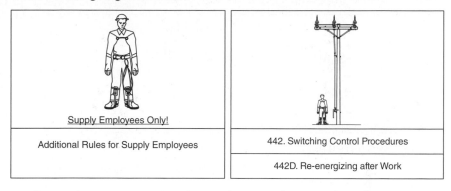

Supply Employees Only!

Additional Rules for Supply Employees

442. Switching Control Procedures

442D. Re-energizing after Work

- Instructions to re-energize after work is complete are to be given by the designated person after the employees who requested the de-energization have reported clear.
- Employees who requested de-energization of lines for other employees or crews cannot request re-energization until the other employees or crews have reported clear.
- The above procedure must be used when more than one location is involved.

442E. Tagging Electric Supply Circuits Associated with Work Activities

Supply Employees Only!

Additional Rules for Supply Employees

442. Switching Control Procedures

442E. Tagging Electric Supply Circuits Associated with Work Activities

- De-energized and grounded circuits must be tagged at all points where the circuit or equipment can be energized.
- When reclosers or circuit breakers are put on one-shot, a tag must be placed at the reclosing device location.
- Exceptions exist for SCADA-controlled systems at both the SCADA operating point and the reclosing device location.
- Tags shall clearly identify the equipment or circuits being worked on.

442F. Restoration of Service after Automatic Trip

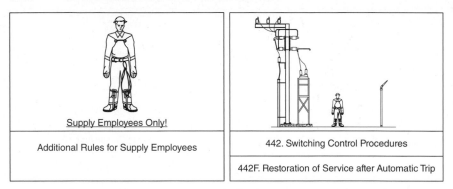

Supply Employees Only!

Additional Rules for Supply Employees

442. Switching Control Procedures

442F. Restoration of Service after Automatic Trip

- Tagged circuits that open automatically (i.e., a circuit with a recloser set to one-shot) must not be closed without authorization.
- When circuits open automatically, local operating rules determine how they may be closed back in with safety.

442G. Repeating Oral Messages

Supply Employees Only!

Additional Rules for Supply Employees

442. Switching Control Procedures

442G. Repeating Oral Messages

- Oral messages associated with line switching must be repeated back to the sender and the sender's identity must be obtained by the receiver.
- The sender of an oral message associated with line switching must require that the message be repeated back by the receiver and must obtain the receiver's identify.

443. WORK ON ENERGIZED LINES AND EQUIPMENT

443A. General Requirements

Supply Employees Only!

Additional Rules for Supply Employees

GENERAL INFORMATION

443. Work on Energized Lines and Equipment

443A. General Requirements

- When working on energized lines, one of the following must be done:
  - ✓ Insulate the employee from energized parts.
  - ✓ Isolate or insulate the employee from ground and other voltages other than the one being worked on.
- Treat covered conductors (i.e., tree wire that is covered with nonrated insulation) as bare energized conductors.
- Consider the effect of a higher voltage line on a lower voltage line. Verify that the line being worked on is free from dangerous leakage and induction voltages or verify that it is effectively grounded.
- Insulated supply cables that cannot be positively identified as de-energized must be pierced or severed with an appropriate tool.
- Consider the operating voltage and take appropriate precautions before cutting an insulated energized supply cable.
- When the insulating covering on an energized cable must be cut, an appropriate tool must be used.
- When the insulating coupling on an energized cable must be cut, suitable eye protection and insulating gloves with protectors must be used.
- When the insulating covering on an energized cable must be cut, extreme care must be taken to prevent short-circuiting conductors.
- Metal-measuring tapes and tapes or ropes containing metal must not be used closer to energized parts than the minimum approach distances in **NESC** Tables 441-1 or 441-5.
- Metal-measuring tapes and tapes or ropes containing metal must be used with care near energized lines due to the effect of induced voltage.
- Metallic equipment or material that is not bonded to an effective ground and could approach energized parts closer than the minimum approach distances must be treated as if it were energized at the voltage to which it is exposed.

443B. Requirement for Assisting Employee

Supply Employees Only!

Additional Rules for Supply Employees

443. Work on Energized Lines and Equipment

443B. Requirement for Assisting Employee

- Employees shall not work alone in bad weather or at night on systems of more than 750 V.
- An exception applies permitting one employee to do certain tasks.

443C. Opening and Closing Switches

Supply Employees Only!

Additional Rules for Supply Employees

443. Work on Energized Lines and Equipment

443C. Opening and Closing Switches

- A smooth continuous motion must be used to close manual switches and disconnects.
- Care must be applied to opening switches to avoid serious arcing.

443D. Working Position

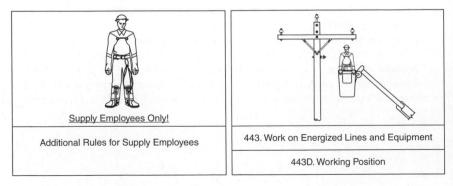

Supply Employees Only!

Additional Rules for Supply Employees

443. Work on Energized Lines and Equipment

443D. Working Position

- Avoid working positions in which a shock or slip will bring the worker's body toward energized parts at a potential different from the employee's body.
- The work position should generally be from below.

443E. Protecting Employees by Switches and Disconnectors

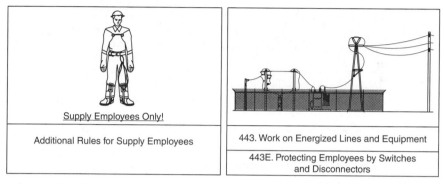

Supply Employees Only!

Additional Rules for Supply Employees

443. Work on Energized Lines and Equipment

443E. Protecting Employees by Switches and Disconnectors

- Load break switches must be opened before non-load-break disconnects.
- Non-load-break disconnects must be closed before closing the load break switch.

443F. Making Connections

Supply Employees Only!

Additional Rules for Supply Employees

443. Work on Energized Lines and Equipment

443F. Making Connections

- When connecting de-energized lines or equipment to energized lines, the jumper wire should first be attached to the de-energized part.
- When disconnecting a line or equipment from an energized line, the source end of the jumper should be removed first.
- Loose conductors (i.e., jumpers) should be kept away from energized parts.

443G. Switchgear

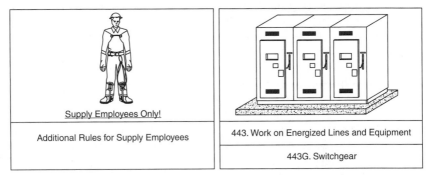

Supply Employees Only!	443. Work on Energized Lines and Equipment
Additional Rules for Supply Employees	443G. Switchgear

- Switchgear must be de-energized and grounded before doing work that involves removing protective barriers unless other safety means are used.
- The switchgear safety features must be replaced after work is completed.

443H. Current Transformer Secondaries

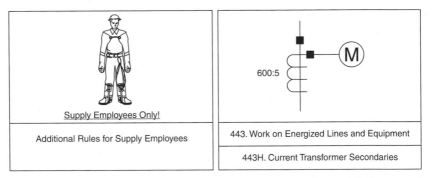

Supply Employees Only!	443. Work on Energized Lines and Equipment
Additional Rules for Supply Employees	443H. Current Transformer Secondaries

- A current transformer secondary must be de-energized before working on it.
- If the current transformer secondary cannot be properly de-energized, the secondary circuit must be bridged (shorted) so that the secondary will not be opened.

443I. Capacitors

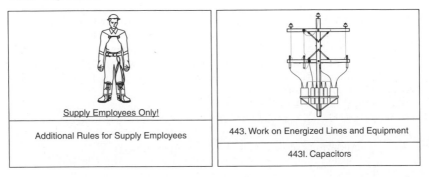

Supply Employees Only!	443. Work on Energized Lines and Equipment
Additional Rules for Supply Employees	443I. Capacitors

- Before working on capacitors, they must be disconnected from the energized source, shorted, and grounded.
- Any line that has capacitors must be short-circuited and grounded before it is considered de-energized.
- Before capacitors are handled, each unit must be shorted between all insulated terminals and the capacitor tank due to series-parallel operation.
- Where capacitors are installed on ungrounded racks, the racks must be grounded before working on the capacitors.
- The internal resistor of a capacitor must not be depended on to discharge the capacitor.

443J. Gas-Insulated Equipment

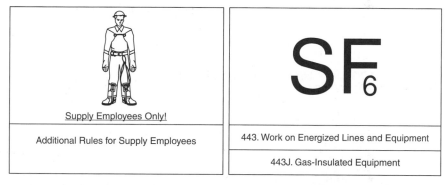

Supply Employees Only!

Additional Rules for Supply Employees

443. Work on Energized Lines and Equipment

443J. Gas-Insulated Equipment

- Employees must be instructed on the special precautions related to handling SF_6 gas.
- The by-products resulting from arcing in SF_6 gas are generally toxic and irritant.

443K. Attendant on Surface

Supply Employees Only!

Additional Rules for Supply Employees

443. Work on Energized Lines and Equipment

443K. Attendant on Surface

- When one employee is in a manhole, another employee must be on the surface to render assistance as required.
- The employee on the surface may enter the manhole to provide short-term assistance.
- An exception permits working alone to do certain tasks.

443L. Unintentional Grounds on Delta Circuits

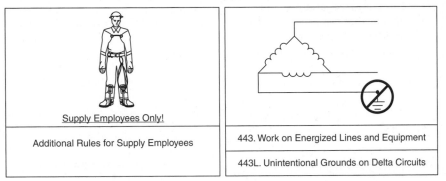

Supply Employees Only!	443. Work on Energized Lines and Equipment
Additional Rules for Supply Employees	443L. Unintentional Grounds on Delta Circuits

- Unintentional grounds on delta circuits must be removed as soon as practical.

444. DE-ENERGIZING EQUIPMENT OR LINES TO PROTECT EMPLOYEES

444A. Application of Rule

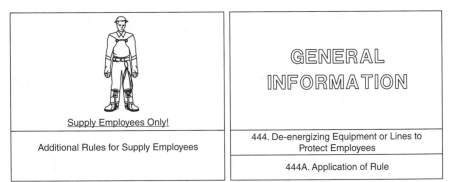

Supply Employees Only!	444. De-energizing Equipment or Lines to Protect Employees
Additional Rules for Supply Employees	444A. Application of Rule

- When employees depend on others to operate switches to de-energize circuits or when employees must secure authorization to operate a switch, Rules 444A through 444H must be followed in order.
- If an employee is in sole charge of a circuit section and is responsible for directing the disconnecting of the section, the portions of Rules 444A through 444H that deal with a designated person may be omitted.
- Records must be kept on utility interactive systems, and these systems must be capable of being visibly disconnected.

444B. Employee's Request

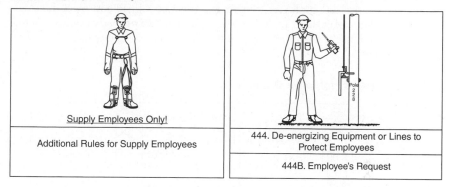

Supply Employees Only!

Additional Rules for Supply Employees

444. De-energizing Equipment or Lines to Protect Employees

444B. Employee's Request

- The employee in charge of the work must make a request to the designated switching person to have the line or equipment de-energized.
- The switching request must properly identify the switch or line section by position, letter, color, number, or other means.

444C. Operating Switches, Disconnectors, and Tagging

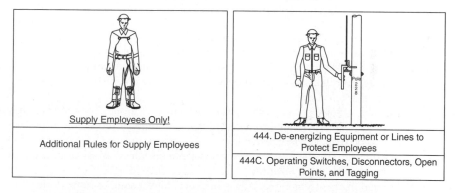

Supply Employees Only!

Additional Rules for Supply Employees

444. De-energizing Equipment or Lines to Protect Employees

444C. Operating Switches, Disconnectors, Open Points, and Tagging

- The designated switching person must direct the operation of switches and disconnects to de-energize the circuit.
- The designated switching person must request that the switches and disconnect be rendered inoperable and tagged.
- If switches that are controlled automatically or remotely can be rendered inoperable, they must be tagged at the switch location.
- If switches that are controlled automatically or remotely cannot be rendered inoperable, then they must be tagged at all points of control.
- The following information must be recorded when placing a tag:
  ✓ Time of disconnection.
  ✓ Name of person making the disconnection.
  ✓ Name of person who requested the disconnection.
  ✓ Name or title or both of the designated switching person.

- When air gaps (i.e., cut or open jumpers) created for de-energizing equipment or lines to protect employees, the air gap must be tagged and meet the minimum clearances in **NESC** Table 444-1 or separated by a properly rated insulator.
- Sources of backfeed must be considered when de-energizing lines and equipment.

444D. Employee's Protective Grounds

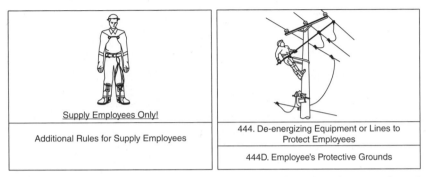

Supply Employees Only!

Additional Rules for Supply Employees

444. De-energizing Equipment or Lines to Protect Employees

444D. Employee's Protective Grounds

- After switching, rendering inoperable where practical, and tagging have been completed and the employee in charge has been given permission by the designated switching person to proceed, protective grounds must be applied.
- Minimum approach distances must be maintained for making voltage tests.
- Minimum approach distances must be maintained for applying grounds.
- Temporary protective grounds must be placed and arranged to protect employees from hazardous differences in electrical potential (e.g., worksite grounds and/or other methods).
- Note that hazardous touch and step potentials may exist around grounded equipment or between separately grounded systems, additional protection measures may be needed (e.g., barriers, insulation, work practices, isolation, or ground mats).
- The minimum approach distances must be maintained from ungrounded conductors at the work location.
- Where making protective grounds is impractical or creates a hazardous condition, grounds may be omitted by special permission.
- An exception applies which allows alternative work methods under special conditions.

444E. Proceeding with Work

Supply Employees Only!

Additional Rules for Supply Employees

444. De-energizing Equipment or Lines to Protect Employees

444E. Proceeding with Work

- After lines are de-energized and grounded, the employee in charge may direct the work to be done.
- Lines may be re-energized for testing under the supervision of the employee in charge and authorization of the designated switching person.
- Additional employees in charge desiring the same equipment or lines to be de-energized and grounded must follow these same procedures.

444F. Reporting Clear—Transferring Responsibility

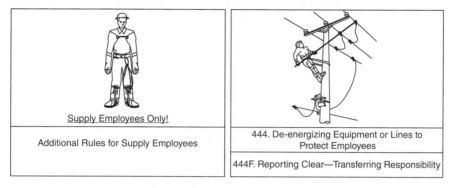

Supply Employees Only!

Additional Rules for Supply Employees

444. De-energizing Equipment or Lines to Protect Employees

444F. Reporting Clear—Transferring Responsibility

- Once work is completed, the employee in charge must verify that the crew assigned to the employee in charge is in the clear.
- Once work is completed, the employee in charge must remove protective grounds.
- Once work is completed, the employee in charge must report to the designated switching person that tags may be removed.
- The employee in charge may transfer responsibilities to another employee after receiving permission from the designated person and personally informing the affected persons of the transfer.

444G. Removal of Tags

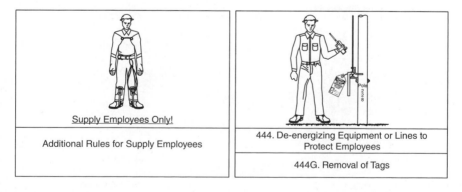

Supply Employees Only!

Additional Rules for Supply Employees

444. De-energizing Equipment or Lines to Protect Employees

444G. Removal of Tags

- The designated switching person must direct the removal of tags.
- The removal of tags must be reported back to the designated switching person by the person removing them.
- Upon tag removal, the record keeping must include the following:
  - ✓ Name of the person requesting removal.
  - ✓ Time of removal.
  - ✓ Name of the person removing the tag.
- The name of the person requesting removal must match the name of the person who requested placement of the tag unless responsibility was properly transferred.

444H. Sequence of Re-energizing

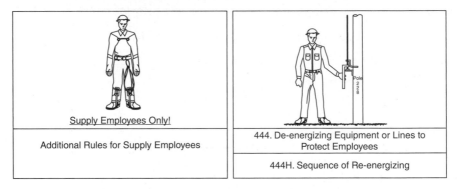

Supply Employees Only!

Additional Rules for Supply Employees

444. De-energizing Equipment or Lines to Protect Employees

444H. Sequence of Re-energizing

- The designated switching person may direct the re-energization of the line only after the removal of protective grounds and tags.

445. PROTECTIVE GROUNDS

445A. Installing Grounds

Supply Employees Only!

Additional Rules for Supply Employees

445. Protective Grounds

445A. Installing Grounds

- Extreme caution must be used to assure that the proper sequence is used when installing and removing protective grounds.
- A specific sequence must be used when installing protective grounds to a previously energized part.
- An exception to the sequence may apply to some high-voltage towers.
- 1st, grounding conductors and devices must be sized to carry anticipated fault currents.
- 2nd, one end of the grounding device must be connected to an effective ground. Grounding switches may be used.
- 3rd, a voltage test must be done utilizing appropriate (insulated) tools and minimum approach distances.
- 4th, if the line shows no voltage, grounding may be completed. If voltage is present, the source must be determined. To complete grounding, the ground device must be brought into contact with the previously energized part using insulating handles and securely attached. Where bundled conductors exist, each conductor of the bundle should be grounded. Only after completion of the fourth step can the employee approach closer than the minimum approach distances or proceed with work on the parts as grounded parts.

445B. Removing Grounds

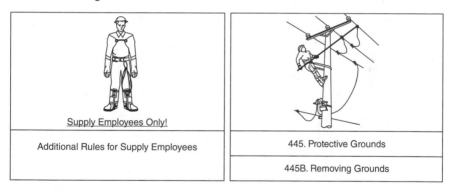

Supply Employees Only!

Additional Rules for Supply Employees

445. Protective Grounds

445B. Removing Grounds

- Extreme caution must be used to assure that the proper sequence is used when installing and removing protective grounds.
- Grounding devices must first be removed from the de-energized parts using tools with insulated handles.
- If multiple ground cables are connected to the same grounding point, all phase connections must be removed before removing any of the ground connections.
- An exception applies to removing all of the phase connections before removing any of the ground connections if this produces a hazard such as unintentional contact of the ground with ungrounded parts. In this case, the grounds may be removed individually from each phase and ground connection.
- The connection of the protective ground to the effective ground must be removed last.
- Note that a hazard may exist when de-energized lines and equipment are in proximity to energized circuits.
- Note that IEEE Std 1048 and IEEE Std 1246 contain additional information on personal protective grounding.

446. LIVE WORK

446A. Training

* Live work practices require adherence to the following rules in addition to other rules in Secs. 42 and 44.
* The minimum approach distances in **NESC** tables 441-1 or 441-2 shall be maintained from all grounded objects and from other conductors, lines and equipment at a different potential from the parts being worked on using the live work method.
* Employees must be trained to use rubber gloves, hot sticks, or the barehand method before using these techniques on energized lines.

446B. Equipment

* Insulated bucket trucks, ladders, and other support equipment must be evaluated for performance at the voltages involved. Tests must be done to ensure the equipment's integrity.
* Insulated bucket trucks and other insulated aerial devices used in bare-hand work must be tested before work is started.
* IEEE Standards and ANSI Standards must be referenced for equipment operation and testing.
* Bucket trucks and other insulated aerial devices must be kept clean.
* Tools and other equipment must not be used in a manner that would reduce the insulating strength of the insulated bucket truck or other insulated aerial device.

446C. When Working on Insulators...

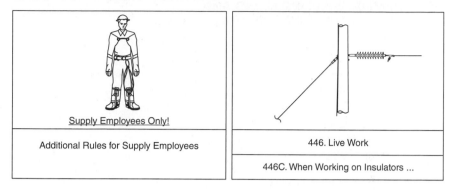

Supply Employees Only!

Additional Rules for Supply Employees

446. Live Work

446C. When Working on Insulators ...

- When insulators are worked on using live-line procedures, the clear insulation distance must not be less than the minimum approach distances of **NESC** Tables 441-1(AC) or 441-5 (DC).

446D. Bonding and Shielding for Bare-Hand Method

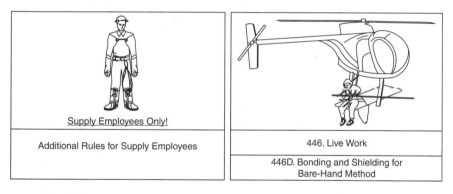

Supply Employees Only!

Additional Rules for Supply Employees

446. Live Work

446D. Bonding and Shielding for Bare-Hand Method

- A conductive bucket liner or other suitable conducting device must be provided to bond the insulated bucket or other insulated aerial device to the energized line.
- The employee must be bonded to the insulated bucket or other insulated aerial device by using conducting shoes, leg clips, or other means.
- Protective clothing for electrostatic shielding must be used where necessary. IEEE Std 516 provides additional information on clothing designed for electrostatic shielding.
- The aerial device must be bonded to the energized conductor before the employee contacts the energized part.

447. PROTECTION AGAINST ARCING AND OTHER DAMAGE WHILE INSTALLING AND MAINTAINING INSULATORS AND CONDUCTORS

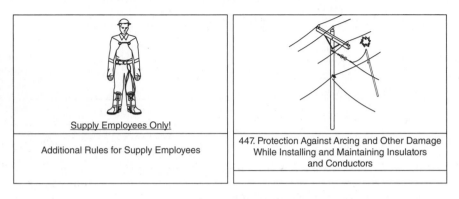

Supply Employees Only!

Additional Rules for Supply Employees

447. Protection Against Arcing and Other Damage While Installing and Maintaining Insulators and Conductors

- When installing and maintaining insulators and conductors, use precautions to limit damage that would cause the conductors or insulators to fall.
- Use precautions to prevent arcs from forming.
- If an arc does form, use precautions to prevent injuring or burning any parts of the supporting structure, insulators, or conductors.

Appendix

Photographs of NESC Applications

The photos shown correspond to figures in the text with the following icon:

See
Photo(s)

Photo 011-2. Typical dividing lines between the NESC and the NEC (see Fig. 011-2).

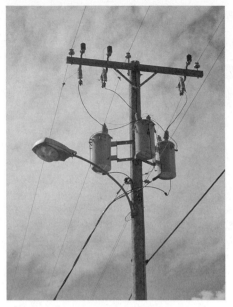

Photo 011-3(1). Examples of NESC and NEC street lighting systems (see Fig. 011-3).

Photo 011-3(2). Examples of NESC and NEC street lighting systems (see Fig. 011-3).

Photo 011-3(3). Examples of NESC and NEC street lighting systems (see Fig. 011-3).

Photo 017-1(1). Example of nominal values (see Fig. 017-1).

Photo 017-1(2). Example of nominal values (see Fig. 017-1).

Photo 092-4. Grounding connections for nonshielded cables over 750 V (see Fig. 092-4).

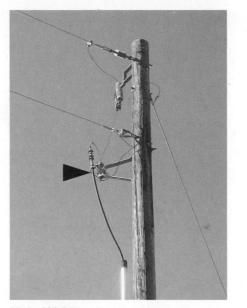

Photo 092-5. Surge arrester cable—shielding inter-connection (see Fig. 092-5).

Photo 092-6. Grounding points for a shielded cable without an insulating jacket (see Fig. 092-6).

Photo 092-7. Grounding points for a shielded cable with an insulating jacket (see Fig. 092-7).

Photo 092-9. Grounding of messenger wires (see Fig. 092-9).

Photo 092-12. Example of conductive electric supply station fence grounding (see Fig. 092-12).

Photo 093-1(1). Example of a copper grounding conductor (pole ground) and a structural metal grounding conductor (steel pole) (see Fig. 093-1).

Photo 093-1(2). Example of a copper grounding conductor (pole ground) and a structural metal grounding conductor (steel pole) (see Fig. 093-1).

Photo 093-2. Connection of grounding conductor to grounded conductor (see Fig. 093-2).

Photo 093-6. Example of pole ground ampacity (see Fig. 093-6).

Photo 093-7(1). Requirements for grounding conductors with or without guards (see Fig. 093-7).

Photo 093-7(2). Requirements for grounding conductors with or without guards (see Fig. 093-7).

Photo 093-10. Example of aluminum grounding conductor transitioning to copper for burial (see Fig. 093-10).

Photo 093-12. Example of common grounding conductor for neutral and equipment (see Fig. 093-12).

Photo 094-4. Made electrodes-driven ground rods (see Fig. 094-4).

Photo 095-1. Connection of grounding conductor to grounding electrode (see Fig. 095-1).

Photo 097-3. Example of a common neutral with single grounding (see Fig. 097-3).

Photo 097-5. Bonding of communication systems to electric supply systems on a joint-use (power and communication) structure (see Fig. 097-5).

Photo 110-1. Example of how Rule 110A applies to Part 1, Electric Supply Stations, or Part 2, Overhead Lines (see Fig. 110-1).

Photo 110-2. Example of how Rule 110A applies to Part 1, Electric Supply Stations, or Part 3, Underground Lines (see Fig. 110-2).

Photo 110-4. Examples of ANSI Z535 safety signs (see Fig. 110-4).

Photo 110-5. Barrier requirements (see Fig. 110-5).

Photo 110-6. VIOLATION! Barrier requirements (see Fig. 110-6).

Photo 110-7. VIOLATION! (unless concurrence by the substation owner) Neighboring fence requirements (see Fig. 110-7).

Photo 110-8. Example of accepted good practice for connecting a neighboring fence to a substation fence (see Fig. 110-8).

Photo 110-9. Example of how to apply the safety clearance zone to a chain-link fence per NESC Table 110-1 and NESC Fig. 110-1 (see Fig. 110-9).

Photo 110-10. Example of how to apply the safety clearance zone to an impenetrable fence per NESC Table 110-1 and NESC Fig. 110-2 (see Fig. 110-10).

Photo 110-16(1). Supporting and securing heavy equipment (see Fig. 110-16).

Photo 110-16(2). Supporting and securing heavy equipment (see Fig. 110-16).

Photo 111-1. Illumination under normal conditions (see Fig. 111-1).

Photo 124-1. Example of how to apply vertical clearances per NESC Table 124-1, NESC Fig. 124-1, and Rule 124A3 (see Fig. 124-1).

Photo 124-2. Clearance measurements made to a permanent supporting surface (see Fig. 124-2).

Photo 124-4. Absence of Code rule related to bus-to-bus clearances (see Fig. 124-4).

Photo 124-5. Example of how to apply NESC Table 124-1 and NESC Fig. 124-2 (see Fig. 124-5).

Photo 124-6. Guarding a substation conductor with insulation (see Fig. 124-6).

Photo 125-3. VIOLATION! Storage materials must not be in the working space (see Fig. 125-3).

Photo 128-1(1). Example of identification in a substation (see Fig. 128-1).

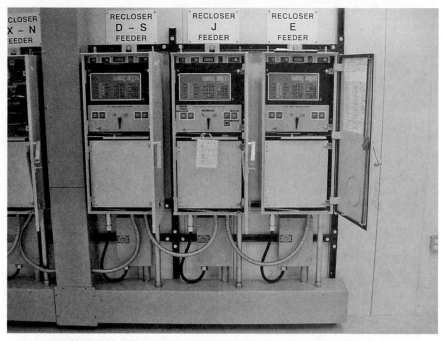

Photo 128-1(2). Example of identification in a substation (see Fig. 128-1).

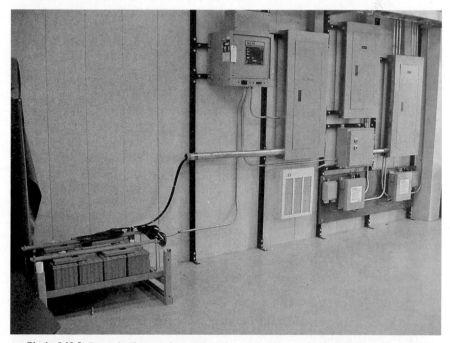

Photo 140-1. Example of storage batteries used in a substation control building (see Fig. 140-1).

Photo 150-1. Example of protecting current-transformer secondary circuits (see Fig. 150-1).

Photo 153-1. Choices for short-circuit protection of power transformers (see Fig. 153-1).

Photo 162-1(1). Substation conductor mechanical protection and support (see Fig. 162-1).

Photo 162-1(2). Substation conductor mechanical protection and support (see Fig. 162-1).

Photo 171-1. Example of a switch position indicator (see Fig. 171-1).

Photo 177-1. Arrester installation (see Fig. 177-1).

Photo 180-2. Metal-enclosed power switchgear (see Fig. 180-2).

Photo 180-4. Adequate clearance for reading meters on control switchboards (see Fig. 180-4).

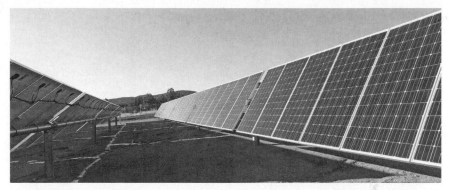

Photo 190-1(1). Examples of photovoltaic (PV) generation installations covered and not covered by Sec. 19 (see Fig. 190-1).

Photo 190-1(2). Examples of photovoltaic (PV) generation installations covered and not covered by Sec. 19 (see Fig. 190-1).

Photo 201-1. Overlap between Overhead Lines (Part 2) and Underground Lines (Part 3) (see Fig. 201-1).

Photo 201-2. Overlap between Overhead Lines (Part 2) and Electric Supply Stations (Part 1) for loading and strength (see Fig. 201-2).

Photo 214-1. VIOLATION! Inspection and tests of lines and equipment when in and out of service (see Fig. 214-1).

Photo 215-8. Example of a guy insulator used to limit galvanic corrosion (see Fig. 215-8).

Photo 216-1. Arrangement of overhead line switches (see Fig. 216-1).

Photo 217-1. Protection of supporting structures (see Fig. 217-1).

Photo 217-2(1). Readily climbable supporting structures (see Fig. 217-2).

Photo 217-2(2). Readily climbable supporting structures (see Fig. 217-2)

Photo 217-2(3). Readily climbable supporting structures (see Fig. 217-2)

Photo 217-3. Permanently mounted and temporary steps on supporting structures (see Fig. 217-3).

Photo 217-4. Arrangement of riser standoff brackets (see Fig. 217-4).

Photo 217-6. Identification of supporting structures (see Fig. 217-6).

Photo 217-7(1). VIOLATION! (remove before climbing) Attachments, decorations, and obstructions on supporting structures (see Fig. 217-7).

Photo 217-7(2). VIOLATION! (unless concurrence by the pole owner) Attachments, decorations, and obstructions on supporting structures (see Fig. 217-7).

Photo 217-7(3). VIOLATION! (unless concurrence by the pole owner) Attachments, decorations, and obstructions on supporting structures (see Fig. 217-7).

Photo 217-7(4). VIOLATION! (unless concurrence by the pole owner) Attachments, decorations, and obstructions on supporting structures (see Fig. 217-7).

Photo 217-7(5). VIOLATION! (unless concurrence by the pole owner) Attachments, decorations, and obstructions on supporting structures (see Fig. 217-7).

Photo 217-10(1). Examples of protection and marking of guys (see Fig. 217-10).

Photo 217-10(2). Examples of protection and marking of guys (see Fig. 217-10).

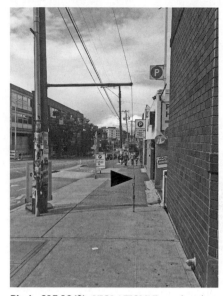

Photo 217-10(3). VIOLATION! Examples of protection and marking of guys (see Fig. 217-10).

Photo 217-10(4). VIOLATION! Examples of protection and marking of guys (see Fig. 217-10).

Photo 218-1(1). General vegetation management requirements (see Fig. 218-1).

Photo 218-1(2). General vegetation management requirements (see Fig. 218-1).

Photo 218-2. VIOLATION! Vegetation management at line, railroad, limited access highway, and navigable waterway crossings (see Fig. 218-2).

Photo 220-1. Standardization of levels of supply and communication conductors (see Fig. 220-1).

Photo 220-2. Relative levels of supply lines of different voltages at crossings (see Fig. 220-2).

Photo 220-3. Relative levels of supply lines of different voltages at structure conflict locations (see Fig. 220-3).

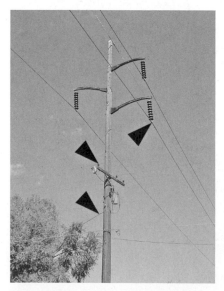

Photo 220-4. Relative levels of supply circuits of different voltages owned by one utility (see Fig. 220-4).

Photo 220-6(1). Example of identification of overhead conductors (see Fig. 220-6).

Photo 220-6(2). Example of identification of overhead conductors (see Fig. 220-6).

Photo 221-1(1). Avoiding conflict between two separate lines (see Fig. 221-1).

Photo 221-1(2). Avoiding conflict between two separate lines (see Fig. 221-1).

Photo 222-1. Example of joint use of structures (see Fig. 222-1).

Photo 225-1. Examples of common terms used in electric railway and trolley systems (see Fig. 225-1).

Photo 230-3. Example of cables laid on grade for emergency conditions (see Fig. 230-3).

Photo 230-6(1). Construction of 230C1 supply cable (see Fig. 230-6).

Photo 230-6(2). Construction of 230C1 supply cable (see Fig. 230-6).

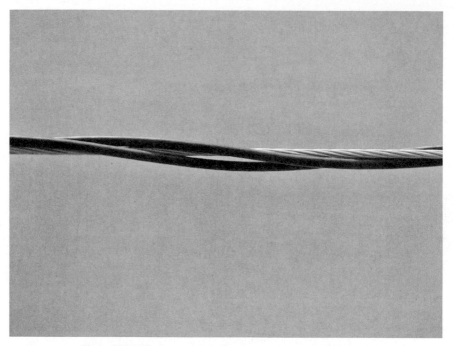

Photo 230-6(3). Construction of 230C3 supply cable (see Fig. 230-6).

Photo 230-7. Covered conductors (see Fig. 230-7).

Photo 230-8. Neutral conductors (see Fig. 230-8).

Photo 230-9. Example of an All-Dielectric Self-Supporting (ADSS) cable (see Fig. 230-9).

Photo 231-1. Clearance of a supporting structure (including anchor guys and attached equipment) from a fire hydrant (see Fig. 231-1).

Photo 231-2(1). VIOLATION! Clearance of a supporting structure (including support arms, anchor guys, braces, and attached equipment) from a curb (see Fig. 231-2).

Photo 231-2(2). Clearance of a supporting structure (including support arms, anchor guys, braces, and attached equipment) from a curb (see Fig. 231-2).

Photo 231-2(3). VIOLATION! Clearance of a supporting structure (including support arms, anchor guys, braces, and attached equipment) from a curb (see Fig. 231-2).

Photo 231-2(4). Clearance of a supporting structure (including support arms, anchor guys, braces, and attached equipment) from a curb (see Fig. 231-2).

Photo 231-3. Clearance of a supporting structure (including support arms, anchor guys, braces, and attached equipment) from a roadway without a curb (see Fig. 231-3).

Photo 231-4. Absence of a Code Rule related to placing a pole in a sidewalk (see Fig. 231-4).

Photo 231-5. Clearance of a supporting structure from railroad tracks (see Fig. 231-5).

Photo 232-7. Common clearance values from NESC Table 232-1 (see Fig. 232-7).

Photo 232-8(1). Common clearance values from NESC Table 232-1 (see Fig. 232-8).

Photo 232-8(2). VIOLATION! Common clearance values from NESC Table 232-1 (see Fig. 232-8).

Photo 232-11. Common clearance values from NESC Table 232-1 (see Fig. 232-11).

Photo 232-14. Common clearance values from NESC Table 232-1 (see Fig. 232-14).

Photo 232-17. Example of clearance to equipment cases and unguarded rigid live parts (see Fig. 232-17).

Photo 232-18(1). Examples of clearance to equipment cases (see Fig. 232-18).

Photo 232-18(2). Examples of clearance to equipment cases (see Fig. 232-18).

Photo 232-19. Example of clearance to street lighting (see Fig. 232-19).

Photo 232-20. Vertical clearance to land surface (see Fig. 232-20).

Photo 232-23. Federal Aviation Administration (FAA) requirements (see Fig. 232-23).

Photo 233-2. Conductor movement envelope (see Fig. 233-2).

Photo 233-4(1). Example of horizontal clearance between wires carried on different supporting structures (see Fig. 233-4).

Photo 233-4(2). Example of horizontal clearance between wires carried on different supporting structures (see Fig. 233-4).

Photo 233-6. Example of vertical clearance between wires carried on different supporting structures (see Fig. 233-6).

Photo 233-7. Example of vertical clearance between wires carried on different supporting structures (see Fig. 233-7).

Photo 233-8. Example of the exception to vertical clearance between wires that are electrically interconnected at a crossing (see Fig. 233-8).

Photo 234-1. Conductor temperature and loading conditions for measuring vertical and horizontal clearance to buildings and other installations (see Fig. 234-1).

Photo 234-4. Example of clearance of a conductor to a street lighting pole (see Fig. 234-4).

Photo 234-5(1). Example of clearance of a conductor to a traffic signal pole (see Fig. 234-5).

Photo 234-5(2). VIOLATION! Example of clearance of a conductor to a traffic signal pole (see Fig. 234-5).

Photo 234-6(1). Application of Rules 234B, 231B, and 232B4 to bent street lighting poles (see Fig. 234-6).

Photo 234-6(2). Application of Rules 234B, 231B, and 232B4 to bent street lighting poles (see Fig. 234-6).

Photo 234-8. Example of skip-span construction (see Fig. 234-8).

Photo 234-9. Example of horizontal clearance of a conductor to a building (see Fig. 234-9).

Photo 234-11. Example of vertical clearance of a supply service drop conductor over a building but not serving it (see Fig. 234-11).

Photo 234-12. Example of clearance of a conductor to a billboard (see Fig. 234-12).

Photo 234-13. Example of horizontal clearance of a conductor to a flagpole and flag (see Fig. 234-13).

Photo 234-15. Example of clearance between a power line and irrigation equipment (see Fig. 234-15).

Photo 234-17. VIOLATION! Example of clearance of supply conductors (service drops) attached to buildings (see Fig. 234-17).

Photo 234-18(1). Example of clearance of supply conductors (service drops) attached to buildings (see Fig. 234-18).

Photo 234-18(2). Example of clearance of supply conductors (service drops) attached to buildings (see Fig. 234-18).

Photo 234-20. VIOLATION! Example of clearance of supply conductors (service drops) attached to buildings (see Fig. 234-20).

Photo 234-21(1). Concerns that are addressed when providing clearance to bridges (see Fig. 234-21).

Photo 234-21(2). Concerns that are addressed when providing clearance to bridges (see Fig. 234-21).

Photo 234-23(1). Concerns that are addressed when providing clearance to grain bins (see Fig. 234-23).

Photo 234-23(2). Concerns that are addressed when providing clearance to grain bins (see Fig. 234-23).

Photo 234-25. Example of a pole, conductors, unguarded rigid live parts, and grounded equipment case adjacent to a building (see Fig. 234-25).

Photo 235-8. Example of vertical clearance between line conductors (see Fig. 235-8).

Photo 235-11. Example of vertical clearance between joint-use (supply and communication) conductors (see Fig. 235-11).

Photo 235-12. Example of vertical clearance between joint-use (supply and communication) conductors (see Fig. 235-12).

Photo 235-13. Example of vertical clearance between joint-use (supply and communication) conductors (see Fig. 235-13).

Photo 235-14. Example of vertical clearance between joint-use (supply and communication) conductors (see Fig. 235-14).

Photo 235-17. Example of clearance of a conductor to an anchor guy and a conductor to a support (see Fig. 235-17).

Photo 235-19. Examples of clearance between supply circuits of different voltages on the same support arm (see Fig. 235-19).

Photo 235-20. Conductor spacing on vertical racks or separate brackets (see Fig. 235-20).

Photo 235-22. Clearance and spacing between communication lines (see Fig. 235-22).

Photo 236-1. Example of climbing space between conductors (see Fig. 236-1).

Photo 236-3. Example of obstructions at the base of the climbing space (see Fig. 236-3).

Photo 236-5. Example of using standoff brackets for equipment (see Fig. 236-5).

Photo 237-2. Example of working clearances from energized equipment (see Fig. 237-2).

Photo 238-1. Definition of equipment as it applies to communication and supply facilities on the same structure (see Fig. 238-1).

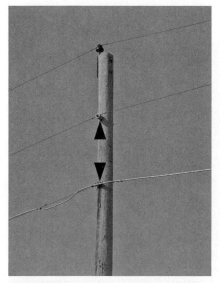

Photo 238-4. Example of vertical clearance between supply and communication equipment on the same structure (see Fig. 238-4).

Photo 238-5. Example of vertical clearance between supply and communication equipment on the same structure (see Fig. 238-5).

Photo 238-6. Example of vertical clearance between supply and communication equipment on the same structure (see Fig. 238-6).

Photo 238-7. Example of vertical clearance between supply and communication equipment on the same structure (see Fig. 238-7).

Photo 238-8(1). Example of vertical clearance between supply and communication equipment on the same structure (see Fig. 238-8).

Photo 238-8(2). Example of vertical clearance between supply and communication equipment on the same structure (see Fig. 238-8).

Photo 238-11. Example of mounting a solar (photovoltaic) panel on a pole (see Fig. 238-11).

Photo 238-12. Example of locating a luminaire on a pole (see Fig. 238-12).

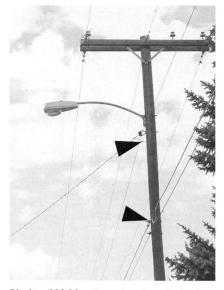

Photo 238-13. Example of vertical clearance between a luminaire or a drip loop feeding a luminaire and communication equipment (see Fig. 238-13).

Photo 238-14. Example of vertical clearance between a luminaire or a drip loop feeding a luminaire and communication equipment (see Fig. 238-14).

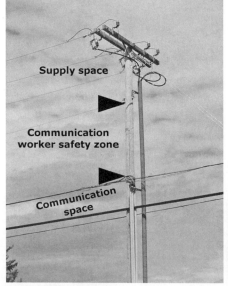

Photo 238-17. Communication worker safety zone (see Fig. 238-17).

Photo 238-18(1). Examples of clearance between supply lines and antennas in the supply space (see Fig. 238-18).

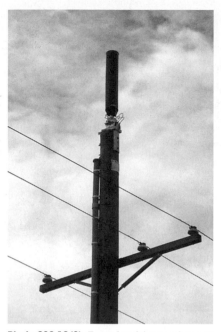

Photo 238-18(3). Examples of clearance between supply lines and antennas in the supply space (see Fig. 238-18).

Photo 238-18(2). Examples of clearance between supply lines and antennas in the supply space (see Fig. 238-18).

Photo 238-18(4). Examples of clearance between supply lines and antennas in the supply space (see Fig. 238-18).

Photo 238-19(1). Example of an antenna mounted in the communication space (see Fig. 238-19).

Photo 238-19(3). Example of an antenna mounted in the communication space (see Fig. 238-19).

Photo 238-19(2). Example of an antenna mounted in the communication space (see Fig. 238-19).

Photo 239-1. Examples of vertical and lateral conductors (see Fig. 239-1).

Photo 239-3(1). Guarding and protection near ground (see Fig. 239-3).

Photo 239-3(2). Guarding and protection near ground (see Fig. 239-3).

Photo 239-3(3). Guarding and protection near ground (see Fig. 239-3).

Photo 239-4. Location of vertical and lateral supply conductors on supply-line structures or within supply space on jointly used structures (see Fig. 239-4).

Photo 239-5. Location of vertical and lateral communication conductors on communication line structures (see Fig. 239-5).

Photo 239-6. Location of vertical supply conductors and cables passing through the communication space on jointly used line structures (see Fig. 239-6).

Photo 239-7. Example of vertical supply conductors on a joint-use (supply and communication) pole (see Fig. 239-7).

Photo 239-8. VIOLATION! Example of vertical supply conductors on a joint-use (supply and communication) pole (see Fig. 239-8).

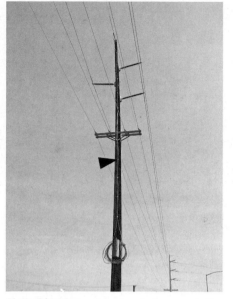

Photo 239-10. Location of vertical communication conductors passing through the supply space on jointly used structures (see Fig. 239-10).

Photo 241-3. Lines considered to be at crossings (see Fig. 241-3).

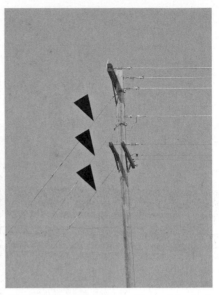

Photo 252-6. Longitudinal loads at a change in grade of construction (see Fig. 252-6).

Photo 252-7(1). Longitudinal loads at deadends (see Fig. 252-7).

Photo 252-7(2). Longitudinal loads at deadends (see Fig. 252-7).

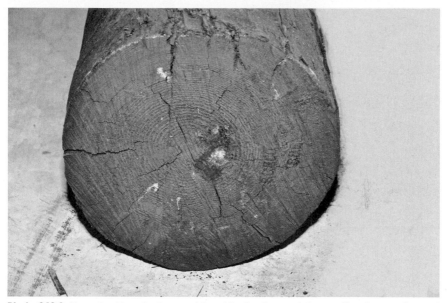

Photo 260-3. Example of how loads, material, and load and strength factors affect the size of a wood pole (see Fig. 260-3).

Photo 260-5(1). Examples of wood pole deterioration (see Fig. 260-5).

Photo 260-5(2). Examples of wood pole deterioration (see Fig. 260-5).

Photo 260-5(3). Examples of wood pole deterioration (see Fig. 260-5).

Photo 260-7(1). Examples of a pole tag and pole brand (see Fig. 260-7).

Photo 260-7(2). Examples of a pole tag and pole brand (see Fig. 260-7).

Photo 260-7(3). Examples of a pole tag and pole brand (see Fig. 260-7).

Photo 260-7(4). Examples of a pole tag and pole brand (see Fig. 260-7).

Photo 261-1(1). Strength requirements of supporting structure (see Fig. 261-1).

Photo 261-1(2). Strength requirements of supporting structure (see Fig. 261-1).

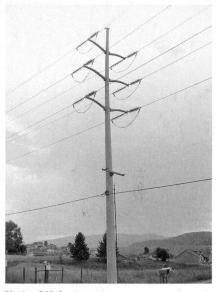

Photo 261-3. Strength requirements of metal, prestressed, and reinforced concrete structures (see Fig. 261-3).

Photo 261-4. Strength requirements of wood structures (see Fig. 261-4).

Photo 261-7(1). Examples of spliced and reinforced wood poles (see Fig. 261-7).

Photo 261-7(2). Examples of spliced and reinforced wood poles (see Fig. 261-7).

Photo 261-7(3). Examples of spliced and reinforced wood poles (see Fig. 261-7).

Photo 261-8. Strength requirements of fiber-reinforced polymer structures (see Fig. 261-8).

Photo 261-15. Strength of concrete and metal crossarms and braces (see Fig. 261-15).

Photo 261-16. Strength of wood crossarms and braces (see Fig. 261-16).

Photo 261-18. Strength of fiber-reinforced polymer crossarms and braces (see Fig. 261-18).

Photo 261-19(1). Example of wood and steel pins (see Fig. 261-19).

Photo 261-19(2). VIOLATION! Example of wood and steel pins (see Fig. 261-19).

Photo 261-20. Strength requirements for armless construction (see Fig. 261-20).

Photo 261-24. Example of open-wire communication conductors (see Fig. 261-24).

Photo 264-1(1). Examples of guying and bracing (see Fig. 264-1).

Photo 264-1(2). Examples of guying and bracing (see Fig. 264-1).

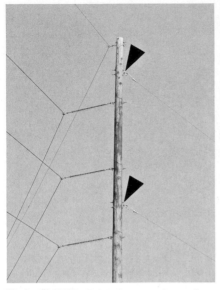

Photo 264-3(1). Point of guy attachment (see Fig. 264-3).

Photo 264-3(2). Point of guy attachment (see Fig. 264-3).

Photo 279-2. Properties of guy insulators (see Fig. 279-2).

Photo 279-4. Example of a guy insulator used to limit galvanic corrosion (see Fig. 279-4).

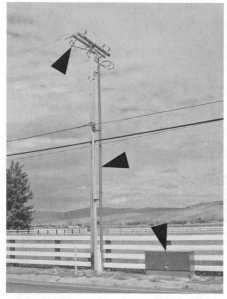

Photo 311-1. Example of worker locating underground lines (see Fig. 311-1).

Photo 301-1. Overlap between Underground Lines (Part 3) and Overhead Lines (Part 2) (see Fig. 301-1).

Photo 312-1(1). Example of the need for adequate working space (see Fig. 312-1).

Photo 312-1(2). Example of the need for adequate working space (see Fig. 312-1).

Photo 312-2. Example of a notification sign used on pad-mounted equipment (see Fig. 312-2).

Photo 314-1. Conductive parts to be grounded (see Fig. 314-1).

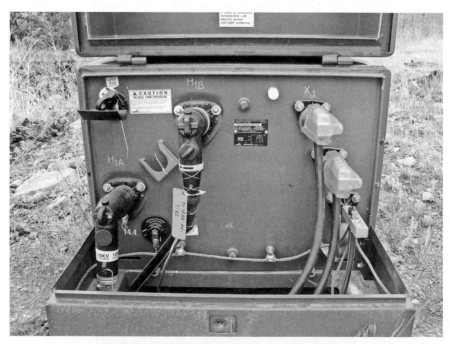

Photo 314-2. Grounding of circuits (see Fig. 314-2).

Photo 320-2. Conduit bending radius (see Fig. 320-2)

Photo 320-4. Conduit routing under highways and streets (see Fig. 320-4).

Photo 320-10. VIOLATION! Radial separation from gas or other lines that transport flammable material (see Fig. 320-10).

Photo 323-1. Loads on manholes, handholes, and vaults (see Fig. 323-1).

Photo 323-2(1). Examples of live loads in roadways and live loads in areas not subject to vehicular loading (see Fig. 323-2).

Photo 323-2(2). Examples of live loads in roadways and live loads in areas not subject to vehicular loading (see Fig. 323-2).

Photo 323-6. Manhole access diameter for a round manhole opening in a manhole containing supply cables (see Fig. 323-6).

Photo 323-13. Example of a vault (see Fig. 323-13).

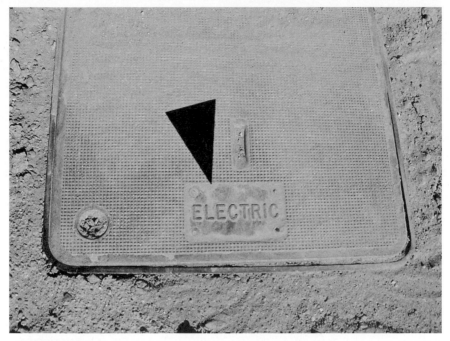

Photo 323-15(1). Requirements for identification of manhole and handhole covers (see Fig. 323-15).

Photo 323-15(2). Requirements for identification of manhole and handhole covers (see Fig. 323-15).

Photo 323-15(3). Requirements for identification of manhole and handhole covers (see Fig. 323-15).

Photo 331-1. Example of a sheath, jacket, and shield on a typical URD supply cable (see Fig. 331-1).

Photo 333-1(1). Examples of cable accessories and joints (see Fig. 333-1).

Photo 333-1(2). Examples of cable accessories and joints (see Fig. 333-1).

Photo 333-1(3). Examples of cable accessories and joints (see Fig. 333-1).

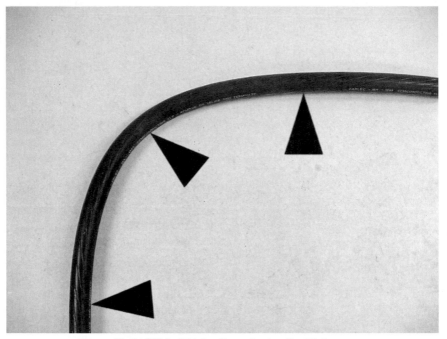

Photo 341-1. Cable bending radius (see Fig. 341-1).

Photo 350-3. Cables below 600 V to ground buried in close proximity (see Fig. 350-3).

Photo 352-1. Trenching requirements for direct-buried cable (see Fig. 352-1).

Photo 352-3. Plowing requirements for direct-buried cables (see Fig. 352-3).

Photo 352-4. Boring requirements for direct-buried cables (see Fig. 352-4).

Photo 354-1. Supply line and steam, gas, or flammable material line in random separation (see Fig. 354-1).

Photo 360-1. General requirements for risers (see Fig. 360-1).

Photo 362-2(1). Examples of using stand-off brackets to organize conduit (duct) risers (see Fig. 362-2).

Photo 362-2(2). Examples of using stand-off brackets to organize conduit (duct) risers (see Fig. 362-2).

Photo 370-1(1). Examples of supply cable terminations (see Fig. 370-1).

Photo 370-1(2). Examples of supply cable terminations (see Fig. 370-1).

Photo 370-1(3). Examples of supply cable terminations (see Fig. 370-1).

Photo 372-1. Example of supply cable identification at termination points (see Fig. 372-1).

Photo 372-2. Absence of Code rule related to underground warning ribbons or tapes and cable route markers (see Fig. 372-2).

Photo 380-1. Examples of supply and communication equipment (see Fig. 380-1).

Photo 380-2. Clearance of pad-mounted equipment to fire hydrants (see Fig. 380-2).

Photo 380-3(1). Absence of Code rule related to underground pad-mounted equipment location adjacent to roads and buildings (see Fig. 380-3).

Photo 380-3(2). Absence of Code rule related to underground pad-mounted equipment location adjacent to roads and buildings (see Fig. 380-3).

Photo 381-3(1). Safety sign requirements for pad-mounted equipment (see Fig. 381-3).

Photo 381-3(2). Safety sign requirements for pad-mounted equipment (see Fig. 381-3).

Photo 384-1. Bonding above ground metallic supply and communications enclosures (see Fig. 384-1).

Photo 390-1. Example of a utility tunnel between buildings (see Fig. 390-1).

Index

Page numbers followed by *f* denote figures or illustrations.